# Springer Series on Naval Architecture, Marine Engineering, Shipbuilding and Shipping

## Volume 9

**Series Editor**

Nikolas I. Xiros, University of New Orleans, New Orleans, LA, USA

The Naval Architecture, Marine Engineering, Shipbuilding and Shipping (NAMESS) series publishes state-of-art research and applications in the fields of design, construction, maintenance and operation of marine vessels and structures. The series publishes monographs, edited books, as well as selected PhD theses and conference proceedings focusing on all theoretical and technical aspects of naval architecture (including naval hydrodynamics, ship design, shipbuilding, shipyards, traditional and non-motorized vessels), marine engineering (including ship propulsion, electric power shipboard, ancillary machinery, marine engines and gas turbines, control systems, unmanned surface and underwater marine vehicles) and shipping (including transport logistics, route-planning as well as legislative and economical aspects).

The books of the series are submitted for indexing to Web of Science.

All books published in the series are submitted for consideration in Web of Science.

More information about this series at http://www.springer.com/series/10523

Karl Dietrich von Ellenrieder

# Control of Marine Vehicles

 Springer

Karl Dietrich von Ellenrieder
Faculty of Science and Technology
Free University of Bozen-Bolzano
Bolzano, Italy

ISSN 2194-8445         ISSN 2194-8453   (electronic)
Springer Series on Naval Architecture, Marine Engineering, Shipbuilding and Shipping
ISBN 978-3-030-75023-7         ISBN 978-3-030-75021-3   (eBook)
https://doi.org/10.1007/978-3-030-75021-3

This Springer imprint is published by the registered company Springer Nature Switzerland AG
The registered company address is: Gewerbestrasse 11, 6330 Cham, Switzerland

*To Antonella, Filippo & Lia*

# Preface

This textbook presents an introduction to the fundamental principles of the control of marine vehicles, at an introductory/intermediate level. The book targets 4th or 5th year undergraduate students, masters students and beginning Ph.D. students and is intended to serve as a bridge across these three academic levels. It will hopefully also be of use to practitioners in the field, who want to review the foundational concepts underpinning some of the recent advanced marine vehicle control techniques found in the literature, which they might find useful for their own applications.

The idea for this text arose from the author's experiences in the research and teaching of marine robotics in the Ocean Engineering Program at Florida Atlantic University in the USA, and his subsequent research and teaching in the areas of automatic control and mobile robotics at the Libera Università di Bolzano in Italy.

This book consists of an introductory chapter followed by two parts. The first part covers the control of linear time invariant systems and is intended to provide a basis and an intuition for the understanding of a system's dynamic response, feedback control and stability. The second part covers the stability and control of nonlinear systems. The end of each chapter includes a list of references and a set of illustrative exercises, which focus on the control of marine vehicles.

It is anticipated that Chaps. 1–6 could be covered in a one semester 4th or 5th year undergraduate course. A one semester M.S. level course could be based on Chaps. 1 and 6, selected parts of Chaps. 7–12. Lastly, if covered in depth, the material in Part II, especially the nonlinear stability analysis methods, feedback linearization and other aspects related to geometric control, would provide a suitable basis for a one semester Ph.D.-level course.

Much of the material presented in the text was originally developed and presented as course notes at both Florida Atlantic University and at the Libera Università di Bolzano. The author is grateful to the many students and colleagues who contributed to the book through their stimulating discussions of the material and their supportive suggestions. A special thanks to Helen Henninger and Travis Moscicki, who provided valuable feedback on early versions of some of the chapters. Fernanda Dalla Vecchia and Maja Fluch created some of the drawings.

Finally, the author would like to thank the U.S. National Science Foundation and the U.S. Office of Naval Research for supporting his research on the use of nonlinear control methods for unmanned surface vessels.

Bolzano, Italy                                                        Karl Dietrich von Ellenrieder
February 2021

# Contents

**Part II   Nonlinear Methods**

# About the Author

**Karl Dietrich von Ellenrieder**  received his B.S. degree in aeronautics and astronautics, with a specialty in avionics, from the Massachusetts Institute of Technology (USA) in 1990, and the M.S. and Ph.D. degrees in aeronautics and astronautics from Stanford University, (USA) in 1992 and 1998, respectively. Since 2016, he has been a Full Professor of Automation in the Faculty of Science and Technology at the Free University of Bozen-Bolzano (Italy). From 2003 to 2016, he was with the Department of Ocean & Mechanical Engineering at Florida Atlantic University (USA) where, after being promoted through the ranks from Assistant Professor to Associate Professor, he ultimately served as a Full Professor of Ocean Engineering and as the Associate Director of the SeaTech Institute for Ocean Systems Engineering. His research interests include automatic control, the development of robotic unmanned vehicles, human-robot interaction and the experimental testing of field robots.

# Chapter 1
# Introduction

## 1.1 Overview

Approximately 70% of the Earth's surface is covered by water. The range of environmental conditions that can be encountered within that water is broad. Accordingly, an almost equally broad range of marine vehicles has been developed to operate on and within conditions as disparate as the rough, cold waters of the Roaring Forties and the Furious Fifties in the Southern Hemisphere; the relatively calm, deep waters of the Pacific; and the comparatively warm, shallow, silty waters of the Mississippi and the Amazon rivers.

Here, we will refer to any system, manned or unmanned, which is capable of self-propelled locomotion or positioning at the water's surface, or underwater, as a *marine vehicle*.[1] Thus, this term will be used interchangeably to mean boat, vessel, ship, semi-submersible, submarine, autonomous underwater vehicle (AUV), remotely operated vehicle (ROV), etc.

The main problem in marine vehicle control is to find a way to actuate a vehicle so that it follows a desired behavior as closely as possible. The vehicle should approximate the desired behavior, even when affected by uncertainty in both its ability to accurately sense its environment and to develop control forces with its actuators (e.g. propellers or fins), as well as, when the vehicle is acted upon by external disturbances, such as current, waves and wind.

## 1.2 Automatic Control

A system is called a *feedback* system when the variables being controlled (e.g. position, speed, etc.) are measured by sensors, and the information is fed back to the controller such that it can process and influence the controlled variables.

---

[1] Some authors similarly use the term *marine craft* in deference to the use of the term *vehicle* to represent a land-borne system.

© Springer Nature Switzerland AG 2021
K. D. von Ellenrieder, *Control of Marine Vehicles*, Springer Series on Naval Architecture, Marine Engineering, Shipbuilding and Shipping 9, https://doi.org/10.1007/978-3-030-75021-3_1

**Example 1.1** Figure 1.1 shows a boat trying to maintain a fixed position and heading on the water (station keeping). Imagine that we set up the system to keep the boat as close as possible to the desired position. Without any control, either human control or machine control, the boat is likely to deviate further and further away from the desired position, possibly because of the action of wind, currents or waves. Thus, this system is inherently unstable. The system being controlled, which is the boat in this example, is often called either the *plant* or the *process*. If a human boat captain were present, you could say that the captain is playing the role of the *controller*. One of the purposes of a controller is to keep the controlled system stable, which in this case would mean to keep the boat near the point $\eta_d$. The task of keeping the combined system consisting of a plant and controller stable is called *stabilization*.

Two possible approaches to stabilizing the system include *open–loop control* and *closed–loop control*:

- Open loop control: This would correspond to: (1) performing a set of detailed experiments to determine the engine throttle and steering wheel positions that keep the boat located at $\eta_d$ for the expected environmental conditions; (2) setting the boat to $\eta_d$, and (3) setting the engine throttle and steering wheel to their experimentally determined positions and hoping that the boat maintains the correct position and heading, even when the environmental conditions vary. As sudden or temporary changes in the environmental conditions (disturbances), such as wind gusts, cannot be predicted, this approach is unlikely to work well. In general, one can never stabilize an unstable plant using open–loop control.
- Closed loop control: This would correspond to using some sort of feedback, such as a human captain monitoring a GPS receiver, to obtain a measure of how well the boat is maintaining the desired position and setting the engine throttle or turning the steering wheel, as needed, to keep the boat in the correct location. As a human captain would use information from the output of the system (the boat's actual position relative to the desired position) to adjust the system input (the positions of the engine throttle and steering wheel), closed–loop control is also called feedback control.

*Automatic control* is simply the use of machines as controllers. As mentioned above, the plant is the system or process to be controlled. The means by which the plant is influenced is called the *input*, which could be a force or moment acting on the boat, for example. The response of the plant to the input is called the *output*, which would be the deviation of the boat from the desired position and heading. In addition to the input, there are other factors that can influence the output of the plant, but which cannot be predicted, such as the forces and moments generated by wind gusts or waves. These factors are called *disturbances*. In order to predict the behavior of the plant, a model of the plant dynamics is needed. This model may be generated using physical principles, by measuring the output response of the plant to different inputs in a process known as *system identification*, or from a combination of these approaches.

In the human feedback system the captain has to perform several subtasks. He/she has to observe the boat's position and compare it to the measured GPS location, make

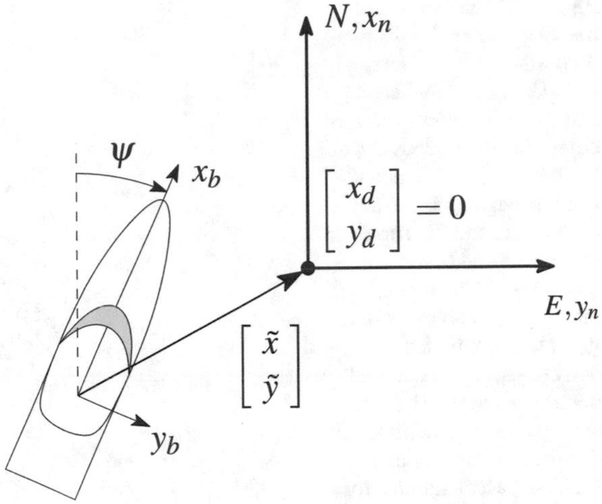

**Fig. 1.1** Station–keeping
about the desired pose
(position and orientation)
$\eta_d = [x_d \ y_d \ \psi_d]^T = 0$.
Here, $\eta := [x_n \ y_n \ \psi]^T$,
where $x_n$ and $y_n$ are the
North and East components,
respectively, of the vessel's
position and $\psi$ is the angle
between $x_b$ and $N$. The error
in position and heading is
$\tilde{\eta} = \eta_d - \eta$. When $\eta_d = 0$,
the pose error is $\tilde{\eta} = -\eta$

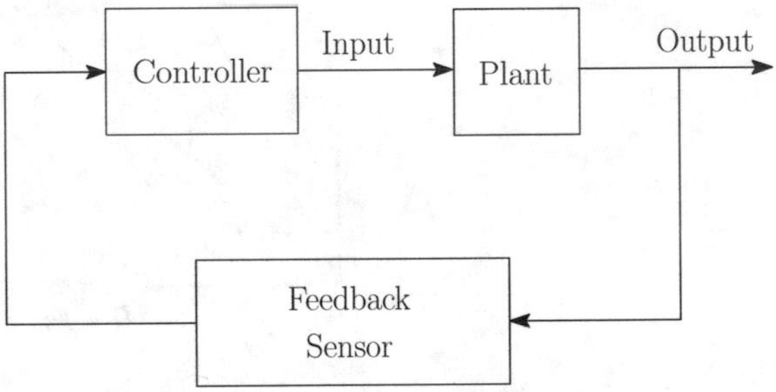

**Fig. 1.2** Structure of a
feedback system for
stabilization purposes

decisions about what actions to take, and exercise his/her decision to influence the
motion of the boat. Normally, these three stages are called *sensing*, *control* and
*actuation*, respectively. If we replace the human captain with a machine, we need
a sensor, a controller and an actuator. As with the human controlled system, the
sensor measures a physical variable that can be used to deduce the behavior of the
output, e.g. the deviation of the boat from the desired position and heading. However,
with machine control, the output of the sensor must be a signal that the controller
can accept, such as an electrical signal. The controller, often a computer or electrical
circuit, takes the reading from the sensor, determines the action needed and sends the
decision to the actuator. The actuator (e.g. the propeller and rudder) then generates
the quantities which influence the plant (such as the forces and moments needed push
the boat towards the desired position). A block diagram of a stabilizing closed–loop
controller is shown in Fig. 1.2. □

**Example 1.2** Consider another example of maneuvering a boat, as shown in Fig. 1.3.
Here, the boat is following a desired, time-dependent, trajectory against a persistent
wind so that its speed follows an external command, such as a varying speed limits
along a coastal marine channel. There are two subtasks in this example. The first is

**Fig. 1.3** A surface vessel
tracking a desired trajectory
$\eta_d$ (dashed curve). Here,
$\eta := [x_n\ y_n\ \psi]^T$, where $x_n$
and $y_n$ are the North and East
components, respectively, of
the vessel's position and the
heading angle $\psi$ is the angle
between $x_b$ and $N$. The error
in position and heading is
$\tilde{\eta} = \eta_d - \eta$. The trajectory
may be time-parameterized
$\eta_d(t)$ with continuous
derivatives, which could also
define desired translational
and rotational velocities, and
desired translational and
rotational accelerations, for
example

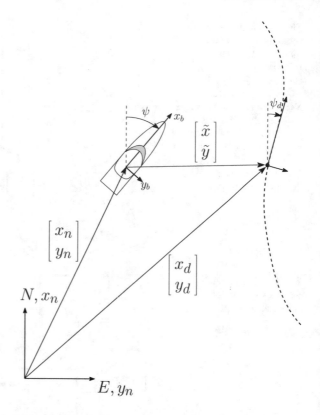

following the desired trajectory without the wind. This problem is called *tracking*.
The second is the reduction (or elimination) of the effects of the wind on the speed
and position of the boat. This problem is called *disturbance rejection*. The overall
problem, encompassing both tracking and disturbance rejection, is called *regulation*.

- Open–loop control: This could correspond to adjusting the positions of the steering wheel and throttle according to a precomputed position profile and the wind
  conditions along different sections of the trajectory, obtained from an accurate
  weather forecast. One can imagine that the scheme will not work well, since any
  error in the precomputed position profile or the weather forecast will cause errors
  in the actual trajectory of the boat.
- Closed–loop control: This could correspond to adjusting the steering and engine
  throttle according to the measured position and speed of the boat. Since we can
  steer, accelerate or decelerate the boat in real time, we should be able to control
  the position, speed and heading of the boat to within a small margin of error. In
  this case it is not necessary to precisely know the environmental conditions.

□

One can see that the main difference between regulation and stabilization is that
there is a command signal, also called a *reference input*, in the regulation problem.

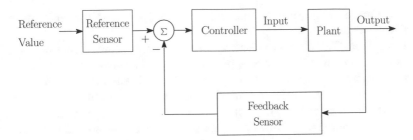

**Fig. 1.4** Structure of a feedback controller for regulation

The reference input needs to be processed, in addition to other signals, by the stabilizing controller. The structure of a feedback system for regulation is shown in Fig. 1.4. Another difference between regulation and stabilization is that in the regulation problem, the disturbance is often assumed to be persistent and has some known features, such as being piecewise constant. Whereas in the stabilization problem, the disturbance is assumed to be unknown and temporary in nature. The study of stabilization is important because there are many stabilization problems of importance for marine engineering, such as the station keeping problem above, but also because it is a key step in achieving regulation.

## 1.3 Background Ideas

Automatic control systems can be found almost everywhere one looks. With the development of embedded computing, small, inexpensive, low–power controllers can be found in almost all electronic devices. The following terms related to automatic control are in common use today:

- *Autonomous*: In the past, systems employing automatic control were often referred to as "autonomous systems". Examples include Autonomous Surface Vehicles (ASVs), Autonomous Underwater Vehicles (AUVs), Autonomous Ground Vehicles (AGVs), etc. However, as advanced artificial intelligence techniques from computer science (machine learning, decision making, etc.) are more and more often employed on unmanned vehicles, an autonomous vehicle has come to mean a vehicle with the following five properties:

  1. Sensing—an autonomous vehicle must be able to sense its environment. Advanced sensing techniques include the use of sensor fusion, which involves combining the use several different types of sensors (e.g. vision—including color and thermal imaging; LiDAR; GPS—including Real–Time Kinematic, RTK GPS; Sonar or Ultrasound) to build up an understanding of a system's environment. Advanced systems also involve the use of Simultaneous Localization and Mapping (SLAM) techniques for building a map of an environment while,

at the same time, using the incomplete or changing map for positioning and navigation.

2. Perception—The ability to discern or identify objects in the environment. For example, in order to properly perform a collision avoidance maneuver an autonomous car might need to understand whether it was avoiding a fixed object, such as a concrete barrier or a moving object, such as another car or a human pedestrian. The geometry of the avoidance maneuver, its urgency and the margin of safety required (for both the occupants of the car and the obstacle) can strongly depend on what the vehicle perceives the obstacle to be.

3. Decision Making and Planning—As mentioned above, one can consider an automatic control system to be making a decision, as it must determine the appropriate output response to an input signal. However, in the fields of robotics and unmanned vehicles, decision–making may also require an autonomous system to develop a list of the tasks required to perform a mission on its own, plan the sequence in which those tasks should be executed, and allocate the responsibility for performing those tasks to different subsystems, other robots, or even humans. As conditions change, the plan may have to be dynamically reformulated and the tasks reallocated in real–time. Further, additional concepts from artificial intelligence may be involved, such as the segmentation of planning into deliberative and reactive tasks. In deliberative planning those (slower) tasks where some sort of optimization of the plan can be performed to minimize time, energy consumption, risk, etc. can be conducted. Reactive planning typically involves decision–making in time–critical situations, where there may not be sufficient time for an optimization of the outcome of a decision to be calculated, such as an impending collision between an autonomous boat and an obstacle. The type of decision-making to be implemented can have a strong influence on the type of software architecture and electronic hardware used. In situations where there is a potential risk to life, the environment or to economically sensitive property, the autonomous vehicle may be required to present its decision to a human operator before taking action.

4. Action—This is the ability to physically implement the outcome of a decision.

5. Learning—Learning from the outcome of an action and applying this experiential learning to future decisions. There are virtually an infinite number of different situations that an autonomous vehicle could find itself within, especially when it operates in an unstructured or uncontrolled environment. Thus, it is virtually impossible to anticipate and preprogram for every possible scenario the system could encounter. Recent research has focused on developing automatically-generated behaviors, instead. These behaviors can be trained through robotic learning techniques in both simulation, before a vehicle is taken into the field, and online, while the vehicle is in operation.

- *Classical Control*: Techniques developed for the analysis and control of primarily linear time invariant (LTI) systems, including root locus and frequency response methods. Excellent textbooks on Classical Control techniques include [19, 33].

- *Digital Control*: Techniques developed for the implementation of control systems on digital computers, such as the discrete representation of dynamic systems, modeling of time delays, and coping with issues related to the quantization, resolution and finite precision that results from the discretization of continuous signals. A good reference for digital control is [20].
- *Modern Control*: The phrase "Modern Control" has been used to describe a collection of control techniques (primarily for linear systems), which usually includes: Classical Control, Digital Control, State Space Control and Robust Control. Good reference texts on Modern Control include [15, 23, 32].
- *State Space Control*: Involves the representation of higher order processes and plants as a collection of first order systems and the use of linear algebra techniques for examining the stability and control of such systems. In comparison to Classical Control techniques, the state space representation of systems can provide a more convenient means to analyze systems with multiple inputs and multiple outputs (MIMO). A widely used text on State Space control is [21]; textbooks containing very nice chapters on state space techniques include [19] and [4].
- *Nonlinear Control*: Classical control techniques are largely based on the analysis and control of linear systems, where the equations describing the dynamics of the plant have been linearized about a desired operating condition. However, most processes or plants exhibit nonlinear behavior, especially when used under conditions sufficiently different from the operating condition. Nonlinear control techniques include approaches to removing nonlinearity (such as feedback linearization) or making systems more robust to existing nonlinearity (such as sliding mode control) and the use of Lyapunov stability methods for analysis and control system design. Good reference works on nonlinear control include [26, 41].
- *Adaptive Control*: An adaptive controller is a controller that can modify its behavior in response to changes in the dynamics of the plant and the character of the disturbances acting on the plant. The main difference between an adaptive controller and ordinary feedback is that an adaptive controller utilizes adjustable control parameters and includes a mechanism for adjusting the parameters, whereas the control parameters in an ordinary feedback controller are constant. In general, adaptive controllers are nonlinear and time-varying because of the parameter adjustment mechanism. A good textbook on adaptive control is [3]; [41] also includes a chapter on adaptive control.
- *Robust Control*: This is an approach to controller design to deal with uncertainty. Robust control methods function properly, provided that uncertain parameters or disturbances remain within some bounded set. Early state-space methods tended to lack robustness; robust control techniques were developed in response. In contrast to adaptive control, robust controllers are designed to work assuming that certain disturbances and plant parameters will be unknown but bounded; the controller parameters remain constant. A very useful text on Robust Control is [49].
- *Model Predictive Control (MPC)*: MPC controllers use an open–loop model of a process or plant and perform simulations over a future finite-time-horizon to predict the response of the system. At each time step, the controller uses the model to identify the best sequence of control inputs to use over a future horizon

of $N$ steps. However, only the first optimal control input is used, the rest of the sequence is thrown away. At the following time step a new measurement of the output is obtained. In this way feedback is incorporated into the process, and the optimization is repeated. MPC has been used in the chemical process control industry since approximately the early 1980s (mainly in oil refining and chemical plants), but it has gained popularity recently for use in the automotive industry, for use in power systems and in the aerospace industry (for the guidance and control of rockets during powered descent during landings). Useful references for MPC include [5, 13, 25, 30, 48].

- *Supervisory Switching Control (SSC)*: Real systems must often perform many different types of tasks and the dynamics of the same plant can vary substantially depending on the specific task the system is performing. For example, the dynamic model of an aircraft performing maneuvers at subsonic speeds is quite different from the dynamic model of the same aircraft performing maneuvers at supersonic speeds. In such situations a control technique called *Gain Scheduling* is often used in which the system selects a controller from among a bank of possible controllers, based upon its operating condition. Supervisory switching control involves the use of multiple controllers for controlling a system that must operate across a diverse range of operating conditions. It differs from gain scheduling in that online optimizations are performed to select a controller from among a bank of possible controllers; advanced nonlinear control techniques, such as Lyapunov falsification and hysteresis switching are used to ensure the stability of the controlled system. SSC differs from MPC in that the controller itself is selected based on the online optimization, not just the control signal. An excellent monograph detailing SSC is [29]; an implementation of the method on a real system can be found in [9].

## 1.4   Typical Guidance, Navigation and Control Architectures of Marine Vehicles

Systems designed to govern the motion of marine vehicles (as well as air-/space-craft, cars, etc.) are typically organized along the functional lines of guidance, navigation and control (GNC). A block diagram of a typical GNC system for marine vehicles is shown in Fig. 1.5. The navigation system uses sensors to estimate the vehicle's position, velocity and acceleration, in addition to other knowledge about the state of the vehicle, such as the available propulsive power. Based upon information from the navigation system, and with knowledge of the vehicle's desired operating condition, the guidance system generates a trajectory. When planning a route some guidance systems may use data on weather conditions, sea states, or other information that could affect the vehicle's progress that has been obtained through communication with satellites or remote ground stations. Lastly, the vehicle's control system uses the data input from both the guidance and navigation systems to generate output

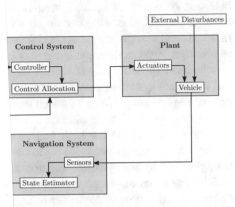

...ntrol System used for marine vehicles

...llers, fins, etc.) so that the vehicle follows
...e.

...tem determines a set of waypoints (inter-
...should follow. The guidance system also
...which is sometimes also called a trajectory
...rajectory through the desired waypoints
...s current position and orientation, obsta-
...r available.

...tion systems of marine vehicles include
..., inertial measurement units (IMUs), and
...typically include vehicle position, speed
...measurement axes available, IMUs can
...ns of the vehicle, and also its rotational
...noisy or have drop-outs where a signal
... underwater vehicles can only acquire
...where they can receive signals from GPS
...ater and when operating in rough seas
...e Earth's magnetic fields around bridges
... be distorted, affecting the outputs of
...ass systems. To mitigate these effects,
...*ators*, also called *observers*, that use a
...r computational models of the vehicle's
...e vehicle. The vehicle state information
...sed as a feedback signal to the vehicle

...nd control allocation system. The input
...ired position, speed, heading, acceler-
...cipated forces and moments that must

be applied to the vehicle, as shown in Fig. 1.5) in order for it to achieve the desired behavior. A *control allocation* system is then used to determine which available actuators (propellers, fins, etc.) should be powered to produce the required forces and moments and to translate the desired forces/moments into actuator commands, such as propeller RPM or fin angle. As it may be possible for more than one combination of actuators to produce the same set of required control forces and moments, the control allocation system often optimizes the choice of actuators to minimize the power or time required to perform a desired maneuver. Note that several different types of controllers may need to be used on the same vehicle, depending upon its operating condition. For example, a vehicle may be required to transit from one port to another, during which time an autopilot (speed and heading controller) is employed. Later it might need to hold position (station–keep) while waiting to be moored, during which time a station–keeping controller may be employed. Systems developed to switch from one type of controller to another, which may have a different dynamic structure, are called *switched-* or *hybrid-control* systems. The system used to select the appropriate controller is known as a *supervisor*.

### 1.4.1 Multilayered Software Architectures for Unmanned Vehicles

In recent robotics and control research involving marine systems, the guidance system is sometimes referred to as a *high-level planner* or *high-level controller* and the block labeled "control system" in Fig. 1.5 is sometimes referred to as a *low-level controller*. Hierarchical models of marine vehicle control systems, such as the one shown in Fig. 1.6, are often constructed along similar lines, see for example [12]. In Fig. 1.6, high-level control, here termed strategic control, involves trajectory planning and what has been traditionally termed *guidance*. The intermediate layer, here termed *tactical control*, involves the controller and control allocation code. The lowest layer, here denoted as *execution control* involves the control of individual actuators, closed loop controllers that regulate propeller RPM or fin angle, for example. In general, one can characterize the top-most layer as requiring the highest computational power, but the lowest bandwidth for communications with other parts of the control system. Conversely, one can categorize the lowest layer as requiring the least computational power, but needing the most bandwidth for communication with actuators.

If one thinks of the functions performed by each of the control layers, then one must be careful to develop an architecture that facilitates the computational speeds required for the higher-level decision-making and trajectory-planning functions, as well as, the higher communication bandwidths of the low–level functions. Thus far, it has been found that that the more tractable software architectures for unmanned marine vehicles are those based on layers of abstraction, where decision making, sensing and actuation are separated in a way that minimizes the number of internal states of the system [1, 11, 14, 22, 35]. The architecture of individual unmanned marine systems tends to have three or four layers of abstraction [1, 14, 37].

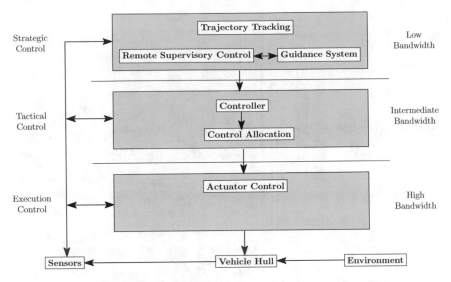

**Fig. 1.6** A typical Guidance, Navigation and Control System used for marine vehicles

An example of a such a system architecture is shown in Fig. 1.7. Here, the guidance system has been split into a *Task Planning Layer*, a *Behavior Layer* and a *Trajectory Planning Layer*. Similarly, the control system is split into a *Trajectory Tracking Layer* and a *Feedback Control Layer*. Based on a mission model the task planning layer determines which task the vehicle should currently perform and queues future tasks that should occur as the vehicle performs its mission. As unmanned vehicles can encounter an infinite number of possible scenarios when operating in the real world, rather than trying to preprogram a vehicle to react to operational conditions in very specific ways, unmanned vehicles can be programmed to behave is more general ways known as *behaviors*. For example, when operating in the vicinity of manned vehicles, it is preferable if an unmanned vehicle can behave as a manned vehicle might so that the human operators of other boats can predict and appropriately react to the behavior of the unmanned vehicle. In many ports vehicles are expected to follow the Coast Guard 'rules of the road', abbreviated as *COLREGs*, that govern which vehicles have the right of way or how vehicles can pass one another. USVs can be programmed to exhibit COLREGs behaviors [7, 8, 38, 39, 47].

### 1.4.2 Inter-Process Communications Methods

As mentioned above, the software architecture for the implementation of multilayered GNC architectures requires reliable communication paths between the layers with sufficient bandwidth, so that vehicle state information and trajectories can be computed and implemented in real-time. Generally, separate software modules are

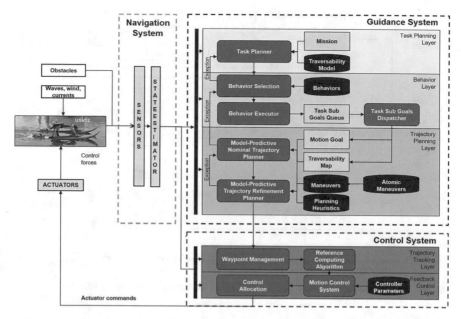

**Fig. 1.7** A multilayered guidance and control architecture for unmanned surface vehicles. Adapted from [46] by permission from Springer © (2012)

written for vehicle state estimation, control and path/trajectory planning. Additionally, each hardware sensor and each actuator will have a software driver that manages communication, exchanging data with the software modules. During implementation, each of the software modules and hardware drivers is run as a separate process on the vehicle's computer system. In order for the vehicle to operate in real-time, some, or all, of these processes may need to be running simultaneously. Communication between running software modules and drivers is known as *inter-process communication* (IPC).

Early unmanned marine vehicle architectures often relied upon customized interrupt programming and multithreading to execute multiple processes concurrently, e.g. [42]. A disadvantage of such approaches is that process scheduling, through the use of semaphores and timers, can become very important (and difficult to set up) for preventing multiple threads from interfering with one another when sharing hardware resources. An additional problem is that hardware support for multithreading is more visible to software, sometimes requiring lengthy modifications to both application programs and operating systems when sensors or actuators are added, removed or reconfigured.

In general, modularity is crucial for handling the complexity of real-time operations. Whenever possible, the software architecture should be modular to permit the addition or removal of systems and subsystems, as required. More recent unmanned vehicle software architectures bypass some of the of the issues mentioned above by taking advantage of multiprocess operating systems, such as the Robot Operating

System [34], the Mission Oriented Operating System [6], or the Lightweight Communications and Marshalling systems [11]. These systems permit large amounts of data to be transferred between the software modules in a distributed guidance, navigation and control system through the use of special software libraries that encapsulate the tasks required for message passing between processes. The communication protocols used in these operating systems relies on a publish/ subscribe model to pass messages between processes and threads. For example, LCM uses the User Datagram Protocol (UDP) to simultaneously broadcast messages to all clients that have subscribed to read them from a given process. At the same time, the process that publishes the message does not need to wait to receive confirmation from subscribing processes that its message has been correctly received (handshaking). This asynchronous, peer-to-peer, multicasting approach simplifies IPC by negating the requirement for a main "server" to arbitrate between subscribers and publishers. It is scalable and permits message passing at higher bandwidths.

An example of a software architecture implemented on a set of unmanned vehicles is shown in Fig. 1.8. The system was used to experimentally test the automatic generation of COLREGs-compliant unmanned vehicle behaviors in [7, 8] in a marine waterway. Each sensor (GPS, electronic compass and inertial measurement unit) is handled by a separate driver operating in parallel and has an exclusive LCM channel. The drivers parse the sensor information, condense the data into an LCM message, and distribute it to the appropriate LCM channel. All sensor information is handled by a state estimator, which collects and organizes the filtered state data into a separate message, which is then transmitted directly to the high-level planner via a separate LCM channel. Along with vehicle telemetry, the vehicle state LCM message includes an overall goal for the high-level planner. This goal may be either static or dynamic, depending on the type of COLREGs behavior being utilized. The end goal consists of a desired position in the $NED$ coordinate system, as well as a desired endpoint heading. Once the estimated state and goal are established and transmitted, the high-level planner produces a dynamically feasible trajectory from a dynamic model of the vehicle. The trajectories received by the low-level control system are broken down into discrete sets of waypoints and a line of sight maneuvering system is employed to guide the vehicle to within a user-set minimum acceptable distance of each waypoint.

## 1.5  Dynamic Modeling of Marine Vehicles

As discussed in Sect. 1.2, the goal of feedback control is to either stabilize the output of a controlled dynamic system about a desired equilibrium point, or to cause a controlled system to follow a desired reference input accurately, despite the path that the reference variable may take, external disturbances and changes in the dynamics of the system. Before designing a feedback controller, we must construct a mathematical model of the system to be controlled and must understand the dynamic responses expected from the closed–loop system, so that we reasonably predict the resulting

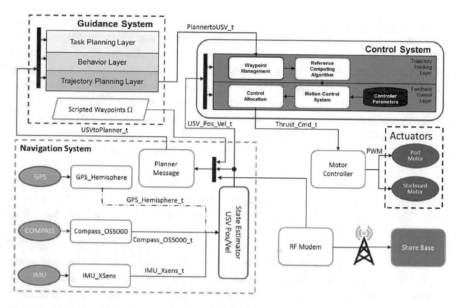

**Fig. 1.8** A multilayered guidance and control architecture for unmanned surface vehicles. Reprinted from [8] © (2015), with permission from Elsevier

performance. Thus, the design of any feedback system fundamentally depends on a sufficiently accurate dynamic model of the system and an appropriate estimate of the closed–loop system's dynamic response.

Thus, the first step in the design of a controller is the development of a suitable dynamic model of the system to be controlled. In addition to physical modeling based on first principles, such as Newton's 2nd Law or the use of Lagrange's Energy Methods for mechanical systems, there are also System Identification techniques that utilize experimental data to determine the parameters in, and sometimes the structure of, the dynamic model of a system.

In this text, we will usually assume that a dynamic model of the system to be controlled has already been developed and focus on how the equations that govern the system's response can be analyzed. However, especially when working with non-linear and/or underactuated systems, an understanding of the underlying dynamics of a system can assist with controller design. In this vein, some of the basic principles involved in modeling marine vehicle systems are summarized in this section. More detailed information on the dynamic modeling of marine systems can be found in the following references [2, 17, 18, 27, 28].

It is traditional to split the study of the physics of motion of particles and rigid bodies into two parts, kinematics and kinetics. Kinematics generally refers to the study of the geometry of motion and kinetics is the study of the motion of a body resulting from the applied forces and moments that act upon its inertia.

## 1.5.1 Kinematics of Marine Vehicles

The key kinematic features of interest for the control of marine vessels are the vehicle's position and orientation; linear and angular velocities; and linear and angular accelerations. Controlling the trajectory of a system is an especially important objective in robotics and unmanned vehicle systems research, where the position and orientation of a vehicle are often referred to as its *pose*. When developing a dynamic model of a marine vehicle, it is common to specify its position with respect to a North-East-Down ($NED$) coordinate system, which is taken to be locally tangent to the surface of the Earth. As shown in Fig. 1.9a, the $NED$ system is fixed at a specific latitude and longitude at the surface of the Earth and rotates with it. Note that the NED frame is not an inertial frame, as it is rotating. Thus, strictly speaking, there are forces arising from the centripetal and Coriolis accelerations caused by the Earth's rotation that act on the vehicle. However, at the speeds that marine vehicles typically travel, these forces are generally much smaller than gravitational, actuator and hydrodynamic forces (such as drag) and so are typically ignored in the equations of motion. However, as we shall see below, there are centripetal and Coriolis effects that arise from the vehicle's rotation about its own axes. Under certain operational conditions, for example when forward and rotational speeds of a vehicle are significant, these effects must be included in a vessel's equations of motion.

The position of the vehicle in NED coordinates is given by

$$\mathbf{p}_{b/n}^n = \begin{bmatrix} x_n \\ y_n \\ z_n \end{bmatrix}, \tag{1.1}$$

where we use the superscript $n$ to indicate that the position is in *NED* coordinates and the subscript $b/n$ to indicate that we are measuring the position of the vehicle (body) with respect to the $NED$ '$n$-frame'. Similarly, the orientation of the vehicle is given by

$$\Theta_{nb} = \begin{bmatrix} \phi \\ \theta \\ \psi \end{bmatrix}, \tag{1.2}$$

where $\phi$ is defined as the *roll* angle about the $x_b$ axis, $\theta$ is the *pitch* angle about the $y_b$ axis, and $\psi$ is the *yaw* angle about the $z_b$ axis. These angles are often referred to as the *Euler angles* and are a measure of the cumulative angular change over time between the $NED$ and body-fixed coordinate systems (Fig. 1.10). Note that the origin of the body-fixed coordinate systems shown here has been taken to be the center of gravity $G$ (center of mass) of the vehicle. In some applications it may be more convenient to locate the origin of the coordinate systems elsewhere, such as at the center of buoyancy $B$ of a vehicle or at a position where measurements of a vehicle's motion

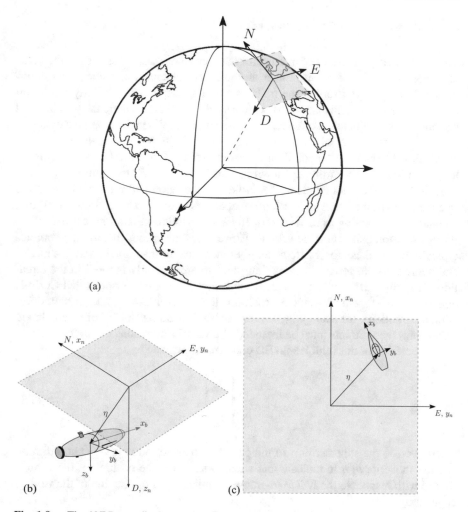

**Fig. 1.9  a** The $NED$ coordinate system referenced to an Earth–fixed, Earth–centered coordinate system. **b** An underwater vehicle referenced with respect to the NED coordinate system. **c** A surface vehicle referenced with respect to the NED coordinate system. Note the motion of surface vessels can typically be assumed to lie within the plane of the $N$–$E$ axes

are collected, such as the location of an IMU. According to D'Alembert's Principle, the forces acting on rigid body can be decomposed into the linear forces applied at the center of gravity of the body and rotational torques applied about the center of gravity [24]. Because of this, when the origin of the coordinate system is not located at the center of gravity additional terms must be included in the equations of motion to determine the local accelerations and velocities in the body-fixed frame, which arise from externally applied forces and moments.

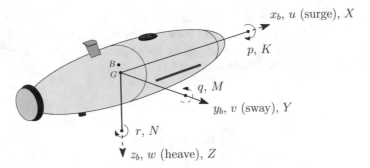

**Fig. 1.10** A body-fixed coordinate system with its origin located at the center of gravity $G$ of an Autonomous Underwater Vehicle (AUV)

As shown in Fig. 1.10, the coordinate along the longitudinal (*surge*) axis of the vehicle is $x_b$, the coordinate along the transverse (*sway*) axis is $y_b$ and the coordinate along the remaining perpendicular (*heave*) axis is $z_b$. The velocity of the vehicle measured in the body fixed frame is given by

$$\mathbf{v}_{b/n}^b = \begin{bmatrix} u \\ v \\ w \end{bmatrix}, \tag{1.3}$$

and the angular velocity of the vehicle in the body–fixed frame is

$$\boldsymbol{\omega}_{b/n}^b = \begin{bmatrix} p \\ q \\ r \end{bmatrix} = \begin{bmatrix} \dot{\phi} \\ \dot{\theta} \\ \dot{\psi} \end{bmatrix}. \tag{1.4}$$

Here, $p$ is the *roll rate* about the $x_b$ axis, $q$ is the *pitch rate* about the $y_b$ axis and $r$ is the *yaw rate* about the $z_b$ axis. As also shown in Fig. 1.10, the external linear forces acting on a marine vehicle are written as

$$\mathbf{f}_b^b = \begin{bmatrix} X \\ Y \\ Z \end{bmatrix} \tag{1.5}$$

and the corresponding external moments are given by

$$\mathbf{m}_b^b = \begin{bmatrix} K \\ M \\ N \end{bmatrix}. \tag{1.6}$$

The generalized motion of an unmanned vehicle has six degrees of freedom (DOF), three translational (linear) and three rotational. We can write the full six DOF equations in vector form by introducing the generalized position $\eta$, velocity $\mathbf{v}$ and force $\tau$, each of which has six components

$$\eta = \begin{bmatrix} \mathbf{p}_{b/n}^n \\ \Theta_{nb} \end{bmatrix} = \begin{bmatrix} x_n \\ y_n \\ z_n \\ \phi \\ \theta \\ \psi \end{bmatrix}, \tag{1.7}$$

$$\mathbf{v} = \begin{bmatrix} \mathbf{v}_{b/n}^b \\ \boldsymbol{\omega}_{b/n}^b \end{bmatrix} = \begin{bmatrix} u \\ v \\ w \\ p \\ q \\ r \end{bmatrix}, \tag{1.8}$$

and

$$\tau = \begin{bmatrix} \mathbf{f}_b^b \\ \mathbf{m}_b^b \end{bmatrix} = \begin{bmatrix} X \\ Y \\ Z \\ K \\ M \\ N \end{bmatrix}. \tag{1.9}$$

**Remark 1.1** The use of mathematical symbols for making definitions or assumptions shorter is very common in the robotics and control literature. A short list of some of the symbols commonly encountered is given in Table 1.1.

**Remark 1.2** In robotics and control research it is often important to keep track of the dimensions and ranges of the components of each vector involved.

**Table 1.1** Common mathematical symbols used in robotics and control.

| Symbol | Meaning |
|---|---|
| $\in$ | An element of |
| $\subset$ | A subset of |
| $\forall$ | For every |
| $\exists$ | There exists |
| $\mathbb{R}$ | The set of real numbers |
| $\mathbb{R}^+$ | The set of positive real numbers |
| $\mathbb{Z}$ | The set of integers |
| $\Rightarrow$ | Implies |

Typically, the components of vectors, such as speed and position, are defined along the set of both positive and negative real numbers and so the dimensions of such vectors are specified as $\mathbf{p}_{b/n}^n \in \mathbb{R}^3$, $\mathbf{v}_{b/n}^b \in \mathbb{R}^3$, $\mathbf{f}_b^b \in \mathbb{R}^3$, and $\mathbf{m}_b^b \in \mathbb{R}^3$, for example. However, the principal values of angular measurements have a limited range, e.g. $0 \leq \phi < 2\pi$, and so are often specified as $\Theta_{nb} \in \mathcal{S}^3$, where $\mathcal{S} := [0, 2\pi)$.

**Coordinate system conventions for surface vehicles**

Unmanned and manned underwater vehicles, such as AUVs and submarines, typically employ the body–fixed coordinate system convention shown in Fig. 1.10. However, the study of the motion of surface vehicles also includes an examination of their response to wave and wind conditions at the water's surface. In the fields of Naval Architecture and Marine Engineering, these surface conditions are often known as a *seaway*. Traditionally, the dynamic modeling of surface vehicles has been split into separate disciplines known as *seakeeping* and *maneuvering* (e.g. [17, 27, 28]). Maneuvering calculations generally examine the dynamic response of a vessel when moving through calm, flat seas and with an eye towards understanding its course-keeping ability and trajectory. Maneuvering calculations do include the wave-making resistance of the vessel (i.e. the drag on the hull generated when the vehicle moves through calm water and creates waves). Seakeeping calculations on the other hand typically examine a vessel's linear and angular displacements from a nominal trajectory when disturbed by the waves already present in a seaway. Seakeeping calculations are typically split into Froude-Kryloff forces caused by the pressure field of waves impacting the hull of a ship, as well as, the 'radiation forces' developed from the ship's own motion [16].

Although, there have been some recent efforts to merge the analyses performed in these two similar disciplines, e.g. [40, 44, 45], the nomenclature and modeling conventions used have traditionally differed slightly, which can lead to confusion. Maneuvering studies have often used a body-fixed coordinate system analogous to that used with underwater vehicles, such as in Fig. 1.11a, with its origin at $G$, the

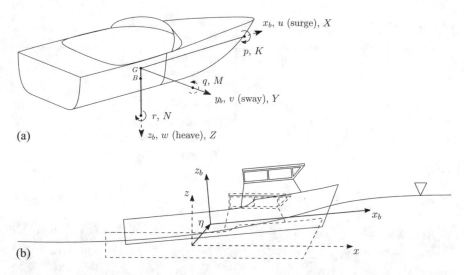

**Fig. 1.11** **a** A traditional maneuvering body–fixed coordinate system. **b** A conventional sea–keeping body–fixed frame

positive sway axis pointing out of the starboard side of the vessel, and the positive heave axis pointing towards the bottom of the vehicle.

However, in place of the $NED$ plane, seakeeping studies often use a coordinate system with its origin at the centroid of the vessel's waterplane and translating at the vessel's nominal constant forward speed (the dashed $x$–$y$–$z$ frame in Fig. 1.11b). A second, body–fixed frame, is used to specify the wave-induced displacement of the vessel $\eta$ from the $x$–$y$–$z$ frame (the $x_b$–$y_b$–$z_b$ frame in Fig. 1.11b). In the $x_b$–$y_b$–$z_b$ coordinate system the positive heave axis points upward, the positive surge axis points forward and the positive sway axis points towards the port side of the vessel. In this text, we shall always use the traditional maneuvering body–fixed coordinate system convention shown in Figs. 1.10 and 1.11a.

### Coordinate System Transformations
When the orientation and position of a marine vehicle changes, the representation of its inertial properties, the inertia tensor, is constant if expressed in a body–fixed coordinate system. Thus, the use of body–fixed coordinate systems can substantially simplify the modeling and dynamic simulation of a vehicle. However, one of the main objectives of vehicle simulations is to predict the time–dependent position and orientation of a vehicle in a real operating space, which is usually specified in a $NED$ coordinate system. In order to determine position and orientation in the $NED$ frame, the velocity of the vehicle in body–fixed coordinates must be converted to $NED$ coordinates and integrated in time. Therefore, it is important to understand how one can perform coordinate transformations to express the velocity vector components in multiple reference frames. Coordinate system transformation techniques are equally important in other instances, for example when converting the representation of

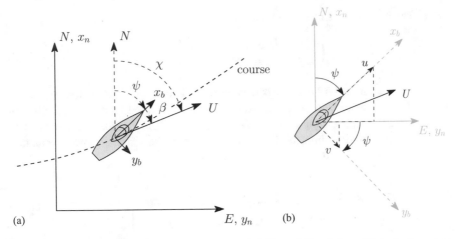

**Fig. 1.12** **a** 3 DOF maneuvering coordinate system definitions. **b** Resolution of the velocity vector components in the $NED$ and body–fixed frames

forces/moments generated by a vectored thruster from a thruster–fixed coordinate system, where the relation between the orientation of the flow entering the thruster and the forces/moments generated is most easily modeled, to a vehicle body–fixed coordinate system, where the thruster forces/moments are best represented during dynamic simulation.

Consider the motion of a boat in the $NED$ coordinate system, as shown in Fig. 1.12a. Here we will assume that the motion is confined to the $N$–$E$ plane and restricted to three degrees of freedom, which include rotation about the $D$ axis measured with respect to $N$ and linear translations in the $N$–$E$ plane. Here the *course* (trajectory) of the boat is represented as a dashed curve. At each time instant the velocity vector $\mathbf{U}$ is tangent to the course. The *course angle* $\chi$, is the angle between $\mathbf{U}$ and $N$. Note that, in general, the bow of the vessel does not need to point in the same direction as the velocity vector. The angle between the $x_b$ axis of the vessel, which usually points through its bow, and $N$ is called the *heading angle* $\psi$. When $\chi$ and $\psi$ differ, the angle between them is known as the drift angle $\beta$ (sometimes also called either the *sideslip angle* or, especially in sailing, the *leeway angle*), and the component of velocity perpendicular to the longitudinal axis of the boat $v$ (sway speed) is called the *sideslip*.

An examination of the velocity vector $\mathbf{U}$ in both the body–fixed and the $NED$ coordinate systems, as shown in Fig. 1.12b, allows one to see how the velocity vector expressed in components along $x_b$ and $y_b$ can be converted to components in the $NED$ frame. Writing the $NED$ frame velocity components $\dot{x}_n$ and $\dot{y}_n$ in terms of the $u$ and $v$ components of the body–fixed frame gives

$$\dot{x}_n = u \cos \psi - v \sin \psi, \tag{1.10}$$

$$\dot{y}_n = u \sin \psi + v \cos \psi, \tag{1.11}$$

which can be rewritten in vector form as

$$\begin{pmatrix} \dot{x}_n \\ \dot{y}_n \end{pmatrix} = \begin{bmatrix} \cos\psi & -\sin\psi \\ \sin\psi & \cos\psi \end{bmatrix} \begin{pmatrix} u \\ v \end{pmatrix}. \tag{1.12}$$

Here, in equation 1.12 the matrix in square brackets transforms vectors expressed in $NED$ coordinates into a vector expressed in body–fixed frame coordinates. In this relatively simple example, one can see that the coordinate transformation can be thought of as a rotation about the $D$ axis. In three dimensions, the same transformation matrix can be expressed as

$$R_{z_b,\psi} = \begin{bmatrix} \cos\psi & -\sin\psi & 0 \\ \sin\psi & \cos\psi & 0 \\ 0 & 0 & 1 \end{bmatrix}, \tag{1.13}$$

where the subscript $z, \psi$ is used to denote a rotation in heading angle $\psi$ about the $z_b$ axis.

Generally, a three dimensional transformation between any two coordinate systems can be thought of as involving a combination of individual rotations about the three axes of a vehicle in body–fixed coordinates. In this context, the heading angle $\psi$, pitch angle $\theta$ and roll angle $\phi$ of the vehicle are Euler angles, and the transformations corresponding to rotation about the pitch and roll axes can be respectively represented as

$$R_{y_b,\theta} = \begin{bmatrix} \cos\theta & 0 & \sin\theta \\ 0 & 0 & 1 \\ -\sin\theta & 0 & \cos\theta \end{bmatrix}, \tag{1.14}$$

and

$$R_{x_b,\phi} = \begin{bmatrix} 1 & 0 & 0 \\ 0 & \cos\phi & -\sin\phi \\ 0 & \sin\phi & \cos\phi \end{bmatrix}. \tag{1.15}$$

In order to determine the transformation between any arbitrary total rotation between the body–fixed coordinate system and the $NED$ frame, we can use the product of the transformation matrices. For example, the complete transformation for an orientation that can be represented as first a rotation in heading and then a rotation in pitch would be given by the product $R_{y_b,\theta} R_{z_b,\psi}$. As shown in Example 1.3 below, the order in which the rotations are performed is important.

**Example 1.3** Consider the successive rotations of a surface vessel, as shown in Fig. 1.13. The final orientation of the vessel after successive $+90°$ rotations in roll first and then pitch, will be different from the final orientation of the vessel after successive $+90°$ rotations in pitch first and then roll. From a mathematical standpoint, this is the result of the fact that matrix operations do not commute, such that in general $R_{y_b,\theta} R_{z_b,\psi} \neq R_{z_b,\psi} R_{y_b,\theta}$. □

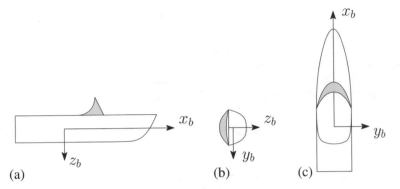

**Fig. 1.13** Successive $+90°$ rotations of the boat shown in1.11a: **a** Initial orientation. **b** Final orientation after roll first and then pitch. **c** Final orientation after pitch first and then roll

**Remark 1.3** As the order of the sequence of rotations is important, in both the fields of marine vehicle maneuvering and aeronautical/astronautical engineering it has become accepted practice for the heading angle $\psi$ to be taken first, the pitch angle $\theta$ to be taken secondly, and lastly the roll angle $\phi$ to be taken.

Thus, the transformation matrix corresponding to a general three dimensional rotation between two coordinate systems is defined as

$$\mathbf{R}_b^n(\Theta_{nb}) := R_{z_b,\psi} R_{y_b,\theta} R_{x_b,\phi}$$

$$= \begin{bmatrix} c\psi c\theta & -s\psi c\phi + c\psi s\theta s\phi & s\psi s\phi + c\psi s\theta c\phi \\ s\psi c\theta & c\psi c\phi + s\psi s\theta s\phi & -c\psi s\phi + s\psi s\theta c\phi \\ -s\theta & c\theta s\phi & c\theta c\phi \end{bmatrix}, \tag{1.16}$$

where $c(\cdot)$ and $s(\cdot)$ are used to represent the $\cos(\cdot)$ and $\sin(\cdot)$ functions, respectively.

Despite the complicated appearance of the rotation matrices, they have some nice mathematical properties that make them easy to implement, and which can be used for validating their correct programming in simulation codes. Rotation matrices belong the *special orthogonal group of order 3*, which can be expressed mathematically as $\mathbf{R}_b^n(\Theta_{nb}) \in SO(3)$, where

$$SO(3) := \{\mathbf{R} | \mathbf{R} \in \mathbb{R}^{3\times3}, \text{ where } \mathbf{R} \text{ is orthogonal and } \det(\mathbf{R}) = 1\}. \tag{1.17}$$

The term $\det(\mathbf{R})$ represents the determinant of $\mathbf{R}$. In words, this means that rotation matrices are orthogonal and their inverse is the same as their transpose $\mathbf{R}^T = \mathbf{R}^{-1}$, so that $\mathbf{R}\mathbf{R}^T = \mathbf{R}\mathbf{R}^{-1} = \mathbf{1}$, where $\mathbf{1}$ is the identity matrix

$$\mathbf{1} := \begin{bmatrix} 1 & 0 & 0 \\ 0 & 1 & 0 \\ 0 & 0 & 1 \end{bmatrix}. \tag{1.18}$$

Use of the transpose can make it very convenient to switch back and forth between vectors in the body–fixed and the $NED$ frames, e.g.

$$\dot{\mathbf{p}}_{b/n}^n = \mathbf{R}_b^n(\Theta_{nb})\mathbf{v}_{b/n}^b, \tag{1.19}$$

and

$$\mathbf{v}_{b/n}^b = \mathbf{R}_b^n(\Theta_{nb})^T \dot{\mathbf{p}}_{b/n}^n. \tag{1.20}$$

When the Euler angles are limited to the ranges

$$0 \le \psi < 2\pi, \quad -\frac{\pi}{2} < \theta < \frac{\pi}{2}, \quad 0 \le \phi \le 2\pi, \tag{1.21}$$

any possible orientation of a vehicle can be attained. When $\theta = \pm\pi/2$, $\psi$ and $\phi$ are undefined, as no unique set of values for them can be found. This condition, wherein the angular velocity component along the $x_b$ axis cannot be represented in terms of the Euler angle rates [24], is known as *gimbal lock*. Marine surface vehicles are unlikely to encounter scenarios in which gimbal lock can occur, however, it is possible for it to occur with ROVs, AUVs or submarines undergoing aggressive maneuvers. In such cases, the kinematic equations could be described by two Euler angle representations with different singularities and the singular points can be avoided by switching between the representations. Alternatively, the use of unit quaternions for performing coordinate transformations, instead of Euler angles, can be used to circumvent the possibility of gimbal lock (see [18]).

**Definition 1.1** [Time derivative of $\mathbf{R}_b^n$ and the cross product operator $\mathbf{S}(\lambda)$] When designing controllers for marine vehicles it is often necessary to compute the time derivative of the rotation matrix $\mathbf{R}_b^n$. A convenient way to do this is to use the cross product operator. Let $\lambda := [\lambda_1 \ \lambda_2 \ \lambda_3]^T$ and $\mathbf{a}$ be a pair of vectors. The cross product between $\lambda$ and $\mathbf{a}$ can be expressed as

$$\lambda \times \mathbf{a} = \mathbf{S}(\lambda)\mathbf{a}, \tag{1.22}$$

where $\mathbf{S}$ is

$$\mathbf{S}(\lambda) = -\mathbf{S}^T(\lambda) = \begin{bmatrix} 0 & -\lambda_3 & \lambda_2 \\ \lambda_3 & 0 & -\lambda_1 \\ -\lambda_2 & \lambda_1 & 0 \end{bmatrix}. \tag{1.23}$$

Then, the time derivative of the rotation matrix that transforms vectors between the body-fixed and the NED reference frames is

$$\dot{\mathbf{R}}_b^n = \mathbf{R}_b^n \mathbf{S}(\omega_{b/n}^b), \tag{1.24}$$

where

$$S(\omega_{b/n}^b) = -S^T(\omega_{b/n}^b) = \begin{bmatrix} 0 & -r & q \\ r & 0 & -p \\ -q & p & 0 \end{bmatrix}. \tag{1.25}$$

☐

Since the orientation between the NED frame and the body–fixed coordinate system can change as a vehicle maneuvers, the relation between the rate of change of the Euler angles $\dot{\Theta}_{nb}$ and the angular velocity of the vehicle $\omega_{b/n}^b$ requires the use of a special transformation matrix $T_{\Theta}(\Theta_{nb})$. As shown in [24], the angular velocities associated with changes in the Euler angles $\dot{\phi}$, $\dot{\theta}$ and $\dot{\psi}$ are not perpendicular to one another. Another way of thinking about this is that the Euler rotations are not taken about orthogonal body axes, but about axes which change during the rotation process [27]. Projecting $\dot{\Theta}_{nb}$ along the directions of the components of the angular velocity vector $\omega_{b/n}^b$, gives

$$\begin{aligned} p &= \dot{\phi} - \dot{\psi} \sin \theta \\ q &= \dot{\theta} \cos \phi + \dot{\psi} \sin \phi \cos \theta \\ r &= -\dot{\theta} \sin \phi + \dot{\psi} \cos \phi \cos \theta, \end{aligned} \tag{1.26}$$

or in vector form $\omega_{b/n}^b = T_{\Theta}^{-1}(\Theta_{nb})\dot{\Theta}_{nb}$, where

$$T_{\Theta}^{-1}(\Theta_{nb}) = \begin{bmatrix} 1 & 0 & -s\theta \\ 0 & c\phi & c\theta s\phi \\ 0 & -s\phi & c\theta c\phi \end{bmatrix}. \tag{1.27}$$

Inverting the relations in equations 1.26, gives

$$\begin{aligned} \dot{\phi} &= p + (q \sin \phi + r \cos \phi) \tan \theta \\ \dot{\theta} &= q \cos \phi - r \sin \phi \\ \dot{\psi} &= (q \sin \phi + r \cos \phi) \sec \theta, \end{aligned} \tag{1.28}$$

or in vector form $\dot{\Theta}_{nb} = T_{\Theta}(\Theta_{nb})\omega_{b/n}^b$ so that

$$T_{\Theta}(\Theta_{nb}) = \begin{bmatrix} 1 & s\phi t\theta & c\phi t\theta \\ 0 & c\phi & -s\phi \\ 0 & \dfrac{s\phi}{c\theta} & \dfrac{c\phi}{c\theta} \end{bmatrix}, \tag{1.29}$$

where $t(\cdot) := \tan(\cdot)$. Note that $\mathbf{T}^T \neq \mathbf{T}^{-1}$.

**The Kinematic Equations**
Putting it all together, the full six DOF kinematic equations relating the motion of a marine vehicle in the NED coordinate system to its representation in the body–fixed coordinate system can be written compactly in vector form as

$$\dot{\boldsymbol{\eta}} = \mathbf{J}_\Theta(\boldsymbol{\eta})\mathbf{v}, \tag{1.30}$$

which, in matrix form, corresponds to

$$\begin{pmatrix} \dot{\mathbf{p}}_{b/n}^n \\ \dot{\Theta}_{nb} \end{pmatrix} = \begin{bmatrix} \mathbf{R}_b^n(\Theta_{nb}) & \mathbf{0}_{3\times3} \\ \mathbf{0}_{3\times3} & \mathbf{T}_\Theta(\Theta_{nb}) \end{bmatrix} \begin{pmatrix} \mathbf{v}_{b/n}^b \\ \boldsymbol{\omega}_{b/n}^b \end{pmatrix}, \tag{1.31}$$

where $\mathbf{0}_{3\times3}$ is a $3 \times 3$ matrix of zeros,

$$\mathbf{0}_{3\times3} := \begin{bmatrix} 0 & 0 & 0 \\ 0 & 0 & 0 \\ 0 & 0 & 0 \end{bmatrix}. \tag{1.32}$$

When studying the stability of submarines, it is common to assume that the motion is constrained to a vertical plane, with only surge, heave and pitch motions. Similarly, in many instances the motion of surface vehicles can be assumed to be constrained to the horizontal plane with only three degrees of freedom: surge, sway and yaw. Maneuvering stability for these cases is presented in [27], for example. For surface vehicles, the three DOF kinematic equations reduce to

$$\dot{\boldsymbol{\eta}} = \mathbf{R}(\psi)\mathbf{v}, \tag{1.33}$$

where $\mathbf{R}(\psi) = R_{z_b,\psi}$ (see equation 1.13) and

$$\boldsymbol{\eta} = \begin{pmatrix} x_n \\ y_n \\ \psi \end{pmatrix}, \quad \text{and} \quad \mathbf{v} = \begin{pmatrix} u \\ v \\ r \end{pmatrix}. \tag{1.34}$$

**Velocity and acceleration in NED and body-fixed frames**
Consider a vector $\mathbf{r}$ in the $N-E$ plane, rotating about the origin as shown in Fig. 1.14. The body-fixed frame is selected so that it rotates with the same angular velocity $\dot{\psi}$ as $\mathbf{r}$. Let the unit vectors along the $x_b$, $y_b$ and $z_b$ axes be denoted by $\hat{i}_b$, $\hat{j}_b$ and $\hat{k}_b$, respectively. Then,

$$\mathbf{r} = r_x \hat{i}_b + r_y \hat{j}_b \quad \text{and} \quad \boldsymbol{\omega} = \dot{\psi}\hat{k}_b. \tag{1.35}$$

The time derivative of $\mathbf{r}$ is

**Fig. 1.14** Relative rotation between the NED frame and a body-fixed frame in two dimensions. The $z_n$ axis of the NED frame coincides with the $z_b$ axis of the body-fixed frame. The body-fixed frame rotates with the vector **r**, which has components $r_x$ and $r_y$

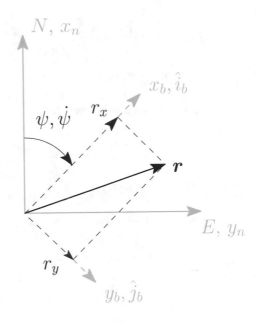

**Fig. 1.15** The unit direction vectors $\hat{i}_b$ and $\hat{j}_b$ correspond to the directions of the body-fixed axes $x_b$ and $y_b$, respectively. After a small angular displacement $\Delta\psi$, the associated changes in $\hat{i}_b$ and $\hat{j}_b$ are given by $\Delta\hat{i}_b$ and $\Delta\hat{j}_b$, respectively

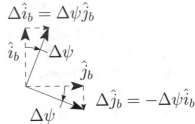

$$\dot{\mathbf{r}} = \dot{r}_x\hat{i}_b + r_x\frac{d\hat{i}_b}{dt} + \dot{r}_y\hat{j}_b + r_y\frac{d\hat{j}_b}{dt}, \tag{1.36}$$

where the time derivatives of the unit vectors $\hat{i}_b$ and $\hat{j}_b$ arises because their directions change as the body fixed frame rotates. From Fig. 1.15 it can be seen that after a small positive angular change $\Delta\psi$, the corresponding changes in $\hat{i}_b$ and $\hat{j}_b$ are given by

$$\Delta\hat{i}_b = |\hat{i}_b|\Delta\psi\,\hat{j}_b, \qquad \Delta\hat{j}_b = -|\hat{j}_b|\Delta\psi\hat{i}_b,$$
$$\text{and} \tag{1.37}$$
$$= \Delta\psi\,\hat{j}_b, \qquad\qquad = -\Delta\psi\hat{i}_b.$$

The time derivatives can then be computed as

$$\lim_{\Delta t \to 0} \frac{\Delta \hat{i}_b}{\Delta t} = \lim_{\Delta t \to 0} \frac{\Delta \psi}{\Delta t} \hat{j}_b, \qquad \lim_{\Delta t \to 0} \frac{\Delta \hat{j}_b}{\Delta t} = -\lim_{\Delta t \to 0} \frac{\Delta \psi}{\Delta t} \hat{i}_b,$$

and

$$\frac{d\hat{i}_b}{dt} = \dot{\psi} \hat{j}_b \qquad\qquad \frac{d\hat{j}_b}{dt} = -\dot{\psi}\hat{i}_b. \tag{1.38}$$

From Fig. 1.15 it can also be seen that these derivatives can also be rewritten as

$$\frac{d\hat{i}_b}{dt} = \boldsymbol{\omega} \times \hat{i}_b, \quad \text{and} \quad \frac{d\hat{j}_b}{dt} = \boldsymbol{\omega} \times \hat{j}_b. \tag{1.39}$$

Using (1.39), (1.36) can be written as

$$\begin{aligned}
\dot{\mathbf{r}} &= \dot{r}_x \hat{i}_b + \dot{r}_y \hat{j}_b + r_x \boldsymbol{\omega} \times \hat{i}_b + r_y \boldsymbol{\omega} \times \hat{j}_b, \\
&= \dot{r}_x \hat{i}_b + \dot{r}_y \hat{j}_b + \boldsymbol{\omega} \times \left( r_x \hat{i}_b \right) + \boldsymbol{\omega} \times \left( r_y \hat{j}_b \right), \\
&= \dot{r}_x \hat{i}_b + \dot{r}_y \hat{j}_b + \boldsymbol{\omega} \times \left( r_x \hat{i}_b + r_y \hat{j}_b \right), \\
&= \dot{r}_x \hat{i}_b + \dot{r}_y \hat{j}_b + \boldsymbol{\omega} \times \mathbf{r}.
\end{aligned} \tag{1.40}$$

As we can see, the time derivative of $\mathbf{r}$ comes from two parts:

(a) The first part corresponds to the time rate of change of $\mathbf{r}$ as measured in the body-fixed frame. Denote this with an overscript $b$ as

$$\overset{b}{\mathbf{r}} = \dot{r}_x \hat{i}_b + \dot{r}_y \hat{j}_b. \tag{1.41}$$

(b) The second part $\boldsymbol{\omega} \times \mathbf{r}$ arises because of the relative rotation $\boldsymbol{\omega}$ between the body-fixed and NED frames.

Let us denote the total time rate of change of $\mathbf{r}$ in the inertial frame with an overscript $n$ as

$$\overset{n}{\mathbf{r}} = \overset{b}{\mathbf{r}} + \boldsymbol{\omega} \times \mathbf{r}. \tag{1.42}$$

Now, let us extend this to the case that includes relative motion between the origins of the NED and body-fixed frames, as shown in Fig. 1.16.

Let $\mathbf{R}$ be a vector from the origin of the NED frame to the origin of the body fixed frame, $\mathbf{r}$ be a vector from the origin of the body-fixed frame to a point $P$ and $\boldsymbol{\rho}$ be a vector from the origin of the NED frame to the same point $P$. Let $\boldsymbol{\omega}$ be the relative rotation between the body-fixed and NED frames.

**Assumption 1.1** The body-fixed frame is selected so that the rotation of the vector $\mathbf{r}$ is the same as the relative rotation between the body-fixed and NED frames.

Then,

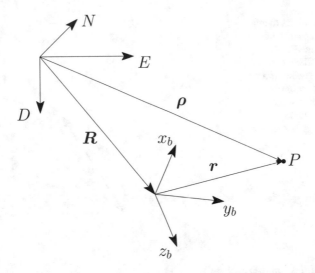

**Fig. 1.16** Relative motion between the NED frame and a body-fixed frame in three dimensions. The motion includes displacement of the origin of the body fixed frame with respect to the NED origin by the vector **R** and the relative rotation rate $\omega$ between the two frames. The vector **r**, which extends from the origin of the body-fixed frame to a point $P$, rotates with respect to the NED frame at same relative rotation rate as the body-fixed frame. The vector $\rho$ extends from the origin of the NED frame to $P$

$$\rho = \mathbf{R} + \mathbf{r}. \tag{1.43}$$

Then the velocity of the point $P$ is given by time derivative of $\rho$, which is

$$\overset{n}{\rho} = \overset{n}{\mathbf{R}} + \overset{n}{\mathbf{r}}, $$

$$= \overset{n}{\mathbf{R}} + \overset{b}{\mathbf{r}} + \omega \times \mathbf{r}. \tag{1.44}$$

Then, using Assumption 1.1 and (1.42), the total acceleration is

$$\overset{nn}{\rho} = \overset{nn}{\mathbf{R}} + \overset{bn}{\mathbf{r}} + \overset{n}{\omega} \times \mathbf{r} + \omega \times \overset{n}{\mathbf{r}}, $$

$$= \overset{nn}{\mathbf{R}} + \overset{bb}{\mathbf{r}} + \omega \times \overset{b}{\mathbf{r}} + \overset{n}{\omega} \times \mathbf{r} + \omega \times \left( \overset{b}{\mathbf{r}} + \omega \times \mathbf{r} \right), $$

$$= \overset{nn}{\mathbf{R}} + \overset{bb}{\mathbf{r}} + \omega \times \overset{b}{\mathbf{r}} + \overset{n}{\omega} \times \mathbf{r} + \omega \times \overset{b}{\mathbf{r}} + \omega \times (\omega \times \mathbf{r}), \tag{1.45}$$

$$= \overset{nn}{\mathbf{R}} + \underbrace{\overset{bb}{\mathbf{r}}}_{\substack{\text{Radial} \\ \text{Acceleration}}} + \underbrace{\omega \times (\omega \times \mathbf{r})}_{\substack{\text{Centripetal} \\ \text{Acceleration}}} + \underbrace{\overset{n}{\omega} \times \mathbf{r}}_{\substack{\text{Tangential} \\ \text{Acceleration}}} + \underbrace{2\omega \times \overset{b}{\mathbf{r}}}_{\substack{\text{Coriolis} \\ \text{Acceleration}}}. $$

**Fig. 1.17** Use of an ordered pair $(\mathbf{p}_{b/n}^n, \mathbf{R})$ specify the motion of a point $P$. The vector $\mathbf{p}_{b/n}^n$ is taken to be fixed in the body-fixed frame so that the motion of the point corresponds to the motion of a rigid body

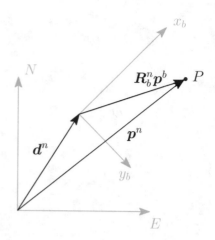

## Homogeneous transformations

The motion of a rigid body can be thought of as an ordered pair $(\mathbf{p}_{b/n}^n, \mathbf{R})$. Since $\mathbf{p}_{b/n}^n \in \mathbb{R}^3$ and $\mathbf{R} \in SO(3)$, the group of all rigid body motions is known as the *special Euclidean group*, $SE(3) := \mathbb{R}^3 \times SO(3)$. Consider the motion of the point $P$, as illustrated in Fig. 1.17.

The motion can be written as the sum of a displacement $\mathbf{d}^n$ and a rotation $\mathbf{R}_b^n$,

$$\mathbf{p}^n = \mathbf{R}_b^n \mathbf{p}^b + \mathbf{d}^n. \tag{1.46}$$

We can compose multiple motions by repeated application of translations and rotations. For example, consider a second translation and rotation from the $B$ frame to the 2 frame

$$\mathbf{p}^b = \mathbf{R}_2^b \mathbf{p}^2 + \mathbf{d}^b. \tag{1.47}$$

The composite motion is given by

$$\begin{aligned} \mathbf{p}^n &= \mathbf{R}_b^n \mathbf{p}^b + \mathbf{d}^n, \\ &= \mathbf{R}_b^n \left( \mathbf{R}_2^b \mathbf{p}^2 + \mathbf{d}^b \right) + \mathbf{d}^n, \\ &= \mathbf{R}_b^n \mathbf{R}_2^b \mathbf{p}^2 + \left( \mathbf{R}_b^n \mathbf{d}^b + \mathbf{d}^n \right), \\ &= \mathbf{R}_2^n \mathbf{p}^2 + \mathbf{d}_2^n, \end{aligned} \tag{1.48}$$

where $\mathbf{R}_2^n := \mathbf{R}_b^n \mathbf{R}_2^b$ and $\mathbf{d}_2^n := \mathbf{R}_b^n \mathbf{d}^b + \mathbf{d}^n$.

If a long sequence of rigid motions is performed, one can see from (1.48) that computation of the final position and orientation will become somewhat complicated. Fortunately, it is possible to represent rigid body motions in a simple matrix form that permits the composition of rigid body motions to be reduced to matrix multiplications, as it was for the composition of rotations in (1.16) above. Rigid body motion can be represented by the set of $4 \times 4$ matrices of the form

$$\mathbf{H} = \begin{bmatrix} \mathbf{R} & \mathbf{d} \\ \mathbf{0}_{1 \times 3} & 1 \end{bmatrix}, \quad \mathbf{R} \in SO(3), \quad \mathbf{d} \in R^3. \tag{1.49}$$

Transformation matrices of the form (1.49) are called *homogeneous transformations*. Since $\mathbf{R}_b^n(\Theta_{nb})$, the inverse transformation is

$$\mathbf{H}^{-1} = \begin{bmatrix} \mathbf{R}^T & -\mathbf{R}^T \mathbf{d} \\ \mathbf{0}_{1 \times 3} & 1 \end{bmatrix}. \tag{1.50}$$

To use $\mathbf{H}$ we must add a component to the representation of the position vector, e.g.

$$\mathbf{p} \rightarrow \mathbf{P} = \begin{bmatrix} \mathbf{p} \\ 1 \end{bmatrix}. \tag{1.51}$$

Then,

$$\mathbf{P}^n = \mathbf{H}_b^n \mathbf{P}^n, \tag{1.52}$$

for example. Six homogeneous transformations can be used to generate $SE(3)$

$$\text{Trans}_{x,a} := \begin{bmatrix} 1 & 0 & 0 & a \\ 0 & 1 & 0 & 0 \\ 0 & 0 & 1 & 0 \\ 0 & 0 & 0 & 1 \end{bmatrix}, \quad \text{Rot}_{x,\phi} := \begin{bmatrix} 1 & 0 & 0 & 0 \\ 0 & c\phi & -s\phi & 0 \\ 0 & s\phi & c\phi & 0 \\ 0 & 0 & 0 & 1 \end{bmatrix}, \tag{1.53}$$

$$\text{Trans}_{y,b} := \begin{bmatrix} 1 & 0 & 0 & 0 \\ 0 & 1 & 0 & b \\ 0 & 0 & 1 & 0 \\ 0 & 0 & 0 & 1 \end{bmatrix}, \quad \text{Rot}_{y,\theta} := \begin{bmatrix} c\theta & 0 & s\theta & 0 \\ 0 & 1 & 0 & 0 \\ -s\theta & 0 & c\theta & 0 \\ 0 & 0 & 0 & 1 \end{bmatrix}, \tag{1.54}$$

$$\text{Trans}_{z,c} := \begin{bmatrix} 1 & 0 & 0 & 0 \\ 0 & 1 & 0 & 0 \\ 0 & 0 & 1 & c \\ 0 & 0 & 0 & 1 \end{bmatrix}, \quad \text{Rot}_{y,\psi} := \begin{bmatrix} c\psi & -s\psi & 0 & 0 \\ s\psi & c\psi & 0 & 0 \\ 0 & 0 & 1 & 0 \\ 0 & 0 & 0 & 1 \end{bmatrix}. \tag{1.55}$$

After a series of motions, the most general form of the homogeneous transformation matrix can be represented as

$$\mathbf{H}_b^n := \begin{bmatrix} n_x & s_x & a_x & p_x \\ n_y & s_y & a_y & p_y \\ n_z & s_z & a_z & p_z \\ 0 & 0 & 0 & 1 \end{bmatrix}, \qquad (1.56)$$

where

$$\mathbf{n} := \begin{bmatrix} n_x \\ n_y \\ n_z \end{bmatrix}, \quad \mathbf{s} := \begin{bmatrix} s_x \\ s_y \\ s_z \end{bmatrix}, \quad \text{and} \quad \mathbf{a} := \begin{bmatrix} a_x \\ a_y \\ a_z \end{bmatrix} \qquad (1.57)$$

are the directions of $x_b$, $y_b$ and $z_b$, respectively, in the $NED$ frame. Additional information about the formulation and use of the homogeneous transformation matrix, especially for applications involving robotic manipulators, can be found in [43]. The use of homogeneous transformations can be particularly useful for the trajectory planning and control of underwater vehicles [10, 36], as one can avoid the singularities associated with Euler angle descriptions of vehicle kinematics.

## 1.5.2  Kinetics of Marine Vehicles

From a control design standpoint, the main goal of developing a dynamic model of a marine vehicle is to adequately describe the vehicle's dynamic response to environmental forces from waves, wind or current and to the control forces applied with actuators, such as fins, rudders, and propellers. The complete equations of motion of a marine vehicle moving in calm, flat water can be conveniently expressed in body-fixed coordinates using the robot-like model of Fossen [18]

$$\dot{\boldsymbol{\eta}} = \mathbf{J}(\boldsymbol{\eta})\mathbf{v} \qquad (1.58)$$

and

$$\mathbf{M}\dot{\mathbf{v}} + \mathbf{C}(\mathbf{v})\mathbf{v} + \mathbf{D}(\mathbf{v})\mathbf{v} + \mathbf{g}(\boldsymbol{\eta}) = \boldsymbol{\tau} + \mathbf{w}_d, \qquad (1.59)$$

where the first equation describes the kinematics of the vehicle (as above) and the second equation represents its kinetics. The terms appearing in these equations are defined in Table 1.2. Here $n$ is the number of degrees of freedom. In general, a marine craft with actuation in all DOFs, such as an underwater vehicle, requires a $n = 6$ DOF model for model-based controller and observer design, while ship and semi-submersible control systems can be designed using an $n = 3$, or 4 DOF model.

The terms $\mathbf{M}$, $\mathbf{C}(\mathbf{v})$ and $\mathbf{D}(\mathbf{v})$ have the useful mathematical properties shown in Table 1.3. These symmetry properties can be exploited in controller designs and stability analyses.

**Table 1.2** Variables used in (1.58) and (1.59)

| Term | Dimension | Description |
|------|-----------|-------------|
| $\mathbf{M}$ | $\mathbb{R}^n \times \mathbb{R}^n$ | Inertia tensor (including added mass effects) |
| $\mathbf{C(v)}$ | $\mathbb{R}^n \times \mathbb{R}^n$ | Coriolis and centripetal matrix (including added mass effects) |
| $\mathbf{D(v)}$ | $\mathbb{R}^n \times \mathbb{R}^n$ | Hydrodynamic damping matrix |
| $\mathbf{g}(\eta)$ | $\mathbb{R}^n$ | Gravity and buoyancy (hydrostatic) forces/moments |
| $\mathbf{J}(\eta)$ | $\mathbb{R}^n \times \mathbb{R}^n$ | Transformation matrix |
| $\mathbf{v}$ | $\mathbb{R}^n$ | Velocity/angular rate vector |
| $\eta$ | $\mathbb{R}^n$ | Position/attitude vector |
| $\tau$ | $\mathbb{R}^n$ | External forces/moments (e.g. wind, actuator forces) |
| $\mathbf{w}_d$ | $\mathbb{R}^n$ | Vector of disturbances |

**Table 1.3** Mathematical properties of $\mathbf{M}$, $\mathbf{C(v)}$ and $\mathbf{D(v)}$

$$\mathbf{M} = \mathbf{M}^T > 0 \Rightarrow \mathbf{x}^T \mathbf{M}\mathbf{x} > 0, \forall \mathbf{x} \neq 0$$
$$\mathbf{C(v)} = -\mathbf{C}^T(\mathbf{v}) \Rightarrow \mathbf{x}^T \mathbf{C(v)}\mathbf{x} = 0, \forall \mathbf{x}$$
$$\mathbf{D(v)} > 0 \Rightarrow \tfrac{1}{2}\mathbf{x}^T [\mathbf{D(v)} + \mathbf{D}^T(\mathbf{v})]\mathbf{x} > 0, \forall \mathbf{x} \neq 0$$

**Remark 1.4** Notice that the velocity, inertial terms and forces in (1.59) are computed in the body-fixed coordinate system. This substantially simplifies the modeling and dynamic simulation of marine vehicle motion because the representation of the inertia tensor is constant when expressed in a body–fixed coordinate system.

An important advantage of representing the equations of motion in this form is that it is similar to that used for robotic manipulators (see [43], for example), which permits one to draw upon a wealth of control and stability analysis techniques developed for more general robotics applications.

**Remark 1.5** The inertia tensor $\mathbf{M}$ includes the effects of *added mass*. These are additional terms added to the true mass, and mass moments of inertia, of a marine vehicle to account for the pressure-related effects acting on the submerged portions of its hull when it accelerates. These pressure-related effects cause forces, which are proportional to the acceleration of the vehicle, and are hence conceptualized to be associated with an effective mass of fluid surrounding the vehicle, which is also accelerated, when the vehicle is accelerated [31].

When a vehicle is operating in a current, or in waves, the equations of motion must be modified slightly. In a current, the hydrodynamic forces (added mass and drag) become dependent on the relative velocity between the vehicle and the surrounding water. When a surface vehicle operates in waves, some of the added mass and hydrodynamic damping terms are dependent on the encounter frequency of the vessel with the waves and must be altered accordingly. As mentioned above, we will normally assume that the dynamic model of the system to be controlled (1.59) has already been developed and focus on how the equations that govern the system's response can be analyzed to design a controller. Detailed descriptions of how maneuvering models can be developed for marine vehicles are presented in the following texts [2, 16–18, 27, 28].

## Problems

**1.1** Some common feedback control systems are listed below.

(a)  Manual steering of a boat in a harbor.
(b)  The water level controlled in a tank by a float and valve.
(c)  Automatic speed control of an ocean–going ship.

For each system, draw a component block diagram and identify the following:

- the process
- the actuator
- the sensor
- the reference input
- the controlled output
- the actuator output
- the sensor output

**1.2** Sailboats can't sail directly into the wind. The closest they can head into the wind is roughly $\psi_{min} \approx 30°$ (Fig. 1.18). In order to move upwind, they must follow a zig-zag shaped course by tacking (changing directions). A sailor's decisions of when to tack and where to go can determine the outcome of a race. Describe the process of tacking a sailboat in a channel as shown in Fig. 1.18. Sketch a block diagram depicting this process.

**1.3** AUVs can only acquire GPS signals when they are at the surface. When underwater, they often use a method known as *dead-reckoning* to estimate their location (when high precision acoustic positioning measurements are not available). The technique involves combining measurements of their orientation with measurements, or estimates, of their speed. When operating near the seafloor, speed can be measured using a sensor known as a Doppler velocimetry logger (DVL). When beneath the surface, but too far above the seafloor for a DVL to function, the speed can be estimated based on knowledge of the correlation between open loop propeller commands and steady state speed. AUVs surface periodically, as their missions permit, to correct the dead-reckoning position estimate using a GPS measurement. Sketch a block diagram depicting this process.

**Fig. 1.18**  A sailboat moving
upwind in a channel

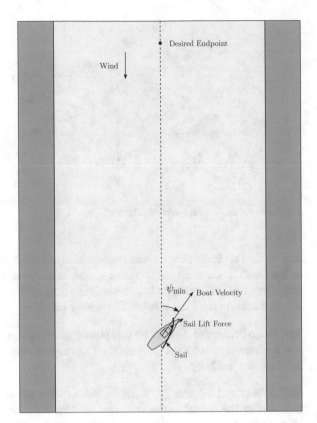

**Fig. 1.19**  Two tugboats
used to dock a ship

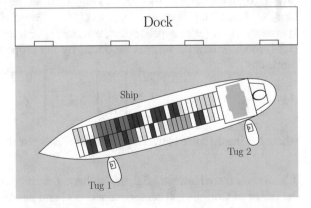

**1.4**  Multiple tugboats are often used to maneuver a ship into place when docking.
Fig. 1.19 shows a typical configuration with one tugboat near the stern and one near
the bow of a ship. Develop a block diagram describing the actions of the two tugboat
captains, the position of each tugboat, and the position of the ship.

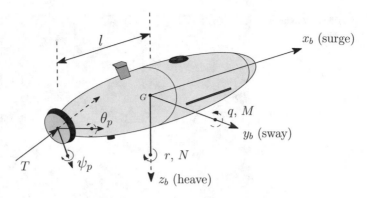

**Fig. 1.20** An AUV with a vectored thruster

**1.5** A ship is performing a circle maneuver about the origin of the $NED$ coordinate system with a constant surge speed $u$ and at a constant radius $R$. Take the time derivative of (1.33) to calculate the acceleration of the ship in the $NED$ frame. In which direction does the acceleration vector point?

**1.6** Redo Problem 1.5 above by computing the vector product $\boldsymbol{\omega}^b_{b/n} \times \dot{\boldsymbol{\eta}}$, where $\boldsymbol{\omega}^b_{b/n}$ and $\dot{\boldsymbol{\eta}}$ are determined as in (1.4) and (1.33), respectively. Do you get the same thing? Why or why not?

**1.7** A vectored thruster AUV uses a gimbal-like system to orient its propeller so that part of the propeller thrust $T$ can be directed along the sway and heave axes (Fig. 1.10) of the vehicle. As shown in Fig. 1.20, let the pitch angle of the the gimbal be $\theta_p \ll 1$ and the yaw angle be $\psi_p \ll 1$. If the thrust of the propeller lies along the symmetry axis of the propeller duct, use the Euler transformation matrices (1.13)–(1.14) to show that the torques produced by the propeller in the vehicle reference frame are

$$\boldsymbol{\tau} = \begin{bmatrix} K \\ M \\ N \end{bmatrix} \approx -lT \begin{bmatrix} 0 \\ \theta_p \\ \psi_p \end{bmatrix},$$

where $l$ is the distance between the center of mass of the AUV and the propeller.

**1.8** As mentioned in Sect. 1.5.2, the mathematical properties of the inertial tensor $\mathbf{M}$, the tensor of Coriolis and centripetal terms $\mathbf{C}$ and the drag tensor $\mathbf{D}$, have important symmetry properties, which are often exploited to show the stability of closed loop marine vehicle control systems. Prove the following relations, which are shown in Table 1.3:

(a) $\mathbf{M} = \mathbf{M}^T > 0 \Rightarrow \mathbf{x}^T \mathbf{M} \mathbf{x} > 0, \forall \mathbf{x} \neq 0,$
(b) $\mathbf{C}(\mathbf{v}) = -\mathbf{C}^T(\mathbf{v}) \Rightarrow \mathbf{x}^T \mathbf{C}(\mathbf{v})\mathbf{x} = 0, \forall \mathbf{x},$
(c) $\mathbf{D}(\mathbf{v}) > 0 \Rightarrow \frac{1}{2}\mathbf{x}^T[\mathbf{D}(\mathbf{v}) + \mathbf{D}^T(\mathbf{v})]\mathbf{x} > 0, \forall \mathbf{x} \neq 0.$

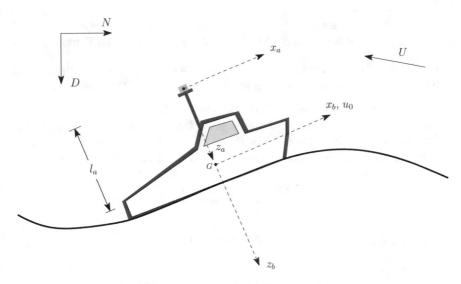

**Fig. 1.21** Wind speed measurements from USV moving over waves

**1.9** A USV is used to measure the wind velocity near the surface as it moves over waves (Fig. 1.21). Assume the vessel is moving at a constant surge speed $u_0$. The Euler angles $(\phi, \theta, \psi)$ and angular rates $(p, q, r)$ of the USV are measured with on board sensors. An anemometer (wind velocity sensor) is mounted on a mast above the center of gravity of the USV, so that its position in the body fixed frame is $\mathbf{r}_a = [0 \ 0 \ -l_a]^T$. What is the true wind velocity $\mathbf{U}$, if the wind velocity measured in the sensor fixed frame is

$$
\mathbf{u}_a = \begin{bmatrix} u_a \\ v_a \\ w_a \end{bmatrix} ? \tag{1.60}
$$

Take the $x_a$ axis to be parallel to the $x_b$ axis, the $y_a$ axis to be parallel to the $y_b$ axis and the $z_a$ axis to be collinear with the $z_b$ axis. Don't forget that even when $\mathbf{U} = 0, \mathbf{u}_a \neq 0$ because the USV is moving. The process of removing the motion-induced effects of a moving platform from sensor measurements is known as *motion compensation*.

**1.10** Inertial measurement units (IMUs) measure the linear accelerations and angular rates along the $x_s$, $y_s$ and $z_s$ axes of the sensor. In addition to motion-induced accelerations, the IMU also outputs a measure of the local gravitational acceleration vector $\mathbf{g} = g\hat{k}_n$, where $\hat{k}_n$ is a unit vector along the D axis of the NED coordinate system. Assume that an IMU is mounted in an unmanned underwater vehicle such that its $x_s$ axis is parallel to the $x_b$ axis, its $y_s$ axis is parallel to the $y_b$ axis and its $z_s$ axis is parallel to the $z_b$ axis. If the vehicle is stationary, what is the sensor output after the vehicle is rotated first in yaw by 45°, then in pitch by 30° and finally in roll by 20°?

**1.11** Find the rotation matrix corresponding to the Euler angles $\phi = \pi/2$, $\theta = 0$ and $\psi = \pi/4$. What is the direction of the $x_b$ axis relative to the NED frame?

**1.12** Use Simulink to create a dynamic simulation of an unmanned surface vessel (USV) moving in a straight line along the horizontal plane of the water's surface.

(a) By simplifying (1.58) and (1.59), write the equations of motion of the vehicle in the surge direction. Assume the origin of the body-fixed frame is located at the center of mass of the USV, there are no exogenous (externally-generated) disturbances, and the $\hat{i}_b$ components of $\mathbf{D(v)v}$ and $\boldsymbol{\tau}$ are given by

$$\mathbf{D(v)v} \cdot \hat{i}_b = d_{nl}u^2 + d_l u$$

and

$$\boldsymbol{\tau} \cdot \hat{i}_b = \tau_x(t),$$

respectively.

(b) Let $m = 10$ kg, $d_{nl} = 1.5$ kg/m, $d_l = 0.5$ kg/s and assume that the USV is at rest for $t < 0$.

  (i) Plot the open loop response over the time interval $0 \le t \le 10$ s for $\tau_x(t) = 100 \cdot 1(t)$ N (e.g. a step input of 100 N).
  (ii) Plot the open loop response over the time interval $0 \le t \le 10$ s when

$$\tau_x(t) = \begin{cases} 0, & t < 0 \\ 20 \cdot t, & 0 \le t \le 5 \text{ s} \\ 100, & t > 5 \text{ s}. \end{cases}$$

# References

1. Aguiar, A.P., Almeida, J., Bayat, M., Cardeira, B.M., Cunha, R., Häusler, A.J., Maurya, P., Pascoal, A.M., Pereira, A., Rufino, M., et al.: Cooperative control of multiple marine vehicles: theoretical challenges and practical issues. In: Manoeuvring and Control of Marine Craft, pp. 412–417 (2009)
2. Allmendinger, E.E., (ed.).: Submersible Vehicle Systems Design, vol. 96. SNAME, Jersey City (1990)
3. Astrom, K.J., Wittenmark, B.: Adaptive Control, 2nd edn. Dover Publications Inc, New York (2008)
4. Åström, K.J., Murray, R.M.: Feedback Systems: An Introduction for Scientists and Engineers. Princeton University Press, Princeton (2010)
5. Bemporad, A., Rocchi, C.: Decentralized hybrid model predictive control of a formation of unmanned aerial vehicles. IFAC Proc. **44**(1), 11900–11906 (2011)
6. Benjamin, M.R., Schmidt, H., Newman, P.M., Leonard, J.J.: Nested autonomy for unmanned marine vehicles with moos-ivp. J. Field Robot. **27**(6), 834–875 (2010)

7. Bertaska, I.R., Alvarez, J., Sinisterra, A., von Ellenrieder, K., Dhanak, M., Shah, B., Švec, P., Gupta, S.K.: Experimental evaluation of approach behavior for autonomous surface vehicles. In: ASME 2013 Dynamic Systems and Control Conference, pp. V002T32A003–V002T32A003. American Society of Mechanical Engineers (2013)
8. Bertaska, I.R., Shah, B., von Ellenrieder, K., Švec, P., Klinger, W., Sinisterra, A.J., Dhanak, M., Gupta, S.K.: Experimental evaluation of automatically-generated behaviors for usv operations. Ocean Eng. **106**, 496–514 (2015)
9. Bertaska, I.R., von Ellenrieder, K.D.: Supervisory switching control of an unmanned surface vehicle. In: OCEANS'15 MTS/IEEE Washington, pp. 1–10. IEEE (2015)
10. Biggs, J., Holderbaum, W.: Optimal kinematic control of an autonomous underwater vehicle. IEEE Trans. Autom. Control **54**(7), 1623–1626 (2009)
11. Bingham, B.S., Walls, J.M., Eustice, R.M.: Development of a flexible command and control software architecture for marine robotic applications. Marine Technol. Soc. J. **45**(3), 25–36 (2011)
12. Breivik, M., Hovstein, V.E., Fossen, T.I.: Straight-line target tracking for unmanned surface vehicles. MIC—Model. Identif. Control **29**(4), 131–149 (2008)
13. Camacho, E.F., Alba, C.B.: Model Predictive Control. Springer Science & Business Media, Berlin (2013)
14. de Sousa, J.B., Johansson, K.H., Silva, J., Speranzon, A.: A verified hierarchical control architecture for co-ordinated multi-vehicle operations. Int. J. Adapt. Control Signal Process. **21**(2–3), 159–188 (2007)
15. Dorf, R.C., Bishop, R.H.: Modern Control Systems. Pearson (Addison-Wesley), London (1998)
16. Faltinsen, O.: Sea Loads on Ships and Offshore Structures, vol. 1. Cambridge University Press, Cambridge (1993)
17. Faltinsen, O.M.: Hydrodynamics of High-Speed Marine Vehicles. Cambridge University Press, Cambridge (2005)
18. Fossen, T.I.: Handbook of Marine Craft Hydrodynamics and Motion Control. Wiley, New York (2011)
19. Franklin, G.F., Powell, J.D., Emami-Naeini, A.: Feedback Control of Dynamic Systems, 3rd edn. Addison-Wesley, Reading (1994)
20. Franklin, G.F., Powell, J.D., Workman, M.L.: Digital Control of Dynamic Systems, vol. 3. Addison-Wesley, Menlo Park (1998)
21. Friedland, B.: Control System Design: An Introduction to State-space Methods. Courier Corporation (2012)
22. Gat, E., et al.: On three-layer architectures (1998)
23. Goodwin, G.C., Graebe, S.F., Salgado, M.E.: Control System Design. Prentice Hall Englewood Cliffs, New Jersey (2001)
24. Greenwood, D.T.: Principles of Dynamics. Prentice-Hall Englewood Cliffs, New Jersey (1988)
25. Howard, T.M., Pivtoraiko, M., Knepper, R.A., Kelly, A.: Model-predictive motion planning: several key developments for autonomous mobile robots. IEEE Robot. Autom. Mag. **21**(1), 64–73 (2014)
26. Khalil, H.K.: Nonlinear Systems, 3rd edn. Prentice Hall Englewood Cliffs, New Jersey (2002)
27. Lewandowski, E.M.: The Dynamics of Marine Craft: Maneuvering and Seakeeping, vol. 22. World Scientific, Singapore (2004)
28. Lewis, E.V.: Principles of Naval Architecture, vol. III, 2nd edn. SNAME, New Jersey City (1989)
29. Liberzon, D.: Switching in Systems and Control. Springer Science & Business Media, Berlin (2012)
30. Mayne, D.Q.: Model predictive control: recent developments and future promise. Automatica **50**(12), 2967–2986 (2014)
31. Newman, J.N.: Marine Hydrodynamics. MIT Press, Cambridge (1997)
32. Ogata, K.: Modern Control Engineering, 5th edn. Pearson, London (2009)
33. Qiu, L., Zhou, K.: Introduction to Feedback Control. Prentice Hall Englewood Cliffs, New Jersey (2009)

34. Quigley, M., Conley, K., Gerkey, B., Faust, J., Foote, T., Leibs, J., Wheeler, R., Ng, A.Y.: Ros: an open-source robot operating system. In: ICRA Workshop on Open Source Software, vol. 3, p. 5. Kobe, Japan (2009)
35. Ridao, P., Batlle, J., Amat, J., Roberts, G.N.: Recent trends in control architectures for autonomous underwater vehicles. Int. J. Syst. Sci. **30**(9), 1033–1056 (1999)
36. Sanyal, A., Nordkvist, N., Chyba, M.: An almost global tracking control scheme for maneuverable autonomous vehicles and its discretization. IEEE Trans. Autom. Control **56**(2), 457–462 (2010)
37. Sellner, B., Heger, F.W., Hiatt, L.M., Simmons, R., Singh, S.: Coordinated multiagent teams and sliding autonomy for large-scale assembly. Proc. IEEE **94**(7), 1425–1444 (2006)
38. Shah, B.C., Švec, P., Bertaska, I.R., Klinger, W., Sinisterra, A.J., von Ellenrieder, K., Dhanak, M., Gupta, S.K.: Trajectory planning with adaptive control primitives for autonomous surface vehicles operating in congested civilian traffic. In: 2014 IEEE/RSJ International Conference on Intelligent Robots and Systems, pp. 2312–2318. IEEE (2014)
39. Shah, B.C., Švec, P., Bertaska, I.R., Sinisterra, A.J., Klinger, W., von Ellenrieder, K., Dhanak, M., Gupta, S.K.: Resolution-adaptive risk-aware trajectory planning for surface vehicles operating in congested civilian traffic. Autonom. Robot. 1–25 (2015)
40. Skejic, R., Faltinsen, O.M.: A unified seakeeping and maneuvering analysis of ships in regular waves. J. Marine Sci. Technol. **13**(4), 371–394 (2008)
41. Slotine, J.-J.E., Li, W.: Applied Nonlinear Control. Prentice-Hall Englewood Cliffs, New Jersey (1991)
42. Song, F., An, P.E., Folleco, A.: Modeling and simulation of autonomous underwater vehicles: design and implementation. IEEE J. Oceanic Eng. **28**(2), 283–296 (2003)
43. Spong, M.W., Hutchinson, S., Vidyasagar, M.: Robot Modeling and Control. Wiley, New York (2006)
44. Subramanian, R., Beck, R.F.: A time-domain strip theory approach to maneuvering in a seaway. Ocean Eng. **104**, 107–118 (2015)
45. Sutulo, S., Soares, C.G.: A unified nonlinear mathematical model for simulating ship manoeuvring and seakeeping in regular waves. In: Proceedings of the International Conference on Marine Simulation and Ship Manoeuvrability MARSIM (2006)
46. Svec, P., Gupta, S.K.: Automated synthesis of action selection policies for unmanned vehicles operating in adverse environments. Autonom. Robot. **32**(2), 149–164 (2012)
47. Svec, P., Shah, B.C., Bertaska, I.R., Alvarez, J., Sinisterra, A.J., von Ellenrieder, K., Dhanak, M., Gupta, S.K.: Dynamics-aware target following for an autonomous surface vehicle operating under colregs in civilian traffic. In: 2013 IEEE/RSJ International Conference on Intelligent Robots and Systems (IROS), pp. 3871–3878. IEEE (2013)
48. Weiss, A., Baldwin, M., Erwin, R.S., Kolmanovsky, I.: Model predictive control for spacecraft rendezvous and docking: Strategies for handling constraints and case studies. IEEE Trans. Control Syst. Technol. **23**(4), 1638–1647 (2015)
49. Zhou, K., Doyle, J.C., Glover, K., et al.: Robust and Optimal Control, vol. 40. Prentice Hall, Upper Saddle River (1996)

# Part I
# Linear Methods

In general, most marine systems are nonlinear, time–varying and of higher order. In Part I, we will focus on approximating the response of nonlinear systems using a linear model. For these models the output of the system represents a small departure from a desired operating point (equilibrium condition). In addition, we will assume that any variation in the parameters contained within the system model occurs much more slowly than the speed of any of the system's actuators, as well as the response of the plant to variations of the input. Using this approach, the first part of this course will consider systems that can be considered to be *Linear and Time–Invariant* (LTI).

The dynamic response of LTI systems will be analyzed in three domains: the *complex s-plane*, the *frequency domain* and *state space*. The basic mathematical approach with the $s$–plane is to transform the linear, constant ordinary differential equations of our process model into algebraic equations using Laplace Transform techniques. By examining the locations of the poles and zeros of the system in the $s$–plane it is possible to guide the design of a feedback controller without having to explicitly solve for the response of the system in the time domain, which can sometimes be quite difficult. Frequency domain techniques are similar, but are associated with the response to purely sinusoidal inputs. A benefit of these methods is that the output amplitude and phase shift responses of a system to a sinusoidal input signal are relatively simple to measure in a laboratory, which makes it possible to characterize the stability and design a controller for the system in the absence of an explicit mathematical model of its dynamics. Lastly, state space techniques involve the formulation of a system's equations of motion as a system of first-order equations, which permits the use of mathematical techniques from linear algebra to understand the system's dynamic response and to design a state feedback controller. State space techniques are well-suited for implementation on microprocessors and for systems that involve multiple input signals and/or multiple output signals. The state space representation of a system is also used in nonlinear control design, and thus knowledge of state space techniques is fundamentally important for the analysis and control of nonlinear systems.

# Chapter 2
# Stability: Basic Concepts and Linear Stability

## 2.1 The Stability of Marine Systems

One of the most important questions about the properties of a controlled system is whether or not it is stable: does the output of the system remain near the desired operating condition as the commanded input condition, the characteristics of any feedback sensors and the disturbances acting on the system vary?

For marine systems the question of stability can be especially critical. For example, unmanned underwater vehicles are often used to make measurements of the water column over several hours. When they are underwater, it can be difficult to localize and track them precisely. If their motion is not robustly stable to unknown disturbances, e.g. salinity driven currents under the polar ice caps, strong wind driven currents, etc. the error in their trajectories could grow substantially over time, making their recovery difficult or sometimes impossible. Anyone who has witnessed a team of oceanographers nervously pacing the deck of a research vessel while anxiously watching for a 2 m UUV to return from the immense blackness of the ocean, with significant research funds and countless hours of preparation and time invested its return, can appreciate the importance of knowing that the vehicle's motion is predictable and that one can be reasonably confident in its safe retrieval.

As another example, consider the motion of a large ship. When moving at design speed, some large ships can take more than 10 minutes and require several kilometers to stop or substantially change direction. The ability to precisely control a ship near populated coastlines, where there could be underwater hazards such as reefs and rocks, is extremely important for ensuring the safety of the lives of the people aboard the ship, as well as for ensuring that the risks of the substantial economic and environmental damage that could occur from a grounding or collision are minimized.

© Springer Nature Switzerland AG 2021
K. D. von Ellenrieder, *Control of Marine Vehicles*, Springer Series on Naval Architecture, Marine Engineering, Shipbuilding and Shipping 9,
https://doi.org/10.1007/978-3-030-75021-3_2

## 2.2 Basic Concepts in Stability

Many advanced techniques have been developed to determine the stability of both linear and nonlinear systems. However, before embarking on a detailed description of the stability analysis techniques typically applied in control systems engineering practice, we can gain some insight into the stability of controlled systems by first exploring some relatively simple graphical approaches that exploit the simplicity of plotting the trajectories of the state variables in phase space [2, 5, 7].

A state space input–output system has the form

$$\dot{\mathbf{x}} = \mathbf{F}(\mathbf{x}, \mathbf{u}), \quad \mathbf{y} = \mathbf{h}(\mathbf{x}, \mathbf{u}), \tag{2.1}$$

where $\mathbf{x} = \{x_1, \ldots, x_n\} \in \mathbb{R}^n$ is the state, $\mathbf{u} \in \mathbb{R}^p$ is the input and $\mathbf{y} \in R^q$ is the output. The smooth maps $\mathbf{F} : \mathbb{R}^n \times \mathbb{R}^p \to \mathbb{R}^n$ and $\mathbf{h} : \mathbb{R}^n \times \mathbb{R}^p \to \mathbb{R}^q$ represent the dynamics and feedback measurements of the system. Here, we investigate systems where the control signal is a function of the state, $\mathbf{u} = \boldsymbol{\alpha}(\mathbf{x})$. This is one of the simplest types of feedback, in which the system regulates its own behavior. In this way, the equations governing the response of a general $n$th–order controlled system can be written as a system of $n$ first order differential equations. In vector form, this equation can be written as

$$\dot{\mathbf{x}} = \mathbf{F}(\mathbf{x}, \boldsymbol{\alpha}(\mathbf{x})) := \mathbf{f}(\mathbf{x}), \tag{2.2}$$

and in the corresponding component form, as

$$\dot{x}_1 = f_1(x_1, \ldots, x_n)$$

$$\vdots \tag{2.3}$$

$$\dot{x}_n = f_n(x_1, \ldots, x_n).$$

We will start by qualitatively exploring some basic stability concepts using graphical approaches, and then proceed to analytical approaches for analyzing the stability of systems in Sects. 2.6–7.4. The graphical representations in phase space of systems with $n > 2$ can be computationally and geometrically complex, so we will restrict our attention to first order $n = 1$ and second order $n = 2$ systems. Further, we will assume that $\mathbf{f}$ is a smooth, real–valued, function of $\mathbf{x}(t)$ and that $\mathbf{f}$ is an autonomous function $\mathbf{f} = \mathbf{f}(\mathbf{x})$, i.e. it does not *explicitly* depend on time $\mathbf{f} \neq \mathbf{f}(\mathbf{x}, t)$.

## 2.3  Flow Along a Line

To start, consider a one–dimensional system $\dot{x} = f(x)$ and think of $x$ as the position of an imaginary particle moving along the real line, and $\dot{x}$ as the velocity of that particle. Then the differential equation represents a vector field on the line—it dictates the velocity vector at each point $x$. To sketch the vector field we can plot $\dot{x}$ versus $x$ and then add vectors to the real line so that at each point of the x–axis, the direction of the vector at that point is determined by the sign of $\dot{x}$. The vectors give the trajectory of the imaginary particle on the real axis. Points where $\dot{x} = 0$ correspond to positions where the particle does not move, i.e. *fixed points*, which represent equilibrium solutions of the differential equation. An equilibrium solution corresponds to a *stable* fixed point if the velocity vectors on both sides of it point towards the fixed point. Conversely, an equilibrium solution corresponds to an *unstable* fixed point if the velocity vectors on both sides of it point away from the fixed point. Note that if the velocity vectors on either side of the fixed point have the same sign (i.e. point in the same direction) the assumption that $f(x)$ is a smooth function would be violated. Here, we'll represent stable fixed points with filled circles and unstable fixed points with open circles, as in Example 2.1 below (after [7]).

**Example 2.1**  Consider the system $\dot{x} = x^2 - 1$. To find the fixed points and examine the stability of the system, we can plot $x$ versus $\dot{x}$, as shown in Fig. 2.1. We can see that the fixed points occur at the locations $x = \pm 1$. Adding velocity vectors to the x–axis we can see that an imaginary particle would travel to the left for $|x| < 1$ and to the right for $|x| > 1$. Thus, the point $x = +1$ is an unstable fixed point and the point $x = -1$ is a stable fixed point. Note that the notion of stable equilibrium is based on small disturbances. A small disturbance to the system at $x = -1$ will return the system to $x = -1$. However, a disturbance that knocks the system from $x = -1$ to the right of $x = 1$ will send the system out to $x \to +\infty$. Thus, $x = -1$ is *locally stable*.  □

**Example 2.2**  Consider the system $\dot{x} = \sin(x + \pi/4) - x$. An approach to graphically analyzing its stability would be to plot the function $f(x) = \sin(x + \pi/4) - x$ and then sketch the associated vector field. An easier approach is to find the equilibrium point where $\dot{x} = 0$ by separately plotting $y_1 = x$ and $y_2 = \sin(x + \pi/4)$ on the same graph and looking for any intersection points where $\dot{x} = y_2 - y_1 = 0$. When this is done, as in Fig. 2.2, it can be seen that this system has a single equilibrium point. It can also be seen that when the line $y_1 = x$ is above the curve $y_2 = \sin(x + \pi/4)$, $\dot{x} < 0$, and when $y_1 = x$ is below the curve $y_2 = \sin(x + \pi/4)$, $\dot{x} > 0$. Thus, the fixed point is a stable equilibrium point. Here, the fixed point is *globally stable* because it is approached from all initial conditions.  □

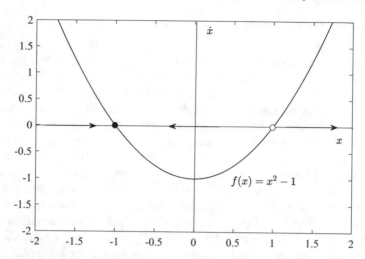

**Fig. 2.1** One dimensional phase diagram of the system $\dot{x} = x^2 - 1$

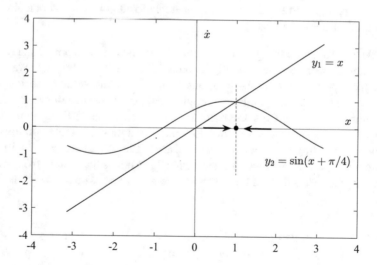

**Fig. 2.2** Graphical approach to finding the stability of the system $\dot{x} = \sin(x + \pi/4) - x$

### 2.3.1  Linear 1D Stability Analysis

Linearization about a desired operating point $x_e$ of a controlled system can be used to determine a more quantitative measure of stability than provided by graphical approaches. To accomplish this we can perform a Taylor series expansion of the differential equation about $x_e$. Let $\eta$ be a small perturbation away from $x_e$, such that $\eta(t) = x(t) - x_e$. Thus, near the point $x_e$, $\dot{\eta} = \dot{x} = f(x) = f(x_e + \eta)$ and using Taylor's expansion gives

$$f(x_e + \eta) = f(x_e) + \eta f'(x_e) + \frac{1}{2} \left.\frac{d^2 f}{dx^2}\right|_{x=x_e} \eta^2 + \cdots + \frac{1}{n!} \left.\frac{d^n f}{dx^n}\right|_{x=x_e} \eta^n, \tag{2.4}$$
$$= f(x_e) + \eta f'(x_e) + O(\eta^2),$$

where "Big–O" notation $O(\eta^2)$ is used to denote quadratically small terms in $\eta$ and $f'(x) := df/dx$. We can think of the desired operating point, or set point of our system as a fixed point of the differential equation governing the controlled system, $f(x_e)$. Assuming that the $O(\eta^2)$ terms are negligible in comparison, we may use the *linear approximation*

$$\dot{\eta} \approx \eta f'(x_e). \tag{2.5}$$

With the initial condition $\eta(t = 0) = \eta_0$, the solution to this equation would be

$$\eta = \eta_0 e^{f'(x_e)t}. \tag{2.6}$$

Thus, we can see that if $f'(x_e) > 0$ the perturbation $\eta$ grows exponentially and decays if $f'(x_e) < 0$. Thus, the sign of the slope $f'(x_e)$ determines whether an operating point is stable or unstable and the magnitude of the slope is a measure of how stable the operating point is (how quickly perturbations grow or decay). When $f'(x_e) = 0$, the $O(\eta^2)$ terms are not negligible and a nonlinear analysis is needed.

**Example 2.3** (*Speed control of a surface vessel*) As discussed in Sect. 1.2, one of the main purposes of an automatic control system is to make a desired operating point an equilibrium point of the controlled system, such that the system is stable at that point. As much as possible, the controlled system should also be stable about the operating point when external disturbances act on the system. Consider the speed control of the boat shown in Fig. 2.3. The equation of motion is given by

$$m\dot{u} = -cu|u| + T, \tag{2.7}$$

where $u$ is speed, $c$ is a constant related to the drag coefficient of the hull and $T$ is the thrust generated by the propeller. Here, we will use the concepts of flow along a line and linearization to explore the stabilization problem of maintaining a desired constant speed $u = u_0$. We will compare the use of both *open loop control* and *closed loop control*, as described in Example 1.1.

Starting with open loop control, suppose we performed a series of careful measurements and found that when conditions are perfect, we can set the thrust to

$$T = cu_0|u_0|,$$

so that at steady state (when $\dot{u} = 0$), $\lim_{t\to\infty} u = u_{ss} \to u_0$. By plotting the flow of (2.7), we can see that it is globally stable (Fig. 2.4). However, we have assumed that the conditions under which the boat is operating are perfect, meaning that there are no external disturbances acting on the system and that we have perfectly characterized the drag-related coefficient $c$.

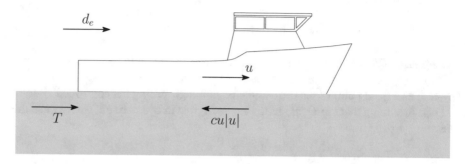

**Fig. 2.3** Speed control of a boat: $u$ is speed, $T$ is propeller thrust, $cu|u|$ is hull drag and $d_e$ is a constant external disturbance force, such as a wind-generated force

Suppose now that there is a very slowly varying external disturbance (for example a strong tailwind could be acting on the boat), which we will characterize as a constant $d_e$. Additionally, assume that there is an uncertainty in the drag-related coefficient, $\delta c$. Then, the open loop thrust will be set to

$$T = (c + \delta c)u_0|u_0| = cu_0|u_0| + \delta cu_0|u_0|$$

and (2.7) becomes

$$m\dot{u} = -cu|u| + cu_0|u_0| + \delta cu_0|u_0| + d_e = -cu|u| + cu_0|u_0| + d, \qquad (2.8)$$

where $d$ is a constant that combines the external disturbances and the uncertainty of the drag-related coefficient. Now, we can solve (2.8) to see that when $\dot{u} = 0$ and $u > 0$, the steady state speed is

$$\lim_{t \to \infty} u = u_{ss} = \sqrt{u_0^2 + d/c}.$$

Owing to $d$, there will always be a speed error and with open loop control there is no way to mitigate it (see Fig. 2.4).

Now, let's explore the use of closed loop control and see if we can improve the speed error caused by $d$. As explained in Example 1.1, closed loop control involves measuring the output of the system and feeding it back to a controller, which produces a signal related to the error between the desired output and the measured output. Here, we want to control the system so that the steady state speed error ($u_{ss} - u_0$) is small. Let the thrust commanded by the closed loop controller be

$$T = cu_0|u_0| + \delta cu_0|u_0| - k_p(u - u_0), \qquad (2.9)$$

where $k_p > 0$ is a controller design parameter called a proportional gain (see Sect. 3.3.1), that can be used to tune the response of the system. Substituting this

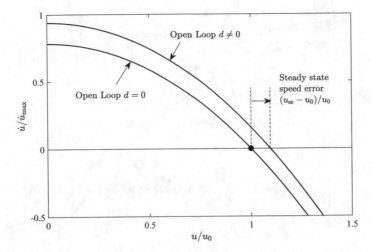

**Fig. 2.4** Open loop speed control without ($d = 0$) and with the effects of external disturbances and parameter uncertainty ($d = cu_0|u_0|/5$)

into (2.7) and accounting for the constant external disturbance $d_e$, (2.7) becomes

$$m\dot{u} = -cu|u| + cu_0|u_0| + \delta cu_0|u_0| - k_p(u - u_0) + d_e,$$
$$= -cu|u| + cu_0|u_0| - k_p(u - u_0) + d. \tag{2.10}$$

For $u > 0$, the term $u|u|$ can be linearized about the desired operating speed $u_0$, to get

$$u|u| = u^2 = u_0^2 + 2u_0(u - u_0) + (u - u_0)^2 \approx u_0^2 + 2u_0(u - u_0).$$

Substituting this into (2.10) gives

$$m\dot{u} \approx -c[u_0^2 + 2u_0(u - u_0)] + cu_0^2 - k_p(u - u_0) + d,$$
$$\approx -(k_p + 2cu_0)(u - u_0) + d. \tag{2.11}$$

Now at steady state ($\dot{u} = 0$), we can solve for $u$ to get

$$\lim_{t \to \infty} u = u_{ss} = u_0 + \frac{d}{k_p + 2cu_0} = u_0 + \left[\frac{d_e + \delta cu_0|u_0|}{k_p + 2cu_0}\right]. \tag{2.12}$$

The advantage of closed loop control, is that the effects of the external disturbance $d_e$ and parameter uncertainty $\delta c$ can be now mitigated by picking a large value of $k_p$, such that

$$\left[\frac{d_e + \delta cu_0|u_0|}{k_p + 2cu_0}\right] \to 0$$

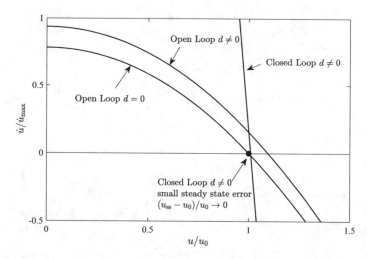

**Fig. 2.5** Performance comparison of closed loop speed control and open loop speed control with external disturbances and parameter uncertainty ($d = cu_0|u_0|/5$)

and $u_{ss} \rightarrow u_0$. Closed loop control with $k_p/(2cu_0) = 10$ and $d = cu_0|u_0|/5$ is shown in Fig. 2.5. It can be seen that the system is stable and that the effect of $k_p$ is to make the slope of $\dot{u}$ very large near $u/u_0 = 1$ so that the system is strongly driven towards $u = u_0$. Of course $k_p$ cannot be made infinitely large. From (2.9), it can be seen that the thrust commanded from the controller is proportional to $k_p$. From a practical standpoint, there are limits to how much thrust a propeller can generate (*actuator saturation*) and how quickly it can respond to changes in the commanded thrust (*actuator rate limits*). As will be discussed later, it can be important to take these actuator limits into account during controller design and stability analysis to ensure that the controller behaves as expected.                                              □

## 2.4  Phase Plane Analysis

The qualitative nature of nonlinear systems is important for understanding some of the key concepts of stability. As mentioned above, the graphical representations in phase space of systems with $n > 2$ can be computationally and geometrically complex. Here we will restrict our attention to second order $n = 2$ systems, where there are two state variables $\mathbf{x} \in \mathbb{R}^2$.

As with our study of one dimensional systems in Sect. 2.3, we can think of a differential equation

$$\dot{\mathbf{x}} = \mathbf{f}(\mathbf{x}), \tag{2.13}$$

as representing a vector field that describes the velocity of an imaginary particle in the state space of the system. The velocity tells us how $\mathbf{x}$ changes.

For two dimensional dynamical systems, each state corresponds to a point in the plane and $\mathbf{f}(\mathbf{x})$ is a vector representing the velocity of that state. We can plot these vectors on a grid of points in the $\mathbf{x} = \{x_1, x_2\}$ plane, which is often called the *state space* or *phase plane* of the system, to obtain a visual image of the dynamics of the system. The points where the velocities are zero are of particular interest, as they represent the *stationary points* or *fixed points* points: if we start at such a state, we stay at that state. These points are essentially equilibrium points.

A *phase portrait*, or *phase plane diagram* can be constructed by plotting the flow of the vector field corresponding to (2.13). For a given initial condition, this flow is the solution of the differential equation in the phase plane. By plotting the solutions corresponding to different initial conditions, we obtain a phase portrait.

The phase portrait can provide insight into the dynamics of a system. For example, we can see whether all trajectories tend to a single point as time increases, or whether there are more complicated behaviors. However, the phase portrait cannot tell us the rate of change of the states (although, this can be inferred from the length of the vectors in a plot of the vector field).

### 2.4.1 Linear 2D Stability Analysis

More generally, suppose that we have a nonlinear system (2.13) that has an equilibrium point at $\mathbf{x}_e$. As in Sect. 2.3.1, let $\eta$ be a small perturbation away from $\mathbf{x}_e$, such that $\eta(t) = \mathbf{x}(t) - \mathbf{x}_e$. Thus, $\dot{\eta} = \dot{\mathbf{x}} = \mathbf{f}(\mathbf{x}) = \mathbf{f}(\mathbf{x}_e + \eta)$. Computing the Taylor series expansion of the vector field, as in Sect. 2.3.1, we obtain

$$\dot{\eta} = \mathbf{f}(\mathbf{x}_e) + \left.\frac{\partial \mathbf{f}}{\partial \mathbf{x}}\right|_{\mathbf{x}_e} \eta + O(\eta^T \eta). \tag{2.14}$$

Since $\mathbf{f}(\mathbf{x}_e) = 0$, we write the *linear approximation*, or the *linearization* at $\mathbf{x}_e$, to the original nonlinear system as

$$\frac{d\eta}{dt} = \mathbf{A}\eta, \quad \text{where} \quad \mathbf{A} := \left.\frac{\partial \mathbf{f}}{\partial \mathbf{x}}\right|_{\mathbf{x}_e}. \tag{2.15}$$

The fact that a linear model can be used to study the behavior of a nonlinear system near an equilibrium point is a powerful one. For example, we could use a local linear approximation of a nonlinear system to design a feedback law that keeps the system near a desired operating point. Also note that, in general, a given dynamical system may have zero, one or more fixed points $\mathbf{x}_e$. When using a phase portrait to analyze the stability of a system it is important to know where these fixed points are located within the phase plane.

## 2.4.2 Classification of Linear 2D Systems

Here we will classify the possible phase portraits that can occur for a given $\mathbf{A}$ in (2.15). The simplest trajectories in the phase plane correspond to straight line trajectories. To start we will seek trajectories of the form

$$\mathbf{x}(t) = e^{\lambda t}\mathbf{v}, \tag{2.16}$$

where $\mathbf{v} \neq 0$ is a fixed vector to be determined and $\lambda$ is a growth rate, also to be determined. If such solutions exist, they correspond to exponential motion along the line spanned by the vector $\mathbf{v}$.

To find the conditions on $\mathbf{v}$ and $\lambda$, we substitute $\mathbf{x}(t) = e^{\lambda t}\mathbf{v}$ into $\dot{\mathbf{x}} = \mathbf{A}\mathbf{x}$, and obtain $\lambda e^{\lambda t}\mathbf{v} = e^{\lambda t}\mathbf{A}\mathbf{v}$. Cancelling the nonzero scalar factor $e^{\lambda t}$ yields

$$\mathbf{A}\mathbf{v} = \lambda\mathbf{v}. \tag{2.17}$$

The desired straight line solutions exist if $\mathbf{v}$ is an *eigenvector* of $\mathbf{A}$ with corresponding *eigenvalue* $\lambda$.

In general, the eigenvalues of a matrix $\mathbf{A}$ are given by the *characteristic equation* $\det(\mathbf{A} - \lambda\mathbf{1}) = 0$, where $\mathbf{1}$ is the identity matrix. If we define the elements of $\mathbf{A}$, as

$$\mathbf{A} = \begin{bmatrix} a & b \\ c & d \end{bmatrix},$$

the characteristic equation becomes

$$\det(\mathbf{A} - \lambda\mathbf{1}) = \begin{vmatrix} a - \lambda & b \\ c & d - \lambda \end{vmatrix} = 0.$$

Expanding the determinant gives

$$\lambda^2 + \tau\lambda + \Delta = 0, \tag{2.18}$$

where

$$\tau = -\text{trace}(\mathbf{A}) = -(a + d),$$
$$\Delta = \det(\mathbf{A}) \qquad ad - bc. \tag{2.19}$$

Then the solutions of the quadratic characteristic equation (2.18) are

$$\lambda_1 = \frac{-\tau + \sqrt{\tau^2 - 4\Delta}}{2}, \quad \lambda_2 = \frac{-\tau - \sqrt{\tau^2 - 4\Delta}}{2}. \tag{2.20}$$

Thus, the eigenvalues only depend on the trace and the determinant of $\mathbf{A}$.

**Fig. 2.6** Eigenvectors in the
phase plane

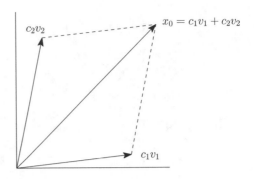

Most often, the eigenvalues are distinct, i.e. $\lambda_1 \neq \lambda_2$. In this case, the corresponding eigenvectors $\mathbf{v}_1$ and $\mathbf{v}_2$ are linearly independent and span the phase plane (Fig. 2.6). Any initial condition $\mathbf{x}_0$ can be written as a linear combination of eigenvectors, $\mathbf{x}_0 = c_1 \mathbf{v}_1 + c_2 \mathbf{v}_2$, where $\mathbf{v}_1$ is the solution to (2.17) when $\lambda = \lambda_1$ and $\mathbf{v}_2$ is the solution to (2.17) when $\lambda = \lambda_2$. Thus, the general solution for $\mathbf{x}(t)$ is

$$\mathbf{x} = c_1 e^{\lambda_1 t} \mathbf{v}_1 + c_2 e^{\lambda_2 t} \mathbf{v}_2. \tag{2.21}$$

The reason that this is the general solution is because it is a linear combination of the solutions to, (2.15), which can be written as $\dot{\mathbf{x}} = \mathbf{A}\mathbf{x}$ and because it satisfies the initial condition $\mathbf{x}(0) = \mathbf{x}_0$, and so by the existence and uniqueness of solutions, it is the only solution.

The types of critical points are classified on the chart shown in Fig. 2.7a: nodes, foci, or saddles can be obtained [3]. The type of trajectory depends on the location of a point defined by $\tau$ and $\Delta$ on the $\tau$—$\Delta$ chart. When both of the eigenvalues are real, either nodes or saddles are produced. When the eigenvalues are complex, a focus is obtained. When $\tau$ and $\Delta$ fall on one of the boundaries indicated by a Roman numeral on the $\tau$—$\Delta$ chart, one of the following degenerate critical points exists (Fig. 2.7b):

Case (I)   For this case $\Delta = 0$. The critical point is known as a node-saddle. The eigenvalues are given by $\lambda_1 = 0$ and $\lambda_2 = \tau$, and the general solution has the form

$$\mathbf{x}(t) = c_1 \mathbf{v}_1 + c_2 \mathbf{v}_2 e^{\lambda_2 t}$$

so that there will be a line of equilibrium points generated by $\mathbf{v}_1$ and an infinite number of straight line trajectories parallel to $\mathbf{v}_2$.

Case (II)   This case occurs when the eigenvalues are repeated, $\lambda_1 = \lambda_2$. If the eigenvectors $\mathbf{v}_1$ and $\mathbf{v}_2$ are independent, the critical point is a star node and the trajectories are straight lines. If the eigenvectors are not linearly independent, the critical point is a node-focus with the solution trajectories approaching/leaving the critical point tangent to the single eigenvector.

Case (III)   The eigenvalues are purely imaginary, so that the trajectories correspond to elliptical orbits about the critical point.

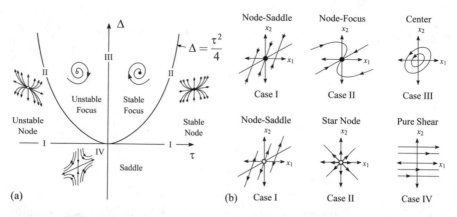

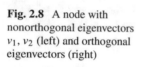

(a)                                                          (b)   Case I                 Case II                 Case IV

**Fig. 2.7** **a** Classification of critical points on the $\tau$—$\Delta$ chart. **b** Degenerate Critical points located on the boundaries I, II, and III and origin IV of the the $\tau$—$\Delta$ chart

**Fig. 2.8** A node with
nonorthogonal eigenvectors
$v_1$, $v_2$ (left) and orthogonal
eigenvectors (right)

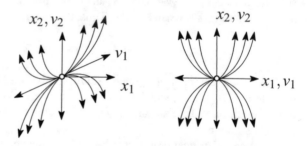

Case (IV)     Finally, in the last degenerate case both $\tau = 0$ and $\Delta = 0$. This corresponds to a system of either the form

$$\begin{bmatrix} \dot{x}_1 \\ \dot{x}_2 \end{bmatrix} = \begin{bmatrix} 0 & a \\ 0 & 0 \end{bmatrix},$$

or

$$\begin{bmatrix} \dot{x}_1 \\ \dot{x}_2 \end{bmatrix} = \begin{bmatrix} 0 & 0 \\ b & 0 \end{bmatrix},$$

where $a$ and $b$ are constants. In this case, the trajectories are either straight lines parallel to the $x_1$-axis given by $x_1 = ac_1t + c_2$, $x_2 = c_1$, or straight lines parallel to the $x_2$-axis given by $x_1 = c_1$, $x_2 = bc_1t + c_2$.

Note that the when the eigenvectors are nonorthogonal, the resulting pattern of trajectories is skewed in the directions of the eigenvectors, when compared to the same case (node or saddle) with orthogonal eigenvectors (Fig. 2.8).

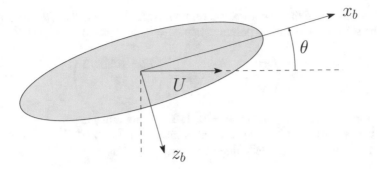

**Fig. 2.9** Pitch stability of an AUV-shaped body moving at speed $U$. The pitch angle is $\theta$

**Example 2.4** Consider the pitch stability of an autonomous underwater vehicle (AUV) moving at a constant speed $U$ (Fig. 2.9). Here we will explore the use of active control to stabilize the system. When the pitch angle is non-zero, a hydrodynamically-induced pitching moment, called the *Munk moment*, can arise [1]. If we assume the motion is confined to the $x_b$–$z_b$ plane of the vehicle and neglect drag, the dynamic equation of motion for pitch is

$$(I_y - M_{\dot{q}})\ddot{\theta} = (I_y - M_{\dot{q}})\dot{q} = -U^2 \sin\theta \cos\theta (Z_{\dot{w}} - X_{\dot{u}}). \qquad (2.22)$$

Here $I_y$ and $M_{\dot{q}}$ are the rigid-body mass and added mass moments of inertia about the sway axis of the AUV. The pitch angle is $\theta$, $q = \dot{\theta}$ is the pitch rate, and $Z_{\dot{w}}$ and $X_{\dot{u}}$ are the added mass in the heave ($z_b$) and surge ($x_b$) directions, respectively. Note that the added mass coefficients are defined such that $M_{\dot{q}} < 0$, $Z_{\dot{w}}$, and $X_{\dot{u}}$, in accordance to the 1950 SNAME nomenclature for the dynamic modeling of underwater vehicles [6]. For long slender vehicles that have a large length-to-diameter ratio, like a typical AUV, $|Z_{\dot{w}}| \gg |X_{\dot{u}}|$. An examination of (2.22) shows that when $0 < \theta < \pi/2$ the Munk moment is proportional to the difference between the added mass coefficients $Z_{\dot{w}}$ and $X_{\dot{u}}$ and that it is a destabilizing moment. Positive pitch angles cause a positive pitching moment, which increases the pitch angle, which causes a larger pitching moment. Similarly, a negative pitch angle, produces a negative pitching moment, which makes the pitch angle more negative, and so on. From a qualitative perspective, we might expect the condition $\theta = 0$ to be unstable and the condition $\theta = \pi/2$ to be stable. First consider the uncontrolled system. We can rewrite (2.22) in vector form. Let

$$\mathbf{x} = \begin{pmatrix} \theta \\ q \end{pmatrix},$$

then

$$\dot{\mathbf{x}} = \begin{pmatrix} q \\ -\dfrac{U^2}{2}\sin(2\theta)\left[\dfrac{Z_{\dot{w}} - X_{\dot{u}}}{I_y - M_{\dot{q}}}\right] \end{pmatrix} = \mathbf{f}, \qquad (2.23)$$

where we have used the trigonometric relation $\sin(2\theta) = 2 \sin\theta \cos\theta$. From (2.23), it can be seen that the critical points, where $\dot{\mathbf{x}} = 0$, are located at

$$\mathbf{x}_{e1} = \begin{pmatrix} n\pi \\ 0 \end{pmatrix} \quad \text{and} \quad \mathbf{x}_{e2} = \begin{pmatrix} (2m+1)\pi/2 \\ 0 \end{pmatrix},$$

where $n = \pm\{0, 1, 2, \ldots\}$ and $m = \pm\{0, 1, 2, \ldots\}$ are integers.

We can start by examining the critical point $\mathbf{x}_{e1}$ for $n = 0$. The matrix $\mathbf{A}$ in (2.15) can be found by taking the *Jacobian* of $\mathbf{f}$ at $\mathbf{x}_{e1}$

$$\mathbf{A} := \left. \frac{\partial \mathbf{f}}{\partial \mathbf{x}} \right|_{\mathbf{x}_{e1}} = \left. \begin{pmatrix} \dfrac{\partial f_1}{\partial x_1} & \dfrac{\partial f_1}{\partial x_2} \\[2mm] \dfrac{\partial f_2}{\partial x_1} & \dfrac{\partial f_2}{\partial x_2} \end{pmatrix} \right|_{\mathbf{x}_{e1}} = \left. \begin{pmatrix} 0 & 1 \\[2mm] -U^2 \cos(2\theta) \left[ \dfrac{Z_{\dot{w}} - X_{\dot{u}}}{I_y - M_{\dot{q}}} \right] & 0 \end{pmatrix} \right|_{\mathbf{x}_{e1}=(0,0)}$$

$$= \begin{pmatrix} 0 & 1 \\[2mm] -U^2 \left[ \dfrac{Z_{\dot{w}} - X_{\dot{u}}}{I_y - M_{\dot{q}}} \right] & 0 \end{pmatrix}.$$

Using (2.19), we can solve for $\tau$ and $\Delta$ to get $\tau = 0$ and

$$\Delta = U^2 \left[ \frac{Z_{\dot{w}} - X_{\dot{u}}}{I_y - M_{\dot{q}}} \right] < 0,$$

(recall that $M_{\dot{q}} < 0$, $Z_{\dot{w}}$, and $X_{\dot{u}}$). Based on Fig. 2.7a, we expect $\mathbf{x}_{e1}$ to be a saddle point.

Using (2.20), we see that the eigenvalues are

$$\lambda_{1,2} = \pm U \sqrt{-\frac{Z_{\dot{w}} - X_{\dot{u}}}{I_y - M_{\dot{q}}}}.$$

The eigenvectors must satisfy (2.17), which can be rewritten as $(\mathbf{A} - \lambda\mathbf{1})\mathbf{v} = 0$, for each eigenvalue. Thus, we can solve the system

$$\begin{pmatrix} -\lambda & 1 \\[2mm] -U^2 \left[ \dfrac{Z_{\dot{w}} - X_{\dot{u}}}{I_y - M_{\dot{q}}} \right] & -\lambda \end{pmatrix} \begin{pmatrix} v_1 \\ v_2 \end{pmatrix} = \begin{pmatrix} 0 \\ 0 \end{pmatrix},$$

to find the eigenvector corresponding to each eigenvalue. Let $\lambda_1$ be the positive eigenvector, which gives

$$\begin{pmatrix} -U\sqrt{\dfrac{Z_{\dot{w}} - X_{\dot{u}}}{I_y - M_{\dot{q}}}} & 1 \\[2ex] -U^2 \left[\dfrac{Z_{\dot{w}} - X_{\dot{u}}}{I_y - M_{\dot{q}}}\right] & -U\sqrt{\dfrac{Z_{\dot{w}} - X_{\dot{u}}}{I_y - M_{\dot{q}}}} \end{pmatrix} \begin{pmatrix} v_1 \\ v_2 \end{pmatrix} = \begin{pmatrix} 0 \\ 0 \end{pmatrix}.$$

A non–trivial solution for $\mathbf{v}_1$ is

$$\mathbf{v}_1 = \begin{pmatrix} v_1 \\ v_2 \end{pmatrix} = \begin{pmatrix} 1 \\[2ex] U\sqrt{-\dfrac{Z_{\dot{w}} - X_{\dot{u}}}{I_y - M_{\dot{q}}}} \end{pmatrix}.$$

Similarly, a solution for the other, negative, eigenvector $\lambda_2$ is

$$\mathbf{v}_2 = \begin{pmatrix} v_1 \\ v_2 \end{pmatrix} = \begin{pmatrix} 1 \\[2ex] -U\sqrt{-\dfrac{Z_{\dot{w}} - X_{\dot{u}}}{I_y - M_{\dot{q}}}} \end{pmatrix}.$$

Thus, the solution to the system (2.23) linearized near $\mathbf{x}_{e1} = [0, 0]^T$ is

$$\mathbf{x} = c_1 \begin{pmatrix} 1 \\[2ex] U\sqrt{-\dfrac{Z_{\dot{w}} - X_{\dot{u}}}{I_y - M_{\dot{q}}}} \end{pmatrix} e^{\lambda_1 t} + c_2 \begin{pmatrix} 1 \\[2ex] -U\sqrt{-\dfrac{Z_{\dot{w}} - X_{\dot{u}}}{I_y - M_{\dot{q}}}} \end{pmatrix} e^{\lambda_2 t},$$

where $c_1$ and $c_2$ are constants, which can be determined using an initial condition for $\mathbf{x}$. As the part of the solution containing $\lambda_1$ is unstable in time, trajectories depart from the critical point along the corresponding eigenvector $\mathbf{v}_1$; the converse is true for trajectories lying along the eigenvector corresponding to $\lambda_2$, $\mathbf{v}_2$.

Turning now to the critical point corresponding to $m = 0$, $\mathbf{x}_{e2} = [0, \pi/2]^T$, we repeat the process above to find that $\tau = 0$ and

$$\Delta = -U^2 \left[\dfrac{Z_{\dot{w}} - X_{\dot{u}}}{I_y - M_{\dot{q}}}\right] > 0.$$

Based on Fig. 2.7a–b, we expect $\mathbf{x}_{e2}$ to be a center. Using (2.20), we see that the eigenvalues are

$$\lambda_{1,2} = \pm jU\sqrt{-\dfrac{Z_{\dot{w}} - X_{\dot{u}}}{I_y - M_{\dot{q}}}},$$

where $j = \sqrt{-1}$ is the imaginary number. A phase diagram of the unstabilized AUV is shown in Fig. 2.10. Owing to the trajectory associated with the positive eigenvalue

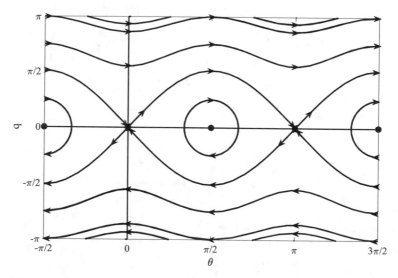

**Fig. 2.10** Pitch angle phase portrait for a AUV with unstabilized Munk moment. Critical points at $(\theta, q) = (n\pi, 0)$, integer $n = \pm\{0, 1, 2, 3, \ldots\}$ are saddle points and critical points at $(\theta, q) = ((2m + 1)\pi/2, 0)$, integer $m = \pm\{0, 1, 2, 3, \ldots\}$ are centers

of the saddle point at $\theta = 0$, the pitch angle is unstable there. Small pitch disturbances or initial conditions with $|\theta| \neq 0$ will grow and exhibit oscillations about $\theta = \pm\pi/2$. The critical point topology is periodic and repeats at other values of $n$ and $m$. Note that the phase portrait is actually best represented on a cylindrical manifold, and simply wraps around the manifold for the other values of $n$ and $m$.

At low speeds, passive design features can be used to create a stabilizing moment when $\theta \neq 0$. For example, by designing the AUV so that its center of buoyancy lies along the $z_b$ axis directly above its center of gravity a stabilizing moment is passively generated when $\theta \neq 0$. Another design strategy is to use fins near the stern of the vehicle. The lift forces generated by the fins when $\theta \neq 0$ will passively generate moments that stabilize the vehicle towards $\theta = 0$. The pitch stability obtained with these approaches is explored in Problem 2.9.

Next, we will explore the use of active control to stabilize pitch. As shown in Problem 1.7, a vectored thruster can be used to generate pitch moments by pivoting the propeller so that its thrust has a component in the $z_b$ direction at the stern (see Fig. 2.11). First, we will explore the use of a controller that produces a restoring torque $-k_p\theta$, which is linearly proportional to the pitch angle, where $k_p > 0$ is a constant. This type of a controller is known as a proportional (P) controller. As will be discussed in Chap. 3, a P controller tends to act like a spring that pushes the response of a system towards a desired equilibrium.

The governing equation for pitch becomes

$$(I_y - M_{\dot{q}})\ddot{\theta} = (I_y - M_{\dot{q}})\dot{q} = -U^2 \sin\theta \cos\theta (Z_{\dot{w}} - X_{\dot{u}}) - k_p\theta,$$

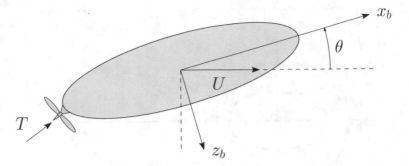

**Fig. 2.11** Pivoting the propeller to generate a pitching moment by tilting the thrust vector

so that

$$\dot{\mathbf{x}} = \begin{pmatrix} q \\ -\dfrac{U^2}{2} \sin(2\theta) \left[ \dfrac{Z_{\dot{w}} - X_{\dot{u}}}{I_y - M_{\dot{q}}} \right] - \dfrac{k_p}{I_y - M_{\dot{q}}} \theta \end{pmatrix} = \mathbf{f}. \qquad (2.24)$$

An inspection of (2.24) shows that an equilibrium point of the system will be at $\mathbf{x}_e = [0, 0]^T$. The Jacobian evaluated at this critical point is

$$\mathbf{A} = \begin{pmatrix} 0 & 1 \\ -U^2 \left[ \dfrac{Z_{\dot{w}} - X_{\dot{u}}}{I_y - M_{\dot{q}}} \right] - \dfrac{k_p}{I_y - M_{\dot{q}}} & 0 \end{pmatrix}$$

so that $\tau = 0$ and

$$\Delta = \frac{k_p + U^2(Z_{\dot{w}} - X_{\dot{u}})}{I_y - M_{\dot{q}}}.$$

From Fig. 2.7a, it can be seen that to prevent the phase portrait from containing saddle points (which have unstable trajectories), we require $\Delta > 0$, so that

$$k_p > -U^2(Z_{\dot{w}} - X_{\dot{u}}). \qquad (2.25)$$

For small angles, when $\sin(\theta) \approx \theta$, a second possible critical point could be located at

$$\mathbf{x} = \begin{bmatrix} \cos^{-1}\left( -\dfrac{k_p}{U^2(Z_{\dot{w}} - X_{\dot{u}})} \right) \\ 0 \end{bmatrix}.$$

However, the constraint (2.25) would make the argument of the $\cos^{-1}(\ )$ term above would be greater than 1, so that a second critical point cannot exist.

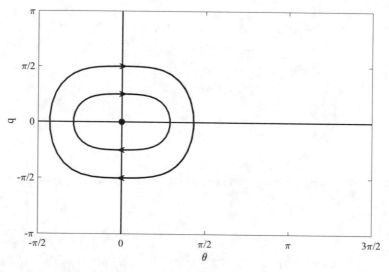

**Fig. 2.12** Pitch angle phase portrait for an AUV with unstable Munk moment counteracted by a proportional controller. Critical points occur at $(\theta, q) = (n\pi, 0)$ and are centers. For initial pitch angles $0 < |\theta| < \pi/2$ solution trajectories oscillate around $\theta = 0$

**Fig. 2.13** Example pitch angle time response for an AUV with unstable Munk moment counteracted by a proportional controller

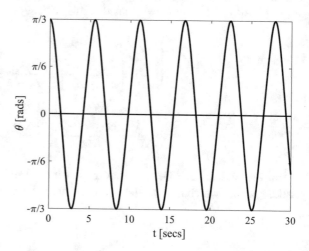

As shown in Fig. 2.12, the phase portrait will consist of a center at $\theta = 0$. If an initial pitch angle is small, the vehicle will exhibit small pitch oscillations within this small initial angle. While this behavior is an improvement over the uncontrolled case, it could be unsuitable for many applications, such as using the AUV to acoustically map a section of the seafloor using a sidescan sonar. The pitch angle time response for a typical AUV with an initial (large) pitch angle of $\pi/3$ is shown in Fig. 2.13. As can be seen, the oscillations persist and would likely be unsuitable in practice.

Lastly, we explore the effectiveness of adding a second feedback term, which is linearly proportional to the derivative of the pitch angle, to the existing controller so that the total control signal is $-k_p\theta - k_d\dot\theta$. This type of a controller is known as a proportional derivative controller. As will be discussed in Chap. 3, the derivative term tends to act like a damping factor. In a situation such as this, it could help to reduce the persistent oscillations observed when using only the $k_p$ term.

The governing equation for pitch becomes

$$(I_y - M_{\dot q})\ddot\theta = (I_y - M_{\dot q})\dot q = -U^2 \sin\theta\cos\theta(Z_{\dot w} - X_{\dot u}) - k_p\theta - k_d\dot\theta,$$

so that

$$\dot{\mathbf{x}} = \begin{pmatrix} q \\ -\dfrac{U^2}{2}\sin(2\theta)\left[\dfrac{Z_{\dot w} - X_{\dot u}}{I_y - M_{\dot q}}\right] - \dfrac{k_p\theta + k_d\dot\theta}{I_y - M_{\dot q}} \end{pmatrix} = \mathbf{f}. \tag{2.26}$$

An inspection of (2.26) shows that the equilibrium point of the system will be at $\mathbf{x}_e = [0, 0]^T$. The Jacobian matrix at this critical point is

$$\mathbf{A} = \begin{pmatrix} 0 & 1 \\ -U^2\left[\dfrac{Z_{\dot w} - X_{\dot u}}{I_y - M_{\dot q}}\right] - \dfrac{k_p}{I_y - M_{\dot q}} & -\dfrac{k_d}{I_y - M_{\dot q}} \end{pmatrix}$$

so that

$$\tau = \frac{k_d}{I_y - M_{\dot q}}$$

and

$$\Delta = \frac{k_p + U^2(Z_{\dot w} - X_{\dot u})}{I_y - M_{\dot q}}.$$

From Fig. 2.7a, it can be seen that if we select $k_d > 0$ and

$$k_p > -U^2(Z_{\dot w} - X_{\dot u})$$

the phase portrait will be a stable focus centered at the critical point $\mathbf{x}_e = [0, 0]^T$ (see Fig. 2.14). The effect of the additional feedback term is to dampen the pitch oscillations (Fig. 2.15).

Note that the main objective of this example is to demonstrate the link between stability and control. As will be seen in the following chapters, many methods exist for designing automatic controllers in order to meet specific performance objectives. The controller gains $k_p$ and $k_d$ are design parameters that could be tuned to produce a desired performance. In this example their values have only been selected to explore some basic stability concepts. □

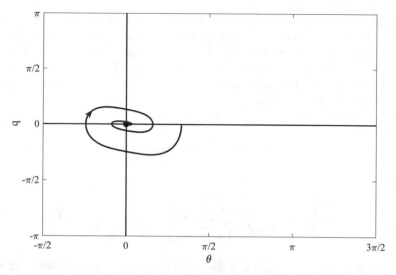

**Fig. 2.14** Pitch angle phase portrait for an AUV with unstable Munk moment counteracted by a proportional-derivative controller. Critical points occur at $(\theta, q) = (n\pi, 0)$ and are stable foci. For initial pitch angles $0 < |\theta| < \pi/2$ solution trajectories approach $\theta = 0$

**Fig. 2.15** Example pitch angle time response for an AUV with unstable Munk moment counteracted by a proportional-derivative controller. For the values of $k_p$ and $k_d$ selected the pitch oscillations are damped out after a couple of oscillation cycles

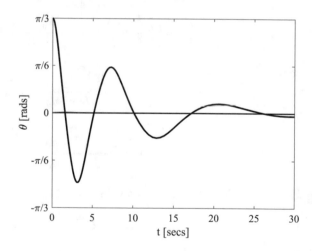

## 2.5  Lyapunov's Indirect (First) Method

As stated in Sect. 2.4.1, it is often possible to study the behavior of a nonlinear system near an equilibrium point using a linearized model of the system. In this vein, Lyapunov's Indirect Method allows one to deduce stability properties of the nonlinear system (2.13), provided that **f** is differentiable and has a continuous derivative, from the stability properties of its *linearization* about an equilibrium point $\mathbf{x}_e$, which is the linear system (2.15) with

$$A = \left.\frac{\partial \mathbf{f}}{\partial \mathbf{x}}\right|_{\mathbf{x}=\mathbf{x}_e}.$$

It can be shown that when $\mathbf{A}$ is nonsingular, i.e. $\det(\mathbf{A}) \neq 0$, the equilibrium point $\mathbf{x}_e$ is an isolated equilibrium point of the nonlinear system (2.13). Note that a system is *asymptotically stable* if the state trajectories $\mathbf{x}(t)$ of the system converge to $\mathbf{x}_e$ for all initial conditions sufficiently close to $\mathbf{x}_e$.

The following theorem can be used to ascertain the stability of the original nonlinear system.

**Theorem 2.1** (Lyapunov's Indirect Stability Criteria)

*(1) If all the eigenvalues of $\mathbf{A}$ have negative real parts (i.e., if $\mathbf{A}$ is Hurwitz, such that the linear approximation (2.15) is asymptotically stable), $\mathbf{x}_e$ is an asymptotically stable equilibrium point of the nonlinear system (2.13).*
*(2) If at least one of the eigenvalues of $\mathbf{A}$ has a positive real part (i.e., if the linear approximation (2.15) is unstable), $\mathbf{x}_e$ is an unstable equilibrium point of the nonlinear system (2.13).*

Recall the definitions of local and global stability from Sect. 2.3. The asymptotic stability of the origin for the linear approximation (which is always global) only implies local asymptotic stability of $\mathbf{x}_e$ for the nonlinear system.

If $\mathbf{A}$ has eigenvalues on the imaginary axis, but no eigenvalues in the open right-half plane, the linearization test is inconclusive. However, in this *critical* case, the system (2.13) cannot be asymptotically stable, since asymptotic stability of the linearization is not only a sufficient, but also a necessary condition, for the (local) asymptotic stability of the nonlinear system.

Lastly, it is noted that the case when $\mathbf{A} = 0$ is analogous to the degenerate case when $f'(x_e) = 0$ for the stability of a flow along a line discussed in Sect. 2.3.1. When this happens, one must use methods from nonlinear stability analysis (Chap. 7) to determine the stability of the system.

## 2.6 Stability of Linear Time Invariant Systems

An important requirement for every feedback control system is that all internal and external states of the system should be well-behaved when all input signals are well-behaved (including reference signals, external disturbances, noise, etc.). One of the simplest definitions of stability is that a signal $x(t)$ is bounded if there is a positive real number $M$, such that

$$|x(t)| \leq M, \quad \forall t \in [0, \infty). \tag{2.27}$$

Here $M$ is said to be a bound of the signal $x(t)$.

**Definition 2.1** (*Bounded input bounded output stability*) A system is bounded-input-bounded-output (BIBO) stable if the output of the system is bounded for every bounded input, regardless of what happens to the internal states of the system.

It is relatively easy to show that a system is unstable if we can identify a single bounded input that makes the output unbounded. However, it is much harder to show that the system is BIBO stable for all bounded inputs. The stability of a LTI system to an input can also be explored by examining its impulse response. For this, we introduce the Laplace Transform.

### 2.6.1  The Laplace Transform

LTI systems have two important properties that form the basis for almost all of the techniques used to analyze them. These are:

- the output responses of different inputs can be superposed to create the overall, composite response; and
- the response is the convolution of the input with the system's unit impulse response.

Let $x_1$ be the output response to an input $u_1$ and $x_2$ be the output response to an input $u_2$. Then, the output of the composite input $\alpha_1 u_1 + \alpha_2 u_2$ is $x = \alpha_1 x_1 + \alpha_2 x_2$.

**Example 2.5** (*Linear system*) Consider the system

$$\dot{x} = -kx + u \tag{2.28}$$

Let $u = \alpha_1 u_1 + \alpha_2 u_2$, and assume that $x = \alpha_1 x_1 + \alpha_2 x_2$. Then $\dot{x} = \alpha_1 \dot{x}_1 + \alpha_2 \dot{x}_2$ since both $\alpha_1$ and $\alpha_2$ are constants. If we substitute these expressions into (2.28), we obtain

$$\alpha_1 \dot{x}_1 + \alpha_2 \dot{x}_2 + k(\alpha_1 x_1 + \alpha_2 x_2) = \alpha_1 u_1 + \alpha_2 u_2.$$

From this it follows that

$$\alpha_1 (\dot{x}_1 + kx_1 - u_1) + \alpha_2 (\dot{x}_2 + kx_2 - u_2) = 0. \tag{2.29}$$

If $x_1$ is the solution with input $u_1$, and $x_2$ is the solution with input $u_2$, then (2.29) is satisfied and our assumption is correct: The response is the sum of the individual responses and superposition holds.                                                                 □

Since many complex signals can be composed by summing impulse functions and/or exponential functions, these two types of functions are often used as elementary input signals to determine the composite response of an LTI system. The Dirac delta function is used to represent an impulse and has the property that

$$\int_{-\infty}^{\infty} f(\tau)\delta(t - \tau)\mathrm{d}\tau = f(t). \tag{2.30}$$

If we replace the function $f(t)$ by $u(t)$, we can think of (2.30) as representing the input $u(t)$ as the integration of impulses of intensity $u(t - \tau)$. To find the response to an arbitrary input we need to find the response to a unit impulse. For an LTI system we can represent the impulse response as $g(t - \tau)$. Equation (2.30) then takes the form

$$x(t) = \int_{-\infty}^{\infty} u(\tau)g(t - \tau)d\tau,$$

$$= \int_{-\infty}^{\infty} u(t - \tau)g(\tau)d\tau.$$

(2.31)

This is known as the *convolution integral* and can also be represented using the symbol $*$, as in $x(t) = u(t) * g(t)$.

One of the properties of (2.31) is that an exponential input function of the form $e^{st}$ results in an output that is also an exponential time function, but changed in amplitude by the function $G(s)$, which is known as the *transfer function*. Note that $s$ may be complex $s = \sigma + j\omega$ and thus, that both the input and output may be complex. If we let $u(t) = e^{st}$ in (2.31), we get:

$$x(t) = \int_{0}^{\infty} u(\tau)g(t - \tau)d\tau,$$

$$= \int_{-\infty}^{\infty} g(\tau)u(t - \tau)d\tau,$$

$$= \int_{-\infty}^{\infty} g(\tau)e^{s(t-\tau)}d\tau,$$

$$= \int_{-\infty}^{\infty} g(\tau)e^{-s\tau}d\tau e^{st},$$

$$= G(s)e^{st}.$$

Here, the integral giving $G(s)$ is the Laplace Transform. In general, if we let $x(t)$ be a signal i.e., a real–valued function defined on the real line $(-\infty, \infty)$, its Laplace transform, denoted by $X(s)$, is a complex function and is defined as

$$X(s) = \mathscr{L}\{x(t)\} = \int_{-\infty}^{\infty} x(t)e^{-st}dt.$$

(2.32)

The set of values of $s$ that makes the integral (2.32) meaningful is called the region of convergence (ROC). The Laplace transforms of some commonly used functions are given in Table 2.1.

The Laplace transform of the unit impulse signal (Dirac Delta Function) is given by

$$\Delta(s) = \mathcal{L}\{\delta(t)\} = \int_{-\infty}^{\infty} \delta(t)e^{-st}dt = 1. \tag{2.33}$$

Additionally, the Laplace transform of the unit step function (Heaviside Step Function) $1(t)$ is given by

$$\Sigma(s) = \mathcal{L}\{1(t)\} = \int_{0}^{\infty} e^{-st}dt = \frac{1}{s}, \tag{2.34}$$

where the ROC is Re$\{s\} > 0$.

**Example 2.6**  Find the transfer function for the system of Example 2.5 when the input is $u = e^{st}$. Start with (2.28)

$$\dot{x} + kx = u = e^{st}.$$

If we assume that we can express $x(t)$ as $G(s)e^{st}$, we have $\dot{x}(t) = sG(s)e^{st}$, and the system equation reduces to

$$sG(s)e^{st} + kG(s)e^{st} = e^{st}.$$

Solving for the transfer function $G(s)$ gives

$$G(s) = \frac{1}{s+k}.$$

$\square$

**Poles and Zeros**
When the equations of a system can be represented as a set of simultaneous ordinary differential equations, the transfer function that results will be a ratio of polynomials, $G(s) = b(s)/a(s)$. We assume that the polynomials $b(s)$ and $a(s)$ are coprime, i.e. that they do not have any common factors. Values of $s$ that make $a(s) = 0$, e.g. the roots of $a(s)$, make $G(s) \to \infty$. These values of $s$ are called the *poles* of the transfer function. Similarly, the values of $s$ that cause $b(s) = 0$, and therefore $G(s) = 0$, are called *zeros*.

**Properties of the Laplace Transform**
Some useful properties of the Laplace transform are listed in Table 2.2.

**Theorem 2.2**  (Initial Value Theorem) *Let* $X(s) = \mathcal{L}\{x(t)\}$. *Then*

$$\lim_{t \to 0} x(t) = \lim_{s \to \infty} sX(s) \tag{2.35}$$

*when the two limits exist and are finite.*

**Table 2.1**  Laplace transform table

| Function name | $x(t), t \geq 0$ | $X(s)$ | ROC |
|---|---|---|---|
| Unit impulse | $\delta(t)$ | $1$ | $\mathbb{C}$ |
| Unit step | $1(t)$ | $\dfrac{1}{s}$ | $\mathrm{Re}\{s\} > 0$ |
| Unit ramp | $t$ | $\dfrac{1}{s^2}$ | $\mathrm{Re}\{s\} > 0$ |
| Unit acceleration | $\dfrac{t^2}{2}$ | $\dfrac{1}{s^3}$ | $\mathrm{Re}\{s\} > 0$ |
| $n$th-order ramp | $\dfrac{t^n}{n!}$ | $\dfrac{1}{s^{n+1}}$ | $\mathrm{Re}\{s\} > 0$ |
| Exponential | $e^{-\alpha t}$ | $\dfrac{1}{s+\alpha}$ | $\mathrm{Re}\{s\} > -\alpha$ |
| $n$th-order exponential | $\dfrac{t^n}{n!}e^{-\alpha t}$ | $\dfrac{1}{(s+\alpha)^{n+1}}$ | $\mathrm{Re}\{s\} > -\alpha$ |
| Sine | $\sin(\omega t)$ | $\dfrac{\omega}{s^2+\omega^2}$ | $\mathrm{Re}\{s\} > 0$ |
| Cosine | $\cos(\omega t)$ | $\dfrac{s}{s^2+\omega^2}$ | $\mathrm{Re}\{s\} > 0$ |
| Damped sine | $e^{-\alpha t}\sin(\omega t)$ | $\dfrac{\omega}{(s+\alpha)^2+\omega^2}$ | $\mathrm{Re}\{s\} > -\alpha$ |
| Damped cosine | $e^{-\alpha t}\cos(\omega t)$ | $\dfrac{s+\alpha}{(s+\alpha)^2+\omega^2}$ | $\mathrm{Re}\{s\} > -\alpha$ |

**Table 2.2**  Properties of Laplace transform. Here it is assumed that the $n$th time derivative of $f(t)$, $f^{(n)}(0) = 0, \forall n$

| Property | Mathematical expression |
|---|---|
| Linearity | $\mathscr{L}\{\alpha x(t) + \beta y(t)\} = \alpha X(s) + \beta Y(s)$ |
| Frequency shift | $\mathscr{L}\{e^{-\alpha t}x(t)\} = X(s+\alpha)$ |
| Time delay | $\mathscr{L}\{x(t-T)\} = e^{-sT}X(s)$ |
| Time scaling | $\mathscr{L}\{x(\alpha t)\} = \dfrac{1}{\alpha}X\left(\dfrac{s}{\alpha}\right)$ |
| Derivative | $\mathscr{L}\{\dot{x}(t)\} = sX(s)$ |
| Higher-order derivative | $\mathscr{L}\{x^{(n)}(t)\} = s^n X(s)$ |
| Integration | $\mathscr{L}\left\{\int_{-\infty}^{t} x(\tau)d\tau\right\} = \dfrac{X(s)}{s}$ |
| Convolution | $\mathscr{L}\{x(t) * y(t)\} = X(s)Y(s)$ |

**Theorem 2.3**  (Final Value Theorem) *Let* $X(s) = \mathscr{L}\{x(t)\}$. *Then*

$$\lim_{t\to\infty} x(t) = \lim_{s\to 0} sX(s) \qquad (2.36)$$

*when the two limits exist and are finite.*

**Inverse Laplace Transform**

The inverse Laplace transform can be used to find a signal $x(t)$ from its transform $X(s)$. Let $s := \sigma + j\omega$, where $\omega \in \mathbb{R}$, be a vertical line in the ROC of $X(s)$. Then, the formula for the inverse Laplace transform is

$$x(t) = \mathcal{L}^{-1}\{X(s)\} = \frac{1}{2\pi j} \int_{\sigma-j\infty}^{\sigma+j\infty} X(s)e^{st}ds. \tag{2.37}$$

In practice, the inverse Laplace transform is rarely used. Instead, tables, such as Tables 2.1 and 2.2, are used together with the Partial Fraction Expansion Technique to determine the inverse Laplace transform of different functions.

**Partial Fraction Expansions**

Here, we begin our discussion with a simple example.

**Example 2.7** Suppose we have calculated a transfer function $G(s)$ and found

$$G(s) = \frac{(s+2)(s+4)}{s(s+1)(s+3)}.$$

Note that the elementary entries in Table 2.1 are of the form $1/(s+a)$, and in order to use $G(s)$ to find the solution to a differential equation, for example, we want to be able to represent it in the form

$$G(s) = \frac{c_1}{s} + \frac{c_2}{(s+1)} + \frac{c_3}{(s+3)},$$

where $c_1$, $c_2$ and $c_3$ are constants. One way to do this would be to equate the two forms of $G(s)$ above and to solve for the three constants. However, an easy way to find the constants is to solve the equations for particular values of $s$. For example, if we multiply both equations by $s$ and set $s = 0$, we find

$$c_1 = \frac{(s+2)(s+4)}{(s+1)(s+3)}\Big|_{s=0} = \frac{8}{3}.$$

Similarly, we evaluate

$$c_2 = \frac{(s+2)(s+4)}{s(s+3)}\Big|_{s=-1} = -\frac{3}{2},$$

and

$$c_3 = \frac{(s+2)(s+4)}{s(s+1)}\Big|_{s=-3} = -\frac{1}{6}.$$

With the partial fractions, the solution for $x(t)$ can be looked up in the tables to be

$$x(t) = \frac{8}{3}1(t) - \frac{3}{2}1(t)e^{-t} - \frac{1}{6}1(t)e^{-3t}$$

$\square$

In the general case, the Partial Fraction Expansion Technique can be carried out as follows. Let $G(s)$ be a strictly proper rational function[1]

$$G(s) = \frac{b(s)}{a(s)} = \frac{b_1 s^{n-1} + b_2 s^{n-2} + \cdots + b_n}{a_0 s^n + a_1 s^{n-1} + \cdots + a_n}, \tag{2.38}$$

where $a_0 \neq 0$. Let the denominator polynomial $a(s)$ have roots $p_1, p_2, \ldots, p_n$, i.e.,

$$a(s) = a_0(s - p_1)(s - p_2) \cdots (s - p_n). \tag{2.39}$$

Here we will assume that $p_1, p_2, \ldots, p_n$ are distinct and that they can be complex. In contrast the $a_0, a_1, \ldots, a_n$ and $b_1, b_2, \ldots, b_n$ are all real. The complex poles among $p_1, p_2, \ldots, p_n$ appear in conjugate pairs. Thus, $G(s)$ can be expanded as

$$G(s) = \frac{c_1}{s - p_1} + \frac{c_2}{s - p_2} + \cdots + \frac{c_n}{s - p_n}, \tag{2.40}$$

where the $c_i$ can be obtained from

$$c_i = \lim_{s \to p_i} (s - p_i)G(s), \tag{2.41}$$

or by solving the polynomial equation

$$b(s) = a(s) \left[ \frac{c_1}{s - p_1} + \frac{c_2}{s - p_2} + \cdots + \frac{c_n}{s - p_n} \right]. \tag{2.42}$$

Notice that the right hand side of (2.42) is indeed a polynomial because of cancellations. The standard way of solving this polynomial equation is to convert it to a set of $n$ linear equations by comparing the coefficients of both sides of the polynomial equation. When the roots $p_i$ and $p_j$ are complex conjugates, so are the coefficients $c_i$ and $c_j$. After the partial fraction expansion is obtained, the inverse Laplace transform can be determined using Table 2.1.

When $G(s)$ has repeated poles (more than one pole at the same location), the partial fraction expansion is more complicated. When repeated poles are present, the denominator can be rewritten as

$$a(s) = a_0(s - p_1)^{n_1} (s - p_2)^{n_2} \cdots (s - p_r)^{n_r}, \tag{2.43}$$

where the $p_1, p_2, \ldots, p_r$ are all distinct and $n_1 + n_2 + \cdots + n_r = n$. Then the partial fraction expansion is of the form

---

[1] A transfer function is *strictly proper* when the degree of the numerator polynomial is less than the degree of the denominator polynomial.

$$G(s) = \frac{c_{11}}{s - p_1} + \frac{c_{12}}{(s - p_1)^2} + \cdots + \frac{c_{1n_1}}{(s - p_1)^{n_1}}$$

$$+ \frac{c_{21}}{s - p_2} + \frac{c_{22}}{(s - p_2)^2} + \cdots + \frac{c_{2n_2}}{(s - p_2)^{n_2}} \tag{2.44}$$

$$+ \cdots + \frac{c_{r1}}{s - p_r} + \frac{c_{r2}}{(s - p_r)^2} + \cdots + \frac{c_{rn_r}}{(s - p_r)^{n_r}},$$

where

$$c_{i1} = \frac{1}{(n_i - 1)!} \lim_{s \to p_i} \frac{d^{n_i - 1}}{ds^{n_i - 1}} \left[ (s - p_i)^{n_i} G(s) \right]$$

$$c_{i2} = \frac{1}{(n_i - 2)!} \lim_{s \to p_i} \frac{d^{n_i - 2}}{ds^{n_i - 2}} \left[ (s - p_i)^{n_i} G(s) \right] \tag{2.45}$$

$$\vdots$$

$$c_{in_i} = \lim_{s \to p_i} (s - p_i)^{n_i} G(s).$$

Alternatively, the coefficients can be obtained by solving the polynomial equation

$$b(s) = a(s) \left[ \frac{c_{11}}{s - p_1} + \cdots + \frac{c_{1n_1}}{(s - p_1)^{n_1}} + \cdots + \frac{c_{r1}}{s - p_r} + \cdots + \frac{c_{rn_r}}{(s - p_r)^{n_r}} \right]. \tag{2.46}$$

The right hand side of this equation is a polynomial since all of the denominators are canceled by $a(s)$. A common way to solve this equation is to convert it to a set of linear equations by comparing coefficients. Now that the Laplace Transform and impulse response function have been defined, we can apply these concepts to explore the stability of LTI systems.

**Theorem 2.4** (Stable Impulse Response) *A LTI system with impulse response function $g(t)$ is stable if and only if*

$$\int_0^\infty |g(t)| dt < \infty. \tag{2.47}$$

A signal is said to be absolutely integrable over an interval if the integral of the absolute value of the signal over the interval is finite. Thus, a linear system is stable if its impulse response is absolutely integrable over the interval $[0, \infty)$. It is still not necessarily an easy task to compute the integral of the absolute value of a function. The following theorem simplifies the task of checking the stability of LTI systems.

**Theorem 2.5** (Stable Transfer Function) *A LTI system with transfer function $G(s)$ is stable if and only if $G(s)$ is proper[2] and all poles of $G(s)$ have negative real parts.*

---

[2]A transfer function is *proper* when the degree of the numerator polynomial does not exceed the degree of the denominator polynomial. A transfer function is *strictly proper* when the degree of the numerator polynomial is less than the degree of the denominator polynomial.

Proofs of Theorems 2.4 and 2.5 can be found in [4]. Using these theorems, we can relate the stability of an impulse response to that of its associated transfer function. When a system is an integrable function, it is stable and its impulse response is also stable. When the transfer function of a system is proper and all of its poles have negative real parts, the transfer function and the system are stable. For an LTI system, the stability of the impulse response and of the transfer function are equivalent concepts. Since the poles of a transfer function $G(s)$ are the roots of its denominator polynomial, according to Theorem 2.5 the stability of an LTI system can be determined by examining the roots of the denominator polynomial. In this regard, we can also discuss the stability of a polynomial.

**Definition 2.2** (*Stable Polynomial*) A polynomial is said to be stable if all of its roots have negative real parts.

## 2.6.2 Routh's Stability Criterion

Often when designing a controller or modeling a dynamic system, some of the coefficients of of the system's characteristic equation may be functions of variable parameters. Computing the roots of the associated *characteristic polynomial*, which is the polynomial in $s$ associated with the Laplace transform of a system's characteristic equation, for all possible parameter variations can be difficult. However, it is generally possible to determine the stability of a polynomial without having to actually compute its roots. Consider a characteristic polynomial of the form

$$a(s) = a_0 s^n + a_1 s^{n-1} + \cdots + a_{n-1}s + a_n, \quad a_0 > 0. \tag{2.48}$$

It can always be factored as

$$a(s) = a_0(s + \gamma_1)(s + \gamma_2) \cdots (s + \gamma_m)[(s + \sigma_1)^2 + \omega_1^2] \cdots [(s + \sigma_l)^2 + \omega_l^2], \tag{2.49}$$

where $\gamma_i$, $i = 1, ..., m$, and $\sigma_i$, $\omega_i$, $i = 1, ..., l$ are all real. Notice that $-\gamma_i$ are the real roots of $a(s)$ and $-\sigma_i \pm j\omega_i$ are the complex pairs of roots of $a(s)$. If $a(s)$ is stable, then $\gamma_i$, $i = 1, ..., m$, and $\sigma_i$, $i = 1, ..., l$, are positive. Since $a_i$, $i = 1, ..., n$, are sums of products of $\gamma_i$, $\sigma_i$, $\omega_i^2$, they must be positive. This leads to the following theorem.

**Theorem 2.6** (Routh stability necessary conditions) *If $a(s)$ is stable, then $a_i > 0$, $i = 1, ..., n$.*

Theorem 2.6 gives a necessary condition for stability, which can be performed by simple inspection of the polynomial (2.48). If one of the coefficients is zero or negative, the polynomial is unstable. However, the condition is not sufficient, except for first-order and second-order polyomials. In general, if all of the coefficients are positive, no immediate conclusion can be reached about the stability.

**Table 2.3** Routh table

| | | | | | |
|---|---|---|---|---|---|
| $s^n$ | $r_{00} = a_0$ | $r_{01} = a_2$ | $r_{02} = a_4$ | $r_{03} = a_6$ | $\cdots$ |
| $s^{n-1}$ | $r_{10} = a_1$ | $r_{11} = a_3$ | $r_{12} = a_5$ | $r_{13} = a_7$ | $\cdots$ |
| $s^{n-2}$ | $r_{20}$ | $r_{21}$ | $r_{22}$ | $r_{23}$ | $\cdots$ |
| $s^{n-3}$ | $r_{30}$ | $r_{31}$ | $r_{32}$ | $r_{33}$ | $\cdots$ |
| $\vdots$ | $\vdots$ | $\vdots$ | $\vdots$ | $\vdots$ | $\vdots$ |
| $s^2$ | $r_{(n-2)0}$ | $r_{(n-2)1}$ | | | |
| $s^1$ | $r_{(n-1)0}$ | | | | |
| $s^0$ | $r_{n0}$ | | | | |

The Routh Criterion can be used to determine the stability of a polynomial without explicitly solving for its roots. To do this, we can start by constructing a Routh table for the polynomial in (2.48) (Table 2.3).

The first two rows come directly from the coefficients of $a(s)$. Each of the other rows is computed from its two preceding rows as:

$$r_{ij} = -\frac{1}{r_{(i-1)0}} \det \begin{bmatrix} r_{(i-2)0} & r_{(i-2)(j+1)} \\ r_{(i-1)0} & r_{(i-1)(j+1)} \end{bmatrix} = \frac{r_{(i-1)0}r_{(i-2)(j+1)} - r_{(i-2)0}r_{(i-1)(j+1)}}{r_{(i-1)0}}.$$

Here, $i$ goes from 2 to $n$ and $j$ goes from 0 to $[(n-i)/2]$. For example,

$$r_{20} = -\frac{1}{r_{10}} \det \begin{bmatrix} r_{00} & r_{01} \\ r_{10} & r_{11} \end{bmatrix} = \frac{r_{10}r_{01} - r_{00}r_{11}}{r_{10}} = \frac{a_1 a_2 - a_0 a_3}{a_1},$$

$$r_{21} = -\frac{1}{r_{10}} \det \begin{bmatrix} r_{00} & r_{02} \\ r_{10} & r_{12} \end{bmatrix} = \frac{r_{10}r_{02} - r_{00}r_{12}}{r_{10}} = \frac{a_1 a_4 - a_0 a_5}{a_1},$$

$$r_{22} = -\frac{1}{r_{10}} \det \begin{bmatrix} r_{00} & r_{03} \\ r_{10} & r_{13} \end{bmatrix} = \frac{r_{10}r_{03} - r_{00}r_{13}}{r_{10}} = \frac{a_1 a_6 - a_0 a_7}{a_1}.$$

When computing the last element of a certain row of the Routh table, one may find that the preceding row is one element short of what is needed. For example, when we compute $r_{n0}$, we need $r_{(n-1)1}$, but $r_{(n-1)1}$ is not an element of the Routh table. In this case, we can simply augment the preceding row by a 0 at the end and keep the computation going. Keep in mind that this augmented 0 is not considered to be part of the Routh table. Equivalently, whenever $r_{(i-1)(j+1)}$ is missing simply let $r_{ij} = r_{(i-2)(j+1)}$. For example, $r_{n0}$ can be computed as

$$
r_{n0} = -\frac{1}{r_{(n-1)0}} \det \begin{bmatrix} r_{(n-2)0} & r_{(n-2)1} \\ r_{(n-1)0} & 0 \end{bmatrix} = r_{(n-2)1}.
$$

**Theorem 2.7** (Routh's Stability Criterion) *The following three statements are equivalent:*

1. *$a(s)$ is stable.*
2. *All elements of the Routh table are positive definite, i.e., $r_{ij} > 0$, $i = 0, 1, ..., n$, $j = 0, 1, ..., [(n-i)/2]$.*
3. *All elements in the first column of the Routh table are positive definite, i.e., $r_{i0} > 0$, $i = 0, 1, ..., n$.*

In general, the Routh table cannot be completely constructed when some elements in the first column are zero. In this case, there is no need to complete the rest of the table since we already know from the Routh criterion that the polynomial is unstable. As pointed out earlier, the most useful application of the Routh criterion is to determine the stability of a polynomial when parameters are involved.

## Problems

**2.1** A CTD (conductivity-temperature-depth) sensor is an oceanographic instrument commonly used to measure profiles of the salinity and temperature of seawater versus depth. A thermistor, a type of resistor whose resistance is dependent on temperature, is often used on CTDs as the temperature sensing element. The temperature response of a thermistor is given by the relation

$$
R = R_0 e^{-0.1T},
$$

where $R$ is resistance, $T$ is temperature in °C and $R_0 = 10 \times 10^3 \Omega$. Find a linear model of the thermister operating at $T = 20$°C for a small range of temperature variations about this operating condition.

**2.2** Use linear stability analysis to classify the fixed points, as either stable or unstable, for the following systems. If linear stability analysis fails because $f'(x) = 0$, use a graphical argument to determine the stability.

(a) $\dot{x} = x(2 - x)$
(b) $\dot{x} = x(2 - x)(4 - x)$
(c) $\dot{x} = \sin x$
(d) $\dot{x} = x^2(2 - x)$

(e) $\dot{x} = ax - x^3$, where $a$ can be: $a > 0$, $a < 0$ or $a = 0$. Discuss all three cases.

**2.3** Analyze the following equations graphically. In each case, sketch the vector field on the real line, find all of the fixed points, classify their stability (as either stable or unstable), and sketch the graph of $x(t)$ for different initial conditions.

(a) $\dot{x} = 2x^2 - 8$
(b) $\dot{x} = 2 + \cos x$
(c) $\dot{x} = e^x - \cos x$ *Hint:* Sketch the graphs of $e^x$ and $\cos x$ on the same axes and look for intersections. You won't be able to find the fixed points explicitly, but you can still find the qualitative behavior.

**2.4** Analyze the stability of the dynamics (corresponding to a mass sinking in a viscous liquid)

$$\dot{u} + 2a|u|u + bu = c, \quad a > 0, \ b > 0.$$

**2.5** For each of the following systems, decide whether the origin is attracting, Lyapunov stable, asymptotically stable, or none of the above.

(a) $\dot{x} = y, \ \dot{y} = -6x$
(b) $\dot{x} = 0, \ \dot{y} = x$
(c) $\dot{x} = -x, \ \dot{y} = -7y$

**2.6** Prove your answers to Problem 2.5 are correct using the different definitions of stability.

**2.7** The system in Fig. 2.16 is a bang-bang heading control system for a small catamaran. The yaw rate is measured using an inertial measurement unit (IMU). Let $\psi$ be the yaw angle measured with respect to North (heading) and $r = \dot{\psi}$ be the yaw rate.

(a) Using the block diagram show that $\ddot{\psi} = -\tau_0 \mathrm{sgn}(\psi)$, where $\mathrm{sgn}(\bullet)$ is the sign function.
(b) Show that

$$\frac{d}{dt}\left(\frac{\dot{\psi}^2}{2}\right) = \dot{\psi}\tau.$$

(c) Integrate the equation found in part b) to obtain a relation between $\dot{\psi}$ and $\psi$.
(d) Draw the phase portrait of the system. Make sure to indicate the direction of each trajectory using arrows.

**2.8** Consider the automatic ship steering system shown in Fig. 2.17.

(a) Write the differential equation relating the heading angle $\psi$ to the rudder angle $\delta$.
(b) Use Routh's Criterion to show whether or not the closed loop system is stable with $K = 1$.

**Fig. 2.16** Bang-bang heading control system

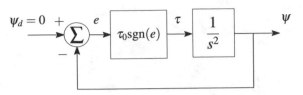

**Fig. 2.17** An automatic ship steering system

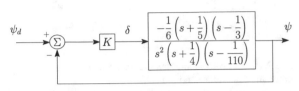

(c) Can the system be made stable with $K < 1$?

**2.9** In Example 2.4 it was shown how active control could be used to stabilize the pitch angle of an AUV by counteracting the Munk moment. It was noted that AUV design features could be used to passively counteract the Munk moment at low speeds. Using Fig. 2.18, address the following items:

(a) By separating the center of buoyancy $B$ and the center of gravity $G$ a hydrostatic moment can be generated to counteract the Munk moment, which is taken to be

$$\tau_\theta = U^2 \sin\theta \cos\theta (Z_{\dot{w}} - X_{\dot{u}}).$$

What is the minimum separation distance $h$ required for pitch stability? Let $m$ be the mass of the AUV. Note that the vehicle is neutrally buoyant.

(b) Assume that the span of the stern plane fins $s$ is the same as that of the rudder fins shown in Fig. 2.18a. The fins actually protrude from both the port and starboard sides of the AUV, but can be modeled as a single wing of span $s$. For the case in which it is not possible to make $h$ large enough, estimate the minimum span of the stern plane fins required to make the pitch stable. Take the location of the center of pressure (where the lift force is centered) to be $x_b = -0.4L$. The lift force generated by the fin can be approximated as

$$L_f = \frac{1}{2}\rho U^2 sl(2\pi\theta),$$

where $l$ is the chord of the fin.

(c) Plot a phase portrait of the pitch dynamics with hydrostatic restoring moment and fins.

(d) Instead of using stern planes to provide pitch stability, it is suggested that fixed bow planes be added so that the center of pressure is located at $x_b = +0.4L$. Is this is a good idea? Use a plot of the corresponding phase portrait to justify your answer.

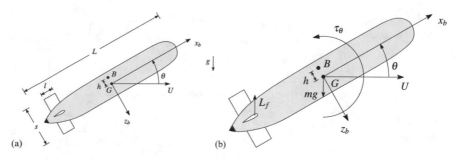

**Fig. 2.18** Passively stabilizing the pitch angle of an AUV. **a** Geometry of the problem. **b** Free body diagram of the forces and moments

**2.10** A boat's cruise control system attempts to maintain a constant velocity in the presence of disturbances primarily caused by wind and current. The controller compensates for these unknowns by measuring the speed of the boat and adjusting the engine throttle appropriately. Let $v$ be the speed of the boat, $m$ the total mass, $T$ the force generated by the propeller, and $F_d$ the disturbance force due to aerodynamic and hydrodynamic drag. The equation of motion of the boat is

$$m\frac{dv}{dt} = T - F_d.$$

The thrust $T$ is generated by the engine, whose torque is proportional to the rate of fuel injection, which is itself proportional to a control signal $0 \leq u \leq 1$ that controls the throttle position. The torque also depends on engine speed $\omega$. Let $P$ be the pitch-to-diameter ratio of the propeller and $d$ the propeller diameter. A simple representation of the torque at full throttle is given by the torque curve

$$Q(\omega) = Q_m\left[1 - \beta\left(\frac{\omega}{\omega_m} - 1\right)^2\right],$$

where the maximum torque $Q_m$ is obtained at engine speed $\omega_m$. Typical parameters are $Q_m = 190$ N-m, $\omega_m = 420$ rad/s (about 4000 RPM) and $\beta = 0.4$. The engine speed is related to the velocity through the expression

$$\omega = \frac{1}{Pd}v := \alpha_n v,$$

and the thrust can be written as

$$T = \frac{nu}{r}Q(\omega) = \alpha_n u\, Q(\alpha_n v).$$

Typical values for $P$ and $\alpha_n$ are $P = 1.2$ and $\alpha_n = 16$. The disturbance force $F_d$ has two major components: $F_w$, the hydrodynamic forces and $F_a$, the aerodynamic drag.

The aerodynamic drag is proportional to the square of the apparent wind speed. Here we will assume that the boat is moving into a headwind $v_w$ so that

$$F_a = \frac{1}{2}\rho C_d A(v + v_w)^2,$$

where $\rho$ is the density of air, $C_d$ is the shape-dependent aerodynamic drag coefficient and $A$ is the frontal area of the boat. Typical parameters are $\rho = 1.3\,\text{kg/m}^3$, $C_d = 0.32$ and $A = 2.4\,\text{m}^2$. Thus, the complete, nonlinear, equation of motion for the boat is

$$m\frac{dv}{dt} = \alpha_n u T(\alpha_n v) - mgC_r\text{sgn}(v) - \frac{1}{2}\rho C_d A(v + v_w)^2 + \frac{1}{2}\rho_w C_{dh} S_w v^2.$$

We'll regulate the boat's speed by using proportional integral (PI) feedback control to determine the control signal $u$. The PI controller has the form

$$u(t) = k_p e(t) + k_i \int_0^t e(\tau)d\tau.$$

This controller can be realized as an input/output dynamical system by defining a controller state $z$ and implementing the differential equation

$$\frac{dz}{dt} = v_r - v, \quad u = k_p(v_r - v) + k_i z,$$

where $v_r$ is the desired (reference) speed. Generate a phase portrait for the closed loop system, in third gear, using a PI controller (with $k_p = 0.5$ and $k_i = 0.1$), $m = 1000\,\text{kg}$ and desired speed 20 m/s. Your system should include the effects of saturating the input between 0 and 1.

# References

1. Faltinsen, O.M.: Hydrodynamics of High-speed Marine Vehicles. Cambridge University Press, Cambridge (2005)
2. Khalil, H.K.: Nonlinear Systems, 3rd edn. Prentice Hall Englewood Cliffs, New Jersey (2002)
3. Perry, A.E., Chong, M.S.: A description of eddying motions and flow patterns using critical-point concepts. Annu. Rev. Fluid Mech. **19**, 125–155 (1987)
4. Qiu, L., Zhou, K.: Introduction to Feedback Control. Prentice Hall Englewood Cliffs, New Jersey (2009)
5. Slotine, J.-J.E., Li, W.: Applied Nonlinear Control. Prentice-Hall Englewood Cliffs, New Jersey (1991)
6. SNAME. Nomenclature for treating the motion of a submerged body through a fluid: Report of the American Towing Tank Conference. Technical and Research Bulletin 1–5, Society of Naval Architects and Marine Engineers (1950)
7. Strogatz, S.H.: Nonlinear Dynamics and Chaos: With Applications to Physics, Biology, Chemistry, and Engineering, 2nd edn. Westview Press, Boulder (2014)

# Chapter 3
# Time Response and Basic Feedback Control

## 3.1 Dynamic Response

The dynamic response of a system is its time-dependent behavior caused by a forcing input, or by a nonequilibrium initial condition. Recall that the transfer function of a linear time invariant system is a function of the complex variable $s = \sigma + j\omega$, where $j = \sqrt{-1}$ is the imaginary number (Sect. 2.6.1). When the equations of motion of a marine system can be treated as LTI, insight about certain features of its dynamic response, and how the response can be manipulated using feedback control, can be obtained by examining the locations of the systems poles and zeros in the complex plane, on which the real part of $s$ is typically plotted on the horizontal axis and the imaginary part of $s$ is plotted on the vertical axis. In control engineering the complex plane is often simply referred to as the $s$-*plane*. Here, we explore simple relationships between the features of a system's dynamic time response and the locations of its poles and zeros.

### 3.1.1 The Impulse Response of 1st and 2nd Order Systems

As the Laplace Transform of the unit impulse function is unity $\mathcal{L}\{\delta(t)\} = 1$ (Table 2.1), the impulse response of a system is given by the inverse Laplace transform of its transfer function and is also known as the system's *natural response*. We first examine how the poles and zeros of a first order system affect its impulse response.

**Example 3.1** Let

$$G(s) = \frac{s+b}{(s+a_1)(s+a_2)}.$$

K. D. von Ellenrieder, *Control of Marine Vehicles*, Springer Series on Naval Architecture, Marine Engineering, Shipbuilding and Shipping 9, https://doi.org/10.1007/978-3-030-75021-3_3

**Fig. 3.1** Pole–zero locations
in the $s$–plane for $G(s)$

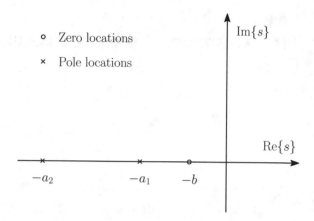

There is a zero located at $s = -b$, a pole at $s = -a_1$ and a pole at $s = -a_2$ (Fig. 3.1).
Expanding by partial fractions, we get

$$G(s) = \frac{A}{s + a_1} + \frac{B}{s + a_2},$$

where

$$A = \frac{b - a_1}{a_2 - a_1} \quad \text{and} \quad B = -\frac{b - a_2}{a_2 - a_1}.$$

The function $g(t)$ corresponding to the inverse Laplace Transform of $G(s)$ is

$$g(t) = \begin{cases} Ae^{-a_1 t} + Be^{-a_2 t}, & t \geq 0, \\ 0, & t < 0. \end{cases}$$

The time dependent components of $g(t)$, $e^{-a_1 t}$ and $e^{-a_2 t}$ depend upon the pole locations, $s = -a_1$ and $s = -a_2$, respectively. The location of the zero $s = -b$ affects
the magnitudes of $A$ and $B$. This is true for more complicated transfer functions as
well. In general, the functional form of the time response is determined by the pole
locations (Fig. 3.3) and the magnitude of the separate components of the response is
governed by the locations of the zeros.                                            □

Next, we look at the impulse response of a second order system with a pair of
complex poles. A classic mathematical representation of such a system in control
engineering is

$$G(s) = \frac{\omega_n^2}{s^2 + 2\zeta\omega_n s + \omega_n^2}. \tag{3.1}$$

Here, $\zeta$ the *damping ratio* and $\omega_n$ is the *undamped natural frequency*. The poles of this
transfer function are located at a radius $\omega_n$ from the origin and at an angle $\theta = \sin^{-1}\zeta$
from the positive imaginary axis in the $s$–plane (Fig. 3.2). In rectangular coordinates,

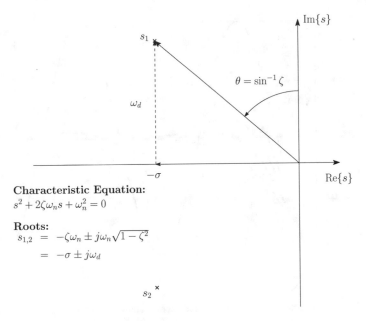

**Characteristic Equation:**
$$s^2 + 2\zeta\omega_n s + \omega_n^2 = 0$$

**Roots:**
$$\begin{aligned} s_{1,2} &= -\zeta\omega_n \pm j\omega_n\sqrt{1-\zeta^2} \\ &= -\sigma \pm j\omega_d \end{aligned}$$

**Fig. 3.2** $s$–plane plot of a pair of complex poles

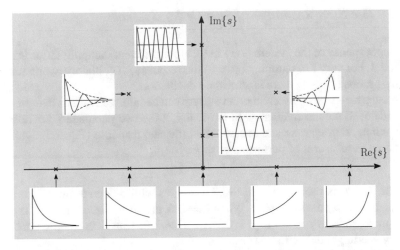

**Fig. 3.3** Impulse response forms associated with various points in the $s$–plane. Note that the time response associated with the pole at the origin ($s = 0$) is constant in time and is often known as the *DC gain* of a system

the poles are located at $s = -\sigma \pm j\omega_d$, where $\sigma = \zeta\omega_n$ and $\omega_d = \omega_n\sqrt{1-\zeta^2}$ is the *damped frequency*. When $\zeta = 0$ there is no damping and the poles are located on the imaginary axis. The qualitative forms of the impulse responses of (3.1) associated with various points in the $s$–plane are shown in Fig. 3.3.

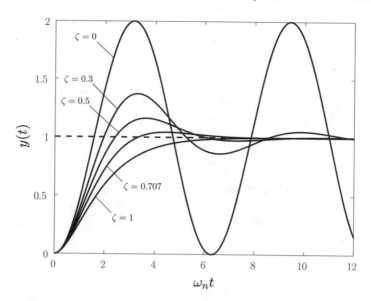

**Fig. 3.4** Characteristic step response of a second order system as the damping ratio $\zeta$ varies

## 3.1.2  The Step Response of 2nd Order Systems

The step response of the system $G(s)$ in (3.1) for different damping ratios is shown in Fig. 3.4. For very low damping ($\zeta \approx 0$) the response is oscillatory, while for large damping ratios ($\zeta \to 1$) there is almost no oscillation.

The step response of a second order system can be inferred from its damping ratio and undamped natural frequency. In particular, these two parameters are related to the *rise time, settling time* and *overshoot* of the step response (Fig. 3.5), which are often used as measures of performance of a system.

(1) *Rise Time*: As can be seen in Fig. 3.4, curves with moderate overshoot rise in approximately the same time. When $\zeta = 0.5$, the normalized time for the output to increase from $y = 0.1$ to $y = 0.9$ is $\omega_n t_r \approx 1.8$. Thus, it is common in control engineering to define the rise time as

$$t_r := \frac{1.8}{\omega_n}. \tag{3.2}$$

(2) *Overshoot*: When $\zeta < 1$, the step response exhibits a peak value greater than $y = 1$ at some *peak time* $t_p$. The difference between the peak value and $y = 1$ is the *peak magnitude* $M_p$ (Fig. 3.5). Since $M_p$ is a local maximum of the step response, the derivative of $y(t)$ is zero at $t_p$. Thus, both $M_p$ and $t_p$ can

**Fig. 3.5** Definitions of rise time $t_r$, settling time $t_s$ and overshoot $M_p$ for unit step response

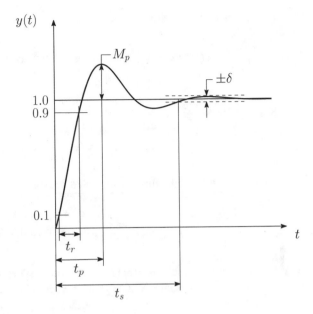

be analytically determined from $y(t)$. The transfer function $G(s)$ provides an algebraic relation between the system's input $u(t)$ and its output $y(t)$

$$G(s) = \frac{Y(s)}{U(s)} = \frac{\omega_n^2}{s^2 + 2\zeta\omega_n s + \omega_n^2}. \tag{3.3}$$

When the input is a unit step function, $u(t) = 1(t)$ and the time response can be found by taking the inverse Laplace transform of

$$Y(s) = \frac{\omega_n^2}{s^2 + 2\zeta\omega_n s + \omega_n^2} U(s) = \frac{\omega_n^2}{s^2 + 2\zeta\omega_n s + \omega_n^2} \cdot \frac{1}{s}.$$

The solution is

$$y(t) = 1 - e^{-\sigma t}\left[\cos(\omega_d t) + \frac{\sigma}{\omega_d}\sin(\omega_d t)\right]. \tag{3.4}$$

Taking the time derivative of this function and setting it equal to zero gives

$$\dot{y}(t) = \sigma e^{-\sigma t}\left[\cos(\omega_d t) + \frac{\sigma}{\omega_d}\sin(\omega_d t)\right] - e^{-\sigma t}\left[-\omega_d t \sin(\omega_d t) + \sigma \cos(\omega_d t)\right],$$

$$= e^{-\sigma t}\left(\frac{\sigma^2}{\omega_d} + \omega_d\right)\sin(\omega_d t) = 0.$$

It can be seen that $\dot{y}(t) = 0$ when $\omega_d t = n\pi$, where $n$ is an integer $n = \pm 0, 1, 2, \ldots$. The peak time corresponds to $n = 1$, so that $t_p = \pi/\omega_d$. Substituting this value for $t_p$ into (3.4), we obtain

$$y(t_p) := 1 + M_p = 1 - e^{-\sigma\pi/\omega_d}\left[\cos\pi + \frac{\sigma}{\omega_d}\sin\pi\right],$$
$$= 1 + e^{-\sigma\pi/\omega_d}.$$

Solving for $M_p$ gives

$$M_p = e^{-\pi\zeta/\sqrt{1-\zeta^2}}, \quad 0 \le \zeta < 1. \tag{3.5}$$

Two frequently used values of $M_p$ used in control engineering practice are

$$M_p = \begin{cases} 0.16, & \text{for } \zeta = 0.5, \\ 0.04, & \text{for } \zeta = 0.707. \end{cases} \tag{3.6}$$

(3) *Settling Time*: The settling time $t_s$ corresponds to the time at which the difference between $y(t)$ and its steady state value becomes bounded by the small quantity $\delta$, such that $|y(t \to \infty) - y(t_s)| = \delta$ (see Fig. 3.5). In control engineering practice it is common to use either a 1% criteria ($\delta = 0.01$) [3], or a 2% criteria ($\delta = 0.02$) [1, 5], to determine $t_s$. Examining (3.4) it can be seen that $t_s$ can be found using the magnitude of the transient exponential in $y(t)$

$$e^{-\sigma t} = e^{-\zeta\omega_n t} = \delta,$$

e.g. when $\zeta\omega_n t_s = -\ln\delta$. Therefore,

$$t_s = \begin{cases} \dfrac{4.6}{\zeta\omega_n} = \dfrac{4.6}{\sigma}, & \delta = 0.01, \\[2mm] \dfrac{4}{\zeta\omega_n} = \dfrac{4}{\sigma}, & \delta = 0.02. \end{cases} \tag{3.7}$$

During the process of control system design, the step response performance measures defined here are often used to quantify design requirements and/or constraints. Typically a maximum overshoot $M_{p\,max}$ and a maximum settling time $t_{s\,max}$, or a minimum rise time $t_{r\,min}$ will be specified. As many systems can be approximated as second order systems, understanding how the locations of poles in the $s$–plane correlate with the values of these performance measures can help to guide controller design. Using (3.2), (3.5) and (3.7) we can specify regions of the $s$–plane where the poles of a second order system should be placed so that the design criteria are satisfied:

$$\omega_n \ge \frac{1.8}{t_{r\,min}}, \tag{3.8}$$

$$\zeta \geq \frac{|\ln(M_{p\,max})|}{\sqrt{[\ln(M_{p\,max})]^2 + \pi^2}}, \tag{3.9}$$

and

$$\sigma \geq \frac{-\ln \delta}{t_{s\,max}}. \tag{3.10}$$

A plot of how these constraints correlate to regions of the $s$-plane is shown in Fig. 3.6. The bounds shown can be used during control system design to help guide the selection of pole locations needed to produce a desired step response.

**Remark 3.1** (*Dominant second order poles*) The time response of many higher order systems is dominated by a single pair of complex poles. This occurs when the real parts of all the other poles (apart from the dominant ones) are located further to the left of the imaginary axis in the complex plane, so that they do not have an appreciable effect on the second order response. For such systems, good low order approximations can be obtained by neglecting the less significant poles and zeros and utilizing the performance formulas (3.8), (3.9) and (3.10) to select suitable values of $\omega_n$ and $\zeta$ based on the dominant poles only. Typically, a closed loop pole or zero will dominate the response of the system when its real part is approximately 1/10th to 1/5th closer to the imaginary axis than the next nearest closed loop pole or zero [5].

### 3.1.3 Effects of Additional Poles and Zeros

Based on the immediately preceding discussion and Fig. 3.6, we can infer the following guidelines, that also loosely apply for systems more complicated than the second order system (3.3):

(1) If the rise time is too slow, the natural frequency should be increased.
(2) If there is too much overshoot in the transient response, damping should be increased.
(3) If the settling time is too long, the poles should be moved towards the left in the $s$–plane.

To explore the effects an additional zero has on a second order system, consider a transfer function with one zero and two complex poles

$$G(s) = \frac{[s/(\alpha\zeta\omega_n) + 1]\omega_n^2}{s^2 + 2\zeta\omega_n s + 1}. \tag{3.11}$$

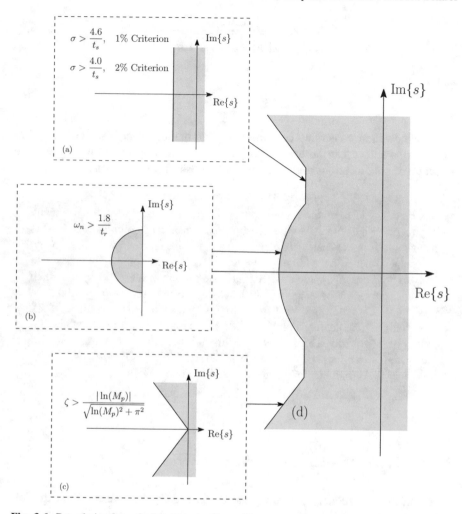

**Fig. 3.6** Boundaries for pole locations in the $s$–plane to produce a desired step response for a second order system, as given by (3.8), (3.9) and (3.10): **a** Settling–time requirements; **b** rise–time requirements; **c** overshoot requirements; and **d** a composite of all three requirements. Poles located to the left of the shaded region will meet the desired requirements

The zero is located at $s = -\alpha \zeta \omega_n = -\alpha \sigma$. By comparing the step response plots in Fig. 3.7 with the step response of a purely second order system shown in Fig. 3.4, it can be seen that when $\alpha$ is large and the zero is far from the poles, it has little effect on the system response. In contrast, when $\alpha = 1$ the zero is located at the value of the real part of the complex poles and has a substantial influence on the response. The main effect of the zero is to increase the overshoot $M_p$. The zero does not affect the settling time.

Expanding $G(s)$ in (3.11) using partial fraction expansion gives

**Fig. 3.7** Step response of a second–order system with an extra zero ($\zeta = 0.5$)

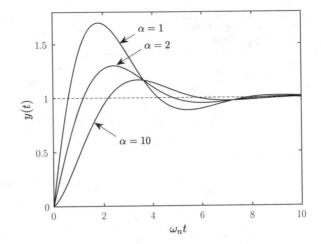

$$G(s) = \frac{\omega_n^2}{s^2 + 2\zeta\omega_n s + 1} + \frac{1}{\alpha\zeta\omega_n} \cdot \frac{s\omega_n^2}{s^2 + 2\zeta\omega_n s + 1}. \tag{3.12}$$

The first term in (3.12) is the original second order transfer function (3.1), without the additional zero, and the second term is the original second order transfer function multiplied by $s$ and scaled by $1/(\alpha\zeta\omega_n)$. As multiplication by $s$ in the $s$-domain corresponds to differentiation in the time domain (see Table 2.2), the second term corresponds to adding the derivative of the original second order transfer function. Let $y_1(t)$ be the step response of the original second order system, $y_2(t)$ be the derivative of the step response of the original second order system and $y(t) = y_1(t) + y_2(t)$ be the total unit step response of (3.12). An examination of the response (Fig. 3.8) shows that the zero increases the overshoot because the derivative of the original term constructively adds to the total response. When $\alpha < 0$, the zero is in the right–half plane and the derivative term is subtracted, rather than added (see Fig. 3.9). This can result in the step response initially starting in the wrong direction before approaching the desired steady state value, a behavior characteristic of nonminimum phase systems. Many marine vehicles exhibit this type response, especially systems that utilize fins and rudders for steering. Consider, for example, ship steering. In order to steer to the left (to port) a ship's rudder is deflected with an angle of attack about the positive yaw axis of the vehicle (see Fig. 1.11a for a definition of ship coordinates). This in turn creates a force in the positive sway direction (to the starboard side) and a negative yaw moment. The small positive sway force from the rudder may initially cause a small displacement of the hull in the positive sway direction, towards the right (starboard) side of the ship. However, the negative yaw moment generated by the rudder also causes the ship to rotate in the negative yaw direction. The resulting attitude of the ship causes hydrodynamic lift to be generated on the main hull that manifests as a side force to the left (port), which ultimately causes the ship to move the left, as desired.

**Fig. 3.8** Second–order step response and its derivative ($\alpha = 2$ and $\zeta = 0.5$)

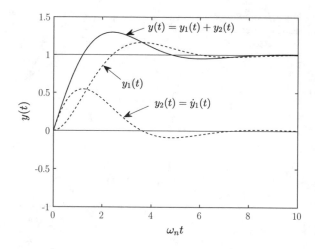

**Fig. 3.9** Second–order step response and its derivative with a right–half plane zero ($\alpha = -2$ and $\zeta = 0.5$). This is known as a nonminimum phase system

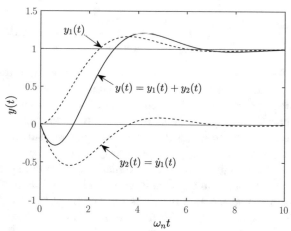

Next consider the effects of an extra pole on the standard second–order step response. The transfer function is

$$G(s) = \frac{\omega_n^2}{[s/(\alpha\zeta\omega_n) + 1][s^2 + 2\zeta\omega_n s + 1]}. \tag{3.13}$$

Plots of the step response for this case are shown in Fig. 3.10 for $\zeta = 0.5$. The pole's main effect is to increase the rise time.

Several qualitative conclusions can be made about the dynamic response of a system based on its pole–zero patterns:

(1) The performance measures of a second order system's response to a step input can be approximated by Eqs. (3.2), (3.5) and (3.7).

**Fig. 3.10** Step response for several third–order systems (extra pole) when $\zeta = 0.5$

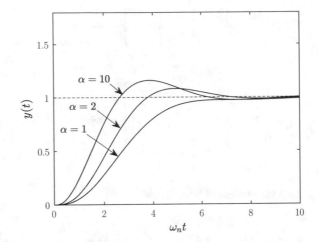

(2) An additional zero in the left–half plane increases overshoot if the zero is near the real part of the complex poles.

(3) An additional zero in the right–half plane depresses overshoot (and may cause the step response to start out in the wrong direction—i.e. nonminimum phase).

(4) An additional pole in the left–half plane increases rise time significantly if the extra pole is near the real part of the complex poles.

## 3.2 Block Diagrams

It is often possible to write the dynamic equations of a controlled system in such a way that they can be grouped into subsystems that do not interact, except that the input to one subsystem may be the output of another subsystem. The subsystems and connections between them can be represented as a *block diagram*. An important benefit of this approach is that use of block diagrams can greatly facilitate the process of solving the equations by permitting them to be graphically simplified.

Three common block diagrams are shown in Fig. 3.11. As mentioned in Sect. 3.1.2, using the Laplace Transform variable $s$, transfer functions provide an algebraic relation between a system's input and its output. Here we will represent each subsystem as a block with the subsystem's transfer function printed on it. The input multiplied by the block transfer function produces its output.

The interconnection of blocks include summing points where multiple signals are added together. These are represented by a circle with the math symbol for summation $\sum$ inside.

In Fig. 3.11a, the block with the transfer function $G_1(s)$ is in series with the block with transfer function $G_2(s)$; the overall transfer function is given by the product $G_1(s)G_2(s)$. In Fig. 3.11b, the two component systems are in parallel with their

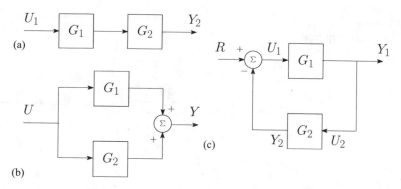

**Fig. 3.11** Elementary block diagrams

outputs added, and the overall transfer function is given by the sum $G_1(s) + G_2(s)$. In Fig. 3.11c, we have a more complicated case. The blocks are connected in a feedback arrangement, so that each feeds into the other. Here, we can examine the equations, solve them and then relate them back to the picture. We have

$$U_1 = R - Y_2,$$
$$Y_2 = G_2 G_1 U_1,$$
$$Y_1 = G_1 U_1.$$

The solution of these equations is

$$Y_1(s) = \frac{G_1}{1 + G_1 G_2} R.$$

The three common block diagrams given in Fig. 3.11 can be used in combination with the block diagram algebra shown in Fig. 3.12 to find an overall transfer function for the complete system. The basic approach is to recursively simplify the system's topology while maintaining exactly the same relations among the remaining variables in the block diagram. Thus, block diagram reduction is essentially a graphical way of solving the dynamic equations by eliminating variables. However, this approach can become tedious and subject to error when the topology of the block diagram is complex. In such cases, the use of Signal Flow Graph techniques or Mason's Rule for Block Diagram Reduction may be useful [1, 3].

## 3.3   Feedback Control

After the dynamic equations of a system plant are formulated, a controller can be designed to compute an input that drives the plant's output towards a desired reference value. Generally, the output will not perfectly track the desired reference value

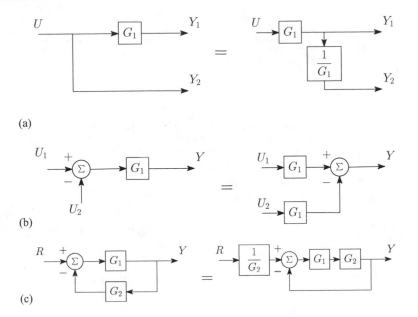

(a)

(b)

(c)

**Fig. 3.12** Block diagram algebra: **a** moving a pick–off point; **b** moving a summing point; **c** conversion to unity feedback

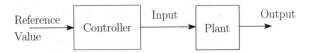

**Fig. 3.13** Open loop control system

because of disturbances, such as wind gusts, currents or waves. Additionally, it is not generally possible to precisely know all of the plant and sensor parameters (either due to their slow drift over time or because of the difficulty in precisely measuring some physical properties, such as hydrodynamic drag, added mass or inertia). Thus, it is almost impossible to exactly predict the effect a given control input may have on the system. As discussed in Sect. 1.2, there are two basic types of control systems, *open-loop* control systems (Fig. 3.13) and *closed loop*, feedback control systems (Fig. 3.14). Closed loop control involves the use of a feedback sensor to measure the output of the system so that it can be compared to the desired reference value. As was shown in Example 2.3, feedback tends to make a control system more *robust* by mitigating the effects of disturbances and parametric uncertainties.

A realization of the closed–loop feedback system shown in Fig. 3.14 using transfer functions is shown in Fig. 3.15. Here, the controller transfer function is $D(s)$, the plant transfer function is $G(s)$, the sensor transfer function is $H(s)$, $R(s)$ is the reference input, $Y(s)$ is the system output and $E(s)$ is the error between the desired output and the measured output $E = R - Y$. The output of the controller $U(s)$ acts on the

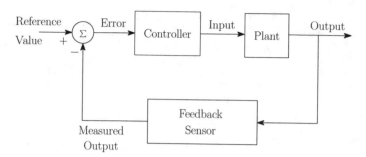

**Fig. 3.14**  Closed loop, feedback control system

**Fig. 3.15**  A closed–loop
feedback control system in
transfer function form

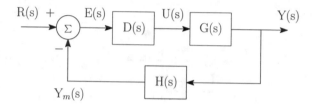

**Fig. 3.16**  A closed–loop
feedback control system with
external disturbance $W_d$

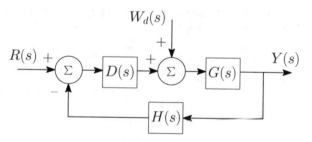

**Fig. 3.17**  A unity feedback
control system

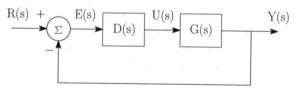

plant to drive it towards the desired reference value. When disturbances also act on
the plant, their effect is combined with the control output, as shown in Fig. 3.16.

In order to simplify the design of feedback control systems, it is common to
analyze controlled systems in *unity feedback* form. Using the graphical block diagram
algebra shown in Fig. 3.12, a non–unity feedback system, such as the one in Fig. 3.15
can be represented in unity feedback form. Here, we will simplify things by taking
$H(s) = 1$, as shown in Fig. 3.17. In this section we will consider how the use of
three common types of controllers affect a systems response when used in a unity
feedback configuration.

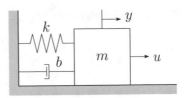

### 3.3.1 *Proportional Feedback Control*

In proportional feedback control the output of the controller is linearly proportional
to the error in the measured output $e = \mathscr{L}^{-1}\{E(s)\}$. The general form of proportional
control is

$$u = k_p e.$$

Thus, the transfer function of the controller $D(s)$ that would appear in the controller
block of Fig. 3.17 is

$$D(s) = k_p.$$

**Example 3.2**  First consider a spring-mass system such as the one shown in Fig. 3.18.
The (unforced) characteristic equation of this system is

$$m\ddot{y} + b\dot{y} + ky = 0,$$

where $y$ is the displacement of the mass from its equilibrium position, $m$ is the mass,
$b$ is the coefficient of damping and $k$ is the spring constant. Taking the Laplace
transform of this equation, and rearranging the coefficients, gives the following form
of the characteristic equation

$$s^2 + \frac{b}{m}s + \frac{k}{m} = 0. \tag{3.14}$$

Note that if the system had no spring, i.e. $k = 0$, there would be no tendency for the
mass to return to its equilibrium position when disturbed.

Next, consider the steering of a surface vessel. The one degree of freedom equation
governing the motion corresponds to the linearized equation of motion for yaw, which
can be written as

$$(I_z - N_{\dot{r}})\ddot{\psi} - N_r\dot{\psi} = N_\delta \delta_R,$$

where $r = \dot{\psi}$ is the yaw rate, $\psi$ is the heading angle, $I_z$ is the mass moment of inertia
of the surface vessel about the yaw axis, $N_{\dot{r}} < 0$ is the yaw added mass, $N_r < 0$ is a
linear drag-related moment, and the product $N_\delta \delta_R$ is the yaw moment applied by a
rudder. The Laplace transform of this equation is

$$(I_z - N_{\dot{r}})s^2\Psi - N_r s\Psi = N_\delta \Delta_R.$$

Thus, its characteristic equation is

$$s^2 + \frac{-N_r}{(I_z - N_{\dot{r}})}s = 0$$

and its open loop transfer function is

$$\frac{\Psi}{\Delta_R} = \frac{-N_\delta/N_r}{s\left[-\dfrac{(I_z - N_{\dot{r}})}{N_r}s + 1\right]}. \tag{3.15}$$

With proportional feedback control using a gain $k_p > 0$ the characteristic equation of the closed loop system is

$$s^2 + \frac{-N_r}{(I_z - N_{\dot{r}})}s + \frac{k_p N_\delta}{(I_z - N_{\dot{r}})} = 0, \quad N_r, \ N_{\dot{r}} < 0. \tag{3.16}$$

Comparing the closed loop characteristic equation to (3.14), it can be seen that proportional feedback provides a control effort that acts somewhat like the spring in the mass-spring-damper system to push the yaw angle towards its desired equilibrium condition.

Let

$$b_\psi = \frac{-N_r}{(I_z - N_{\dot{r}})} \quad \text{and} \quad k_\psi = k_p\frac{N_\delta}{(I_z - N_{\dot{r}})}, \tag{3.17}$$

so that the characteristic equation of the system is written as

$$s^2 + b_\psi s + k_\psi = 0 \tag{3.18}$$

and the associated roots are

$$s_{1,2} = -\frac{b_\psi}{2} \pm \sqrt{\left(\frac{b_\psi}{2}\right)^2 - k_\psi}.$$

From (3.17), it can be seen that as $k_p$ increases, so does $k_\psi$. Plotting these roots in the s-plane (Fig. 3.19) shows that for $0 < k_\psi < (b_\psi/2)^2$, as $k_p$ increases the largest root at

$$s_1 = -\frac{b_\psi}{2} + \sqrt{\left(\frac{b_\psi}{2}\right)^2 - k_\psi}$$

becomes more negative. From (3.7), this implies that the settling time $t_s$ decreases. When $k_\psi > (b_\psi/2)^2$, the roots are complex. As $k_p$ increases, the settling time $t_s$ is constant because the real part of each root $\text{Re}\{s_{1,2}\} = -b_\psi/2$ is constant. However, the imaginary part of each root

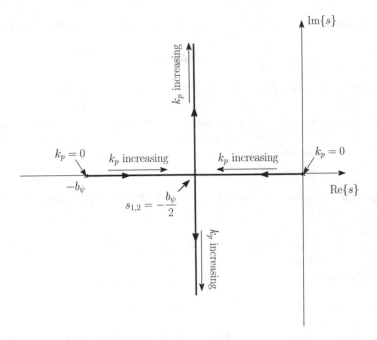

**Fig. 3.19** The locus of roots of (3.18) vs. $k_p$

$$\text{Im}\{s_{1,2}\} = \pm j\sqrt{k_\psi - \left(\frac{b_\psi}{2}\right)^2}$$

increases as $k_p$ increases. From Fig. 3.2, it can be seen that this means the damping ratio $\zeta$ will decrease and the natural frequency $\omega_n$ will increase. Because of this, the overshoot $M_p$ will increase and the rise time $t_r$ will decrease, as can be seen from (3.5) and (3.2). Thus, the selection of $k_p$ involves a compromise between obtaining a sufficiently fast time response and reducing overshoot. In Example 2.3, it was shown how proportional gain is important for reducing the disturbance-induced steady state error of a system. However, as shown here, if the value of $k_p$ is too large, the dynamic response may not be acceptable. There is an upper limit on how much proportional control can be used to reduce the steady state errors caused by disturbances, while still achieving a desired dynamic response.  □

### 3.3.2  Derivative Feedback Control

A discussed in Example 2.4, derivative feedback control terms have the form

$$u(t) = k_d \dot{e} = k_p T_D \dot{e}.$$

Therefore, when $D(s)$ is a derivative feedback controller the transfer function

$$D(s) = k_p T_D s,$$

can be used in the controller block of Fig. 3.17, where $T_D$ is the *derivative time*.

Unless the controlled system has a natural proportional term (e.g. buoyancy in a depth control system), derivative feedback control by itself will not drive the error to zero. Therefore, derivative feedback control is normally used with proportional and/or integral feedback control to improve the stability of a system's response by increasing its effective damping.

**Example 3.3** Let's re-examine the steering of a surface vessel discussed in Example 3.2, but with no proportional control term (no $k_\psi$ term). The characteristic equation (3.18) is now

$$s^2 + \left(b_\psi + k_p T_D\right) s = 0,$$

which has roots at

$$s_{1,2} = \begin{cases} 0, \\ -(b_\psi + k_p T_D). \end{cases}$$

The position of the root at $s = 0$ is independent of the value of $k_p T_D$. As can be seen from Fig. 3.3, the time response associated with a root at the origin is constant in time. Thus, $\psi$ will not go to the desired value for an arbitrary initial condition $\psi_0$. □

### 3.3.3 Integral Feedback Control

An integral feedback controller produces an output signal of the form

$$u(t) = k_i \int_{t_o}^{t} e \, dt = \frac{k_p}{T_I} \int_{t_o}^{t} e \, dt.$$

The transfer function of the controller $D(s)$ that would appear in the controller block of Fig. 3.17 becomes

$$D(s) = \frac{k_i}{s} = \frac{k_p}{T_I s},$$

where $T_I$ is the *integral time*, or *reset time*, and $1/T_I$ is the *reset rate*. In comparison to the proportional controller and the derivative controller, the integral controller can provide a finite value of $u$ when the error–signal $e$ is non-zero. The reason for this is because $u$ is a function of the past values of $e$, rather than only its current value. Past errors $e$ "wind up" the integrator to a value that remains, permitting the controller to continue counteracting disturbances $w_d$ (Fig. 3.16) even after $e = 0$. Thus, an important reason for using integral feedback control is to reduce the steady state error when disturbances are present. However, this benefit may come at the cost of reduced stability.

**Example 3.4** Consider the surface vehicle speed controller discussed in Example 2.3. Let the velocity error be $e$, the linearized equation of motion (2.11) without proportional feedback control can be written for the error as

$$\dot{e} = -ke + w_d,$$

where $k = 2cu_0$ and $w_d = d$. Adding pure integral control to this equation gives

$$m\dot{e} = -ke - \frac{k_p}{T_I s} \int_{t_o}^{t} e \, dt + w_d.$$

Taking the Laplace Transform and rearranging terms yields

$$\left[ s^2 + \frac{k}{m}s + \frac{k_p}{mT_I} \right] \frac{E}{s} = \frac{W_d}{m}. \tag{3.19}$$

Solving for the input output relation between the error and disturbance, we get

$$\frac{E(s)}{W_d(s)} = \frac{\frac{1}{m}s}{s^2 + \frac{k}{m}s + \frac{k_p}{mT_I}}.$$

Recall from Example 2.3 that $w_d$ is a combination of environmental disturbances and uncertainty in the drag parameter. Let $w_d$ have the character of a step function so that $w_d(t) = A\, 1(t)$ and $W_d(s) = A/s$, where $A$ is a constant. The steady state error for the disturbance can be calculated using the Final Value Theorem (2.36),

$$\lim_{t \to \infty} e(t) = e_{ss} = \lim_{s \to 0} s\, E(s) = \lim_{s \to 0} s \left[ \frac{\frac{A}{m}}{s^2 + \frac{k}{m}s + \frac{k_p}{mT_I}} \right] = 0. \tag{3.20}$$

In comparison, from (2.12) it can be seen that for purely proportional control the steady state error is

$$e_{ss} = \left[ \frac{d_e + \delta c u_0 |u_0|}{k_p + 2cu_0} \right],$$

which becomes small as the proportional gain term $k_p$ is made large. However, as mentioned in both Examples 2.3 and 3.2, there is an upper limit to how large the proportional gain can be made and so the steady state error with proportional control will always be finite. In contrast, from (3.20) we can see that purely integral control can drive the steady state error to zero when disturbances and uncertainty are present.

However, the use of integral control can also affect the dynamic stability of the system. From (3.19), we see that the integral-only speed controller has the associated characteristic equation

$$s^2 + \frac{k}{m}s + \frac{k_p}{mT_I} = 0.$$

Comparing this to the canonical form of a second order system (3.1), we can take $k/m = 2\zeta\omega_n$ and $\omega_n^2 = k_p/(mT_I)$. With these relations, we can then solve for the damping ratio

$$\zeta = \frac{k}{2}\sqrt{\frac{T_I}{m\,k_p}}.$$

As the control gain $k_p/T_I$ increases, the damping ratio decreases, which in turn increases the overshoot of the response. Thus, a possible limitation of integral control is that the controlled system may become less dynamically stable when the gain is large.                                                                                      □

### 3.3.4   PID Feedback Control

A PID controller combines proportional, integral and derivative feedback control. The transfer function that would appear in the controller block of Fig. 3.17 is

$$D(s) = k_p\left(1 + \frac{1}{T_I s} + T_D s\right). \tag{3.21}$$

Owing to their robustness and ease of use, PID controllers are standard in many commercial marine systems. Controller design largely consists of *tuning* the three constants in (3.21) to achieve an acceptable performance. As explained in [3], the basic tuning approach exploits the ideas that increasing $k_p$ and $1/T_I$ tends to reduce system errors, but may not produce adequate stability, while stability can be improved by increasing $T_D$.

**Example 3.5**  Figure 3.20 shows the response of P (proportional), PD (proportional-derivative), and PID feedback controllers to a unit step disturbance $W_d(s) = 1/s$ for the second-order plant

$$G(s) = \frac{1}{(4s + 1)(s + 1)}.$$

The PD controller exhibits less oscillatory behavior than the P controller (i.e. including a derivative term dampens the response) and that adding the integral term to the PD controller increases the oscillatory behavior of the response, but reduces the magnitude of the steady state error.                                                                □

**Fig. 3.20** Response of a
second-order system to a unit
step disturbance with
$k_p = 25$, $T_I = 2.5$ and
$T_D = 0.2$

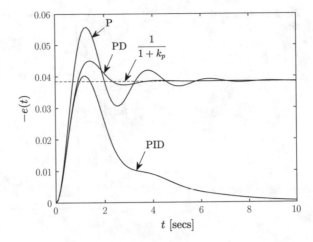

**Fig. 3.21** Process reaction
curve

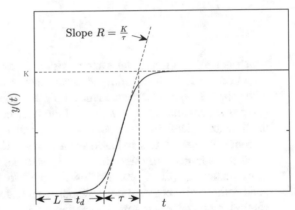

## Ziegler–Nichols Tuning of PID Regulators

Most controller design techniques require a complete model of the plant dynamics.
However, owing to the complex situations and environments in which marine systems are deployed, the development and (especially) the validation of such a model
can be particularly difficult. Ziegler–Nichols tuning methods provide a means of
using simple experiments to determine first-cut estimates of the three PID constants
in (3.21) that will yield acceptable performance, without the need for a complete
plant model. Ziegler and Nichols developed two methods for experimentally tuning
controllers [7, 8]:

(1) The first method assumes that the step response of most systems can be approx-
    imated by the *process reaction curve* in Fig. 3.21 and that the curve can be
    generated by performing experiments on the plant. The the plant input–output
    behavior is be approximated by the transfer function

**Fig. 3.22** Quarter decay ratio

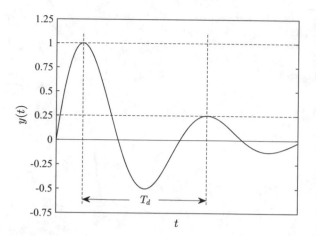

$$\frac{Y(s)}{U(s)} = \frac{K e^{-t_d s}}{\tau s + 1},\tag{3.22}$$

which is simply a first-order system with a time delay of $t_d$. The constants in (3.22) can be determined from the experimentally measured unit step response of the plant. The slope of the tangent at the inflection point of the process reaction curve is approximately $R = K/\tau$, and the intersection of the tangent line with the time axis identifies the time delay (also called the *lag*) $L = t_d$. The choice of controller parameters is based on a decay ratio of approximately 1/4, i.e. when the dominant transient in the response decays by 1/4 over one period of oscillation $T_d$ (Fig. 3.22). This corresponds to $\zeta = 0.21$ and is generally a good compromise between quick response and adequate stability margins. The controller parameters required to achieve this, as determined by Ziegler and Nichols, are shown in Table 3.1.

(2) In Ziegler–Nichols' second method, the criteria for adjusting controller parameters are based on experimentally evaluating the plant at the limit of stability. Using only proportional control, the gain is increased until continuous oscillations are observed, i.e. when the system becomes marginally stable. The corresponding gain $k_u$ (the *ultimate gain*) and the period of oscillation $T_u$ (the *ultimate period*) are measured as shown in Fig. 3.23. The measurements should be performed when the amplitude of oscillation is small. The recommended controller parameters can then be determined from the measurements using the relations given in Table 3.2.

Controller parameters determined using the Ziegler–Nichols tuning methods provide reasonable closed-loop performance for many systems. However, additional manual fine tuning of the controller is typically conducted to obtain the "best" control.

*Integrator Antiwindup*

In Sect. 1.2 it was explained that a controller is typically some sort of electrical circuit or computer that reads a signal from a sensor and then outputs a signal to an actuator,

**Table 3.1** Ziegler–Nichols tuning for $D(s) = k_p \left(1 + \dfrac{1}{T_I s} + T_D s\right)$ for a decay ratio of 0.25

| Type of controller | Optimum gain |
|---|---|
| P | $k_p = \dfrac{1}{RL}$ |
| PI | $k_p = \dfrac{0.9}{RL}, T_I = \dfrac{L}{0.3}$ |
| PID | $k_p = \dfrac{1.2}{RL}, T_I = 2L, T_D = 0.5L$ |

**Fig. 3.23** A marginally stable system

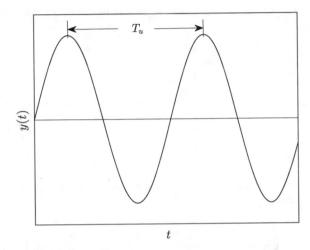

**Table 3.2** Ziegler–Nichols tuning for $D(s) = k_p \left(1 + \dfrac{1}{T_I s} + T_D s\right)$ based on a stability boundary

| Type of controller | Optimum gain |
|---|---|
| P | $k_p = \dfrac{1}{2}k_u$ |
| PI | $k_p = 0.45k_u, T_I = \dfrac{1}{1.2}T_u$ |
| PID | $k_p = 0.6k_u, T_I = \dfrac{1}{2}T_u, T_D = \dfrac{1}{8}T_u$ |

which is then used to generate some sort of effort (a force, a moment, etc.) that acts upon the plant. In this chapter, the presence of the actuator has thus far been missing from our discussion of feedback control. The block diagram of a closed loop feedback system in Fig. 3.14 can be extended to include an actuator, as shown in Fig. 3.24.

The output range of a real actuator is always finite. For example, the stall angle of most symmetric hydrofoils is about $\pm 15°$ [6]. Beyond this angle most hydrofoil

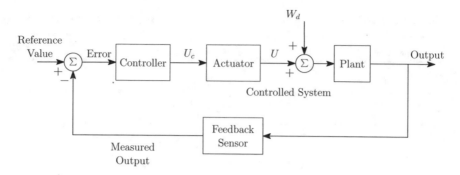

**Fig. 3.24**  Feedback system with actuator block ($W_d$ is a disturbance)

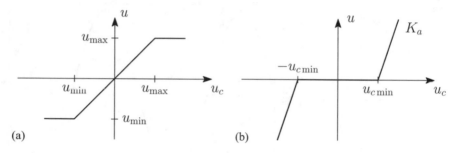

**Fig. 3.25  a** Actuator saturation and **b** actuator deadband with gain $K_a$. Actuator input is $u_c$ and output is $u$

sections are not very effective at producing any additional lift, bounding the available turning force of fins and rudders. For this reason, the angle of attack of the fins and rudders on most commercially available underwater vehicles tends to be limited to about $\pm 15°$. As another example, consider an underwater thruster. Owing to a limit on the output power of the motor turning the propeller, there is a bounds on the thrust that can be generated in the forward and reverse directions. Even if an extremely powerful motor is used to drive the propeller, when a propeller is operated at very high rotational speeds, cavitation (or ventilation when operated near the waterline) will limit the thrust that can be produced. This upper and lower bounds on the output effort of an actuator is known as *actuator saturation* and is often approximated with a saturation function, as sketched in Fig. 3.25a.

A second type of nonlinearity that can occur with real actuators known as a *deadband* is sketched in Fig. 3.25b. Many actuators do not respond until the input command reaches a minimum value. For example, owing to friction in motor windings and shafts (starting torque), a thruster may not start to rotate until its input current and voltage reach minimum values. As another example, there is often some play in gearbox systems. When the direction of rotation of the input shaft is changed, it may need to be turned through some small angle before the gear on the input shaft re-engages with the gear on the output shaft, causing a small range of input shaft

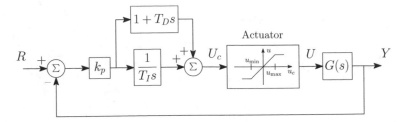

**Fig. 3.26** A PID feedback system with actuator saturation

motion for which the output shaft does not turn. As will be shown below, a deadband element can be used as part of an antiwindup scheme.

If actuator saturation is not taken into account during controller design, a phenomenon known as *integrator windup* can occur when the controller includes an integral term. If the error is large, the controller will produce a large actuator command $u_c$. However, since the actuator output $u$ is bounded, the real system will be slower than anticipated in driving the output of the plant towards the reference value. As a result, the integral of the error will become large over time and will persist, even if the real system finally achieves the desired reference value. This can in turn cause a large overshoot and error. Consider the PID feedback system shown in Fig. 3.26. Suppose a large reference step causes the actuator to saturate at $u_{max}$. The integrator will continue integrating the error $e$, and the signal $u_c$ will keep growing, even though the actuator has reached its saturation limit (i.e., $u > u_{max}$ or $u < u_{min}$). Serious performance degradation can result as the increase in the magnitude of $u_c$ does not cause the output of the plant to change. The integrator output may become very large if the actuator is saturated for a long time and can require a large or persistent error of the opposite sign to discharge the integrator to its proper value.

The solution to the integrator windup problem is to "turn off" the integral action when the actuator is saturated. This can be done fairly easily with computer control. Two *antiwindup* schemes are shown in Fig. 3.27a, b using a dead-zone nonlinearity with a PID controller. As soon as the actuator saturates, the feedback loop around the integrator drives the input to the integrator $E_1$ to zero. The slope of the dead-zone nonlinearity $K_a$ should be large enough so that the antiwindup circuit can follow $E$ and keep the output from saturating. The effect of antiwindup is to reduce the error overshoot, as well as the control effort. In practice, it may be difficult to measure the output of the actuator. Thus, the approach shown in Fig. 3.27b is more widely used.

**Example 3.6** Consider the first-order plant

$$G(s) = \frac{3(s+2)}{s(s+12)}$$

with an integral controller

$$D(s) = \frac{15}{s}$$

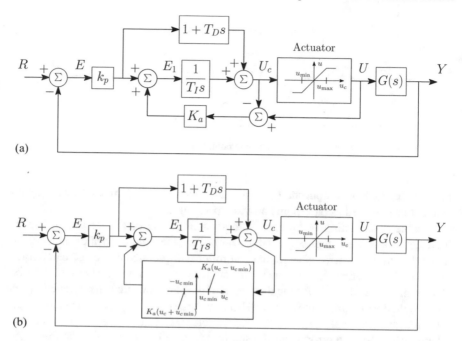

**Fig. 3.27** PID controller with two different antiwindup schemes: **a** actuator output feedback; **b** control signal feedback

and an actuator that saturates at $u_{max} = 1$ and $u_{min} = -1$. With the system at rest, a unit step input causes the actuator to saturate and integral action to begin. The integrator output will grow even after the plant response exceeds the reference input, as shown by the dashed line in Fig. 3.28a. An antiwindup mechanism of the type shown in Fig. 3.27b with a deadband of $|u_{c\,min}| = 1$ and a gain of $K_a = 1$ is implemented. The solid lines in Fig. 3.28a, b show the response and corresponding control effort $u$ with antiwindup. The antiwindup system causes the actuator to be saturated for less time and produces a response with less overshoot and a shorter settling time.    □

## 3.4  Steady State Response

An important performance measure often used for controller design is the steady state error of the closed loop system, $E(s) = R(s) - Y(s)$. For the feedback control system in Fig. 3.15, we have

$$E(s) = R(s) - \frac{D(s)G(s)}{1 + D(s)G(s)H(s)} R(s) = \frac{1 + D(s)G(s)H(s) - D(s)G(s)}{1 + D(s)G(s)H(s)} R(s).$$

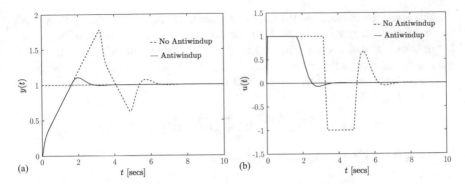

**Fig. 3.28** Step response of system with integral controller: **a** output response and **b** actuator output with antiwindup (—) and without (−−) antiwindup

With unity feedback (Fig. 3.17), $H(s) = 1$ and the error is

$$E(s) = \frac{R(s)}{1 + D(s)G(s)}.$$

Using the Final Value Theorem (2.36), we can see that the steady–state error is given by

$$e_{ss} = \lim_{t \to \infty} e(t) = \lim_{s \to 0} sE(s) = \lim_{s \to 0} \frac{sR(s)}{1 + D(s)G(s)}. \tag{3.23}$$

We can study the general steady–state error of a system by taking the reference input to be a polynomial in time of the form

$$r(t) = \frac{t^k}{k!} 1(t).$$

The corresponding Laplace transform of the reference input is

$$R(s) = \frac{1}{s^{k+1}}. \tag{3.24}$$

When $k = 0$, the input is a step of unit amplitude; when $k = 1$, the input is a ramp with unit slope; and when $k = 2$, the input is a parabola with unit second derivative. Historically, these inputs are known as position, velocity and acceleration inputs, respectively. An examination of the steady state error of unity feedback systems for these three "standard" test inputs can be used to define the type of a system and a corresponding error constant. These error constants can be used as a quantitative measure of steady–state system performance and in the controller design process, a designer may attempt to increase the error constants, while maintaining an acceptable transient response.

As we will see, the steady–state error of a system will be zero for input polynomials below a certain degree $k$, constant for inputs of degree $k$ and unbounded for degrees larger than $k$. Stable systems can be classified according to the value of $k$ for which the steady–state error is constant—this classification is known as the *system type*.

From (3.23) and (3.24) we can see that the steady–state error for unity feedback is

$$e_{ss} = \lim_{s \to 0} \frac{1}{s^k[1 + D(s)G(s)]}. \tag{3.25}$$

For a step input $k = 0$, when the system is of type 0, we get

$$e_{ss} = \lim_{s \to 0} \frac{1}{[1 + D(s)G(s)]} = \frac{1}{1 + K_p},$$

where $K_p$ is the *position error constant* given by

$$K_p = \lim_{s \to 0} D(s)G(s).$$

For a ramp input $k = 1$ and the steady state error is

$$e_{ss} = \lim_{s \to 0} \frac{1}{s[1 + D(s)G(s)]} = \lim_{s \to 0} \frac{1}{s D(s)G(s)}.$$

For this limit to exist, the product $D(s)G(s)$ must have at least one pole at $s = 0$. When there is exactly one pole at $s = 0$, the system is type 1 and the error is

$$e_{ss} = \frac{1}{K_v},$$

where $K_v$ is known as the *velocity error constant*, given by

$$K_v = \lim_{s \to 0} s D(s)G(s).$$

Similarly, systems of type 2 must have two poles located at $s = 0$ and the error is

$$e_{ss} = \frac{1}{K_a},$$

where $K_a$ is the *acceleration error constant*,

$$K_a = \lim_{s \to 0} s^2 D(s)G(s).$$

For the special case of unity feedback, the error constants can be computed as

$$\lim_{s \to 0} s^k D(s)G(s).$$

**Table 3.3**  Correspondence between steady–state error and system type for unity feedback systems

| Type | Step | Ramp | Parabola |
|------|------|------|----------|
| 0 | $\dfrac{1}{1+K_p}$ | $\infty$ | $\infty$ |
| 1 | 0 | $\dfrac{1}{K_v}$ | $\infty$ |
| 2 | 0 | 0 | $\dfrac{1}{K_a}$ |

The steady–state error for systems of type 0, 1 and 2 are summarized in Table 3.3. The definition of system type helps one to quickly identify the ability of a system to track polynomials at steady state. System types and their associated error constants for non-unity feedback systems are discussed in [1, 5].

## 3.5  Additional Performance Measures

As discussed above, a system's rise time, settling time, peak overshoot (Sect. 3.1.2) and its steady–state error constants (Sect. 3.4) can be used to assess the performance of a controller and are often used in the controller design process. Modern control systems sometimes also consider the following additional metrics: the *Integral Square of the Error* (ISE), the *Integral of the Absolute magnitude of the Error* (IAE), the *Integral of Time multiplied by Absolute Error* (ITAE), and the *Integral of Time multiplied by the Squared Error* (ITSE).

The general form of these metrics is given by

$$I = \int_0^T f(e(t), r(t), y(t), t)dt, \tag{3.26}$$

where $f$ is a function of the error $e(t)$, reference input $r(t)$, output $y(t)$ and time $t$. The upper limit of the integral $T$ may be chosen somewhat arbitrarily, but is typically taken so the integral approaches a steady–state value. To be useful a performance index must be a number that is always positive or zero. These metrics can be used to design an *optimum control system* where the system parameters are adjusted so that a selected performance index reaches an extremum value, commonly a minimum value. Then the best, "optimum", system is defined as the system that minimizes this index. The minimization of IAE and ISE is often of practical significance, such as for the minimization of fuel consumption for aircraft and space vehicles.

ISE is given by

$$ISE = \int_0^T e^2(t)dt. \tag{3.27}$$

This criterion is often used to discriminate between excessively overdamped and excessively underdamped systems.

IAE is given by

$$\text{IAE} = \int_0^T |e(t)| dt, \tag{3.28}$$

and has been found to be particularly useful when used in simulation studies.

The metric ITAE was developed to reduce the large contribution of the initial error to the value of the performance integral and to emphasize errors that occur later in the time response of a system

$$\text{ITAE} = \int_0^T t|e(t)| dt. \tag{3.29}$$

ITSE is a similar metric

$$\text{ITSE} = \int_0^T t e^2(t) dt. \tag{3.30}$$

Generally, ITAE provides the best selectivity of the performance indices defined here because its minimum value tends to be easily discernible as system parameters are varied. Numerical or analytical techniques can be used to select controller parameters, e.g. $k_p$, $T_D$ and $T_I$ by minimizing one or more of these measures of performance. Further information about these metrics and how they can be used for controller design can be found in [1].

# Appendix

**Example 3.7** Use Matlab to find the response of a second order system with a natural frequency of $\omega_n = 1$ and damping ratios of $\zeta = \{0.3, 0.5, 0.7, 0.9\}$ to a unit step input.

Answer: The transfer function of a second order system is given by (3.3), where the numerator is given by $\omega_n^2$ and the denominator is $s^2 + 2\zeta\omega_n s + \omega_n^2$. Using the following code we can repeatedly plot the response of the system for the different values of $\zeta$ on the same plot using the "hold on" command; it is helpful to incorporate the code within an M-file. Note that comments within the code start with the symbol %. The plotted response is shown in Fig. 3.29.

```
%% Code for plotting the step response of a second order system
omegan = 1.0;
zeta = [0.3, 0.5, 0.7, 0.9];

num = omegan^2; % system numerator

for i = 1:length(zeta),
    den = [1 2*zeta(i)*omegan omegan^2]; % system denominator

    G = tf(num, den); % construct the transfer function

    step(G) % plots the step response
    hold on
end
hold off
```

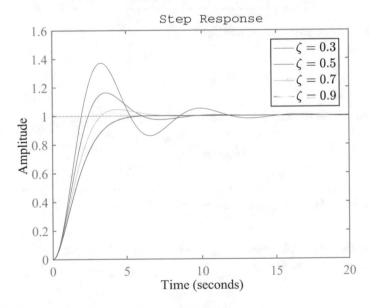

**Fig. 3.29** Unit step response of second order system with $\omega_n = 1$

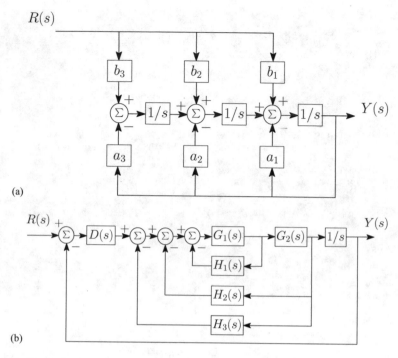

**Fig. 3.30** Block diagrams for Problem 3.2

## Problems

**3.1** Consider the second–order system with the transfer function

$$G(s) = \frac{2}{s^2 + 1s - 2}.$$

(a) Determine the DC gain for this system (see Fig. 3.3).
(b) What is the final value of the step response of $G(s)$?

**3.2** Find the transfer functions $Y(s)/R(s)$ of the block diagrams in Fig. 3.30a, b using the rules of Fig. 3.12a, b.

**3.3** Find the transfer function $Y(s)/W_d(s)$ for the system in Fig. 3.31 when $R(s) = 0$.

**3.4** The DUKW-Ling is an amphibious USV designed to traverse the surf zone of a beach [4]. A Small-Waterplane Area Twin Hull (SWATH) design [2] is employed

**Fig. 3.31** A system with
disturbance input $W_d(s)$

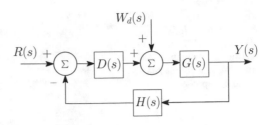

to mitigate the wave-induce motion of the vehicle (Fig. 3.32a). When a system is
excited by external forces near one of its resonant frequencies, the amplitude of the
response can be quite pronounced. Thus, it is desirable for the natural frequencies
of the USV to be different from those of the waves that will be encountered. When
excited by a small external impulse, such as a brief downward push on one of its
hulls, the natural roll response of the DUKW-Ling can be measured using an inertial
measurement unit (IMU). As shown in Fig. 3.32b, the natural roll response resembles
that of a classic damped second order system. Assume that the roll response has the
form

$$\phi(t) = \phi_0 e^{-\zeta \omega_n t} \cos(\omega_d t + \psi), \tag{3.31}$$

where $\phi$ is the roll angle, $\phi_0$ is the amplitude of the response, $\zeta$ is the damping ratio,
$\omega_n$ is the natural frequency, $t$ is time,

$$\omega_d = \omega_n \sqrt{1 - \zeta^2}$$

is the damped frequency and $\psi$ is a phase angle.

(a) Note that the period of oscillation in the measured data $T_d$ will be that associated
with the damped frequency $\omega_d$. Examine the roll response at two times $t_n$ and $t_1$,
which are separated by an integer number of periods $n$, so that $(t_n - t_1) = n T_d$.
Let

$$\delta := \frac{1}{n} \ln \left[ \frac{\phi_1}{\phi_n} \right],$$

where $\phi_1 := \phi(t_1)$ and $\phi_n := \phi(t_n)$. Using (3.31), show that

$$\zeta = \frac{1}{\sqrt{1 + \left( \dfrac{2\pi}{\delta} \right)^2}}.$$

(b) Use the data in Fig. 3.32b to estimate the DUKW-Ling's damping ratio $\zeta$ in roll.
(c) Use the data in Fig. 3.32b to estimate the DUKW-Ling's natural frequency $\omega_n$
in roll.

**3.5** A UUV surge speed integral control system is shown in Fig. 3.33.

**Fig. 3.32  a** The
Amphibious DUKW-Ling
USV, ©[2009] IEEE,
reprinted, with permission,
from [4], **b** measured roll
impulse response

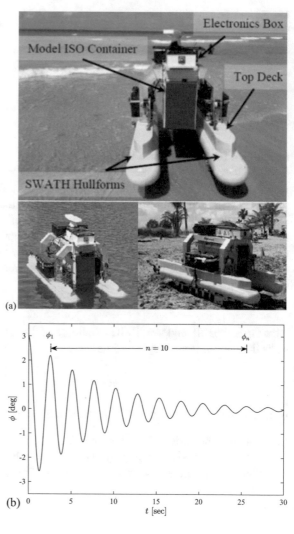

**Fig. 3.33** UUV
speed-control system

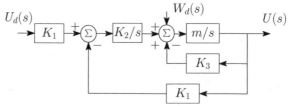

(a) Give the transfer function relating the output $u$ to the current disturbance $w_d$ when the desired surge speed is $u_d = 0$.
(b) What is the steady state response of $u$ if $w_d$ is a unit ramp function?
(c) What is the system type in relation to $u_d$? What are the values of the steady state error constants $K_p$ and $K_v$?
(d) What is the system type in relation to the disturbance $w_d$? What is the error constant due to the disturbance?

**3.6** A DC motor is used to control the speed of an underwater thruster. The system is described by the differential equation

$$\dot{\omega} + \frac{1}{2}\omega = \frac{3}{4}v_a - 50q,$$

where $\omega$ is the motor shaft speed, $v_a$ is the commanded input motor voltage, and $q$ is the load torque from the propeller. Let

$$v_a = -K\left(e + \frac{1}{T_I}\int^t e\,d\tau\right),$$

where the speed error is $e := \omega_d - \omega$ and $\omega_d$ is the desired reference speed.

(a) Compute the transfer function from $Q(s) := \mathscr{L}\{q(t)\}$ to $\Omega(s) := \mathscr{L}\{\omega(t)\}$ as a function of $K$ and $T_I$.
(b) Compute the values of $K$ and $T_I$ so that the characteristic equation of the closed–loop system will have roots at $s_{1,2} = -4 \pm 4j$.

**3.7** A CTD is a device used to obtain profiles of the conductivity, temperature, density and sound speed versus depth. CTDs are often lowered and raised from the side of a boat using a davit (small crane) with a cable reel to let out and pull in the CTD cable. Consider a proportional controller used to control the speed of the cable, which uses a tachometer as a feedback sensor. A block diagram of the system is shown in Fig. 3.34. Let the radius of the reel $r$ be 0.3 m when full and 0.25 m when empty. The rate of change of the radius is

$$\frac{dr}{dt} = -\frac{d_c^2\omega}{2\pi w},$$

where $d_c$ is the diameter of the cable, $w$ is the width of the reel and $\omega$ is the angular velocity of the reel. The profiling speed of the cable is $v(t) = r\omega$. CTDs are often designed to be used with profiling speeds in the range $0.5 \le |v| \le 2$ m/s, with speeds of about 1 m/s generally being the best compromise between data quality and profile resolution. Thus, take the desired cable speed to be $v_d = 1$ m/s. Simulate this system and obtain the step response of the speed over 100 s when the cable is let out for three values of gain $K = 0.2$, 0.3, and 0.4. Note that the relation between reel angular velocity and the input torque from the controller $\tau$ is

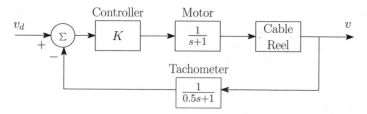

**Fig. 3.34** CTD cable reel control system

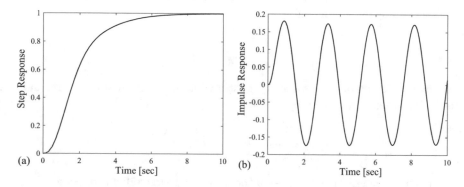

**Fig. 3.35** **a** Step response and **b** impulse response of the ROV depth control system

$$I\frac{d\omega}{dt} = \tau,$$

and that the inertia $I$ changes when the cable is wound or unwound. The equation $I = 50r^4 - 0.1$ can be used within the simulation to account for the changing inertia. Which gains have the smallest overshoot and the fastest response?

Assume $w = 0.5$ m, $d_c = 0.01$ m, and $r = 0.3$ m at $t = 0$ s.

**3.8** A vertical tunnel thruster based depth control system is being developed for a semi-automatic underwater remotely operated vehicle (ROV) using Ziegler–Nichols tuning methods.

(a) The open loop, unit-step response of the system is shown in Fig. 3.35a. The time–delay and the steepest slope may be determined from this transient response. Find the P-, PI-, and PID-controller parameters using the Ziegler–Nichols transient response method.

(b) The system with proportional feedback has the unit–impulse response shown in Fig. 3.35b. The gain $K_u = 5.75$ corresponds to the system being on the verge of instability. Determine the P-, PI-, and PID-controller parameters according to the Ziegler–Nichols ultimate sensitivity method.

**3.9** A certain control system has the following specifications: rise time $t_r \le 0.01$ s and overshoot $M_p \le 0.17$; and steady–state error to unit ramp $e_{ss} \le 0.005$.
(a) Sketch the allowable region of the s–plane for the dominant second–order poles of an acceptable system.
(b) If $y/r$ for this problem is $G/(1 + G)$, what is the behavior of $G(s)$ near $s = 0$ for an acceptable system?

**3.10** A negative unity feedback control system has the plant

$$G(s) = \frac{K}{s(s + \sqrt{2K})}.$$

(a) Determine the percentage of overshoot and settling time due to a unit step input.
(b) For what range of $K$ is the settling time less than 1 s?

**3.11** A second order control system has the closed–loop transfer function $T(s) = Y(s)/R(s)$. The system specifications for a step input are as follows:
   (1) $M_p \le 0.05$.
   (2) $t_s < 4$ s.
   (3) Peak time $t_p < 1$ s (e.g. Fig. 3.5).
Show the permissible area of the s–plane for the poles of $T(s)$ required to achieve the desired response.

**3.12** Consider the plant

$$G(s) = \frac{1}{4 - s^2}.$$

Design a unity feedback controller $D(s)$, such that the position error constant $K_p$ is infinity and the closed-loop poles are located at $s_{1,...,4} = -2, -4, -6, -8$. What is the velocity error constant $K_v$?

**3.13** A proportional controller is used to steer a large boat, as represented in Fig. 3.36. Take the desired course deviation to be $R(s) = 0$ and let the transfer function of the boat be

$$G(s) = \frac{0.25}{s^2 + 0.5s + 0.25}.$$

The output of the controller, $U(s)$ would generally represent the control input to an actuator, such as a rudder, which causes steering forces to be generated along the sway axis of the vessel to correct the course deviation. Actually, the rudder generates a small yaw moment, which causes the hull of the vessel to turn. Owing to its new orientation with respect to the oncoming flow, lift-like hydrodynamic forces along the hull are produced in the sway direction. Here, we will idealize the process and take $U(s)$ to be a steering force. When the wind force is a step input $W_d(s) = 1/s$, plot both $u(t)$ and the corresponding response of the vessel $y(t)$ for $0 \le t \le 30$ s using gains of $K = 1$ and $K = 20$. Show that $u(t)$ works to drive $y(t)$ towards zero. How do the steady state responses differ when $K = 1$ and $K = 20$?

**Fig. 3.36** Proportional
steering control of a large
boat

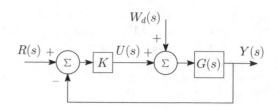

**3.14** Steady–state errors for reference and disturbances inputs. Compute the steady–state error of the system in Problem 3.6(b) for the following inputs:

(a)  a unit step reference input.
(b)  a unit ramp reference input.
(c)  a unit step disturbance input.
(d)  a unit ramp disturbance input.

**3.15** An attitude control system for a small UUV uses a sternplane to regulate the pitch angle of the vehicle (Fig. 3.37). Use Matlab to simulate the dynamic response of the system for times $0 \leq t \leq 60\,\mathrm{s}$.

(a)  Plot the time response of the system for a step input of $\theta_d(t) = 5°1(t)$ when a proportional (P) controller is used

$$D(s) = k_p = \frac{1}{3}.$$

What is the pitch angle error at 30 s?
(b)  If an integral term is added to the controller, it may be possible to reduce the steady state error. Using a proportional integral (PI) controller of the form

$$D(s) = k_p \left(1 + \frac{1}{T_I s}\right) = \frac{1}{3}\left(1 + \frac{1}{6s}\right),$$

repeat the step response simulation in (a) with the PI controller. What is the pitch angle error at 30 s?
(c)  Compare the steady–state tracking error of the P controller with that of the PI controller.

**Fig. 3.37** Block diagram of
a UUV attitude control
system

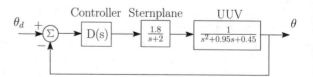

# References

1. Dorf, R.C., Bishop, R.H.: Modern Control Systems. Pearson (Addison-Wesley), London (1998)
2. Faltinsen, O.M.: Hydrodynamics of High-Speed Marine Vehicles. Cambridge University Press, Cambridge (2005)
3. Franklin, G.F., Powell, J.D., Emami-Naeini, A.: Feedback Control of Dynamic Systems, 3rd edn. Addison-Wesley, Reading (1994)
4. Marquardt, J.G., Alvarez, J., von Ellenrieder, K.D.: Characterization and system identification of an unmanned amphibious tracked vehicle. IEEE J. Ocean. Eng. **39**(4), 641–661 (2014)
5. Qiu, Li., Zhou, Kemin: Introduction to Feedback Control. Prentice Hall, Englewood Cliffs (2009)
6. von Ellenrieder, K.D., Dhanak, M.R.: Hydromechanics. In *Springer Handbook of Ocean Engineering*, pp. 127–176. Springer, Berlin (2016)
7. Ziegler, J.G., Nichols, N.B.: Process lags in automatic control circuits. Trans. ASME **65**(5), 433–443 (1943)
8. Ziegler, J.G., Nichols, N.B.: Optimum settings for automatic controllers. Trans. ASME **64**(11) (1942)

# Chapter 4
# Root Locus Methods

## 4.1 Introduction

As shown in Sect. 3.1.2, the time response of a second order system is largely determined by the locations of its dominant closed loop poles. Finding the locations of the closed loop poles is fairly straightforward when the values of the system's parameters are precisely known. However, there will generally be some degree of uncertainty associated with each parameter and so it is important to understand whether the system will perform as desired as the parameters change. Additionally, understanding how control parameters affect system performance is important for controller design. A classical technique for studying how pole locations vary with parameter changes is the Root-Locus Method. The root locus of a system is the locus of its closed loop roots in the $s$-plane. A root locus diagram was presented in Example 3.2 (see Fig. 3.19), although it was not explicitly defined as such. In general, it is usually more important to know how to quickly sketch a root locus in order to see the trend of the root loci, rather than solving for the exact root locus plot (which can be done using a computer). Here, we will see how the root locus can be used as a tool for system analysis and controller design.

## 4.2 Root–Locus Diagrams

Recall some of the basic mathematical characteristics of complex variables. As shown in Fig. 4.1, a complex variable

$$s := \sigma + j\omega \qquad (4.1)$$

has a real component $\sigma$ and an imaginary component $\omega$ and can be thought of as a vector with magnitude

$$r = |s| = \sqrt{\sigma^2 + \omega^2},$$

© Springer Nature Switzerland AG 2021
K. D. von Ellenrieder, *Control of Marine Vehicles*, Springer Series on Naval Architecture, Marine Engineering, Shipbuilding and Shipping 9,
https://doi.org/10.1007/978-3-030-75021-3_4

**Fig. 4.1** Vector nature of a
complex variable

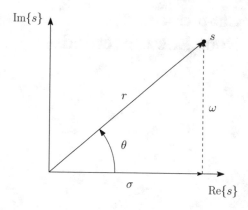

and direction

$$\theta = \tan^{-1}\left(\frac{\omega}{\sigma}\right).$$

Complex variables can also be expressed in polar form, e.g.

$$s = re^{j\theta}.$$

Sums of complex variables

$$\begin{aligned}
s_3 &= s_1 + s_2, \\
&= (\sigma_1 + j\omega_1) + (\sigma_2 + j\omega_2), \\
&= (\sigma_1 + \sigma_2) + j(\omega_1 + \omega_2), \\
&= (\sigma_3 + j\omega_3), \\
&= r_3 e^{j\theta_3},
\end{aligned}$$

products of complex variables

$$\begin{aligned}
s_3 &= s_1 \cdot s_2, \\
&= r_1 e^{j\theta_1} \cdot r_2 e^{j\theta_2}, \\
&= r_1 r_2 e^{j(\theta_1 + \theta_2)}, \\
&= r_3 e^{j\theta_3},
\end{aligned}$$

and ratios of complex variables

$$\begin{aligned}
s_3 &= \frac{s_1}{s_2}, \\
&= \frac{r_1 e^{j\theta_1}}{r_2 e^{j\theta_2}}, \\
&= \frac{r_1}{r_2} e^{j(\theta_1 - \theta_2)}, \\
&= r_3 e^{j\theta_3},
\end{aligned}$$

are also complex variables. Thus, we can see that the open loop transfer function of a LTI system, which is the ratio of two polynomials in the complex variable $s$, will also be a complex variable that possesses both magnitude and direction. As will be shown below, this fundamental concept is the basis for plotting root locus diagrams.

**Example 4.1**  Consider the transfer function

$$G(s) = \frac{(s+2)(s^2 + 2s + 9)}{s(s+3)(s+5)}.$$

By factoring the second order polynomial in the numerator, $G(s)$ can be rewritten as

$$G(s) = \frac{(s+2)(s+1 + j\sqrt{8}/2)(s+1 - j\sqrt{8}/2)}{s(s+3)(s+5)}.$$

If we evaluate this transfer function at the point $s = 2 + 3j$, we can see that the transfer function essentially represents the ratio of the product of the vectors from the poles and zeros of $G(s)$ to the point $s$ (Fig. 4.2). Thus,

$$
\begin{aligned}
G(s)|_{s=2+3j} &= \left. \frac{(s+2)(s+1 + j\sqrt{8}/2)(s+1 - j\sqrt{8}/2)}{s(s+3)(s+5)} \right|_{s=2+3j}, \\
&= \frac{s_1 \cdot s_2 \cdot s_3}{s_4 \cdot s_5 \cdot s_6}, \\
&= \frac{r_1 e^{j\psi_1} \cdot r_2 e^{j\psi_2} \cdot r_3 e^{j\psi_3}}{r_4 e^{j\phi_4} \cdot r_5 e^{j\phi_5} \cdot r_6 e^{j\phi_6}}, \\
&= \frac{r_1 r_2 r_3}{r_4 r_5 r_6} e^{j[(\psi_1 + \psi_2 + \psi_3) - (\phi_4 + \phi_5 + \phi_6)]}.
\end{aligned}
$$

The magnitude is

$$|G(s)|_{s=2+3j} = \frac{r_1 r_2 r_3}{r_4 r_5 r_6},$$

and the angle is

$$\angle\, G(s)|_{s=2+3j} = [(\psi_1 + \psi_2 + \psi_3) - (\phi_4 + \phi_5 + \phi_6)].$$

□

*Complex Conjugate:* Let the complex conjugate of $s$ be denoted as $\bar{s}$. If $s$ is given by (4.1), then

$$\bar{s} := \sigma - j\omega, \tag{4.2}$$

in polar form, we have

$$\bar{s} = r e^{-j\theta},$$

and the magnitude squared of $s$ is

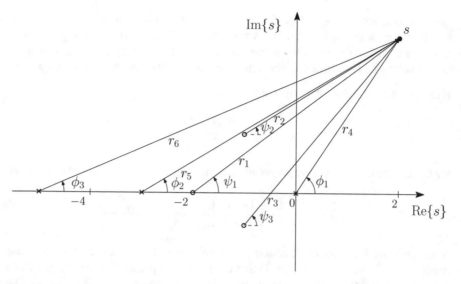

**Fig. 4.2** Vector diagram of $G(s)$ evaluated at $s = 2 + 3j$

$$|s|^2 = s \cdot \bar{s} = r^2.$$

### 4.2.1   Constructing a Root–Locus

Consider the unity feedback system shown in Fig. 3.17. The closed loop transfer function is

$$\frac{Y(s)}{R(s)} = \frac{D(s)G(s)}{1 + D(s)G(s)}, \tag{4.3}$$

where $R$ is the reference input and $Y$ is the output. For simplicity, define the loop transfer function as

$$L(s) := D(s)G(s)$$

and assume that $L(s)$ has the form

$$
\begin{aligned}
L(s) &= K\frac{b(s)}{a(s)} \\
&= K\frac{s^m + b_1 s^{m-1} + \cdots + b_m}{s^n + a_1 s^{n-1} + \cdots + a_n} \\
&= K\frac{(s - z_1)(s - z_2)\cdots(s - z_m)}{(s - p_1)(s - p_2)\cdots(s - p_n)}.
\end{aligned}
\tag{4.4}
$$

Here, $b(s)$ is the numerator polynomial of the open loop system, $a(s)$ is the denominator polynomial of the open loop system, $z_1, \ldots, z_m$ are the open loop zeros, $p_1, \ldots, p_n$ are the open loop poles, and $K$ is a variable gain. With this assumed form of $L(s)$, (4.3) could also be written as

$$\frac{Y(s)}{R(s)} = \frac{Kb(s)}{a(s) + Kb(s)}. \tag{4.5}$$

Thus, the denominator polynomial of the closed loop system is

$$a_{CL}(s) = a(s) + Kb(s), \tag{4.6}$$

and the characteristic equation of the closed loop system could be expressed as either

$$1 + D(s)G(s) = 1 + K\frac{b(s)}{a(s)} = 0, \tag{4.7}$$

or, equivalently,

$$a_{CL}(s) = a(s) + Kb(s) = 0. \tag{4.8}$$

The goal is to study how the locations of the closed loop poles, which are the roots of the closed loop characteristic equation, change when $K$ varies within the range $0 \leq K \leq \infty$ (*complementary* root locus diagrams can also be plotted for $K < 0$, see [3, 5]). Our goal is to find the locus of all points $s$ that satisfy

$$1 + L(s) = 0 \quad \Longleftrightarrow \quad L(s) = -1. \tag{4.9}$$

As discussed above, the transfer function $L(s)$ is a complex variable with magnitude and direction. When $s$ is on the root locus, two conditions from (4.9) must hold:

**MagnitudeCondition** : $\qquad\qquad\qquad\qquad |L(s)| = 1, \quad$ (4.10)

**PhaseCondition** : $\qquad \angle L(s) = (2k+1)\pi, \quad k = 0, \pm 1, 2, \ldots \quad$ (4.11)

The magnitude condition can be satisfied by an appropriate $K \geq 0$. However, the phase condition does not depend on $K$,

$$\angle L(s) = \sum_{i=1}^{m} \angle(s - z_i) - \sum_{j=1}^{n} \angle(s - p_j) = (2k+1)\pi, \ k = \pm 0, 1, 2, \ldots.$$

To construct a root locus, we need to identify the points in the $s$-plane that satisfy the phase condition.

## 4.2.2  Properties of the Root Locus

Some properties of the root locus can be determined by examining (4.8). We will assume that the closed loop transfer function is proper, so that if $a(s)$ has degree $n$ and $b(s)$ has degree $m$, the *pole excess*, $n - m \geq 0$.

*Branches*: Since the polynomial $a_{CL}(s)$ has degree $n$, there will be $n$ roots of the closed loop characteristic equation. Thus, the root locus plot will have $n$ branches and each branch will correspond to one of the closed loop roots.

*Starting points*: As can be seen from (4.6), when $K = 0$ the closed loop poles are equal to the open loop poles $a_{CL}(s) = a(s)$. Thus, the root locus starts at the open loop poles.

*Ending points*: From (4.8), it can be seen that for $K \to \infty$, the roots of the closed loop poles are located at $Kb(s) = 0$, or $b(s) = 0$. Thus, some of the branches will end at the open loop zeros. Other branches will go off towards asymptotes as $s \to \infty$.

*Asymptotes*: Far from the origin of the $s$-plane, as $s \to \infty$, the effects of $m$ poles will tend to cancel out the effects of the $m$ zeros. The cluster of roots will act as $n - m$ poles centered at a point $s_0$, where using (4.4), we can see that (4.7) asymptotically approaches the expression

$$1 + \frac{K}{(s - s_0)^{n-m}} = 0. \tag{4.12}$$

As will be shown below, $s_0$ is the centroid of the poles and zeros of the open loop system. Let $(s - s_0) = Re^{j\theta}$, then far from the origin, points on the root locus will occur when the angle condition

$$\angle \frac{K}{(s - s_0)^{n-m}} = \angle -1,$$
$$\angle \frac{K}{R^{n-m}} e^{-j(n-m)\theta} = -(2k + 1)\pi, \quad k = \pm 0, 1, 2, \ldots \tag{4.13}$$
$$\theta = \frac{(2k + 1)\pi}{n - m}, \quad k = \pm 0, 1, 2, \ldots$$

is satisfied. The asymptotes are thus $(n - m)$ lines that radiate from $s = s_0$. When $K > 0$, the lines have the angles $(2k + 1)\pi/(n - m)$, with respect to the real line. Figure 4.3 shows the asymptotes of the root locus for large $K$ for different values of $(n - m)$.

A mathematical property of monic polynomials of degree $n$ is that the coefficient of the $(n - 1)$th term will be the negative of the sum of the roots. Thus, for the open loop poles $a(s)$

$$a_1 = -\sum_1^n p_k,$$

and, similarly, for the open zeros $b(s)$,

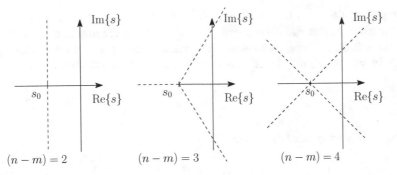

**Fig. 4.3** Asymptotes of the root locus for systems with $(n-m) = 2, 3$ and 4. The $(n-m)$ asymptotes radiate from $s_0$. The angles between asymptotes are $(2k+1)\pi/(n-m), k = \pm0, 1, 2, \ldots$

$$b_1 = -\sum_1^m z_k.$$

As shown above, when $K = 0$, the closed loop poles are equal to the open loop poles $a_{CL}(s) = a(s)$. From (4.8) this implies that for $K = 0$ the sum of the closed loop roots is the same as the sum of the open loop poles

$$-\sum_1^n r_k = -\sum_1^n p_k = a_1.$$

The coefficient $a_1$ is constant, independent of $s$. When both $K \to \infty$ and $s \to \infty$, the term $Kb(s)$ in (4.8) is significant. In this case, the negative sum of the roots associated with the asymptotic open loop poles in (4.12) and those roots associated with the open loop zeros in (4.8) must also be equal to $a_1$, so that

$$-(n-m)s_0 - \sum_1^m z_k = a_1 = -\sum_1^n p_k.$$

Solving for $s_0$, gives

$$s_0 = \frac{\sum_1^n p_k - \sum_1^m z_k}{(n-m)},\tag{4.14}$$

which corresponds to the centroid of the open loop poles and zeros.

*Breakaway/breakin points*: Breakaway points and breakin points are points (normally) along the real axis, where two branches meet and then split apart as $K$ is increased. As example of this can be seen in Fig. 3.19, where a branch of the root locus originating at the pole at $s = 0$ meets a branch of the root locus that originates at the pole $s = -b_\psi$ at the point $s = -b_\psi/2$. The two branches then split from the real axis at an angle of $\pi/2$ and tend towards $s \to \infty$ as $K \to \infty$ along vertical asymptotes of $\theta = \pm\pi/2$. Note that along the real axis $K$ increases on one side of the breakaway point and decreases on the other side of the breakaway point. Thus, $K$ passes through a local maximum or minimum at the breakaway point. Using this property, we can identify breakaway points as those locations where

$$\frac{dK}{ds} = 0.$$

From (4.4), we see that $L(s)$ has the form

$$L(s) = K G_{OL}(s), \tag{4.15}$$

where $G_{OL}(s)$ is the open loop forward transfer function (the product of the controller and plant transfer functions). Using (4.9), we can see that

$$K = -\frac{1}{G_{OL}(s)}$$

so that breakaway points occur when

$$\frac{dK}{ds} = -\frac{d}{ds}\left(\frac{1}{G_{OL}(s)}\right) = \frac{1}{G_{OL}(s)^2}\frac{d}{ds}G_{OL}(s) = 0,$$

or more simply

$$\frac{d}{ds}G_{OL}(s) = 0. \tag{4.16}$$

In general, there will normally be more than one value of $s$ which satisfies (4.16). However, by constructing the other parts of a root locus, one can generally determine which values of $s$ are the correct breakaway or breakin points. In some cases, there may only be one breakaway or breakin point.

In summary, the root locus has $n$ branches that start at the open loop poles and end at either the open loop zeros, or at $s \to \infty$. The branches that end at $s \to \infty$ have star-patterned asymptotes centered at $s_0$ and separated by angles given by (4.13). An immediate consequence is that open loop systems with right half plane zeros or a pole excess $(n - m) > 2$ will always be unstable for sufficiently large gain, $K$. Breakaway and breakin points can be identified by determining where $K$ passes through a local maximum or minimum so that $dG_{OL}(s)/ds = 0$ when $L(s) = K G_{OL}(s)$. As shown in the following note, there are a few simple rules for sketching root loci. Excellent references for use of root locus techniques include [3, 5]

### Guidelines for Constructing a Root Locus with $0 \le K \le \infty$

1. The root locus is symmetric about the real axis.

2. The root loci start from $n$ poles (when $K = 0$) and approach the $m$ zeros ($m$ finite zeros $z_i$ and $n - m$ infinite zeros when $K \to \infty$).
3. The root locus includes all points on the real axis to the left of an odd number of open loop real poles and zeros.
4. As $K \to \infty$, $n - m$ branches of the root-locus approach asymptotically $n - m$ straight lines (called *asymptotes*) with angles

$$\theta = \frac{(2k + 1)\pi}{n - m}, k = 0, \pm 1, \pm 2, \ldots$$

and the starting point of all asymptotes is on the real axis at

$$s_0 = \frac{\sum_{i=1}^{n} p_i - \sum_{j=1}^{m} z_j}{n - m} = \frac{\sum \text{poles} - \sum \text{zeros}}{n - m}.$$

5. The *breakaway points* (where the root loci meet and split away, usually on the real axis) and the *breakin points* (where the root loci meet and enter the real axis) are among the roots of the equation: $dG_{\text{OL}}(s)/ds = 0$. (On the real axis, only those roots that satisfy Rule 3 are breakaway or breakin points.)
6. The *departure angle* $\phi_k$ (from a pole at $s = p_k$) is given by

$$\phi_k = \sum_{i=1}^{m} \angle(p_k - z_i) - \sum_{j=1, j\neq k}^{n} \angle(p_k - p_j) \pm \pi.$$

When a pole is *repeated* $l$ times at a location $s = p_k$ (i.e. when multiple poles exist at the same point $p_k$), the departure angle is

$$\phi_k = \frac{1}{l} \left[ \sum_{i=1}^{m} \angle(p_k - z_i) - \sum_{j=1, j\neq k}^{n} \angle(p_k - p_j) \pm \pi \right].$$

The *arrival angle* $\psi_k$ (at a zero, $z_k$) is given by

$$\psi_k = - \sum_{i=1, i\neq k}^{m} \angle(z_k - z_i) + \sum_{j=1}^{n} \angle(z_k - p_j) \pm \pi.$$

When a zero is *repeated* $l$ times at a location $s = z_k$ (i.e. when multiple zeros exist at the same point $z_k$), the arrival angle is

$$\psi_k = -\frac{1}{l}\left[\sum_{i=1,i\neq k}^{m}\angle(z_k - z_i) + \sum_{j=1}^{n}\angle(z_k - p_j) \pm \pi\right].$$

A quick sketch of the root-locus can often be obtained using only Rules 1–4; rules 5 and 6 are less frequently used nowadays since the exact root-locus can be readily generated using a computer.

**Example 4.2** Angles of departure and breakin points are illustrated here. Consider the trajectory-tracking of a four-wheel underwater sampling rover (Fig. 4.4). It is assumed that the wheels roll on the seafloor without slipping and that a linear Dubins model of the car's kinematics can be used [1]. When the trajectory of the rover consists of small lateral deviations $Y$ from a straight line along the $X$-direction, the transfer function $G(s)$ between $Y$ and the steering angle of the front wheels $\delta$ is

$$G(s) = \frac{\mathscr{Y}}{\Delta} = \frac{av_0}{b}\frac{(s + v_0/a)}{s^2}, \tag{4.17}$$

where $a$ is the distance from the rear axle to the center of gravity of the car, $b$ is the distance between the front and rear axles, $v_0$ is the translational speed of the rear axle with respect to the seafloor, $\mathscr{Y} = \mathscr{L}\{Y\}$ and $\Delta = \mathscr{L}\{\delta\}$.

The system has a pair of repeated poles ($n = 2$) at $p_{1,2} = 0$ and one zero ($m = 1$) at $z_1 = -v_0/a$, so there is only one asymptote at $\pi$ (Root Locus Rule 4).

Owing to Root Locus Rule 3, there will be a breakin point on the real axis between the zero at $s = -v_0/a$ and $s \to -\infty$. According to Root Locus Rule 5, the breakin point will occur at one of the roots of

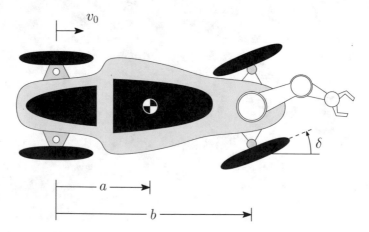

**Fig. 4.4** The underwater rover

$$\frac{d}{ds} G_{OL}(s) = \frac{d}{ds} \left[ \frac{av_0}{b} \frac{(s + v_0/a)}{s^2} \right] = \frac{av_0}{b} \left[ \frac{s^2 - 2s(s + v_0/a)}{s^4} \right] = 0.$$

This reduces to

$$s \left( s + \frac{2v_0}{a} \right) = 0.$$

The two roots of $dG_{OL}(s)/ds = 0$ are at

$$s_{1,2} = \begin{cases} 0 \\ -\dfrac{2v_0}{a} \end{cases}.$$

Thus, the breakin point is at $s = -2v_0/a$.

The departure angles from the two poles at the origin satisfy the following equation

$$2\phi_{1,2} = \angle(p_{1,2} - z_1) \pm \pi = \pm\pi,$$

which gives $\phi_{1,2} = \pm\pi/2$.

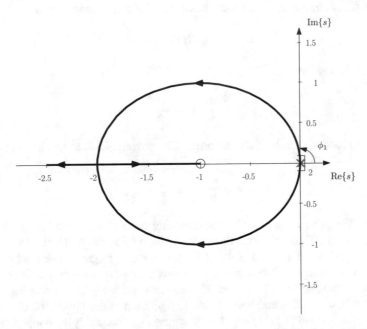

**Fig. 4.5** Root locus of the underwater rover for $v_0 = 0.5\,$m/s, $a = 0.5\,$m and $b = 1\,$m. There is a zero at $s = -1$, a breakin point at $s = -2$, and two poles at $s = 0$ (indicated using a bracket and the subscript 2). The departure angle $\phi_1$ of one of the branches of the root locus departing the poles at the origin is labeled

The root locus of $G(s)$ for the underwater rover, with $v_0 = 0.5\,\text{m/s}$, $a = 0.5\,\text{m}$ and $b = 1\,\text{m}$, is shown in Fig. 4.5. The two branches of the root locus leave the two poles at the origin at angles of $\pm\pi/2$, they intersect at the breakin point $s = -2v_0/a = -2$ and, as $K \to +\infty$, one branch asymptotically approaches the zero at $s = -1$, while the other branch extends towards $s \to -\infty$ along the real axis.  □

## 4.3   Root Locus Controller Design Methods

### 4.3.1   Selecting Gain from the Root Locus

The root locus is a plot of all possible locations of the closed loop roots of (4.7) versus $K$, for all positive real $K$ (e.g. $\forall K \in \mathbb{R}_{>0}$). The simplest root locus controller design technique is to select a value of $K$ that satisfies a given set of transient and steady-state time response requirements. Note that, in general, $K$ could be a loop gain, such as a proportional controller gain, or a system parameter.

**Example 4.3** Returning to the underwater rover discussed in Example 4.2, let $D(s) = K$. Then, the characteristic equation of the closed loop system is

$$1 + KG(s) = 0,$$

which can be rewritten as

$$s^2 + \frac{Kav_0}{b}s + \frac{Kv_0^2}{b} = 0. \tag{4.18}$$

The closed loop system is of second order. Comparing (4.18) to (3.1), we find that

$$\omega_n = v_0\sqrt{\frac{K}{b}} \quad \text{and} \quad \zeta = \frac{a}{2}\sqrt{\frac{K}{b}}.$$

The gain $K$ can be chosen so that the damping ratio of the closed loop system is $\zeta = 1/\sqrt{2} \approx 0.707$, which generally tends to give minimal overshoot and a relatively fast response (see Sect. 3.1.2). Solving for the value $K$ to get this desired damping ratio gives $K = 2b/a^2$. The location of the resulting complex pair of closed loop roots corresponding to this value of $K$ are shown on the root locus in Fig. 4.6 for $v_0 = 0.5\,\text{m/s}$, $a = 0.5\,\text{m}$ and $b = 1\,\text{m}$. As can be seen in the figure, the closed loop roots will be located at $s_1, \bar{s}_1 = -1 \pm j$, where $\bar{s}_1$ is the complex conjugate of $s_1$. The resulting motion of the rover as it follows a sine wave reference trajectory with a 1 m amplitude and a 40 s period is shown in Fig. 4.7. The displacement of the vehicle in the $X$ direction is linearly approximated as $X(t) = v_0t$, where $t$ is time.  □

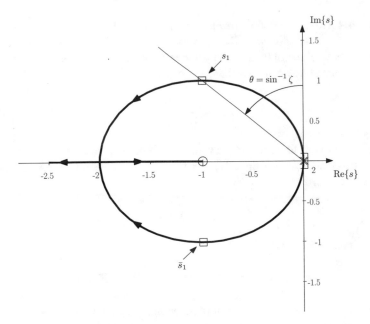

**Fig. 4.6** Root locus of the underwater rover for $v_0 = 0.5\,\mathrm{m/s}$, $a = 0.5\,\mathrm{m}$ and $b = 1\,\mathrm{m}$. When $K = 2b/a^2$ the closed loop poles are located at $s_1, \bar{s}_1 = -1 \pm j$ and the resulting damping ratio line in the complex $s$-plane occurs at an angle of $\theta = \sin^{-1}\zeta = \pi/4$ with respect to the imaginary axis

**Fig. 4.7** Trajectory of the underwater rover for $0 \le t \le 100\,\mathrm{s}$, $v_0 = 0.5\,\mathrm{m/s}$, $a = 0.5\,\mathrm{m}$ and $b = 1\,\mathrm{m}$

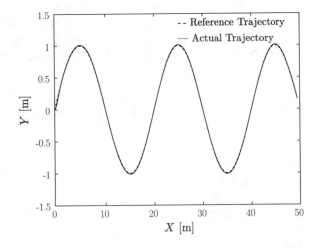

**Example 4.4** (*Root locus for parameter variation*)  Thus far we have seen how the root locus can be plotted as the loop gain of the system is varied. However, as mentioned above, the root locus also provides a powerful tool for understanding how the performance of a system can change if one of its parameters varies. Consider the system

$$L(s) = \frac{s+1}{s(s+K)},$$

where $K$ is now used to represent the value of an uncertain system parameter (which could be associated with a linear drag of an AUV, for example). As before, the characteristic equation is given by

$$1 + L(s) = 0,$$

which can be rewritten as

$$s(s+K) + s + 1 = (s^2 + s + 1) + sK = 0.$$

Dividing by $(s^2 + s + 1)$, the characteristic equation can then be written in the form of (4.9),

$$1 + K\frac{s}{(s^2 + s + 1)} = 1 + L_1(s) = 0,$$

and all the normal guidelines for constructing a root locus can be applied to the new loop transfer function $L_1(s)$ to understand how variations of the uncertain parameter $K$ affect system performance. The root locus can also be used to study the simultaneous effects of multiple uncertain parameters, see [2]. □

## 4.3.2  Compensation by Adding or Moving Poles and Zeros

When transient and steady-state time response requirements cannot be met by adjusting the gain alone, the use of a more complex controller design is necessary, such as one of the PD-, PI- or PID-controllers presented in Chap. 3. These more complex controllers are often called *compensators*, as they are designed to compensate for the dynamics of the plant. They effectively alter the root locus of the system so that its closed loop time response satisfies the given requirements. Adding, or moving, poles and zeros to a system affects its root locus and can be exploited for controller design.

- **Adding open loop poles:** Open loop poles tend to repel the root locus, generally pushing it towards the right half of the $s$-plane ($\text{Re}\{s\} > 0$). When there are more than 2 poles, at least two branches of the root locus will be pushed towards the right half plane. An example of this can be seen in Fig. 4.8.

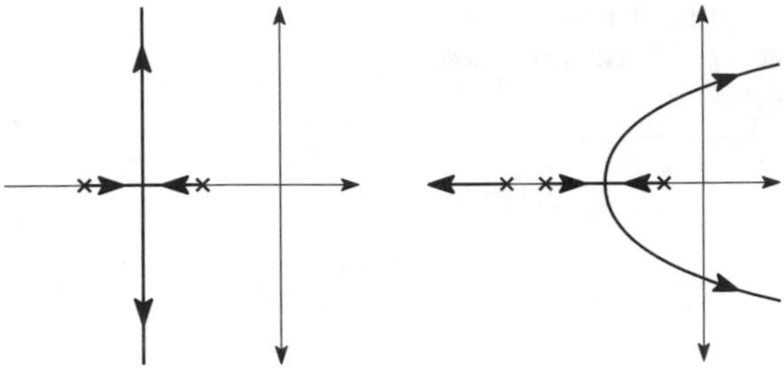

**Fig. 4.8** The effect of adding an open loop pole to the root locus—the two vertical branches of the original root locus (left) are pushed into the right half of the *s*-plane (right)

**Fig. 4.9** The effect of adding an open loop zero to the root locus—the two vertical branches of the original root locus (left) are attracted to the zero (right)

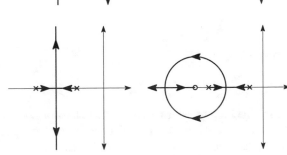

- **Adding open loop zeros:** Open loop zeros attract the root locus; suitably placed zeros can move the root loci towards a desired region of the *s*-plane. The effect of adding an additional zero can be seen in Fig. 4.9.

When designing compensators, it is useful to keep in mind that the control effort required is related to how far from the open loop poles the closed loop poles are moved by the feedback. Additionally, it should be remembered that considerable control effort may be required to move a closed loop pole away from an open loop zero, as open loop zeros attract the closed loop poles. A controller design philosophy that aims to *only* fix the undesirable characteristics of the open loop response will allow smaller control efforts (and smaller actuators) than an approach that arbitrarily picks the closed loop poles in a given location without regard for the locations of the open loop poles.

### 4.3.3 Phase Lag Controllers

A first-order phase lag controller has the form

$$D(S) = \frac{K(s+b)}{s+a}, \quad b > a > 0. \tag{4.19}$$

The pole-zero configuration of the controller is shown in Fig. 4.10. It is known as a phase lag controller because the phase angle of $D(s)$ is less than zero for all $s$ in the upper half of the *s*-plane. Thus, it always contributes a negative phase (phase

**Fig. 4.10** Pole/zero configuration of a phase lag controller

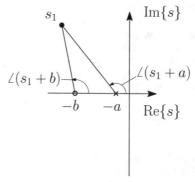

$$\angle D(s) = \angle(s_1 + b) - \angle(s_1 + a) < 0$$

lag) to the loop transfer function. Phase lag controllers are generally used when the dominant closed loop poles (root locus) of a system already pass through a region of the $s$-plane that satisfies a set of desired transient response characteristics (e.g. rise time, settling time and peak overshoot), but when the steady state error of the system is too large. Thus, a phase lag controller is typically designed so that the original root locus of the system is insignificantly affected at a desired closed loop pole location (i.e. the change in phase from the controller is almost zero there), while providing a substantial gain increase for $s \to 0$ to reduce the steady-state error. For this reason, lag controllers are also sometimes called *gain compensators*. Suppose $s_1$ is a desired closed loop (dominant) pole location, then the ratio $b/a$ is typically chosen so that

$$\frac{s_1 + b}{s_1 + a} \approx 1,$$

i.e.

$$D(s_1) \approx K,$$

and

$$D(s_1)G(s_1) \approx KG(s_1) = -1.$$

This can usually be achieved by choosing the zero $b$ and the pole $a$ far from $s_1$, but relatively close to the origin. Note that $a$ and $b$ do not need to be small in the absolute sense. At the same time, note that $D(0) = Kb/a$. By suitably selecting the ratio $b/a$, a phase lag controller can be used to substantially increase the steady-state gain constant (see Sect. 3.4).

**Example 4.5** Let $s_1 = -100 + 100j$, $b = 1$ and $a = 0.1$, then

$$D(s_1) = \frac{K(s_1 + b)}{s_1 + a} = \frac{K(-99 + 100j)}{-99.9 + 100j} \approx K,$$

and $D(0) = Kb/a = 10K$. □

**Phase lag controller design steps**

Step 1 Construct the root locus plot of the uncompensated system $KG(s)$.

Step 2 Locate suitable dominant root locations $s_1$ and $\bar{s}_1$, where $\bar{s}_1$ is the complex conjugate of $s_1$, on the uncompensated root locus that satisfy the desired transient response and find the corresponding gain $K$. Denote this gain as $K_0$.

Step 3 Calculate the value of $K$ needed to obtain the desired steady-state response (steady-state error constant) and denote it as $K_s$.

Step 4 Compare the uncompensated error constant with the desired error constant, the pole-zero ratio of the compensator will be selected to achieve the required increase.

Step 5 The pole and zero should be located near the origin of the $s$-plane in comparison to $\omega_n$. Pick a number $b$, where $b > a$, that is much smaller than $|s_1|$, so that $(s_1 + b)/(s_1 + a) \approx 1$ and let $a = K_0 b/K_s$.

Step 6 Verify the controller design by simulation with

$$D(s) = \frac{K_0(s + b)}{s + a}.$$

Note that $D(0) = K_s$.

*PI Controllers*: A PI controller has the transfer function

$$D(s) = \frac{K_0(s + b)}{s}, \quad b > 0,$$

and can be regarded an extreme form of the lag controller with the pole placed at the origin. The lag controller design procedure can also be used to design PI controllers by taking $K_s \to \infty$, so that $D(0) \to \infty$. Note that a PI controller will increase the system type by 1.

**PI controller design steps**

Step 1 Construct a root locus plot of $KG(s)$.

Step 2 Find the closed loop (dominant) poles $s_1$ and $\bar{s}_1$ on the root locus plot that will give the desired transient response and find the corresponding $K$ value, call this $K_0$. Note that the transient response characteristics are usually specified in terms of rise time, settling time and overshoot. Use of (3.8), (3.9) and (3.10) can be used to translate them into a desired $\zeta$ and $\omega_n$, and thereby to a desired point in the $s$-plane $s_1$.

Step 3 Pick a number $b$ that is much smaller than $|s_1|$, so that $(s_1 + b)/s_1 \approx 1$.

Step 4 Verify the controller design by simulation with

$$D(s) = \frac{K_0(s + b)}{s}.$$

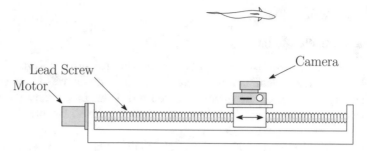

**Fig. 4.11**  Motion tracking underwater imaging system

**Example 4.6**  An underwater imaging system is designed to track the motion of a certain species of marine organisms, which are known to habitually swim past the system's location (Fig. 4.11). A camera is positioned along a track using a lead-screw traversing mechanism. The transfer function between the lead-screw motor and camera position is given by

$$G(s) = \frac{250}{s(s + 10)}.$$

It is desired that the closed system has less that 5% overshoot to a unit step input and that the steady-state error to a unit ramp input be no greater than $e_{ss} = 2.5\%$.

According to (3.6), a damping ratio of $\zeta = 0.707$ should give an overshoot of 4.33%. From the root locus of the system shown in Fig. 4.12 that, it can be seen that it is possible to achieve the desired damping ratio at the point $s_1 = -5 + 5j$. Given that the branches of the root locus do not need to be moved, we will explore the design of a lag controller. The loop gain at $s_1$ is

$$K_0 = \frac{1}{|G(s_1)|} = \frac{|s_1||s_1 + 10|}{250} = \frac{\sqrt{(-5)^2 + 5^2}\sqrt{(5)^2 + 5^2}}{250} = \frac{1}{5}.$$

The gain required to achieve the desired steady state error to a ramp input is given by

$$K_s = \frac{1}{e_{ss}} = 40.$$

The zero should be placed along the real axis at a location $b \ll |s_1|$. We will try placing the zero at $b = -\text{Re}\{s_1\}/100 = 0.05$. The location of the pole is then $a = K_0 b/K_s = 2.5 \times 10^{-4}$. As can be seen in Fig. 4.13, away from the compensator pole-zero pair, the original root locus is minimally affected.

In simulation, it is found that the overshoot for a step response is just a little higher than desired (5.35%). The damping ratio can be increased slightly by moving the desired closed loop root a little lower along the vertical branch of the original root locus. We try the location $s_1 = -5 + 4.75j$. With this selection, $K_0 = 0.1902$,

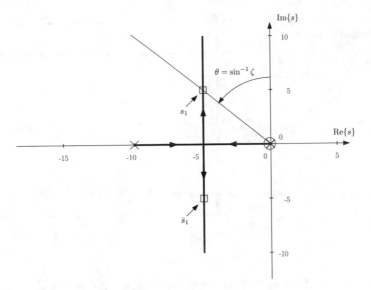

**Fig. 4.12**  Root locus and closed loop roots for motion tracking system with $K_0 = 1/5$

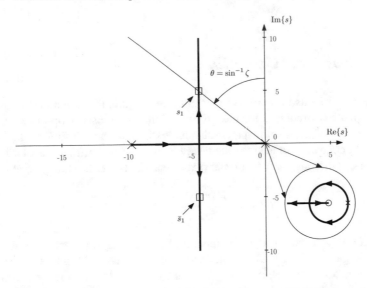

**Fig. 4.13**  Root locus of the lag-compensated system. The compensator only affects the shape of the root locus near the origin (shown as an inset on the lower right)

**Fig. 4.14** Root locus of the lag-compensated system. The affect of the compensator on the shape of the locus is shown as an inset on the lower right

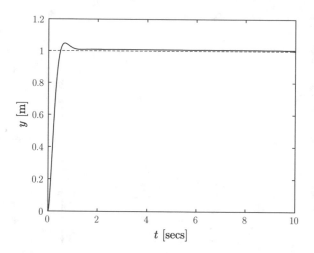

$b = 0.05$ and $a = 2.378 \times 10^{-4}$. The peak overshoot to a step response is now 4.73% and all of the design requirements are satisfied (Fig. 4.14).                                     □

### 4.3.4   Phase Lead Controllers

When the root locus of a system does not already pass through a desired point $s_1$ in the $s$-plane (as required for the system's dominant closed loop poles to satisfy a specified set of transient response characteristics), it may be possible to design a phase lead controller to achieve the desired response by shifting the root locus so that it passes through $s_1$. A first order phase lead controller has the general form

$$D(S) = \frac{K(s+b)}{s+a}, \quad a > b > 0.$$

Since $\angle D(s) > 0$ for any $s$ on the upper half of the complex plane, it contributes a positive angle, or phase lead. In contrast to the phase lag controller, the phase lead controller is a *phase compensator*. The controller provides a positive phase to the loop transfer function in order to move the system's closed loop poles further into the left half plane, or to a desired $s$-plane location.

**Example 4.7**   Consider a system with two poles $p_1$ and $p_2$. Suppose $s_0$ is a point on the root locus where

$$-\phi_1 - \phi_2 = -\pi.$$

In order to move the closed loop pole from $s_0$ to $s_1$, the following phase lead is needed so that $s_1$ satisfies the phase condition

**Fig. 4.15** Phase needed to move from $s_0$ to $s_1$

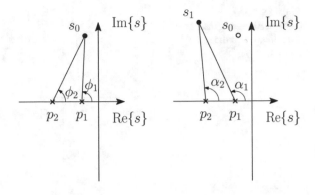

$$\theta = \alpha_1 + \alpha_2 - \phi_1 - \phi_2 > 0.$$

□

**Phase lead controller design steps**

Step 1 Construct a root locus plot of $KG(s)$.

Step 2 Find the closed loop (dominant) poles $s_1$ and $\bar{s}_1$, where $\bar{s}_1$ is the complex conjugate of $s_1$, on the root locus plot that will give the desired transient response. As noted in the lag controller design procedure above, the transient response characteristics are usually specified in terms of rise time, settling time and overshoot. Use of (3.8), (3.9) and (3.10) can be used to translate them into a desired $\zeta$ and $\omega_n$, and thereby to a desired point in the $s$-plane $s_1$.

Step 3 Calculate the angle required so that $s_1$ is on the root locus plot

$$\angle D(s_1) + \angle G(s_1) = (2k + 1)\pi,$$

i.e.

$$\theta = \angle D(s_1) = (2k + 1)\pi - \angle G(s_1) > 0,$$

for an integer $k$ (Fig. 4.15).

Step 4 Find $b$ and $a$ so that

$$\angle(s_1 + b) - \angle(s_1 + a) = \theta,$$

and make sure that $s_1$ is the dominant pole. There are infinitely many choices, so long as the angle between the two vectors is $\theta$ (see Fig. 4.16). A convenient starting point is to first try placing the compensator zero on the real axis directly below $s_1$, or to the left of the first two real poles.

Step 5 Find $K_0$ so that

$$\frac{K_0|s_1 + b|}{|s_1 + a|}|G(s_1)| = 1.$$

**Fig. 4.16** Two
configurations of phase lead
controllers with the same
phase

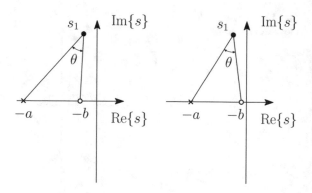

Step 6   Verify the controller design by simulation with

$$D(s) = \frac{K_0(s+b)}{s+a}.$$

If the steady state error achieved is not sufficient, the steps may need to be
repeated (design iteration) trying slightly different values for $s_1$, $b$ and $a$.

*PD Controller:* A PD controller has the form

$$D(S) = K(s+b), \quad b > 0.$$

This controller can be regarded as a special case of the phase lead controller where
$a \to \infty$ and $K/a$ is constant. The design procedure for a PD controller is almost the
same as that of a phase lead controller.

PD controllers are not generally physically realizable, but can be approximated
using the relation

$$D(S) = K\left(\frac{s}{Ts+1} + b\right)$$

where $T > 0$, $T \ll 1$, see [5] for additional details.

As will be discussed in Sect. 5.7.2 below, PD controllers tend to amplify the
existing noise in a system. Their implementation may be suitable when the dynamics
of the plant are heavily damped, but should be avoided otherwise.

**PD controller design steps**

Step 1   Construct a root locus plot of $KG(s)$.

Step 2   Find the closed loop (dominant) poles $s_1$ and $\bar{s}_1$ on the root locus plot that
will give the desired transient response.

Step 3   Calculate the angle required so that $s_1$ is on the root locus plot

$$\angle D(s_1) + \angle G(s_1) = (2k+1)\pi,$$

**Fig. 4.17** Zero location of a
PD controller

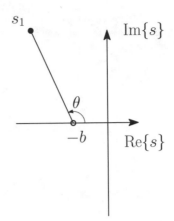

i.e.

$$\theta = \angle D(s_1) = (2k+1)\pi - \angle G(s_1) > 0,$$

for an integer $k$ (Fig. 4.17).

Step 4  Find $b$ so that

$$\angle(s_1 + b) = \theta,$$

and make sure that $s_1$ is the dominant pole.

Step 5  Find $K_0$ so that

$$K_0|s_1 + b||G(s_1)| = 1.$$

Step 6  Verify the controller design by simulation with

$$D(s) = K_0(s + b).$$

**Example 4.8**  A Lagrangian float is a type of underwater buoy often used for data collection (Fig. 4.18). They are typically designed to drift in ocean currents and so their horizontal motion is typically passive, while their vertical motion (depth) may be automatically controlled. Using precise depth control, a Lagrangian float can be used as a camera platform to provide high resolution seafloor imaging to evaluate the health of benthic habitats, such as coral reefs [6]. The linearized equation of motion for the depth of a Lagrangian float is given by

$$\ddot{z} = \frac{F_z}{m},$$

where $m$ is the mass of the float (including added mass), $z$ is float depth and $F_z$ is the control force along the vertical direction, which could be developed using a thruster or buoyancy-control mechanism. Take $u := F_z/m$ to be the control input, the transfer function of the float is

**Fig. 4.18** A Lagrangian
float designed for underwater
imaging

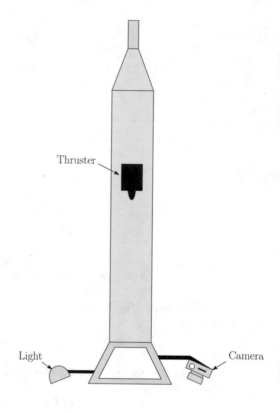

$$G(s) = \frac{1}{s^2}.$$

The associated root locus is shown in Fig. 4.19. From the root locus it can be seen that, for any value of loop gain, the closed loop roots will be located on the imaginary axis. Thus, the closed loop response will be marginally stable and consist of sinusoidal oscillations of constant amplitude. In order to facilitate imaging, the system should be capable of maintaining a desired depth with minimal oscillation. A settling time of about 30 s (2% criterion) and an overshoot of about 20% would permit the system to change depth relatively quickly. From (3.9) it can be seen that $\zeta \geq 0.46$ is required. Taking $\zeta = 0.707$, (3.7) requires that $\omega_n \geq 0.189$ rad/s to satisfy the 2% settling time criterion.

Based on these values we select the desired closed loop pole locations as

$$s_1, \bar{s}_1 = -\zeta\omega_n \pm j\omega_n\sqrt{1 - \zeta^2} = -0.1336 \pm 0.1336j.$$

From Step 3 of the PD controller design procedure the angle needed from the lead compensator is

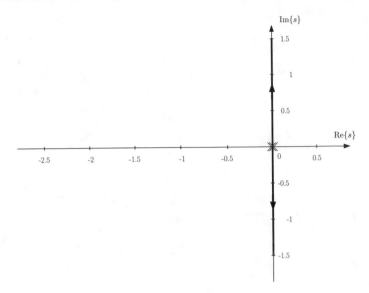

**Fig. 4.19**   The root locus of the uncontrolled Lagrangian float

$$\theta = (2k + 1)\pi - \angle G(s_1) = (2k + 1)\pi + 2 \cdot \frac{3\pi}{4} = \frac{\pi}{2} > 0,$$

when $k = -1$. As a first pick, take $b = |\text{Re}\{s_1\}| = 0.133$, then $a$

$$a = -\text{Re}\{s_1\} + \frac{\text{Im}\{s_1\}}{\tan(\angle(s_1 + b) - \theta)} = 442.$$

Then from Step 5,

$$K_0 = \frac{1}{|s_1 + b||G(s_1)|} = 118.$$

Thus, we have a phase lead controller

$$D(s) = \frac{118(s + 0.133)}{s + 442}.$$

The root locus of $D(s)G(s)$ is shown in Fig. 4.20.
The step response shown in Fig. 4.21 indicates that the design specifications are met.

$\square$

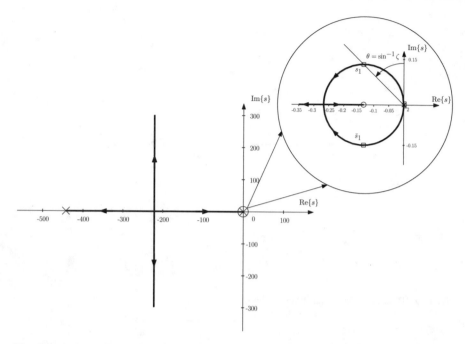

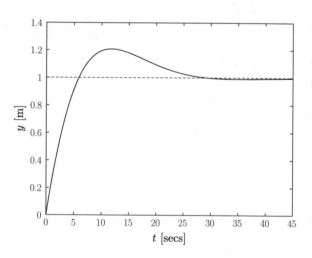

**Fig. 4.20** The root locus of the Lagrangian float with lead compensator. The compensator shifts the vertical axis of the root locus far into the left half of the $s$-plane, but more importantly causes the root locus to pass through the desired complex closed loop poles $s_1$, $\bar{s}_1$ (inset), which are also the dominant poles

**Fig. 4.21** Step response of the Lagrangian float with lead compensator

## 4.4 General Guidelines for Root Locus Controller Design

Bearing in mind that compensator designs which significantly change the root locus of the original plant should be avoided, as they generally require a greater control effort (see Sect. 4.3.2) and consequently can cause more wear and tear on actuators, the following approach can be followed when designing a controller using the Root Locus Method:

(1) Identify the transient and steady-state design requirements for the system.
(2) Determine suitable dominant pole locations based on the transient requirements.
(3) Using the steady-state error requirements, determine the associated steady-state error constants.
(4) If the steady-state error requirements indicate that the system type should be increased, a PI or PID controllers should be used.
(5) Plot the root locus of $KG(s)$.
(6) If the root locus passes through the desired dominant poles, a phase lead, PD, PID or lead-lag controller can be used. Sometimes, a pure phase lag or pure phase lead controller may not provide enough design freedom to satisfy all of the design specifications. In such cases, a combination of phase lag and phase lead, or a PID controller is necessary. The basic idea is to use first use a phase lead to satisfy the transient response and then a phase lag to satisfy the steady-state requirements.

## 4.5 Matlab for Root Locus Analysis and Controller Design

### 4.5.1 Constructing a Root Locus with Matlab

Some systems are unstable for small values of the loop gain $K$. Here we plot the root locus of such a system using Matlab and discuss how to determine $K$ at the stability limit. Consider the system

$$D(s)G(s) = \frac{K}{(s-2)(s+3)}.$$

The closed loop poles are given by the roots of

$$1 + D(s)G(s) = 0,$$

or

$$s^2 + s + K - 6 = 0. \tag{4.20}$$

The following commands can be used to plot the root locus with Matlab

```
num = 1; % numerator coefficients of L(s), excluding K
den = conv([1, −2], [1, 3]); % denominator coefficients of L(s)
L = tf(num, den); % create the transfer function
rlocus(L) % generate a root locus plot
sgrid; % draw damping ratio lines
```

From Routh Stability Theorem 2.6, we know that all of the coefficients of the characteristic polynomial must be positive in order for the closed loop system to be stable. From (4.20), we can see that stability requires $K > 6$. This condition can also be determined graphically using the root locus plot in Matlab. The *rlocfind* command can be used to find the numerical value of any point on the root locus and the corresponding value of the gain.

```
[K, poles] = rlocfind(L) % click on any desired point on the plot
```

To find the value of $K$ when the branch of the root locus starting at the unstable open loop pole at $s = +2$ first enters the left half plane, click on the intersection of the root locus and the imaginary axis. As illustrated in Fig. 4.22, it might not be possible to place the cursor exactly at the point of interest by hand, but one can very close

$$K = 6.0171, \quad \text{selected point } s_1 = -0.0174 - 0.0019j.$$

This shows that the closed loop system will only be stable for $K > 6$.

### 4.5.2   Use of Matlab for Designing a Phase Lag Controller

Recall the lag controller designed for an underwater imaging system presented in Example 4.6. We follow the step-by-step procedure outlined there using Matlab.

Step 1   Construct the root locus plot of the uncompensated system $KG(s)$.

```
% construct zero-pole-gain model of open loop system
G = zpk([], [0, -10], 250);
rlocus(G); % plot its root locus
```

Step 2   Locate suitable dominant root locations $s_1$ and $\bar{s}_1$ on the uncompensated root locus that satisfy the desired transient response and find the corresponding gain $K$. Denote this gain as $K_0$.

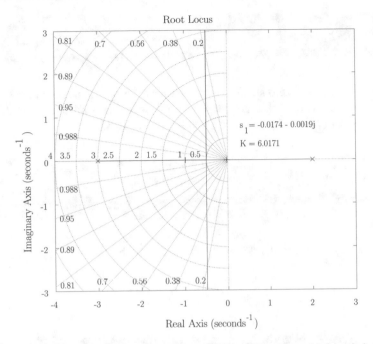

**Fig. 4.22** Use of Matlab to determine the gain required for stability. The red '+' symbols indicate the closed loop roots corresponding to the gain selected

```
s1 = -5 + j*4.75; % the desired dominant pole location
% use the evaluate frequency response function to determine
% the magnitude and phase of G(s1)
Gs1 = evalfr(G,s1);
K0 = 1/abs(Gs1); % gain of uncompensated system at s1
```

**Step 3** Calculate the value of $K$ needed to obtain the desired steady-state response (steady-state error constant) and denote it as $K_s$.

```
ess = 0.025; % max steady state error
Ks = 1/ess; % steady state error constant
```

**Step 4** Compare the uncompensated error constant with the desired error constant, the pole-zero ratio of the compensator will be selected to achieve the required increase.

```
K = Ks/K0; % required gain increase
```

Step 5  The pole and zero should be located near the origin of the $s$-plane in comparison to $\omega_n$. Pick a number $b$, where $b > a$, that is much smaller than $|s_1|$, so that $(s_1 + b)/(s_1 + a) \approx 1$ and let $a = K_0 b / K_s$.

```
b = -real(s1)/100; % pick b much smaller than s1
a = b/K; % calculate a
```

Step 6  Verify the controller design by simulation with

$$D(s) = \frac{K_0(s + b)}{s + a}.$$

```
D = zpk(-b,-a,K0); % construct lag compensator
T = feedback(D*G, 1); % construct CL unity feedback system
step(T); % plot step response of CL system
stepinfo(T) % calculate overshoot and 2% settling time
```

The step response of the system is shown in Fig. 4.14. The output of the Matlab stepinfo command is given below (all time-related data are in seconds and overshoot is in percent).

|  |  |
|---:|:---|
| RiseTime: | 0.3150 |
| SettlingTime: | 0.9670 |
| SettlingMin: | 0.9005 |
| SettlingMax: | 1.0473 |
| Overshoot: | 4.7347 |
| Undershoot: | 0 |
| Peak: | 1.0473 |
| PeakTime: | 0.6665 |

### 4.5.3  Use of Matlab for Designing a Phase Lead Controller

Recall the lead controller designed to control the depth of a Lagrangian float, which was presented in Example 4.8. We follow the step-by-step procedure outlined there using Matlab.

**Step 1** Construct a root locus plot of $KG(s)$.

```
% construct zero-pole-gain model of open loop system
G = zpk([], [0, 0], 1);
rlocus(G); % plot its root locus
```

**Step 2** Find the closed loop (dominant) poles $s_1$ and $\bar{s}_1$ on the root locus plot that will give the desired transient response.

```
ts = 30; % desired settling time
zeta = 0.707; % damping ratio
omegan = 4/zeta/ts; % natural frequency
sigmad = zeta*omegan; % real part of dominant poles
omegad = omegan*sqrt(1 - zeta^2); % imag part of dominant poles
s1 = -sigmad + j*omegad;
```

**Step 3** Calculate the angle required so that $s_1$ is on the root locus plot

$$\angle D(s_1) + \angle G(s_1) = (2k + 1)\pi,$$

i.e.

$$\theta = \angle D(s_1) = (2k + 1)\pi - \angle G(s_1) > 0,$$

for an integer $k$.

```
Gs1 = evalfr(G,s1); % the magnitude and phase of G(s1)
theta = pi - angle(Gs1); % angle required to put s1 on root locus
```

**Step 4** Find $b$ and $a$ so that

$$\angle(s_1 + b) - \angle(s_1 + a) = \theta,$$

and make sure that $s_1$ is the dominant pole.

```
% usually best to try placing b below s1 first
b = sigmad; % zero
% find compensator pole location from angle condition
% note: be careful with discontinuity of tangent near ±90°
a = -real(s1) + imag(s1)/tan(angle(s1+b) - theta); % pole
```

Step 5  Find $K_0$ so that

$$\frac{K_0|s_1 + b|}{|s_1 + a|}|G(s_1)| = 1.$$

```
K0 = abs(s1+a)/abs(Gs1)/abs(s1+b);
```

Step 6  Verify the controller design by simulation with

$$D(s) = \frac{K_0(s + b)}{s + a}.$$

```
D = zpk(-b, -a, K0);% construct lead compensator
T = feedback(D*G, 1); % construct CL unity feedback system
step(T); % plot step response of CL system
stepinfo(T) % calculate overshoot and 2% settling time
```

The step response of the system is shown in Fig. 4.20. The output of the Matlab stepinfo command is given below (all time-related data are in seconds and overshoot is in percent.).

|  |  |
|---:|:---|
| RiseTime: | 4.4855 |
| SettlingTime: | 25.9492 |
| SettlingMin: | 0.9330 |
| SettlingMax: | 1.2080 |
| Overshoot: | 20.8036 |
| Undershoot: | 0 |
| Peak: | 1.2080 |
| PeakTime: | 11.7432 |

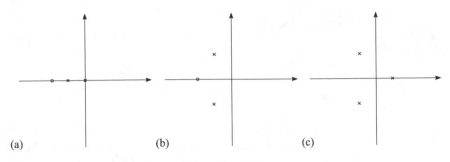

**Fig. 4.23** Open loop pole-zero diagrams

## Problems

**4.1** Consider the root locus for the closed loop characteristic equation

$$1 + \frac{K}{s(s+2)(s+10)} = 0.$$

(a) Construct the root locus of the system by hand, clearly showing:

    i. the real axis segments,
    ii. the asymptotes for $K \to \infty$, and
    iii. any breakaway or breakin points.

(b) Use Routh's Criterion to determine the value of $K$ when a pair of the closed loop roots are located on the imaginary axis (stability bounds).

**4.2** Sketch the root loci for the pole-zero maps shown in Fig. 4.23. Show asymptotes, centroids, and estimates of arrival and departure angles. Each map has a closed loop characteristic equation of the form

$$1 + K \frac{b(s)}{a(s)} = 0.$$

**4.3** Consider the Lagrangian float discussed in Example 4.8, which has the plant transfer function

$$G(s) = \frac{1}{s^2}.$$

It was shown that the lead compensator

$$D(s) = \frac{118(s + 0.133)}{s + 442}$$

**Fig. 4.24** Lagrangian float
depth controller with
feedback sensor

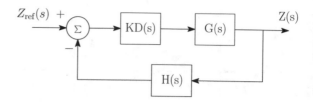

provides a suitable time response when the depth $z$ is perfectly measured (unity feedback). Here, assume that the depth is measured using a depth sensor with a finite time response. The sensor transfer function is

$$H(s) = \frac{1}{\tau s + 1},$$

where $\tau = 0.75\,\mathrm{s}$ is the response time of the sensor. If a loop gain $K$ is added to the system (Fig. 4.24), such that the loop transfer function is

$$L(s) = K D(s) G(s) H(s).$$

Using the root-locus procedure, find the value of $K$ that will maximize the damping ratio.

**4.4** Many marine systems use bilateral teleoperation to provide a human user with a sense of the conditions at a remote location. For example, an underwater manipulator (robotic arm) may measure the forces it exerts when manipulating objects at a remote work site and transmit them back to a control station, where a measure of the forces is imparted to the user's hand through a force feedback joystick. This can allow a user to better regulate the process.

Consider the control of one axis of a force feedback joystick. The plant transfer function of the joystick is given by

$$G(s) = \frac{\Theta}{T} = \frac{1/I}{s(s + b/I)},$$

where $\Theta(s)$ is the output joystick angle, $T$ is the input torque from a motor within the joystick base, $I = 0.762\,\mathrm{kg\text{-}m^2}$ is the mass moment of inertia of the joystick system and $b = 0.254\,\mathrm{N\text{-}m\text{-}s}$ is a frictional drag coefficient.

(a)  Sketch the root locus of the system by hand.
(b)  Design a compensator that can control the joystick handle to follow a step command in angle with a settling time of $t_s \leq 0.3\,\mathrm{s}$ and a peak overshoot of about 4%.
(c)  What is the system type and the steady state error of the closed loop system for the step input?

(d) In order to provide force feedback to a user, assume that the reference position of the joystick is an angle of $\theta_{ref} = 0°$ and that the input from the user will be modeled as a disturbance torque.

    i. What is the steady state error to step disturbance?

    ii. Design an additional compensator, to be used in series with the compensator designed in part (a), to reduce the steady state error to a step disturbance to less than 2%.

**4.5** A transfer function for a simplified, linear one degree of freedom model that relates the output lateral deviation $Y$ of a small unmanned surface vessel (USV) to input changes in its rudder angle $\delta_R$ is

$$\frac{Y}{\Delta_R} = \frac{U/100}{s^2(s + 5/4)},$$

where $U = 1.5\,\text{m/s}$ is the constant surge speed of the USV.

(a) Plot the root locus of the system.

(b) Design a compensator so that the closed loop step response of the system has less than 30% overshoot and a settling time of less than 20 s.

(c) Plot the root locus of the compensated system.

(d) Plot the time response of the system to a step input.

(e) Most symmetric hydrofoils stall at an angle of attack of near $\pm15°$. Therefore, most rudders are ineffective beyond these angles. Assume that the rudder angle is limited (saturated) to $\pm15°$. Plot the time response of the system and explain how and why it differs from the time response plotted in part (d).

**4.6** A spar buoy is a large cylindrical buoy with a circular cross section, which is anchored so that its axis of symmetry remains vertical. In this configuration, the bottom of the spar is generally located deep below the waterline and only a short section of the spar protrudes above the surface. Owing to a small waterplane area, properly-designed spar buoys tend to be minimally disturbed by waves and are used when a stable offshore platform is required, such as for applications involving offshore wind turbines or large cranes. Here we explore the angular positioning of a spar buoy using underwater thrusters. The block diagram of a proposed PD controller is shown in Fig. 4.25. Suppose we require the ratio of $K_1/K_2 = 5$. Using root locus techniques, find the ratios of $K_1/I_z$ and $K_2/I_z$ that provide a settling time $t_s \leq 25\,\text{s}$ (2% criterion) and a step response with a peak overshoot of $M_p \leq 10\%$.

**4.7** AUVs are often used to image the seafloor or underwater objects using side scan sonar or camera systems. As sonars and cameras are often mounted in fixed positions on the AUV, precise roll control of the AUV is sometimes used to ensure that the imaging systems are pointed in the appropriate direction when images are collected (see [4], for example). A rolling moment can be generated either using fins mounted at the aft end of the AUV or by shifting a mass within the AUV, for example.

**Fig. 4.25** Spar angular
positioning with PD
controller

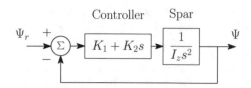

**Fig. 4.25** Spar angular
positioning with PD
controller

The block diagram of such a roll control system is shown in Fig. 4.26, where both
the roll angle and roll rate are measured. It is required that the step response of the
system have an overshoot of $M_p \le 10\%$ and a settling time of $t_s \le 9\,\mathrm{s}$ (2% criterion).
Determine $K$ and $K_g$.

**4.8** A crane is used to deploy unmanned vehicles from the deck of a research vessel.
The dynamics of the crane about its azimuthal axis are given by the transfer function

$$G(s) = \frac{0.25}{s(s+0.2)(s+4)(s+4.5)}.$$

Design a lead compensator so that the system has a steady state error $e_{ss} \le 10\%$ for a
ramp input, and a step response with a rise time is $t_r \le 5\,\mathrm{s}$, an overshoot $M_p \le 20\%$
and a settling time $t_s \le 12\,\mathrm{s}$.

**4.9** The pitch angle of a fast planing hull boat depends on its forward speed. The
flow of water along the hull produces a pitching moment on the vessel. When the
speed is low, the hull will tend to pitch upwards (so the bow is higher than the stern).
The drag on the hull increases when this occurs and so trim tabs are often used to
reduce the pitch angle of the hull at low and moderate speeds. A block diagram of a
trim tab system is shown in Fig. 4.27.

(a) Design a compensator that will provide a step response with an overshoot of
   $M_p \le 20\%$ and a settling time of $t_s \le 2\,\mathrm{s}$.
(b) What is the resulting steady-state error?

**4.10** A semi-automatic control system is used to maintain the depth of a remotely-
operated vehicle (ROV) so that its operator can concentrate on other tasks, such as

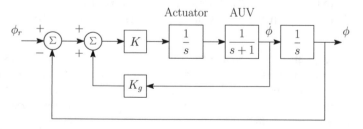

**Fig. 4.26** Roll control of an AUV

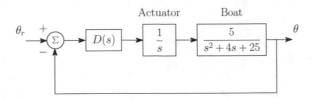

**Fig. 4.27** Pitch control system for a planing hull boat

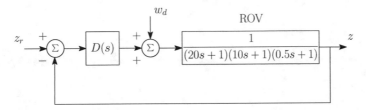

**Fig. 4.28** ROV depth control system

collecting samples with a underwater manipulator. The system uses a tunnel thruster for depth control; a block diagram is shown in Fig. 4.28.

(a) Design a compensator that achieves a steady-state depth error of $e_{ss} \leq 5\%$, an overshoot of $M_p \leq 18\%$ and a settling time of $t_s \leq 5\,\text{s}$ for a unit step input, $z_r(t) = 1(t)$.

(b) When the reference input is $z_r = 0$, plot the time response of the system $z(t)$ to a unit step disturbance, $w_d(t) = 1(t)$.

# References

1. Åström, K.J., Murray, R.M.: Feedback Systems: An Introduction for Scientists and Engineers. Princeton University Press, Princeton (2010)
2. Dorf, R.C., Bishop, R.H.: Modern Control Systems. Pearson (Addison-Wesley), London (1998)
3. Franklin, G.F., Powell, J.D. Emami-Naeini, A.: Feedback Control of Dynamic Systems, 3rd edn. Addison-Wesley, Reading (1994)
4. Hong, E.Y., Chitre, M.: Roll control of an autonomous underwater vehicle using an internal rolling mass. In Mejias, L., Corke, P., Roberts, J. (eds.) Field and Service Robotics: Results of the 9th International Conference, pp. 229–242. Springer International Publishing, Berlin (2015)
5. Qiu, L., Zhou, K.: Introduction to Feedback Control. Prentice Hall, Englewood Cliffs (2009)
6. Snyder, W., Roman, C., Licht, S.: Hybrid actuation with complementary allocation for depth control of a Lagrangian sea-floor imaging platform. J. Field Robot. **35**(3), 330–344 (2017)

# Chapter 5
# Frequency Response Methods

## 5.1 Frequency Domain Analysis

When a stable system is given a sinusoidal input, its steady state response (output) is also sinusoidal. However, with respect to the input signal, the magnitude of the response will be amplified or attenuated by a certain amount and its phase will be shifted. The amplification factor and phase shift vary with the frequency of the input and are respectively called the *magnitude frequency response* and the *phase frequency response* of the system. Stability analyses and controller design can be performed using only the system's frequency response, which can often be obtained experimentally for a stable system, even though an analytic model may not be available. Thus, it can be particularly useful for the design of marine control systems for which precise analytical models and their associated transfer functions are difficult to develop.

**Example 5.1** Consider the open loop transfer function for the steering of a surface vessel (3.15), which was presented in Example 3.2. The transfer function is a relation between the output response of the yaw angle and an input rudder angle. Let the numerical coefficients be selected so that the resulting motion would be representative of a small unmanned surface vessel (USV),

$$G(s) = \frac{\Psi(s)}{\Delta_R(s)} = \frac{\pi/20}{s(s + 1/20)}.$$

If we imagine that we are conducting an experiment in which we sinusoidally oscillate the rudder and measure the output yaw angle of the USV, we can let the input to $G(s)$ be the sinusoidal signal

$$\delta_R(t) = \frac{\pi}{12} \sin(\omega t) 1(t) \rightarrow \Delta_R(s) = \frac{\pi}{12} \cdot \frac{\omega}{(s^2 + \omega^2)}$$

K. D. von Ellenrieder, *Control of Marine Vehicles*, Springer Series on Naval Architecture, Marine Engineering, Shipbuilding and Shipping 9, https://doi.org/10.1007/978-3-030-75021-3_5

where $\omega$ is the frequency of oscillation. This would correspond to sinusoidal oscillations of the rudder with an amplitude of $\pm 15°$. Then,

$$\Psi(s) = G(s)\Delta_R(s) = \frac{\pi}{12} \cdot \frac{\pi/20}{s(s+1/20)} \cdot \frac{\omega}{(s^2 + \omega^2)},$$

$$= \frac{\pi^2}{12}\left\{\frac{1}{\omega s} - \frac{\omega/\left[\omega^2 + (1/20)^2\right]}{(s+1/20)}\right.$$

$$\left. + \frac{|G(j\omega)|}{(s^2 + \omega^2)}\left(s\sin[\angle G(j\omega)] + \omega\cos[\angle G(j\omega)]\right)\right\}.$$

Taking the inverse Laplace Transform of $\Psi(s)$, the corresponding yaw angle time response can be found as

$$\psi(t) = \frac{\pi^2}{12}\left\{\frac{1}{\omega} - \frac{\omega\exp(-t/20)}{\left[\omega^2 + \left(\frac{1}{20}\right)^2\right]} + |G(j\omega)|\sin[\omega t + \angle G(j\omega)]\right\}1(t).$$

As $(t \to \infty)$ the transient part of the time response

$$\psi_t(t) = -\frac{\pi^2\omega}{12\left[\omega^2 + \left(\frac{1}{20}\right)^2\right]}\cdot\exp\left(-\frac{t}{20}\right),$$

gradually decays, leaving the steady-state part of the response

$$\psi_{ss}(t) = \frac{\pi^2}{12}\left\{\frac{1}{\omega} + |G(j\omega)|\sin[\omega t + \angle G(j\omega)]\right\}.$$

Here $1/\omega$ is the constant, *DC component*, of the response. Thus, we expect the output heading angle of the USV to oscillate, with the same frequency as the input rudder angle, about a fixed angle $\pi^2/(12\omega)$. The oscillation frequency is related to its period as $\omega = 2\pi/T$. Some typical time responses for periods of $T = 5, 10, 15$ s are shown in Fig. 5.1. Table 5.1 gives the magnitude and phase shift of the sinusoidal component of the steady-state response for each different period of oscillation.

□

Example 5.1 illustrates a practical implication for the system identification of a stable system. Suppose the system transfer function $G(s)$ is not analytically available. One can still experimentally generate an approximate frequency response model over an interval of frequencies, even if the original system is nonlinear. The approach is to input a sequence of sinusoidal signals $r(t) = A(\omega)\sin(\omega t)$ and measure the steady-state response of the output, $y_{ss}(t) = B(\omega)\sin[\omega t + \phi(\omega)]$ for a given $B(\omega)$

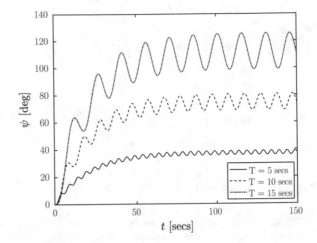

**Fig. 5.1** Time response of USV yaw angle to sinusoidal rudder angle inputs

**Table 5.1** Variation of the magnitude and phase of $G(j\omega)$ with rudder period $T$

| $T$ [sec] | 5 | 10 | 15 |
|---|---|---|---|
| $\omega$ [rad/s] | 1.26 | 0.63 | 0.42 |
| $|G(j\omega)|$ | 0.03 | 0.13 | 0.29 |
| $\angle G(j\omega)$ [deg] | −178 | −175 | −173 |

and $\phi(\omega)$. The measured data can be used to construct empirical models for the *magnitude frequency response*

$$|G(j\omega)| = \frac{B(\omega)}{A(\omega)},$$

and for the *phase frequency response*

$$\angle G(j\omega) = \phi(\omega).$$

Most control system analysis and design can be performed without an analytical model when the magnitude and phase frequency responses of the system are available.

## 5.2  The Nyquist Criterion

The use of frequency response techniques for controller design and analysis requires an understanding of how to characterize the stability of a system in the frequency domain. The Nyquist Criterion can be used to determine the stability of a feedback

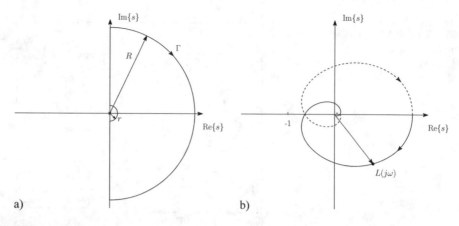

**Fig. 5.2** The Nyquist contour $\Gamma$ and the Nyquist plot. The Nyquist plot is the image of the loop transfer function $L(s)$ when $s$ traverses $\Gamma$ in the clockwise direction. The solid line corresponds to $\omega > 0$, and the dashed line to $\omega < 0$. The gain and phase at the frequency $\omega$ are $g = |L(j\omega)|$ and $\phi = \angle L(j\omega)$

system through analysis of the loop transfer function $L(s) = D(s)G(s)$ (Sect. 4.2). The Nyquist plot is a graphical tool that can be used to ascertain system stability.

### 5.2.1  Nyquist Plots

The Nyquist plot of the loop transfer function $L(s)$ is formed by tracing $s \in \mathbb{C}$ around the Nyquist "D contour," consisting of the imaginary axis combined with an arc at infinity connecting the endpoints of the imaginary axis. This contour, denoted as $\Gamma \in \mathbb{C}$, is illustrated in Fig. 5.2a. The image of $L(s)$ when $s$ traverses $\Gamma$ gives a closed curve in the complex plane and is referred to as the Nyquist plot for $L(s)$, as shown in Fig. 5.2b. Note that if the transfer function $L(s)$ goes to zero as $s$ gets large (the usual case), then the portion of the contour "at infinity" maps to the origin. Furthermore, the portion of the plot corresponding to $\omega < 0$ is the mirror image of the portion with $\omega > 0$.

There is a subtlety in the Nyquist plot when the loop transfer function has poles on the imaginary axis because the gain is infinite at the poles. To solve this problem, we modify the contour $\Gamma$ to include small deviations that avoid any poles on the imaginary axis, as illustrated in Fig. 5.2a (assuming a pole of $L(s)$ at the origin). The deviation consists of a small semicircle to the right of the imaginary axis pole location.

We now state the Nyquist condition for the special case where the loop transfer function $L(s)$ has no poles in the right half plane.

**Theorem 5.1** (Simplified Nyquist criterion) *Let $L(s)$ be the loop transfer function for a negative feedback system and assume that $L(s)$ has no poles in the open right* half-plane (Re$\{s\} \geq 0$). *Then the closed loop system is stable if and only if the closed contour given by* $\Omega = \{L(j\omega) : -\infty < \omega < +\infty\} \subset \mathbb{C}$ *has no net encirclements of the critical point* $s = -1$.

The Nyquist criterion does not require that $|L(j\omega_{\text{cross}})| < 1$ for all $\omega_{\text{cross}}$ corresponding to a crossing of the negative real axis. Rather, it says that the number of encirclements must be zero, allowing for the possibility that the Nyquist curve could cross the negative real axis and cross back at magnitudes greater than 1. One advantage of the Nyquist criterion is that it tells us how a system is influenced by changes of the controller parameters. For example, it is very easy to visualize what happens when the gain is changed since this just scales the Nyquist curve.

**Example 5.2** Consider the transfer function

$$L(s) = \frac{1}{\left(s + \frac{1}{2}\right)^2}.$$

To compute the Nyquist plot we start by evaluating points on the imaginary axis $s = j\omega$, which are plotted in the complex plane in Fig. 5.3, with the points corresponding to $\omega > 0$ drawn as a solid line and $\omega < 0$ as a dashed line. These curves are mirror images of each other. To complete the Nyquist plot, we compute $L(s)$ for $s$ on the outer arc of the Nyquist $D$-contour. This arc has the form $s = Re^{j\theta}$ for $R \to \infty$. This gives

$$L(Re^{j\theta}) = \frac{1}{\left(Re^{j\theta} + \frac{1}{2}\right)^2} \to 0 \quad \text{as} \quad R \to \infty.$$

The outer arc of the $D$-contour maps to the origin on the Nyquist plot. $\qquad\Box$

### 5.2.2 General Nyquist Criterion

Theorem 5.1 requires that $L(s)$ has no poles in the closed right half-plane. In some situations this is not the case and a more general result is required. Nyquist originally considered this general case, which we summarize as a theorem.

**Theorem 5.2** (Nyquist stability criterion) *Consider a closed loop system with the loop transfer function* $L(s)$ *that has* $P$ *poles in the region enclosed by the Nyquist contour. Let* $N$ *be the net number of clockwise encirclements of* $-1$ *by* $L(s)$ *when* $s$ *encircles the Nyquist contour* $\Gamma$ *in the clockwise direction. The closed loop system then has* $Z = N + P$ *poles in the right half-plane.*

The full Nyquist criterion states that if $L(s)$ has $P$ poles in the right half-plane, then the Nyquist curve for $L(s)$ should have $P$ counterclockwise encirclements

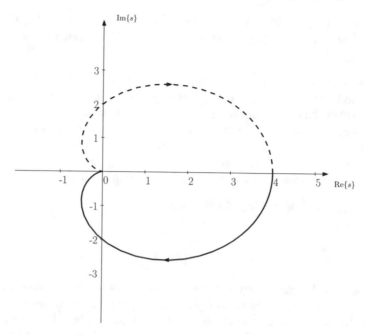

**Fig. 5.3** Nyquist plot of $L(s) = 1/(s + 1/2)^2$

of $-1$ (so that $N = -P$). In particular, this requires that $|L(j\omega_g)| > 1$ for some $\omega_g$ corresponding to a crossing of the negative real axis. Care has to be taken to get the right sign of the encirclements. The Nyquist contour has to be traversed clockwise, which means that $\omega$ moves from $-\infty$ to $\infty$ and $N$ is positive if the Nyquist curve winds clockwise. If the Nyquist curve winds counterclockwise, then $N$ will be negative (the desired case if $P \neq 0$).

As in the case of the simplified Nyquist criterion, we use small semicircles of radius $r$ to avoid any poles on the imaginary axis. By letting $r \to 0$, we can use Theorem 5.2 to reason about stability. Note that the image of the small semicircles generates a section of the Nyquist curve with large magnitude, requiring care in computing the winding number.

**Example 5.3** Consider the pitch stability of an autonomous underwater vehicle (AUV) moving at a constant speed $U$, as presented in Example 2.4. The nonlinear equation describing the pitch dynamics of the AUV (2.22), can be linearized about zero pitch ($\theta = 0$) to get

$$\ddot{\theta} = -\frac{U^2(Z_{\dot{w}} - X_{\dot{u}})}{(I_y - M_{\dot{q}})}\theta + u_c, \tag{5.1}$$

where $u_c$ is a control input. For a 2.5 meter long, 0.5 meter diameter AUV moving at $U = 1.5$ m/s the coefficients in (5.1) can be approximated as

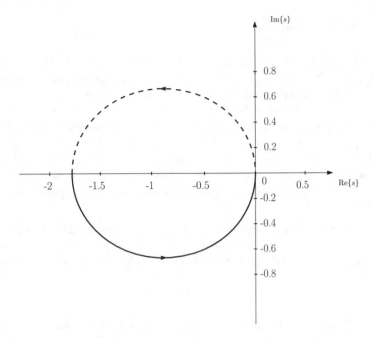

**Fig. 5.4** PD control of an AUV. Nyquist plot of the loop transfer function for gain $k = 4$

$$-\frac{U^2(Z_{\dot{w}} - X_{\dot{u}})}{(I_y - M_{\dot{q}})} \approx 9. \tag{5.2}$$

Thus, the linearized dynamics can be represented with the transfer function $G(s) = 1/(s^2 - 9)$. If we attempt to stabilize the AUV with a proportional-derivative (PD) controller having the transfer function $D(s) = k(s + 4)$. The loop transfer function is

$$L(s) = \frac{k(s + 4)}{s^2 - 9}.$$

The Nyquist plot of the loop transfer function is shown in Fig. 5.4. We have $L(0) = -4k/9$ and $L(\infty) = 0$. If $k > 9/4$, the Nyquist curve encircles the critical point $s = -1$ in the counterclockwise direction when the Nyquist contour $\Gamma$ is encircled in the clockwise direction. The number of encirclements is thus $N = -1$. Since the loop transfer function has one pole in the right half-plane ($P = 1$), we find that $Z = N + P = 0$ and the system is thus stable for $k > 9/4$. If $k < 9/4$, there is no encirclement and the closed loop will have one pole in the right half-plane. Note that systems in which increasing the gain leads from instability to stability, such as this one, are known as a *conditionally stable* systems. □

### 5.2.3 Stability Margins

A key advantage of the Nyquist Stability Criterion is that it not only indicates whether or not a system is stable, but it also (perhaps more importantly) gives quantitative measures of the system's relative stability, i.e., stability margins. These measures describe how far from instability the system is and its robustness to perturbations. The main idea is to plot the loop transfer function $L(s)$. An increase in controller gain simply expands the Nyquist plot radially. An increase in the phase of the controller twists the Nyquist plot. From the Nyquist plot we can easily determine the amount of gain or phase that can be added without causing the system to become unstable. Thus, the stability margins are conventionally characterized using gain margin (GM) and phase margin (PM).

*Gain margin*: The smallest amount the loop gain can be increased before the closed loop system becomes unstable. Suppose the closed loop system with the open loop transfer function $L(s)$ is stable. Let $\omega_g$ be the only *phase crossover frequency*, i.e.,

$$\angle L(j\omega_g) = -180°$$

and let $|L(j\omega_g)| < 1$. Then $L(j\omega_g) = -|L(j\omega_g)|$, and the closed loop system with an open loop transfer function $KL(s)$ will still be stable if $K < 1/|L(j\omega_g)|$ since $KL(j\omega_g) > -1$ and the Nyquist plot of $KL(j\omega)$ will have the same number of encirclements as $L(j\omega)$. On the other hand, the closed loop system with an open loop transfer function $KL(s)$ will become unstable if $K \geq 1/|L(j\omega_g)|$, since $KL(j\omega_g) \leq -1$ and the Nyquist plot of $KL(j\omega)$ will have a different number of encirclements compared to the Nyquist plot of $L(j\omega)$.

The gain margin is given by

$$\text{GM} = \frac{1}{|L(j\omega_g)|}.$$

In the case when the loop transfer function $L(s)$ has more than one phase crossover frequency, then the above argument carries over if $\omega_g$ is taken as the one so that $|L(j\omega_g)|$ is less than 1, but closest to 1.

*Phase margin*: The amount of phase lag required to reach the stability limit of the closed loop system. Again, suppose that the closed loop system with the open loop transfer function $L(s)$ is stable. Let $\omega_c$ be the only *gain crossover frequency*, i.e.,

$$|L(j\omega_c)| = 1.$$

Then $L(j\omega_c) = e^{j\angle L(j\omega_c)}$ and the closed loop system with an open loop transfer function $e^{-j\phi}L(s)$ will still be stable if $\angle L(j\omega_c) - \phi > -180°$ since $e^{-j\phi}L(j\omega_c) = e^{j(\angle L(j\omega_c) - \phi)}$ and the Nyquist plot of $e^{-j\phi}L(j\omega)$ will have the same number of encirclements as $L(j\omega)$. On the other hand, the closed loop system with an open loop transfer function $e^{-j\phi}L(s)$ will become unstable if $\angle L(j\omega_c) - \phi \leq -180°$

**Fig. 5.5** Stability margins
on the Nyquist plot. The gain
margin $GM$ corresponds to
the smallest increase in gain
that creates an encirclement,
and the phase margin $PM$ is
the smallest change in phase
that creates an encirclement.
The stability margin $SM$ is
the shortest distance to the
critical point $-1$. Only the
part for $0 \le \omega \le \infty$ is
shown, the other branch is
symmetric about the real axis

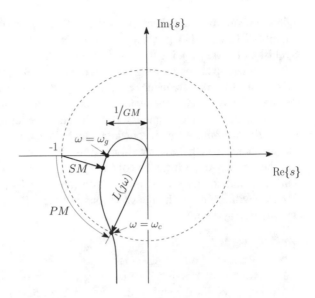

since $e^{-j\phi}L(j\omega_c) = e^{j(\angle L(j\omega_c)-\phi)}$ and the Nyquist plot of $e^{-j\phi}L(j\omega)$ will have a
different number of encirclements compared to the Nyquist plot of $L(j\omega)$.

The phase margin is given by

$$PM = 180° + \angle L(j\omega_c).$$

The gain and phase margins have simple geometric interpretations in the Nyquist
plot, as shown in Fig. 5.5, where we have plotted the portion of the curve corresponding to $\omega > 0$. The gain margin is given by the inverse of the distance to the nearest
point between $-1$ and 0 where the loop transfer function crosses the negative real
axis. The phase margin is given by the smallest angle on the unit circle between $-1$
and the loop transfer function. When the gain or phase is monotonic, this geometric
interpretation agrees with the formulas above.

A drawback with gain and phase margins is that both have to be given to guarantee
that the Nyquist curve is not close to the critical point. An alternative way to express
margins is by a single number, the *stability margin* SM, which is the shortest distance
from the Nyquist curve to the critical point. The gain and phase margins can also be
determined using a Bode plot, as will be shown below.

## 5.3 Bode Diagrams

In Sect. 5.2.1 it was shown that $|L(s)| \to 0$ as $s \to \infty$ (the usual case) so that the
portion of the Nyquist D-contour "at infinity" maps to the origin of the Nyquist
plot and that the portion of the plot for $\omega < 0$ corresponds to the portion for $\omega > 0$

reflected about the real axis of the $s$-plane. Thus, most of the information about the stability of the closed loop system can be determined by examining the portion of the Nyquist plot corresponding to $\omega > 0$. Imagine that the $\omega > 0$ portion of the Nyquist plot is "unwrapped" by plotting the magnitude $|L(j\omega)|$ versus frequency $\omega$ and the phase $\angle L(j\omega)$ versus frequency $\omega$ separately on Cartesian axes. When the frequency response of a system is represented in this way, usually using logarithmic scales for both the magnitude and frequency and a linear scale for the phase, the plot is known as a *Bode diagram*. As with root locus techniques, it is generally beneficial to be able to quickly sketch the approximate Bode diagram of a system by hand to explore how the stability or performance of the system varies as its parameters are changed, rather than obtaining a precise plot, which can be easily performed with a computer. Here, we will discuss techniques for constructing a Bode plot and for analyzing the stability of a system.

As will be seen in the following section, Bode techniques exploit some basic properties of logarithmic functions. The logarithmic function $g(s) = \log_b(s)$ is the inverse of the exponential function $f(s) = b^s$, where the positive constant $b$ is called the *base* of the logarithm. Thus, if

$$y = \log_b(s), \tag{5.3}$$

$$b^y = b^{\log_b(s)} = s. \tag{5.4}$$

Setting $y = 1$ in (5.4) gives $s = b$. Using this result in (5.3) gives $\log_b(b) = 1$. In control theory, the base 10 logarithm ($b = 10$) is typically used. The properties listed here are commonly used to construct Bode diagrams:

(a) Logarithms of a term $s$ raised to a power $n$ can be expressed as

$$\log_{10}\left(s^n\right) = n\log_{10}(s). \tag{5.5}$$

This property implies that

$$\log_{10}\left(10^n\right) = n\log_{10}(10) = n \tag{5.6}$$

and that

$$\log_{10}(1) = \log_{10}\left(10^0\right) = 0. \tag{5.7}$$

(b) The logarithm of the product of two terms is the sum of their logarithms

$$\log_{10}(fg) = \log_{10}(f) + \log_{10}(g). \tag{5.8}$$

This can be applied recursively to show that the logarithm of the product of $n$ terms $g_1, \ldots, g_n$ is simply the sum of the logarithms of each term

$$\log_{10}(g_1 \cdot g_2 \cdots g_{n-1} \cdot g_n) = \log_{10}(g_1) + \cdots + \log_{10}(g_n). \quad (5.9)$$

(c) The logarithm of the ratio of two terms is the difference of their logarithms

$$\log_{10}\left(\frac{f}{g}\right) = \log_{10}(f) - \log_{10}(g). \quad (5.10)$$

This can be applied recursively to show that the logarithm of $f$ divided by product of $n$ terms $g_1, \ldots, g_n$ is

$$\log_{10}\left(\frac{f}{g_1 \cdots g_n}\right) = \log_{10}(f) - \log_{10}(g_1) - \cdots - \log_{10}(g_n). \quad (5.11)$$

## 5.3.1 Constructing A Bode Diagram

Recall that in Sect. 4.2 it was shown that a transfer function of an LTI system is a complex variable that possesses both magnitude and direction (phase angle). Additionally, as shown in Example 4.1, transfer functions can be decomposed into a product of zeros, divided by a product of poles. For example, a transfer function with three zeros and three poles can be represented as

$$
\begin{aligned}
G(s) &= \frac{s_1 \cdot s_2 \cdot s_3}{s_4 \cdot s_5 \cdot s_6}, \\
&= \frac{r_1 e^{j\psi_1} \cdot r_2 e^{j\psi_2} \cdot r_3 e^{j\psi_3}}{r_4 e^{j\phi_4} \cdot r_5 e^{j\phi_5} \cdot r_6 e^{j\phi_6}}, \\
&= \frac{r_1 r_2 r_3}{r_4 r_5 r_6} e^{j[(\psi_1 + \psi_2 + \psi_3) - (\phi_4 + \phi_5 + \phi_6)]}.
\end{aligned}
$$

Note that the magnitude is

$$|G(s)| = \frac{r_1 r_2 r_3}{r_4 r_5 r_6}.$$

If we take the logarithm of the magnitude, we get

$$
\begin{aligned}
\log_{10}|G(s)| &= \log_{10}\left(\frac{r_1 r_2 r_3}{r_4 r_5 r_6}\right) \\
&= \log_{10} r_1 + \log_{10} r_2 + \log_{10} r_3 - \log_{10} r_4 - \log_{10} r_5 - \log_{10} r_6.
\end{aligned}
$$

For historical reasons, it is common in control engineering to represent the log magnitude in *decibels* (often represented with the acronym dB), which are defined as

$$|G|_{dB} := 20 \log_{10} |G|,$$

then
$$|G(s)|_{dB} = 20 \log_{10} r_1 + 20 \log_{10} r_2 + 20 \log_{10} r_3$$
$$- 20 \log_{10} r_4 - 20 \log_{10} r_5 - 20 \log_{10} r_6.$$

Note that the corresponding phase angle of the system is

$$\angle G(s) = [(\psi_1 + \psi_2 + \psi_3) - (\phi_4 + \phi_5 + \phi_6)].$$

Thus, we can see that we can simply add the separate components of the magnitude and phase plots coming from the poles and zeros of a transfer function, as well as from products of transfer functions, in order to construct a Bode diagram. Lastly, it is also common convention in control engineering practice to drop the "10" from $\log_{10}$ and simply use "log" to represent the base 10 logarithm.

*Bode Form:* Thus far, when analyzing the dynamic response (Chap. 3) and the root locus (Chap. 4) of a system it was convenient to represent the loop transfer function in the form

$$L(s) = K \frac{(s + z_1)(s + z_2) \dots}{(s + p_1)(s + p_2) \dots}.$$

However, when constructing Bode diagrams it is more convenient to represent the loop transfer function in "Bode" form

$$L(j\omega) = K_b \frac{(j\omega\tau_1 + 1)(j\omega\tau_2 + 1) \dots}{(j\omega\tau_a + 1)(j\omega\tau_b + 1) \dots}, \tag{5.12}$$

where $s$ is replaced by $j\omega$ and the $K_b$ is $K$ rescaled as

$$K_b = K \frac{z_1 \cdot z_2 \dots}{p_1 \cdot p_2 \dots}.$$

The main advantage of rewriting the loop transfer function in Bode form is that for Type 0 systems, the magnitude of the transfer function in dB is simply $20 \log |K_b|$ at low frequencies ($\omega \to 0$). As Bode diagrams are generally constructed from left to right (low frequency to high frequency), use of the Bode form permits one to avoid having to continuously redraw the plot, shifting it upwards or downwards as new terms are added.

For most types of systems, there are three types of terms, which form the basic building blocks of a Bode diagram:

(a) The first class of terms includes poles and zeros located at the origin. Let

$$G(j\omega) = K(j\omega)^m$$

where $m$ is a positive or negative integer. Then, for $K > 0$

$$|G(j\omega)|_{dB} = 20 \log K + 20m \log |\omega|, \quad \angle G(j\omega) = m\frac{\pi}{2}. \tag{5.13}$$

Note that the magnitude plot is straight line with a slope of $20m$ [dB/decade], where a *decade* is a factor of 10 on the logarithmic scale. A simple approach for drawing the magnitude curve on the Bode diagram is to plot the point $20 \log K$ at $\omega = 1$ and then draw a line with slope $20m$ [dB/decade] through the point. The phase is constant and can be drawn as a horizontal line on the phase diagram. The Bode diagrams for this system with $K = 1$ are illustrated in Fig. 5.6.

(b) The second class of terms includes poles and zeros, which may be repeated $m$ times, located away from the origin. Consider the system

$$G(j\omega) = (j\omega\tau + 1)^m.$$

The magnitude and phase frequency responses are given by

$$|G(j\omega)| = [1 + (\omega\tau)^2]^{m/2} \quad \text{and} \quad \angle G(j\omega) = m \tan^{-1}(\omega\tau).$$

At the *corner frequency* (also called the *breakpoint*), $\omega = 1/\tau$, we have

$$|G(j)|_{dB} = 20 \cdot \frac{m}{2} \log(2) = 3m \text{ [dB]}, \quad \angle G(j) = m \tan^{-1}(1) = m\frac{\pi}{4}.$$

On the Bode diagram, we have

$$|G(j\omega)|_{dB} = 20 \cdot \frac{m}{2} \log[1 + (\omega\tau)^2].$$

For very low frequencies $\omega\tau \ll 1$, $[1 + (\omega\tau)^2] \to 1$, so that

$$\lim_{\omega\tau \to 0} |G(j\omega)|_{dB} \to 0 \quad \text{and} \quad \lim_{\omega\tau \to 0} \angle G(j\omega) \to 0°.$$

For very high frequencies $\omega\tau \gg 1$, $[1 + (\omega\tau)^2] \to (\omega\tau)^2$ making

$$\lim_{\omega\tau \to \infty} |G(j\omega)|_{dB} \to 20m \log(\omega\tau) \quad \text{and} \lim_{\omega\tau \to 0} \angle G(j\omega) \to m\frac{\pi}{2}.$$

Thus, the plot of $|G(j\omega)|_{dB}$ can be approximated by two straight lines along the asymptotes for $\omega\tau \ll 1$ and $\omega\tau \gg 1$, as shown in Fig. 5.7. The intersection (or corner) of these two straight lines is the breakpoint $\omega = 1/\tau$. Similarly, the plot of $\angle G(j\omega)$ can also be approximated using straight lines.

(c) The third class of terms includes non-repeated poles and zeros of the form

$$G(j\omega) = \left[ \left( \frac{j\omega}{\omega_n} \right)^2 + 2\zeta \left( \frac{j\omega}{\omega_n} \right) + 1 \right]^{\pm 1}, \quad 0 \le \zeta \le 1.$$

At low frequencies $\omega/\omega_n \ll 1$, $[(j\omega/\omega_n)^2 + 2\zeta(j\omega/\omega_n) + 1] \to 1$ so that

$$\lim_{\omega/\omega_n \to 0} |G(j\omega)|_{dB} \to 0 \quad \text{and} \quad \lim_{\omega/\omega_n \to 0} \angle G(j\omega) \to 0°.$$

At high frequencies $\omega/\omega_n \gg 1$, $[(j\omega/\omega_n)^2 + 2\zeta(j\omega/\omega_n) + 1] \to (j\omega/\omega_n)^{\pm 2}$ giving

$$\lim_{\omega/\omega_n \to \infty} |G(j\omega)|_{dB} \to \pm 40 \log(\omega/\omega_n) \quad \text{and} \quad \lim_{\omega/\omega_n \to 0} \angle G(j\omega) \to \pm \pi.$$

Here, it can be seen that the corner frequency is the natural frequency $\omega_n$. The low frequency magnitude asymptote is a horizontal line of 0 dB and the high frequency magnitude asymptote is a line with a slope of $+40$ dB/decade for a pair of complex zeros or $-40$ dB/decade for a pair of complex poles. Similarly, the low frequency phase is a horizontal line of 0° and the high frequency phase asymptote is a horizontal line at $+180°$ for a pair of complex zeros or $-180°$ for a pair of complex poles.

Figure 5.8 illustrates the Bode diagrams for a second-order complex pole for various damping ratios. Bode diagrams for transfer functions with complex poles are more difficult to construct since the magnitude and phase responses depend on damping ratio near the corner frequency. The magnitude frequency response has a peak if $0 < \zeta < 1/\sqrt{2}$ and decreases monotonically if $\zeta \ge 1/\sqrt{2}$. At the corner frequency the magnitude is

$$|G(j\omega_n)|_{dB} = 20 \log[(2\zeta)^{\pm 1}].$$

When $\zeta \ne 0$ the frequency of the resonant peak $\omega_r$ differs slightly from the corner frequency and is given by

$$\omega_r = \omega_n \sqrt{1 - 2\zeta^2},$$

as explained in Sect. 5.5.1.

As mentioned above, the separate components of the magnitude and phase plots coming from the poles and zeros of a transfer function, as well as from products of transfer functions, can simply be added to construct a Bode diagram. Sketching the magnitude and phase frequency responses in such cases requires that the separate components be combined into a composite curve.

When plotting the magnitude frequency response, the slope of the asymptotes will be the sum of the slopes of the individual pole and zero components. As can be seen from (5.13), if there are $m$ poles or $m$ zeros at $j\omega = 0$, the slope of the lowest-frequency part of the composite curve will be $20m$ dB/decade and the magnitude at $\omega = 1$ will be $20 \log K$ dB. Thus, the procedure for constructing the magnitude

**Fig. 5.6** Bode diagrams for $G(j\omega) = (j\omega)^m$, when $m = \{+1, -1, -2\}$

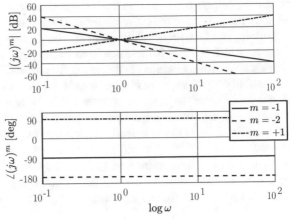

**Fig. 5.7** Bode diagrams for the first order system $(\tau s + 1)$, when $\tau = 1$. Asymptotes are indicated with dashed lines and the solid curves represent the true magnitude and phase. Note that the true magnitude differs from the composite asymptote by 3 dB at the corner frequency and that the true phase differs from the phase of the composite asymptote by 5.7° one decade above and one decade below the corner frequency

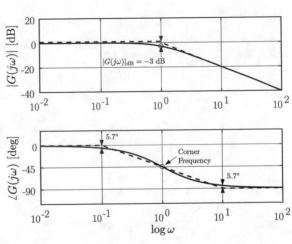

**Fig. 5.8** Frequency response of a second-order system

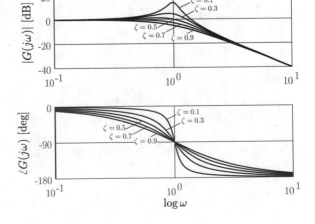

frequency response is to start at the left (low frequency portion) of the diagram first plotting the portion of the curve corresponding to the gain and any poles or zeros at the origin, and then sequentially changing the slope of the asymptote at each corner frequency as frequency is increases towards the right, and then finally sketching the actual curve using the concepts discussed for classes (b) and (c) above.

Similarly, the composite phase curve is the sum of the individual phase curves. A quick sketch of the composite curve can be drawn by starting below the lowest corner frequency and setting the phase equal to $\pm m\pi/2$ for the $m$ poles or zeros at the origin. The phase is then sequentially stepped at each corner frequency by $\pm m\pi/2$ for each $(j\omega\tau + 1)^{\pm m}$ term and by $\pm\pi$ for each complex pole/zero pair from left to right corresponding to increasing frequency. As with the magnitude frequency response, an approximate sketch of actual curve is then drawn using the concepts discussed for classes (b) and (c) above.

### Guidelines for Constructing a Bode Diagram

On the magnitude frequency response plot:

1. Write the transfer function in Bode form, as shown in (5.12).
2. Determine $m$ for the $K(j\omega)^m$ term. Plot the point $20\log K$ at $\omega = 1$ and then draw a line with slope $20m$ [dB/decade] through the point.
3. Create the composite magnitude asymptotes by extending the low frequency asymptote until the first corner frequency, then stepping the slope by $\pm 20m$ dB/decade or by $\pm 40$ dB/decade depending on whether the corner frequency corresponds to a $(j\omega\tau + 1)^m$ term or to a complex second order term in the numerator or denominator, and continuing through all subsequent corner frequencies in ascending order.
4. Sketch the approximate magnitude curve by increasing from the asymptotes by $\pm 3m$ dB for each $(j\omega\tau + 1)^m$ term. At corner frequencies corresponding to complex poles or zeros sketch the resonant peak or valley with $|[(j\omega/\omega_n)^2 + 2\zeta(j\omega/\omega_n) + 1]^{\pm 1}|_{\omega=\omega_n}|_{dB} = 20\log[(2\zeta)^{\pm 1}]$.

On the phase frequency response plot:

1. Plot the low frequency phase asymptote as a horizontal line at $m\pi/2$ corresponding to the $(j\omega)^m$ term.
2. Sequentially step the phase at each corner frequency by $\pm m\pi/2$ for each $(j\omega\tau + 1)^m$ term and by $\pm\pi$ for each complex pole/zero pair from left to right corresponding to increasing frequency.
3. Locate the asymptotes for each individual phase curve so that their phase change corresponds to the steps in phase from the approximate curve in the previous step. Sketch each individual phase curve using the asymptotes as a guide.
4. Graphically add the phase curves.

**Example 5.4** The magnitudes and phases from each transfer function in a product of transfer functions can simply be added to construct a Bode diagram. Sketching the magnitude and phase frequency responses simply requires that the separate components be combined into a composite curve. For example, consider the Langrangian float presented in Example 4.8. A lead compensator was designed to control its depth using root locus design techniques. The loop transfer function of the controlled system is $D(s)G(s)$, thus

$$\log|D(j\omega)G(j\omega)| = \log|D(j\omega)| + \log|G(j\omega)|,$$
$$\angle D(j\omega)G(j\omega) \quad = \angle D(j\omega) + \angle G(j\omega).$$

The Bode diagrams of the controlled system can be obtained by graphically adding the separate Bode diagrams of the controller $D(s)$ and the plant $G(s)$.

The transfer functions

$$D(s) = \frac{118(s + 0.133)}{s + 442}$$
$$G(s) = \frac{1}{s^2}$$

are plotted separately as dashed lines in Fig. 5.9. The composite system is shown as a thick solid curve. It can be seen how the separate magnitude and phase frequency response diagrams from the controller and the plant simply add to give the Bode diagram of the complete system. The construction of a composite Bode diagram can also be done by hand relatively easily. Straight line approximations to each curve, such as those shown in Fig. 5.7, can be graphically added to produce the magnitude frequency response. For example, the approximate magnitude plot of $D(s)G(s)$ can be obtained by first shifting the entire line $G(s) = 1/s^2$ downward by $20 * \log|D(j\omega)|_{\lim \omega \to 0}$, then changing the slope of the resulting line to $-20$ dB/decade at the compensator zero $\omega = 0.133$ rad/s and then changing the slope of the resulting curve again at the compensator pole $\omega = 442$ rad/s back to the original slope of $-40$ dB/decade. Straight line approximations for the phase frequency response can similarly be constructed, but require substantial care when used for design. Intermediate phase angles of the composite curve can be numerically calculated at discrete frequencies where it is expected to change rapidly in order to ensure that the composite phase curve is accurate.                                                                     □

## 5.4 Assessing Closed Loop Stability from the Bode Diagram

The stability of a closed loop system can be determined from the Bode diagram of its loop transfer function $L(j\omega)$. Recall from Sect. 5.2.3 that the gain margin (the smallest factor by which the loop gain can be increased before the closed loop system becomes unstable)

**Fig. 5.9** Constructing a
composite Bode diagram
(dark solid curves) from the
separate frequency responses
(dashed curves) of two
transfer functions
(Example 5.4)

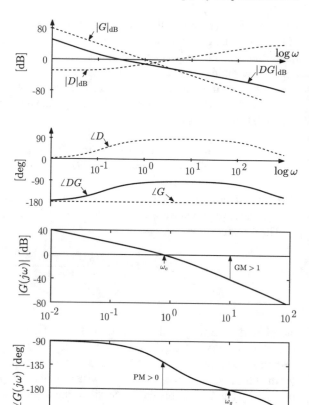

**Fig. 5.10** Stability margins
on a Bode diagram. The
system is stable, as $GM > 1$
and $PM > 0$

$$GM = \frac{1}{|L(j\omega_g)|},$$

is measured at the phase cross over frequency ($\omega_g$, where $\angle L(j\omega_g) = -180°$) and
that the phase margin (the phase lag at which the stability limit is reached)

$$PM = 180° + \angle L(j\omega_c),$$

is measured at the gain cross over frequency ($\omega_c$, where $|L(j\omega_c)|_{dB} = 0$). By identi-
fying these frequencies on the Bode diagram of a system GM and PM can be directly
measured, as shown in Fig. 5.10. An example of the margins for an unstable system
is shown in Fig. 5.11.

However, care must be taken when determining stability margins from the Bode
diagram. The above approach only holds for systems where increasing the loop gain
leads to instability. For conditionally stable systems (unstable systems that can be
stabilized by increasing the loop gain, see Example 5.3), or for systems in which there

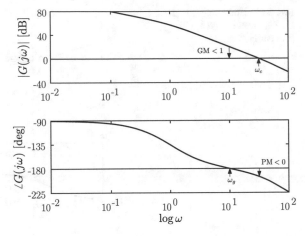

**Fig. 5.11** Stability margins on a Bode diagram. The system is unstable, as $GM < 1$ and $PM < 0$

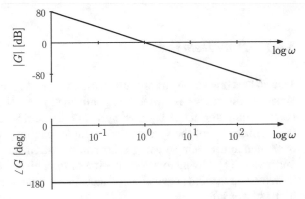

**Fig. 5.12** Bode diagram of the Langrangian float with proportional controller of gain $K = 1$

are multiple resonances and the magnitude crosses 0 dB multiple times, and/or the phase crosses $-180°$ several times, analysis should be performed using the Nyquist Stability Criterion (Sect. 5.2).

**Example 5.5** Returning to the Lagrangian float discussed in Examples 4.8 and 5.4, we can see that if the float were controlled by a simple proportional controller of gain $K$, the Bode diagram of the system would be as shown in Fig. 5.12. The gain crossover frequency is $\omega_c = 1/\sqrt{K}$ and PM$> 0$ for any value of $K$ (the gain margin is said to be infinite, as is can be increased without making the system unstable). Here, the system is actually metastable with a damping ratio of $\zeta = 0$. The output of the system will consist of undamped sinusoidal oscillations. This can also be seen from its root locus Fig. 4.19, where the closed loop roots are on the imaginary axis for all values of $K$.

When the lead compensator designed in Example 4.8 is used, the phase margin is increased to PM= 66° (Fig. 5.13). The gain margin is still infinite. ☐

**Fig. 5.13** Bode diagram of
the Langrangian float with
lead compensator

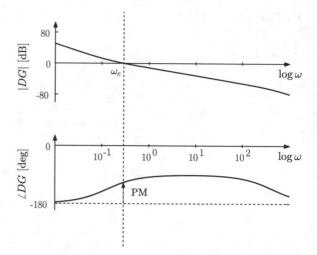

## 5.4.1  Time Delays

Time delays (also called transportation delays) are common in many marine systems.
When a control loop is closed using information received from a remote sensor or
control computer, the delay arising from communications constraints can degrade
the performance of the system. For example, one may want to control the position
of an unmanned vehicle using sensor information that is acoustically transmitted
underwater [1]. Owing to a combination of long distances, the finite speed of sound
in water, and the finite bandwidth of acoustic communications, time delays can occur
in the sensor information.

The analysis of LTI, unity feedback systems with time delays using frequency
response techniques is fairly straightforward. A pure time delay can be modeled as

$$G_d(s) = \exp^{-sT_d},  \tag{5.14}$$

where $T_d$ is the delay time. The associated magnitude and phase frequency response
are given by

$$|G_d(j\omega)| = 1, \quad \text{and} \quad \angle G_d(j\omega) = -\omega T_d.$$

Thus, the magnitude of the time delay is constant, but the delay will contribute a
negative phase to any input signal, which will grow without bound as the frequency
increases $\omega \rightarrow \infty$. A consequence of this is that time delays can negatively affect the
performance of a system by decreasing its phase margin, possibly also destabilizing
the system by making PM< 0.

**Example 5.6** Consider the station keeping of a small unmanned surface vehicle
using an external underwater acoustic positioning sensor system. It is assumed that
the longitudinal axis of the vessel is oriented along the same direction as the current

**Fig. 5.14** Effect of sensor time delays of $T_d = \{0, 0.5, 3\}$ s on the frequency response of the station keeping vessel. As the time delay increases, the phase margin decreases. The system is unstable, PM$< 0$, when $T_d = 3$ s

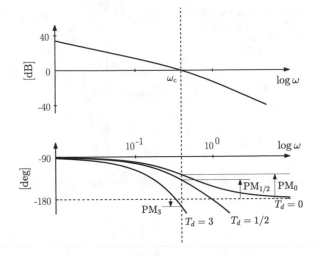

and that the vehicle is maintaining its position along this axis only. The transfer function describing the motion is

$$G(s) = \frac{1/2}{s(s + 1/2)}.$$

In the absence of any communications delays in the sensing system $T_d = 0$ s, a proportional controller with gain $K = 1/2$ can be used to stably control the position of the vessel with a phase margin of PM $= 52°$ (Fig. 5.14) and the time response shown in Fig. 5.15. When the localization sensor is about 3 km from the vessel, communications delays can be as long as 2 s [1]. As the time delay increases, the overshoot in the time response also increases (Fig. 5.15). As can be seen from the Bode diagrams in Fig. 5.14, this corresponds to a decreasing phase margin. The phase of the system can be examined to show that when the time delay is greater than 2.3 s, PM$< 0$ and the closed loop system is unstable. □

## 5.5 Dynamic Response from the Bode Diagram

The step response of a stable first order system is characterized by an exponential rise or decay. The *time constant* of the response is related to the corner frequency on the system's Bode diagram, as

$$\tau = \frac{1}{\omega}.$$

As discussed in Sect. 3.1.2, the time response of second order systems, as well as that of many higher-order systems, is dominated by a single pair of complex poles.

**Fig. 5.15** Time response of
station keeping vessel for
sensor time delays of
$T_d = \{0, 0.5\}$ s

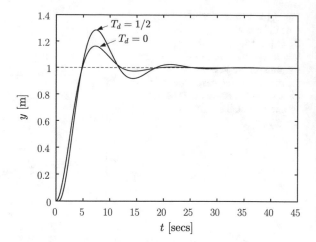

In such cases, the dynamic response of the closed loop system (rise time $t_r$, settling
time $t_s$, and peak overshoot $M_p$) is related to its damping ratio $\zeta$ and the natural
frequency $\omega_n$. Since the damping ratio and natural frequency of a system can also be
ascertained from its frequency response, the step response of a closed loop system
can also be inferred from a Bode diagram of its loop transfer function.

Consider a second-order system with loop transfer function

$$L(s) = \frac{\omega_n^2}{s(s + 2\zeta\omega_n)}. \tag{5.15}$$

Since $\angle L(j\omega) \geq -\pi, \forall\omega$, GM $= \infty$. We can solve the relation $|L(j\omega_c)| = 1$ to get

$$\omega_c = \omega_n\sqrt{\sqrt{1 + 4\zeta^4} - 2\zeta^2}. \tag{5.16}$$

Thus, the PM is given by

$$PM = 180° + \angle L(j\omega_c) = \tan^{-1}\left[\frac{2\zeta}{\sqrt{\sqrt{1 + 4\zeta^4} - 2\zeta^2}}\right].$$

The relationship between PM and $\zeta$ is shown in Fig. 5.16. For $0 \leq \zeta \leq 0.6$, the
damping ratio $\zeta$ varies almost linearly with the PM:

$$\zeta \approx \frac{PM \text{ [deg]}}{100}. \tag{5.17}$$

This formula is often used in frequency-domain controller design to estimate system
damping and overshoot.

**Fig. 5.16** The relationships among phase margin, damping ratio, and percentage overshoot for a second-order system

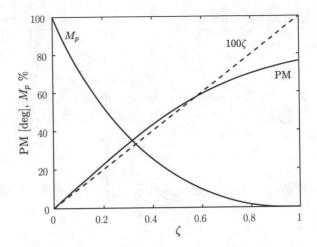

Thus, from a frequency response diagram one can determine $\omega_c$ and PM. In turn, using (5.16) and (5.17), the damping ratio $\zeta$ and natural frequency $\omega_n$ can be calculated. Now equipped with both $\zeta$ and $\omega_n$, one can then use the relations in (3.2), (3.6) and (3.7) to determine the step response of the closed loop system in terms of its rise time, peak overshoot and settling time.

### 5.5.1  Closed Loop Frequency Response

When the loop transfer function in (5.15) is the forward part of a unity feedback system (see Fig. 3.17), the closed loop transfer function from $R$ to $Y$

$$T(s) = \frac{Y(s)}{R(s)} = \frac{L(s)}{1 + L(s)} = \frac{\omega_n^2}{s^2 + 2\zeta\omega_n s + \omega_n^2} \tag{5.18}$$

has the canonical form of a second order system (3.1).

Important information about the closed loop time response of a system can be qualitatively determined from the system's closed loop frequency response. Further, as will be shown shortly, bandwidth of the closed loop system, which is a measure of its ability to track a desired input while also rejecting high frequency noise, can be estimated from the open loop frequency response.

Figure 5.17 illustrates a typical closed loop frequency response $T(j\omega)$. Important features of the response include the resonant peak $M_r$, resonant frequency $\omega_r$, bandwidth $\omega_b$ and cutoff rate, which are defined as follows:

(1) *Resonant Peak $M_r$*: The maximum value of $T(j\omega)$. This is an indicator of the relative stability (or damping) of the system. A relatively large $M_r$ indicates a lightly damped system.

**Fig. 5.17** Closed loop
frequency response

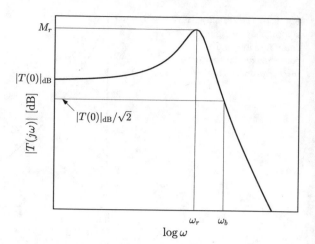

(2) *Peak Frequency $\omega_r$*: The frequency at which $M_r = |T(j\omega_r)|$.

(3) *Bandwidth $\omega_b$*: The smallest frequency such that $|T(j\omega_b)| = |T(j0)|/\sqrt{2}$. This frequency indicates that the system will pass essentially all signals in the band $[0, \omega_b]$ without much attenuation. This is usually used as a qualitative indicator of rise time (as well as settling time which also strongly depends on $M_r$). It also gives an indication of the system's ability to reject high-frequency noise.

(4) *Cutoff rate*: The slope of $|T(j\omega)|$ beyond the bandwidth. A large (negative) slope indicates a good noise rejection capability.

The value of the resonant peak $M_r$ is the maximum of $|T(j\omega)|$. Thus, when the closed loop transfer function in (5.18) is differentiated with respect to $\omega$ and set to zero, e.g.

$$\frac{d|T(j\omega)|}{d\omega} = 0,$$

the resulting equation can be solved to find the resonant frequency

$$\omega_r = \omega_n\sqrt{1 - 2\zeta^2}, \quad 0 < \zeta \le \frac{1}{\sqrt{2}}.$$

The range of $\zeta$ in the solution above is limited because the frequency must be real valued (not a complex number). The closed loop frequency response $|T(j\omega)|$ will not have a resonant peak if $\zeta > 1/\sqrt{2}$. Substituting the solution for $\omega_r$ back into (5.18) gives the value of the resonant peak

$$M_r = \frac{1}{2\zeta\sqrt{1 - \zeta^2}}, \quad 0 < \zeta \le \frac{1}{\sqrt{2}}.$$

**Fig. 5.18** Relationship between the value of the resonant peak $M_r$ and the peak overshoot to a step input $M_p$ for a standard second-order system with $0 < \zeta \le 1/\sqrt{2}$

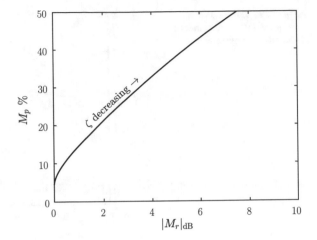

Since the peak overshoot $M_p$ of the closed loop system's step response is also a function of the damping ratio $\zeta$ (Sect. 3.1.2), $M_r$ can be used to determine $M_p$. The relationship is shown in Fig. 5.18.

To find the bandwidth $\omega_b$, (5.18) can be solved for the frequency at which

$$|T(j\omega)| = \frac{|T(0)|}{\sqrt{2}} = \frac{1}{\sqrt{2}},$$

giving

$$\omega_b = \omega_n \sqrt{1 - 2\zeta^2 + \sqrt{4\zeta^4 - 4\zeta^2 + 2}}. \tag{5.19}$$

We can now use (5.19) and (5.16) to develop a relation between the open loop gain crossover frequency $\omega_c$ and the closed loop bandwidth $\omega_b$,

$$\omega_b = \frac{\sqrt{1 - 2\zeta^2 + \sqrt{4\zeta^4 - 4\zeta^2 + 2}}}{\sqrt{\sqrt{1 + 4\zeta^4} - 2\zeta^2}} \cdot \omega_c.$$

The ratio $\omega_b/\omega_c$ is plotted in Fig. 5.19 for a range of $\zeta$; it can be seen that the ratio is almost constant over a fairly large range (roughly $0 < \zeta \le 1/\sqrt{2}$), where

$$\frac{\omega_b}{\omega_c} \approx 1.56.$$

This approximation can be helpful in control system design as it permits one to estimate a closed loop system's bandwidth using the gain crossover frequency from its open loop Bode diagram.

*Note:* For a first order system with closed loop transfer function

**Fig. 5.19** Variation of $\omega_b/\omega_c$ with $\zeta$ for a standard second-order system

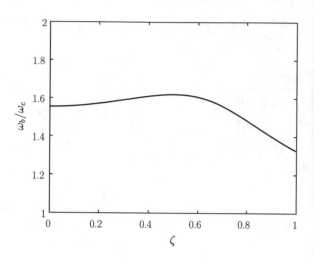

$$T(s) = \frac{1}{\tau s + 1}$$

The bandwidth corresponds to the corner frequency, $\omega_b = 1/\tau$. There is no peak, and the cutoff rate is $-20$ [dB/ decade].

## 5.6   Steady-State Response from the Bode Diagram

Suppose the Bode diagram of a unity feedback system with loop transfer function $L(s) = D(s)G(s)$ is given. Recall from Sect. 3.4, that the steady state error coefficient of a Type $k$ system is given by

$$K = \lim_{s \to 0} s^k D(s)G(s) = \lim_{s \to 0} s^k L(s).$$

Thus, at low frequencies, the loop transfer function can be approximated as

$$\lim_{\omega \to 0} L(j\omega) \sim \frac{K}{(j\omega)^k}.$$

With this approximation, the steady-state error constants $K_p$ ($k = 0$), $K_v$ ($k = 1$), and $K_a$ ($k = 2$) can be determined graphically from the open loop Bode magnitude diagram, as shown in Fig. 5.20.

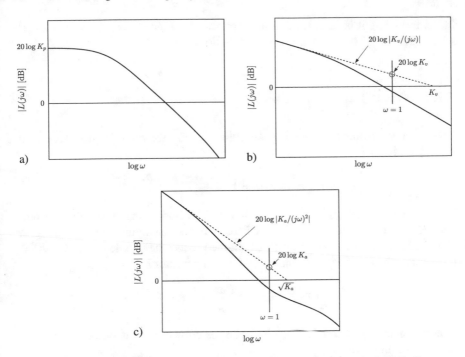

**Fig. 5.20** Determination of steady-state error constants from open loop magnitude Bode diagrams: **a** $K_p$ can be determined by reading $|L(j\omega)|_{dB}$ at low frequency, **b** $K_v$ and c) $K_a$ can be determined from the intersection between the low frequency asymptote of $|L(j\omega)|_{dB}$ (dashed lines) and the 0 dB line

## 5.7   Controller Design in the Frequency Domain

Design techniques using root locus methods were presented in Sect. 4.3. Generally, an advantage to using root locus controller design methods is that one can quickly explore how the closed loop roots of the system (and hence its stability and time response) vary as the loop gain or other system parameters are changed. However, it can sometimes be easier to graphically pick stability margins and to understand the bandwidth limitations of a system using frequency response techniques. In this regard, root locus and frequency response design techniques complement one another. The control designer can choose to apply one method over another, or to use them both, as best suits the problem at hand. The ability to switch between design methods can permit one to reach the best compromise in selecting controller parameters and to gain insight into the ramifications of alternate designs. Of course, when a mathematical model of the plant to be controlled is not available, frequency response controller design techniques can still be applied using an experimentally generated Bode diagram of the open loop system, whereas root locus techniques cannot. Here, we shall explore some simple design techniques using Bode diagrams.

In the following it is assumed that the loop transfer function is $L(s) = D(s)G(s)$, where $D(s)$ is the compensator transfer function.

## 5.7.1 Phase Lag Controllers

For frequency response phase lag controller design, we consider the Bode form analog of (4.19)

$$D(s) = \frac{K(s/b + 1)}{s/a + 1}, \quad 0 < a < b,$$

where $D(0) = K$ and $|D(j\omega)|_{\omega \to \infty} = Ka/b$. The Bode diagram of this controller with $K = 1$ is shown in Fig. 5.21, where only the asymptotes of the controller magnitude frequency response, rather than the precise magnitude curve, are shown for clarity. The phase lag contributed by the compensator is

$$\angle D(j\omega) = \tan^{-1} \frac{\omega}{b} - \tan^{-1} \frac{\omega}{a}.$$

The maximum lag occurs halfway between the corner frequencies of the compensator's pole and zero, e.g. at the location

$$\frac{\log a + \log b}{2} = \frac{1}{2} \log(ab) = \log \sqrt{ab} \tag{5.20}$$

on the logarithmic frequency axis of the Bode diagram (Fig. 5.21).

As pointed out in Sect. 4.3.3, a phase lag controller is a gain compensator. The useful effect of the compensator is not the phase lag, but the fact that the compensation increases the low frequency gain to improve the system's steady state response, while also attenuating the system's response near the desired crossover frequency. If necessary, the attenuation of the magnitude response from the lag controller can be used to reduce the crossover frequency, thereby increasing the phase margin to a satisfactory level and reducing the closed loop bandwidth of the system. The following procedure can be used to design a phase lag controller.

**Phase lag controller design steps**

Step 1    Find $K$ so that the DC gain requirements of the open loop system are satisfied. The controller gain $K$ should be chosen to satisfy steady-state error requirements on tracking, disturbance rejection, etc.

Step 2    Determine the desired crossover frequency $\omega_c$ and $\text{PM}_{\text{des}}$ (Fig. 5.22). (Recall that the crossover frequency is related to the rise time and settling time, and the PM is related to the overshoot of the system response.)

Step 3    Plot the Bode diagram of $KG(s)$.

**Fig. 5.21** Bode diagram of a phase lag controller $D(s) = \frac{s/b+1}{s/a+1}$

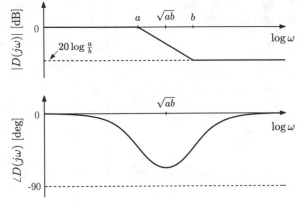

**Fig. 5.22** Phase lag controller design

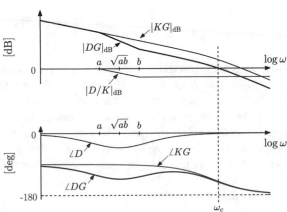

**Step 4** Check the phase plot of $KG(s)$ at the desired crossover frequency $\omega_c$. If $180° + \angle KG(j\omega_c) \geq PM_{des} + 5°$, then a phase lag controller can be designed to satisfy the design specifications. (The reason for adding $5°$ is that we expect the controller $D(s)$ to contribute about $-5°$ at $\omega_c$ when the controller parameters are chosen as below.)

**Step 5** Pick

$$b \approx \omega_c/10, \quad a = \frac{b}{K|G(j\omega_c)|}.$$

**Step 6** Finally, a phase lag controller is given by

$$D(s) = \frac{K(s/b+1)}{s/a+1}.$$

*Proportional-Integral (PI) Control*: Consider a PI controller of the form:

**Fig. 5.23** Bode diagram of a
PI controller $D(s) = \frac{s/b+1}{s}$

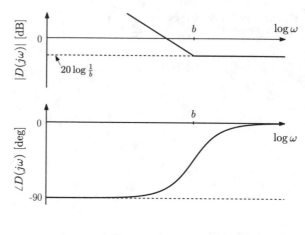

$$D(s) = \frac{K(s/b + 1)}{s}.$$

Note that a PI controller increases the system type by 1 with $D(0) = \infty$ and
$|D(j\omega)|_{\omega \to \infty} = K/b$. The Bode diagram of this controller with $K = 1$ is shown in
Fig. 5.23. A desirable aspect of this controller is that it has infinite gain for $\omega \to 0$,
which causes the steady state error to vanish. This is accomplished at the cost of a
phase decrease below the controller's corner frequency $\omega = b$. In order to prevent
the phase margin of the system from being adversely affected, $b$ is typically selected
to be significantly less than the crossover frequency $b \ll \omega_c$.

**PI controller design steps**

Step 1    Determine the desired crossover frequency $\omega_c$ and phase margin $\mathrm{PM}_{\mathrm{des}}$.
    (Recall that the crossover frequency is related to the rise time and settling time,
    and the PM is related to the overshoot of the system response.)
Step 2    Plot the Bode diagram of $KG(s)$ for any given $K$ (for example, $K = 1$).
Step 3    Check the phase plot of $KG(s)$ at the desired crossover frequency $\omega_c$. If
    $180° + \angle KG(j\omega_c) \geq \mathrm{PM}_{\mathrm{des}} + 5°$, then a PI controller can be designed to satisfy
    the design specifications. (The reason for adding $5°$ is that we expect the con-
    troller $D(s)$ to contribute about $-5°$ at $\omega_c$ when the controller parameters are
    appropriately chosen as below.)
Step 4    Choose

$$b \approx \omega_c/10, \quad K = \frac{b}{|G(j\omega_c)|}.$$

Step 5    Finally, a PI controller is given by

$$D(s) = \frac{K(s/b + 1)}{s}.$$

**Fig. 5.24** Phase lead controller
$$D(s) = \frac{s/b + 1}{s/a + 1}, \ (b < a)$$

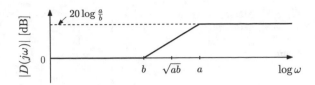

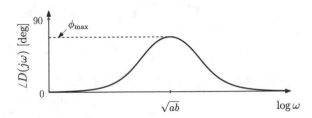

*Note*: When designing lag or PI controllers it is not generally desirable to select a value of $b$ that is significantly smaller than $b \approx \omega_c/10$. Doing so on a real physical systems tends to require the use of large actuators.

### 5.7.2 Phase Lead Controllers

Consider a first-order phase lead controller of the form

$$D(s) = \frac{K(s/b + 1)}{s/a + 1}, \quad 0 < b < a.$$

The Bode diagram of this controller with $K = 1$ is shown in Fig. 5.24. The phase lead contributed by the compensator is

$$\angle D(j\omega) = \tan^{-1}\frac{\omega}{b} - \tan^{-1}\frac{\omega}{a}.$$

The maximum phase (see Fig. 5.25) is given by

$$\phi_{max} = \sin^{-1}\left(\frac{a - b}{a + b}\right) = \sin^{-1}\left(\frac{a/b - 1}{a/b + 1}\right).$$

As with the lag controller, the maximum phase lead angle occurs halfway between the compensator's zero and pole (5.20), at the frequency

$$\omega_{max} = \sqrt{ba}.$$

**Fig. 5.25** Maximum phase lead angle versus pole-zero ratio for a phase lead controller

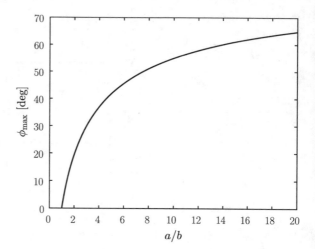

A phase lead controller is a phase compensator. It is used mainly to increase PM without reducing the crossover frequency. The following procedure can be used to design a phase lead controller.

**Phase lead controller design steps**

Step 1    Find $K_1$ so that the DC gain requirements of the open loop system $L(s) = K_1 G(s)$ are satisfied. For example, $K_1$ has to be chosen to satisfy steady-state error requirements on tracking, disturbance rejection, and so on (See Fig. 5.26).

Step 2    Determine the desired crossover frequency $\omega_{des}$, $PM_{des}$, and so on. (Recall that the crossover frequency is related to the rise time and settling time, and the PM is related to the overshoot of the system response.)

Step 3    Plot the Bode diagram of $K_1 G(s)$ and calculate the crossover frequency $\omega_1$.

Step 4    If $\omega_1 \ll \omega_{des}$, let $\omega_c = \omega_{des}$ and

$$\phi_{max} = PM_{des} - \angle K_1 G(j\omega_c) - 180°.$$

Find $a$ and $b$ from

$$\frac{a}{b} = \frac{1 + \sin \phi_{max}}{1 - \sin \phi_{max}}, \, \omega_c = \sqrt{ba}.$$

Find $K$ such that

$$\left| \frac{K(j\omega_c/b + 1)}{j\omega_c/a + 1} \right| |G(j\omega_c)| = 1.$$

If $K \geq K_1$, go to step 6. Otherwise, go to step 5.

Step 5    If $\omega_1 \approx \omega_{des}$ or $\omega_1 \geq \omega_{des}$, let $K = K_1$. Estimate the phase $\phi_{max}$ needed by examining the Bode diagram in the frequency range $\geq \omega_1$ (it is expected that the

**Fig. 5.26** Phase lead controller design

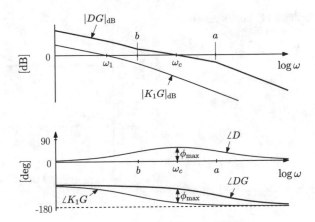

lead compensator will increase the crossover frequency somewhat). Let

$$\frac{a}{b} = \frac{1 + \sin \phi_{max}}{1 - \sin \phi_{max}},$$

and let $\omega_c$ be such that

$$|K_1 G(j\omega_c)| = \sqrt{\frac{b}{a}}.$$

Find $b$ and $a$ by setting

$$\omega_c = \sqrt{ab}.$$

Step 6 Let a lead controller be given by

$$D(s) = \frac{K(s/b + 1)}{s/a + 1}.$$

Step 7 Plot the Bode diagram of $D(s)G(s)$ and check the design specifications. Adjust $K$, $a$ and $b$, if necessary. If the design specifications cannot be satisfied then a lead-lag controller may be needed.

*Proportional-Derivative (PD) Control*: Consider a PD controller of the form

$$D(s) = K(s/b + 1).$$

The Bode diagram of this controller with $K = 1$ is shown in Fig. 5.27. Note that the phase of this compensator is given by

$$\phi = \tan^{-1}\left(\frac{\omega}{b}\right).$$

**Fig. 5.27** Bode diagram of a PD controller $D(s) = s/b + 1$

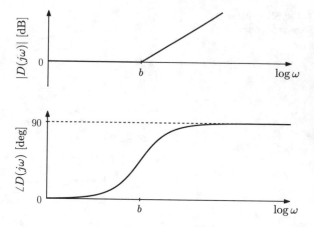

The increase in phase above the controller corner frequency $\omega = b$ stabilizes the system by increasing the PM. However, it can be seen from Fig. 5.27 that the magnitude of the controller's frequency response grows without bound as the frequency increases $|D(j\omega)|_{\omega \to \infty} \to \infty$. This is undesirable, as any high frequency noise present in a real system would be amplified by the controller, especially if the dynamics of the plant are lightly damped. The phase lead controller alleviates the high frequency noise with the introduction of the compensator pole, which occurs at a higher corner frequency than the lead compensator zero ($a > b$).

### PD controller design

Step 1    Find $K_1$ so that the DC gain requirements of the open loop system $L(s) = K_1 G(s)$ are satisfied. For example, $K_1$ has to be chosen to satisfy steady-state error requirements on tracking, disturbance rejection, and so on.

Step 2    Determine the desired crossover frequency $\omega_{des}$, $PM_{des}$, and so on. (Recall that the crossover frequency is related to the rise time and settling time, and the PM is related to the overshoot of the system response.)

Step 3    Plot the Bode diagram of $K_1 G(s)$ and calculate the crossover frequency $\omega_1$.

Step 4    If $\omega_1 \ll \omega_{des}$, let $\omega_c = \omega_{des}$ and

$$\phi_{max} = PM_{des} - \angle K_1 G(j\omega_c) - 180°.$$

Find $b$ from

$$b = \frac{\omega_c}{\tan \phi_{max}}.$$

Find $K$ such that

$$|K(j\omega_c/b + 1)| |G(j\omega_c)| = 1.$$

If $K \geq K_1$, go to step 6. Otherwise, go to step 5.

**Step 5** If $\omega_1 \approx \omega_{des}$ or $\omega_1 \geq \omega_{des}$, let $K = K_1$. Estimate the phase $\phi_{max}$ needed by examining the Bode diagram in the frequency range $\geq \omega_1$ (it is expected that the PD compensator will increase the crossover frequency somewhat). Let $\omega_c$ and $b$ be such that

$$b = \frac{\omega_c}{\tan \phi_{max}}$$

and

$$|K(j\omega_c/b + 1)| |G(j\omega_c)| \approx 1.$$

(This can be done graphically from the Bode diagram.)

**Step 6** Let a PD controller be given by

$$D(s) = K(s/b + 1).$$

**Step 7** Plot the Bode diagram of $D(s)G(s)$ and check the design specifications. Adjust $b$ and $K$, if necessary. If the design specifications cannot be satisfied, a PID or lead-lag controller may be needed.

The design process described above can be regarded only as a starting point for any practical design. Subsequent adjustments to the design parameters are usually highly desirable to produce a satisfactory controller design.

**Example 5.7** The root locus design of a compensator to control the lateral deviation $y$ of a USV to input changes in its rudder angle $\delta_R$ is considered in Problem 4.5. A simplified kinematic model is used to develop the loop transfer function of the system, which is given by

$$\frac{Y}{\Delta_R} = \frac{U/100}{s^2(s + 5/4)},$$

where $U = 1.5$ m/s is the constant surge speed of the USV. Here, we will explore the use of frequency response techniques to design a suitable controller. The system is of Type 2, so the steady state error will be zero for a reference step command in the lateral deviation, thus the DC gain does not need to be selected to satisfy a steady state error requirement (in position or velocity). We will design a compensator so that the closed loop step response of the system has less than 30% overshoot and a settling time of less than 20 s.

The frequency response of the open loop system is shown in Fig. 5.28. The crossover frequency is $\omega_1 = 0.1093$ rad/s and the phase margin is PM $= -5°$. In order to obtain an overshoot of less than 30%, a phase margin of about $PM_{des} \geq 50°$ is needed (see Fig. 5.16), which means that the increase in phase margin required is about $\phi_{max} = 55°$. The desired phase margin corresponds to a damping ratio of about $\zeta \geq 0.4$. Thus, a 2% settling time of $t_s \leq 20$ s requires that the natural frequency of the closed loop system be $\omega_n \geq 0.5$ rad/s. The corresponding desired crossover frequency is calculated using (5.16) to get $\omega_c = 0.4272$ rad/s. Given that the system is of Type 2 we will explore the use of a PD controller; at high frequency the compensated system

**Fig. 5.28** Bode diagram of open loop USV lateral position dynamics

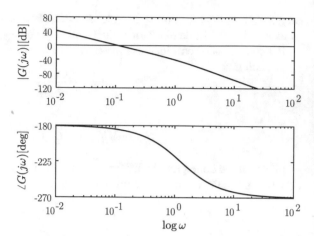

will act somewhat somewhat like a Type 1 system so that the effects of high frequency noise are not likely to be an issue.

The PD compensator zero is selected as

$$b = \frac{\omega_c}{\tan \phi_{max}} = 0.300 \text{ rad/s}$$

and

$$K = \frac{1}{|(j\omega_c b + 1)||G(j\omega_c)|} = 9.222.$$

With these values the step response is found to have a settling time of $t_s = 25$ s and an overshoot of 44%. If the system is tuned slightly by increasing the damping ratio to $\zeta = 0.47$, then $\omega_c = 0.3435$ and $\phi_{max} = 64.7°$, placing the compensator zero at $b = 0.1623$ rad/s and giving a compensator gain of $K = 4.355$. The frequency response and the step response of the compensated system are shown in Figs. 5.29 and 5.30, respectively. The system meets the design requirements by achieving a settling time of 18.7 s and an overshoot of 29.7%.                                    □

### 5.7.3  Lead-Lag or PID Controllers

A lead-lag controller can take the general form

$$D(s) = \frac{K(s/b_1 + 1)}{s/a_1 + 1} \frac{s/b_2 + 1}{s/a_2 + 1}, \quad 0 < a_1 < b_1 < b_2 < a_2.$$

The Bode diagram of this controller with $K = 1$ is shown in Fig. 5.31.

**Fig. 5.29**  Bode diagram of
PD compensated USV lateral
positioning system

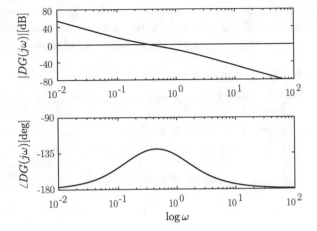

**Fig. 5.30**  Step response of
PD compensated USV lateral
positioning system

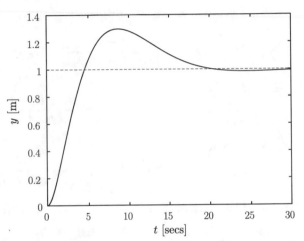

**Fig. 5.31**  Lead-lag
controller
$D(s) = \frac{K(s/b_1+1)}{s/a_1+1} \frac{s/b_2+1}{s/a_2+1}$

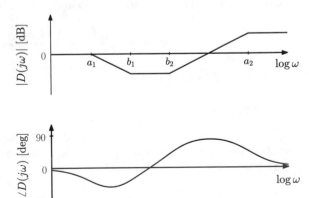

The design of this controller essentially follows a combination of a lead controller design and a lag controller design although it is much more flexible. The following procedure can be used to design a lead-lag controller.

**Lead-lag controller design**

Step 1    Find $K$ so that the DC gain requirements of the open loop system $L(s) = KG(s)$ are satisfied. For example, $K$ has to be chosen to satisfy steady-state error requirements on tracking, disturbance rejection, and so on.

Step 2    Determine the desired crossover frequency $\omega_c$ and $PM_{des}$.

Step 3    Plot the Bode diagram of $KG(s)$ and calculate the phase $\phi_{max}$ needed at $\omega_c$ in order to achieve the desired PM,

$$\phi_{max} = PM_{des} - \angle KG(j\omega_c) - 180° + 5°.$$

Step 4    Choose $a_2$ and $b_2$ such that

$$\frac{a_2}{b_2} = \frac{1 + \sin \phi_{max}}{1 - \sin \phi_{max}}, \quad \omega_c = \sqrt{a_2 b_2}.$$

Let

$$D_{lead}(s) = \frac{K(s/b_2 + 1)}{s/a_2 + 1}.$$

Step 5    Choose

$$b_1 \approx \omega_c/10, \quad a_1 = \frac{b_1}{|D_{lead}(j\omega_c)G(j\omega_c)|}.$$

Step 6    Plot the Bode diagram of $D(s)G(s)$ and check the design specifications.

Similarly, a PID controller can be designed by essentially combining the design process for PD and PI controllers. The Bode diagram of a PID controller

$$D(s) = \frac{K(s/b_1 + 1)(s/b_2 + 1)}{s}$$

with $K = 1$ is shown in Fig. 5.32.

**PID controller design**

Step 1    Determine the desired crossover frequency $\omega_c$ and $PM_{des}$.

Step 2    Plot the Bode diagram of $KG(s)$ and calculate the phase $\phi_{max}$ needed at $\omega_c$ in order to achieve the desired PM,

$$\phi_{max} = PM_{des} - \angle G(j\omega_c) - 180° + 5°.$$

Step 3    Choose $b_2$ such that

**Fig. 5.32** PID controller $D(s) = \frac{(s/b_1+1)(s/b_2+1)}{s}$

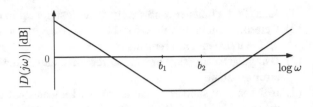

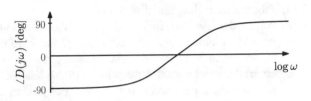

$$b_2 = \frac{\omega_c}{\tan \phi_{\max}}.$$

Let

$$D_{\mathrm{pd}}(s) = s/b_2 + 1.$$

Step 4   Choose

$$b_1 \approx \omega_c/10, \quad K = \frac{b_1}{|D_{\mathrm{pd}}(j\omega_c)G(j\omega_c)|}.$$

Step 5   A PID controller is given by

$$D(s) = \frac{K(s/b_1 + 1)(s/b_2 + 1)}{s}.$$

Step 6   Plot the Bode diagram of $D(s)G(s)$ and check the design specifications.

### 5.7.4  Summary of Compensator Design in the Frequency Domain

The general design guidelines presented for compensator design using root locus methods (Sect. 4.4) also apply to compensator design using frequency response techniques. However, in contrast to root locus methods, where one typically tries to satisfy the transient response requirements first and then the steady state requirements, with frequency response techniques, one first satisfies the steady state requirements and then the transient requirements. A general design procedure is as follows:

(1) Identify the transient and steady-state design requirements for the system.
(2) Determine suitable values for the phase margin and crossover frequency based on the transient requirements.
(3) Using the steady-state error requirements, determine the associated steady-state error constants.
(4) If the steady-state error requirements indicate that the system type should be increased, a PI or PID controller should be used.
(5) Plot the Bode diagram of $KG(s)$.
(6) Adjust the low frequency (DC) gain for $\omega \to 0$ to ensure that the steady state error requirements are satisfied. If the phase margin at the crossover frequency is more than 5° higher than the desired phase margin, a lag controller or a PI controller can be used. Otherwise, a phase lead or PD controller can be used. Sometimes, a pure phase lag or pure phase lead controller may not provide enough design freedom to satisfy all of the design specifications. In such cases, a combination of phase lag and phase lead, or a PID controller is necessary. The basic idea is to use first satisfy the steady state requirements by selecting the appropriate controller gain, using a phase lead to satisfy the transient response at a suitable crossover frequency, and then determining a pole and a zero for a phase lag controller that does not alter the selected crossover frequency.

In summary, the general characteristics of each type of compensator discussed above are:

(a) *Lag compensation*: Can be used to increase the DC gain in order to decrease steady-state errors. Or, when the DC gain is not significantly increased, it can alternatively be used to shift the gain crossover frequency to a lower value in order to achieve an acceptable phase margin. The compensator contributes a phase lag between the two corner frequencies, which must be low enough that it does not excessively degrade the stability of the closed loop system.

(b) *PI control*: Decreases steady-state errors in the system by increasing the magnitude frequency response at frequencies below the zero corner frequency. The compensator also increases the phase lag at frequencies lower than the zero corner frequency. In order to prevent a degradation in the stability of the system, the corner frequency must made be sufficiently low.

(c) *Lead compensation*: Adds phase lead within the range of frequencies between the compensator zero and pole, which are usually selected to bracket the crossover frequency. Unless the DC gain of the system is increased, lead compensation will increase the system's crossover frequency and its speed of response. If the DC gain is reduced to prevent an increase in the crossover frequency, the steady-state errors of the system will increase.

(d) *PD control*: Adds a phase lead at all frequencies higher than the corner frequency. Since its magnitude response increases without bound as the input frequency increases, the controller can be sensitive to high frequency noise. PD compensation increases the crossover frequency and speed of response, unless the low-frequency gain of the controller is suitably reduced.

Here we have discussed the design of first order compensators, or a lead-lag/PID combination of simple first order compensators. Using a technique known as *Loop Shaping*, it is possible to design higher order controllers that satisfy stricter design requirements on the steady state error, bandwidth and noise rejection, resonant response, etc. by using the requirements to specify which regions of the Bode diagram the controlled system's magnitude and phase responses should pass through. Such controllers can be more closely tailored to suit a set of control design requirements. However, the improved performance can come at the cost of controller complexity and they may not always be practical, owing to uncertainties in the plant or actuator models, sensor noise or external disturbances. Additional details about this approach can be found in [3, 4].

## 5.8 Matlab for Frequency Response Analysis and Control Design

### 5.8.1 Nyquist Plots

A Nyquist plot of the system with loop transfer function $L(s) = 1/(1 - 0.5)^2$ (Fig. 5.33) can be generated as follows:

```
L = zpk([],[-0.5,-0.5],1); % create the loop transfer function
% set the frequency range from 10^-3 to 10^2 with 100 sample points.
w = logspace(-3,2,100);
nyquist(L, w);
```

It is not necessary to use the optional frequency argument w in the nyquist function call. However, doing so can provide more resolution and a smoother curve. The data cursor point can be used to interrogate the real and imaginary parts of the system and the corresponding frequency at desired points on the Nyquist curve. A grid showing contours of constant magnitude can be added to the plot using the *grid* command. The Nyquist diagrams of mulitple systems, e.g. L1(s), L2(s) and L3(s) can also be superimposed by inserting multiple transfer functions as arguments to the Matlab *nyquist* function: nyquist(L1, L2, L3, w).

### 5.8.2 Bode Plots

The Bode plot of a system can be obtained using the Matlab *bode* command. The magnitude in dB and the phase in degrees are plotted as a function of frequency,

**Fig. 5.33** Nyquist plot of
$L(s) = 1/(1 - 0.5)^2$

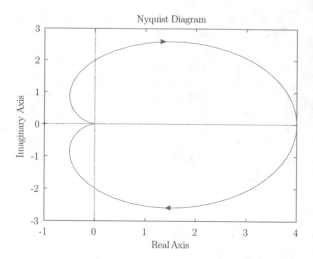

which is plotted on a logarithmic scale. As an example, the frequency response of
the lag controller $D(s) = K(s/b + 1)/(s/a + 1)$ where $a = 1/5$ rad/s, $b = 5$ rad/s
and $K = 1$ (Fig. 5.34) was generated with the following code.

```
a = 1/5; % Compensator pole
b = 1/a; % Compensator zero
K = 1; % Compensator zero
D = zpk(-b, -a, K*a/b); % Compensator transfer function
w = logspace(-3, 3, 100); % Frequency range
bode(D, w); % Create Bode plot
```

The ratio $a/b$ is included in the third (gain) argument to the *zpk* Matlab function
in order to convert the transfer function from Bode format to the normal pole zero
format used by the *zpk* command, e.g.

$$D(s) = \frac{K\left(\dfrac{s}{b} + 1\right)}{\left(\dfrac{s}{a} + 1\right)} = \frac{K\dfrac{1}{b}(s + b)}{\dfrac{1}{a}(s + a)} = \frac{Ka}{b}\frac{(s + b)}{(s + a)}.$$

As with the *nyquist* Matlab command, it is not necessary to use the optional
frequency argument w in the *bode* function call. However, doing so can provide more
resolution and a smoother curve. The data cursor point can be used to interrogate
the both the magnitude and phase frequency responses at desired points (here is
where having more resolution by defining w yourself can be very useful). Grid lines
showing levels of constant magnitude and phase can be added to the plot using the

**Fig. 5.34** Bode plot of
$D(s) =$
$K(s/b + 1)/(s/a + 1)$ for
$a = 1/5$ rad/s, $b = 5$ rad/s
and $K = 1$

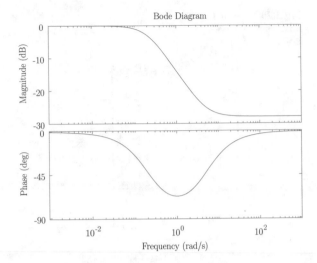

*grid* command. The Bode diagrams of mulitple systems, e.g. L1(s), L2(s) and L3(s) can also be superimposed by inserting multiple transfer functions as arguments to the Matlab *bode* function: bode(L1, L2, L3, w).

*Magnitude and Phase at a Specific Frequency*:
The magnitude and phase of a transfer function at a specific frequency can also be determined using the *bode* command. For example, with the following code

```
wc = 0.933; % frequency of interest
bode(D, wc); % get magnitude and phase at wc
```

Matlab will return

| | | |
|---|---|---|
| mag | = | 0.2132, and |
| phase | = | −67.3312. |

Alternatively, although it is more complicated, one also could use the *evalfr* command to determine the magnitude and phase at a desired frequency. When the second argument of the *evalfr* command is given as a frequency on the imaginary axis, it will return the complex value of the transfer function at that frequency. The corresponding magnitude and phase of the transfer function can be determined using the *abs* and *angle* commands as follows

**Fig. 5.35** Gain and phase margins of $G(s) = 20/[s(s^2 + 7s + 12)]$

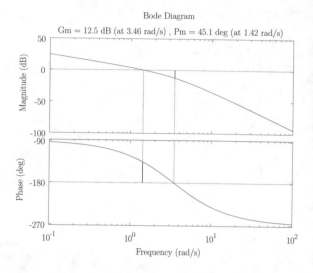

wc = 0.933; % frequency of interest
% get complex value of D at wc,
% where wc*1j gives $j\omega_c$ with $j := \sqrt{-1}$ being the imaginary number
Dwc = evalfr(D,wc*1j);
magDwc = abs(Dwc); % get magnitude of D at wc
% get phase of D at wc and convert from radians to degrees
phaseDwc = angle(Dwc)*180/pi;

as above, Matlab returns

| | | |
|---|---|---|
| magDwc | = | 0.2132, and |
| phaseDwc | = | −67.3312. |

*Determining GM, PM, $\omega_c$ and $\omega_g$:*

The gain margin GM, phase margin PM, gain crossover frequency $\omega_c$ and phase crossover frequency $\omega_g$ of a transfer function can be determined using the *margin* command. When called without equating the output to any variables, *margin* will create a Bode plot of the transfer function and label it with the gain margin, phase margin, gain crossover frequency and phase crossover frequency (Fig. 5.35). The following code illustrates how the command can be used.

```
G = zpk([], [0 −3 −4], 20);% build the transfer function G(s)
margin(G); % plot frequency response of G(s) and label margins
```

When variables are assigned to the output of the *margin* command, no Bode plot is created. The following code shows how to assign the gain margin, phase margin, gain crossover frequency and phase crossover frequency to variables using the *margin* command. The variables can subsequently be used for controller design. In this example omega_c $= \omega_c$ is the gain crossover frequency and omega_g $= \omega_g$ is the phase crossover frequency.

```
G = zpk([], [0 –3 –4], 20);% build the transfer function G(s)
% compute the margins and crossover frequencies and store them for later use
[GM, PM, omega_g, omega_c] = margin(G);
```

*Time Delays*:
The importance of time delays in marine control systems was discussed in Sect. 5.4.1. The Bode diagram of a transfer function with a pure time delay can be generated using the following code (Fig. 5.36).

```
w = logspace(–1, 1, 100);
G = zpk([], [0 –3 -4], 20);
Gd = zpk([], [0 –3 –4], 20, 'InputDelay', 0.1);
bode(G, Gd, w)
grid
legend('G(s)', 'G(s)e^{–0.1T}')
```

Here, the time delay is included in the definition of the transfer function when the *zpk* function is called by using the 'InputDelay' argument. Time delays can be similarly included with the 'InputDelay' argument when defining transfer functions using the *tf* Matlab command.

### 5.8.3  Matlab for Constructing A PD Controller

Let us explore the use of Matlab for designing the PD controller developed in Example 5.7. The loop transfer function of the system is given by

$$\frac{Y}{\Delta_R} = \frac{U/100}{s^2(s + 5/4)},$$

where $U = 1.5$ m/s is the constant surge speed of the USV. The compensator design requirements are a closed loop step response with less than 30% overshoot and a settling time of less than 20 s.

**Fig. 5.36** Bode diagrams of
$G(s) =$
$20/[s(s^2 + 7s + 12)]$ and of
$G(s)$ with time delay, i.e.
$G_d(s) = G(s)e^{-sT}$, where
$T = 0.1$ s

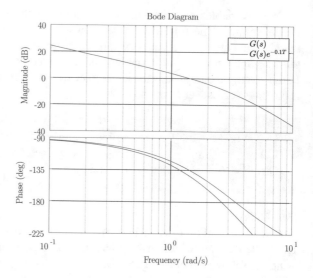

We follow the step-by-step procedure for designing a PD controller outlined in
Sect. 5.7.2.

Step 1    Find $K_1$ so that the DC gain requirements $L(s) = K_1 G(s)$ of the open loop
system are satisfied. For example, $K_1$ has to be chosen to satisfy steady-state
error requirements on tracking, disturbance rejection, and so on. The DC gain
does not need to be selected to satisfy a steady state error requirement (in position
or velocity), as the system is of Type 2. Thus, we select $K_1 = 1$.

Step 2    Determine the desired crossover frequency $\omega_{des}$, $PM_{des}$, and so on (recall
that the crossover frequency is related to the rise time and settling time, and the
PM is related to the overshoot of the system response). After some iteration with
various values of damping ratio $\zeta$, it was found that a value of $\zeta = 0.47$ will satisfy
the design requirements. As each iteration is conducted in exactly the same way,
only the final one is shown here. The following code can be used to calculate the
desired crossover frequency and phase margin from the design requirements.

```
zeta = 0.47;% potentially good damping ratio
ts = 20;% settling time
omegan = 4/zeta/ts;% natural frequency
% desired phase margin from (5.5)
PM_des = 180/pi*atan(2*zeta/sqrt(sqrt(1+4*zeta^4)-2*zeta^2));
% desired crossover frequency from (5.16)
omega_des = omegan*sqrt(sqrt(1+4*zeta^4)-2*zeta^2);
```

Step 3    Plot the Bode diagram of $K_1G(s)$ and calculate the crossover frequency $\omega_1$.

```
U = 1.5; % Define the boat speed
G = zpk([],[0 0 -5/4], U/100); % Define plant transfer function
w = logspace(-2,2,100); % Specify frequency range of interest
K1 = 1; % Set the DC gain
bode(K1*G,w); % Plot the Bode diagram
grid % Add grid to plot
% calculate the crossover frequency
[GM1, PM1, omega_g1, omega_1] = margin(K1*G);
```

The calculation gives $\omega_1 = 0.1093$ rad/s. The corresponding Bode diagram is shown in Fig. 5.37.

Step 4    The crossover frequency $\omega_1$ is about 3 times smaller than $\omega_{des}$, so we will take $\omega_c = \omega_{des}$ and calculate

$$\phi_{max} = PM_{des} - \angle K_1 G(j\omega_c) - 180°,$$

find $b$ using

$$b = \frac{\omega_c}{\tan \phi_{max}},$$

and solve for $K$ with

$$|K(j\omega_c/b + 1)| |G(j\omega_c)| = 1.$$

The following Matlab code can be used to perform these steps.

```
% Magnitude and phase at desired crossover frequency
[magc, phasec] = bode(K1*G,omega_des);
PMc = 180 + phasec;
% Compensator phase required at desired crossover frequency
phi_max = PM_des - PMc;
% Compensator zero location
b = omega_des/tand(phi_max);
% Compensator gain needed for correct gain crossover
Dc = zpk(-b,[], 1/b);
[magd,phased] = bode(Dc,omega_des);
```

```
[mag,phase] = bode(G,omega_des);
K = 1/mag/magd;
```

With the code above, we find that $K = 4.355$, which is greater than $K_1 = 1$, so we proceed to Step 6 of the design.

Step 6    Let a PD controller be given by

$$D(s) = K(s/b + 1).$$

This is accomplished with the following lines of code.

```
% Build the PD final compensator
D = zpk(-b,[], K/b);
```

Step 7    Plot the Bode diagram of $D(s)G(s)$ and check the design specifications. Adjust $b$ and $K$, if necessary. If the design specifications cannot be satisfied, a PID or lead-lag controller may be needed. This can be accomplished with the following commands.

```
% Confirm that design specifications are satisfied
sys = D*G; % Construct the new loop transfer function
% Compute margins and crossover frequencies
[Gm,Pm,Wcg,Wcp] = margin(sys);
figure(2) % Open a new window for the plot
bode(sys,w) % Create the Bode diagram
grid % Add gridlines to the plot
```

We find that the desired crossover frequency and phase margin are achieved with the compensator designed, thus the design requirements are met. The Bode plot with PD compensator is shown in Fig. 5.38. A plot of the step response is shown in Fig. 5.39.

## Problems

**5.1** Draw the Bode plots for the following systems, first by hand, then using Matlab
(a)

$$G(s) = \frac{9000s}{(s + 1)(s + 10)(s^2 + 30s + 900)},$$

**Fig. 5.37** Bode diagram of $K_1 G(s)$

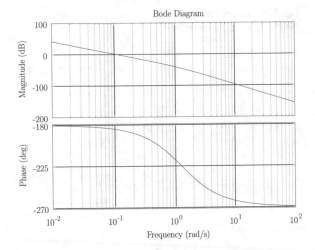

**Fig. 5.38** Bode diagram of the PD compensated system

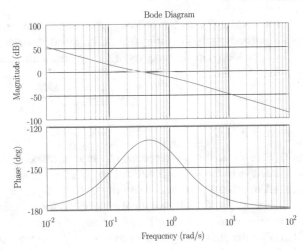

(b)

$$G(s) = \frac{5(s + 10)}{s(s + 1)(s^2 + s + 0.5)}.$$

**5.2** Determine the range of $K$ for which the system

$$KG(s) = \frac{40K}{(s + 10)(s + 2)^2}$$

is stable by making a Bode plot with $K = 1$ and imagining the magnitude plot sliding up or down until the system becomes unstable. Qualitatively verify the stability rules with a root locus plot.

**Fig. 5.39** Step response of the PD compensated system

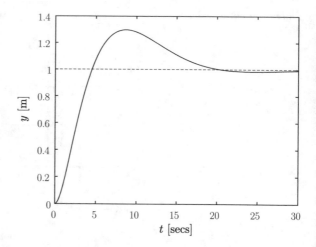

**5.3** Consider the station keeping of a small unmanned surface vehicle using an external underwater acoustic positioning sensor system, as discussed in Example 5.6. The open loop transfer function of the system with a proportional controller is

$$KG(s) = \frac{K/2}{s(s+1/2)},$$

where $K = 1/2$.

(a) Plot the frequency response of the controlled system when there are no time delays in the sensing system. Find the gain crossover frequency $\omega_c$.
(b) Assuming a pure time delay of the form $G_d(s) = \exp(-sT_d)$, where $T_d$ is the time delay, find an analytical expression for the phase margin of the system.
(c) Numerically solve for $T_d$ at the limit of stability.
(d) Plot the phase margin as a function of $T_d$ for the value of $\omega_c$ found in Part (a).

**5.4** A surface vessel steering system has the transfer function

$$G(s) = \frac{K_v(-s/8+1)}{s(3s/8+1)(3s/80+1)}.$$

(a) By hand, sketch the Bode magnitude and phase response of the system with $K_v = 2$.
(b) On your sketch, indicate the following:

    (i) The gain crossover frequency.
    (ii) The phase crossover frequency.
    (iii) The gain margin.
    (iv) The phase margin.

**Fig. 5.40** USV speed
control system

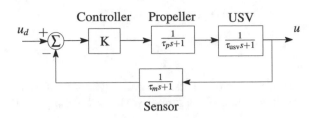

Controller   Propeller      USV

(c) Validate your sketch by plotting the Bode magnitude and phase response of the
system for $K_v = 2$ using Matlab.
(d) Confirm the answers found in (b) using the Matlab "margin" command.

**5.5** An unmanned surface vessel (USV) speed control system is shown in Fig. 5.40.
The speed sensor is slow enough that its dynamics are included in the loop transfer
function. The time constant of the speed measurement is $\tau_m = 0.25$ s, the time
constant of the USV (associated with the drag on its hull and the mass of the vehicle)
is $\tau_{usv} = 7$ s and the time constant of the propeller is $\tau_p = 0.5$ s.

(a) What proportional control gain $K$ is needed in order for the steady state error to
   be less than 7% of the reference speed setting?
(b) Use the Nyquist Criterion to investigate the stability of the system for the gain
   $K$ determined in (a).
(c) Estimate GM and PM. Is this a good system design? Why?

**5.6** Consider the USV station keeping problem studied in Example 5.6 with loop
transfer function

$$KG(s) = \frac{K/2}{s(s + 1/2)}.$$

The system will be operated as a closed loop unity feedback system. We are given a
time response specification that $M_p \leq 30\%$.

(a) Determine the corresponding specification in the frequency domain for the resonant peak value $M_r$ of the closed loop system.
(b) Determine the resonant frequency $\omega_r$ for the maximum value of $K$ that satisfies
   the specifications.
(c) Determine the bandwidth of the closed loop system using the results of (b).

**5.7** Consider the surface vessel steering system studied in Problem 5.4.

(a) Design a compensator so that the controlled system has a velocity constant
   $K_v = 2$, a phase margin of PM$\geq 50°$ and unconditional stability (PM$\geq 0°$ for
   all frequencies $\omega$).
(b) Plot the root locus of the final system design with compensator and indicate the
   location of the closed loop roots.

**5.8** The transfer function

$$G(s) = \frac{\pi}{s(20s+1)(s/3+1)}$$

can be used to model the combined dynamics of a surface vessel and rudder for the heading response of a surface vessel (see Example 5.1).

(a) Design a lead compensator $D(s)$ so that the phase margin with unity feedback is approximately 40° and the DC gain is 1.
(b) What is the approximate bandwidth of the system?

**5.9** An autopilot is used to control the depth of an AUV. The second order linearized relationship between depth *rate* and elevator fin angle is approximated as

$$\frac{sZ(s)}{\Delta(s)} = \frac{32(s+1/4)}{s^2 + 2s + 4} \frac{\text{m/s}}{\text{deg}}.$$

A block diagram of the system is shown in Fig. 5.41. You have been given the task of designing a compensator for the system.

First consider $D(s) = K_0$.

(a) Plot the magnitude and phase versus frequency (Bode plots) of the open loop system for $D(s) = K_0 = 1$.
(b) What value of $K_0$ provides a crossover frequency of 9 rad/s?
(c) For the value of $K_0$ in part (b), would this system be stable if the loop were closed?
(d) What is the PM?
(e) Plot the root locus. Locate the roots for your value of $K_0$ from (b).
(f) What steady-state error would result with $K_0$ from part (b) if the command were a ramp of 1 m/s?

Next, consider the compensator

$$D(s) = K \frac{s/b+1}{s/a+1}.$$

(g) Choose the parameters $K$, $a$ and $b$ so the crossover frequency is 9 rad/s and PM $\geq$ 50°. Plot the compensated system on the same plot as developed in part (a), and label the crossover frequency and PM on the plot.
(h) Plot the root locus. Show the locations of the closed loop roots on the root locus for the value of $K$ you determined in part (g).
(i) What steady-state error would result with $K$ from part (h) if the command were a ramp of 1 m/s?
(j) If the error in part (i) is too large, what type of compensation would you add? Approximately, at what frequencies would you put the breakpoints?

**Fig. 5.41** AUV depth
control system

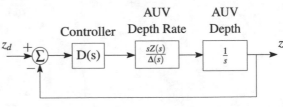

**Fig. 5.42** Block diagram of
the pitch control system

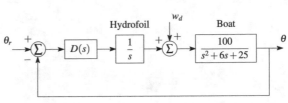

**5.10** At high speeds planing hull boats can experience wave-induced resonant pitch oscillations. The resonant frequency of the oscillations depends on boat speed and boat length [2]. For example, a 10 m boat moving at a speed of about 40 knots will experience resonant pitch oscillations with a frequency of about 5 rad/s when encountering (head on) waves with a wavelength of about 40 m. The block diagram of a notional fast hydrofoil-based pitch control system is shown in Fig. 5.42.

(a) Design a lead controller so that the response of the system to a unit step reference input has a settling time of $t_s \approx 0.25$ s and an overshoot of $M_p \approx 20\%$.
(b) In deep water a wave with a wavelength of 40 m will have a period of about 5 s. Treating the waves as a disturbance input, plot the response of the system for waves of this period when they impose a pitching moment of unity amplitude and the commanded pitch angle is $\theta_r = 0°$.
(c) Plot the open loop response of the system and the response of the system using the lead controller on the same plot. How much is the amplitude of the response attenuated by the pitch control system?

# References

1. Arrichiello, F., Liu, D.N., Yerramalli, S., Pereira, A., Das, J., Mitra, U., Sukhatme, G.S.: Effects of underwater communication constraints on the control of marine robot teams. In: Second International Conference on Robot Communication and Coordination, 2009. ROBOCOMM'09, pp. 1–8. IEEE (2009)
2. Faltinsen, O.M.: Hydrodynamics of High-speed Marine Vehicles. Cambridge University Press, Cambridge (2005)
3. Qiu, Li, Zhou, Kemin: Introduction to Feedback Control. Prentice Hall Englewood Cliffs, New Jersey (2009)
4. Åström, KJ, Murray, R.M.: Feedback Systems: An Introduction for Scientists and Engineers. Princeton University Press, New Jersey (2010)

# Chapter 6
# Linear State Space Control Methods

## 6.1 Introduction

Most practical applications involving marine vehicles typically require one to simultaneously control more than one system state at a time, such as both heading and speed, for example. One might attempt to control each state with a separate single-input single-output (SISO) controller. However, as system states are often coupled, the use of multiple SISO controllers will not generally suffice. State space techniques are more well-suited to the control of marine vehicles than other approaches (such as the Root Locus or Frequency Response Techniques), as they can be easily applied to multiple input multiple output (MIMO) systems. Further, stability analysis and control design for nonlinear and time varying systems is typically performed in the time domain. Thus, the state space techniques presented here can also be considered a bridge linking linear SISO systems techniques and nonlinear techniques, which are very often required for the control of marine vehicles.

### 6.1.1 State Variables

Both the time-domain analysis and the design of control systems utilize the concept of the state of a system. Consider the system shown in Fig. 6.1, where $y_1(t)$ and $y_2(t)$ are the output signals and $u_1(t)$ and $u_2(t)$ are the input signals. A set of variables $(x_1, x_2, \ldots, x_n)$ represents the state of the system if knowledge of the initial values of the variables $[x_1(t_0), x_2(t_0), \ldots, x_n(t_0)]$ at the initial time $t_0$, and of the input signals $u_1(t)$ and $u_2(t)$ for $t \geq t_0$, suffices to determine the values of the outputs and state variables for all $t \geq t_0$. For marine vehicles, the state variables are typically

© Springer Nature Switzerland AG 2021
K. D. von Ellenrieder, *Control of Marine Vehicles*, Springer Series on Naval
Architecture, Marine Engineering, Shipbuilding and Shipping 9,
https://doi.org/10.1007/978-3-030-75021-3_6

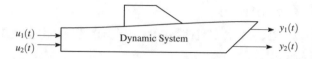

$u_1(t)$

$u_2(t)$

Dynamic System

$y_1(t)$

$y_2(t)$

either the position and orientation of the vehicle (pose) or its velocity and angular
rates. For robotic manipulators the joint angles and their rates are state variables. For
electronic systems the state variables are typically voltages and currents. Note that
path-dependent variables, such as work and power, cannot be state functions and are
known as process functions, instead.

The general form of a state space equation can be written as a set of coupled first
order differential equations. The input–output system has the form

$$\dot{x} = F(x, u), \quad y = h(x, u), \tag{6.1}$$

where $x = \{x_1, \ldots, x_n\} \in \mathbb{R}^n$ is the state, $u \in \mathbb{R}^p$ is the input and $y \in \mathbb{R}^q$ is the
output. The smooth maps $F : \mathbb{R}^n \times \mathbb{R}^p \to \mathbb{R}^n$ and $h : \mathbb{R}^n \times \mathbb{R}^p \to \mathbb{R}^q$ represent the
dynamics and output measurements of the system.

When $F$ is linear in $x$ and $u$, the state equation can be written as a linear combination of the states and inputs, e.g.

$$\begin{aligned}
\dot{x}_1 &= a_{11}x_1 + a_{12}x_2 + \cdots + a_{1n}x_n + b_{11}u_1 + \cdots + b_{1p}u_p, \\
\dot{x}_2 &= a_{21}x_1 + a_{22}x_2 + \cdots + a_{2n}x_n + b_{21}u_1 + \cdots + b_{2p}u_p, \\
&\vdots \\
\dot{x}_n &= a_{n1}x_1 + a_{n2}x_2 + \cdots + a_{nn}x_n + b_{n1}u_1 + \cdots + b_{np}u_p.
\end{aligned} \tag{6.2}$$

This set of simultaneous differential equations can be written in matrix form as

$$\frac{d}{dt} \begin{bmatrix} x_1 \\ x_2 \\ \vdots \\ x_n \end{bmatrix} = \begin{bmatrix} a_{11} & a_{12} & \cdots & a_{1n} \\ a_{21} & a_{22} & \cdots & a_{2n} \\ \vdots & \vdots & \cdots & \vdots \\ a_{n1} & a_{n2} & \cdots & a_{nn} \end{bmatrix} \begin{bmatrix} x_1 \\ x_2 \\ \vdots \\ x_n \end{bmatrix} + \begin{bmatrix} b_{11} & \cdots & b_{1m} \\ b_{21} & \cdots & b_{2m} \\ \vdots & \cdots & \vdots \\ b_{n1} & \cdots & b_{nm} \end{bmatrix} \begin{bmatrix} u_1 \\ u_2 \\ \vdots \\ u_m \end{bmatrix}, \tag{6.3}$$

or more compactly in state equation form as

$$\dot{x} = Ax + Bu. \tag{6.4}$$

Similarly, when $h$ is linear in $x$ and $u$, the output equation can be written as a linear
combination of the states and inputs as

$$y = Cx + Du, \tag{6.5}$$

where $y$ is a vector of output signals. In general, the coefficients in the matrices $A$, $B$, $C$, and $D$ may be time dependent. When they are also constant in time, the system is a *Linear Time Invariant* (LTI) system. In the control engineering literature $A$ is known as the *dynamics matrix*, $B$ is the *control matrix*, $C$ is the *sensor matrix* and $D$ is the *direct term*.

**Example 6.1** As shown in Fig. 6.2, a wave adaptive modular vessel (WAM-V) is a catamaran with inflatable hulls, which uses a set of springs and dampers to mitigate the effects of waves on the motion of its payload [3]. Consider a simplified one dimensional model of the vessel's vertical motion given by

$$M\ddot{d} + b(\dot{d} - \dot{y}) + k(d - y) = u,$$
$$m\ddot{y} + b(\dot{y} - \dot{d}) + k(y - d) = 0, \tag{6.6}$$

where $M$ is the mass of the hull, $m$ is the mass of the payload and superstructure, $b$ is the damping coefficient, $k$ is the spring constant, $d$ is the vertical displacement of the hulls, $y$ is the output vertical displacement of the payload and superstructure, and $u$ is the input wave forcing. The effects of added mass, wave damping and hull flexibility are neglected here. Take the state variables to be the position and velocity of each mass

$$x_1 = y,$$
$$x_2 = \dot{y},$$
$$x_3 = d,$$
$$x_4 = \dot{d}.$$

**Fig. 6.2** A wave adaptive modular vessel (WAMV) USV16. Photo courtesy 2014 FAU-Villanova Maritime RobotX Challenge Team

Then, the equations written in state variable form are

$$\dot{x}_1 = x_2,$$
$$\dot{x}_2 = -\frac{k}{M}x_1 - \frac{b}{M}x_2 + \frac{k}{M}x_3 + \frac{b}{M}x_4 + \frac{u}{M},$$
$$\dot{x}_3 = x_4,$$
$$\dot{x}_4 = \frac{k}{m}x_1 + \frac{b}{m}x_2 - \frac{k}{m}x_3 - \frac{b}{m}x_4.$$

It can be seen that when the matrices $A, B, C$ and $D$ are selected as

$$A = \begin{bmatrix} 0 & 1 & 0 & 0 \\ -\dfrac{k}{M} & -\dfrac{b}{M} & \dfrac{k}{M} & \dfrac{b}{M} \\ 0 & 0 & 0 & 1 \\ \dfrac{k}{m} & \dfrac{b}{m} & -\dfrac{k}{m} & -\dfrac{b}{m} \end{bmatrix}, \quad B = \begin{bmatrix} 0 \\ \dfrac{1}{M} \\ 0 \\ 0 \end{bmatrix},$$

$$C = \begin{bmatrix} 1 & 0 & 0 & 0 \end{bmatrix}, \qquad\qquad D = 0,$$

(6.6) can be rewritten in the state equation form (6.4).                     □

The solution of (6.4) can be obtained in a manner similar to the approach typically utilized for solving a first-order differential equation. Consider the equation

$$\dot{x} = ax + bu, \tag{6.7}$$

where $x(t)$ and $u(t)$ are scalar functions of time. We expect an exponential solution of the form $e^{at}$. Taking the Laplace transform of (6.7), we have

$$sX(s) - x(0) = aX(s) + bU(s),$$

and therefore

$$X(s) = \frac{x(0)}{s-a} + \frac{b}{s-a}U(s). \tag{6.8}$$

The inverse Laplace transform of (6.8) results in the solution

$$x(t) = e^{at}x(0) + \int_0^t e^{a(t-\tau)}bu(\tau)d\tau. \tag{6.9}$$

We expect the solution of the state differential equation to be similar to (6.9) and to be of vector form. Define the matrix exponential function as

$$e^{At} = \exp(At) := 1 + At + \frac{(At)^2}{2!} + \cdots + \frac{(At)^n}{n!} + \cdots , \tag{6.10}$$

which converges for all finite $t$ and any $A$. Then the solution of the state differential equation is found to be

$$x(t) = e^{At} x(0) + \int_0^t e^{A(t-\tau)} Bu(\tau) d\tau. \tag{6.11}$$

Equation (6.11) may be obtained by taking the Laplace transform of (6.4) and rearranging to obtain

$$X(s) = [s\mathbf{1} - A]^{-1} x(0) + [s\mathbf{1} - A]^{-1} BU(s). \tag{6.12}$$

Define the *state transition matrix* as $\Phi(s) := [s\mathbf{1} - A]^{-1}$, where $\Phi(s)$ is simply the Laplace transform of $\phi(t) = \exp(At)$. The inverse Laplace transform of (6.12) gives (6.11). Written in terms of $\phi(t)$, (6.11) is

$$x(t) = \phi(t)x(0) + \int_0^t \phi(t-\tau) Bu(\tau) d\tau. \tag{6.13}$$

## 6.2 Reachability/Controllability

The *reachability* of a system is its ability to reach an arbitrary desired state $x(t)$ in a transient fashion through suitable selection of a control input $u(t)$. The *controllability* of a system is its ability to reach the origin $x(t) = \mathbf{0}$, given a suitable control input $u(t)$. Thus, reachability is a more general form of controllability and for linear systems, through a suitable change of coordinates, could be viewed as the same property. In some texts, the term controllability is used instead of reachability [4, 6]. Here, as in [1], the term reachability shall be used. Reachability is important for understanding whether feedback control can be used to obtain the desired dynamics of a system.

Define the reachable set $\mathcal{R}(x_0, t \leq t_1)$ as the set of all points $x_f$, such that there exists an input $u(t)$, $0 \leq t \leq t_1$ that steers the system from $x(0) = x_0$ to $x(t_1) = x_f$ in the finite time $t_1$ (Fig. 6.3). A system described by the matrices $A \in \mathbb{R}^n \times \mathbb{R}^n$ and $B \in \mathbb{R}^n \times \mathbb{R}^p$ is controllable if there exists an unconstrained control input[1] $u \in \mathbb{R}^p$ that can transfer any initial state $x(0) \in \mathbb{R}^n$ to any desired state $x(t) \in \mathbb{R}^n$. The reachability matrix $W_r \in \mathbb{R}^n \times \mathbb{R}^n$ is defined as

$$W_r := \begin{bmatrix} B & AB & A^2 B & \cdots & A^{n-1} B \end{bmatrix}. \tag{6.14}$$

---

[1] An input signal can be constrained when it is affected by actuator saturation (Sect. 3.3.4) or actuator rate limits (Sect. 10.6), for example.

**Fig. 6.3** The reachable set
$\mathscr{R}(x_0, t \leq t_1)$

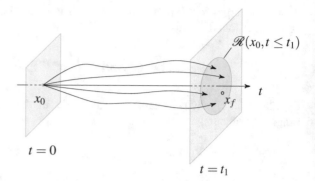

**Theorem 6.1** *A linear system of the form (6.4) is reachable, if and only if, the reachability matrix* $\mathbf{W}_r$ *is invertible (has full rank n), i.e.*

$$\text{rank} \begin{bmatrix} \mathbf{B} & \mathbf{AB} & \mathbf{A}^2\mathbf{B} & \cdots & \mathbf{A}^{n-1}\mathbf{B} \end{bmatrix} = n. \tag{6.15}$$

If the determinant of $\mathbf{W}_r$ is nonzero, the system is reachable/controllable.

**Example 6.2** The pitch dynamics of an AUV are given by $\dot{x} = \mathbf{A}x + \mathbf{B}u$, where

$$\mathbf{A} = \begin{bmatrix} -1 & -\dfrac{1}{2} \\ 1 & 0 \end{bmatrix} \quad \text{and} \quad \mathbf{B} = \begin{bmatrix} 1 \\ 0 \end{bmatrix}.$$

For this system $n = 2$. To determine whether the dynamics are controllable (reachable), we can compute the reachability matrix as

$$\mathbf{W}_r := [\mathbf{B} \quad \mathbf{AB}] = \begin{bmatrix} 1 & -1 \\ 0 & 1 \end{bmatrix}. \tag{6.16}$$

We can see that the rank of $\mathbf{W}_r$ is 2 and that its determinant is nonzero. Thus, the system is controllable.                                                                    □

The stability of an autonomous system can be graphically described as a trajectory in the phase (state) space of a system near a critical point (see Sect. 2.4). The reachability of a system permits us to show whether or not it is possible to reach a desired critical point $x_f$ in the system's state space. Given an initial location $x_0$ and assuming there are no exogenous (externally-generated) disturbances, state feedback permits us to control the system by selecting the unique trajectory its state will follow between $x_0$ and $x_f$.

### 6.2.1 Reachable Canonical Form

The use of canonical forms of the dynamical equations describing a system can greatly simplify the selection of controller gains. When a system is expressed in reachable canonical form, the coefficients of its characteristic equation

$$a(s) = s^n + a_1 s^{n-1} + \cdots + a_{n-1}s + a_n, \tag{6.17}$$

directly appear in the dynamics matrix. To do this, we seek a change of coordinates that transforms the state variable $x$ into a new variable $z$, so that (6.4) and (6.5) can be written in the reachable canonical form

$$\frac{dz}{dt} = A_r z + B_r u$$

$$= \begin{bmatrix} -a_1 & -a_2 & -a_3 & \cdots & -a_n \\ 1 & 0 & 0 & \cdots & 0 \\ 0 & 1 & 0 & \cdots & 0 \\ \vdots & \ddots & \ddots & \ddots & \vdots \\ 0 & 0 & \cdots & 1 & 0 \end{bmatrix} z + \begin{bmatrix} 1 \\ 0 \\ 0 \\ \vdots \\ 0 \end{bmatrix} u, \tag{6.18}$$

$$y = C_r z + du$$

$$= \begin{bmatrix} b_1 & b_2 & b_3 & \cdots & b_n \end{bmatrix} z + du.$$

To see how the transformed coordinates $z$ and the matrices $A_r$, $B_r$ and $C_r$ are defined, consider the transfer function representation of a system with characteristic polynomial $a(s)$ with control input $u$ and output $y$, as shown in Fig. 6.4. Here, $d$ represents the direct term. Define an auxiliary variable $\xi(t)$ with the Laplace transform $\Xi(s)$, where

$$\frac{\Xi(s)}{U(s)} = \frac{1}{a(s)}. \tag{6.19}$$

Then $a(s)\Xi(s) = U(s)$ giving

$$\left( s^n + a_1 s^{n-1} + \cdots + a_{n-1}s + a_n \right) \Xi(s) = U(s). \tag{6.20}$$

**Fig. 6.4** Transfer function representation of the relation between the control input $u$ and the system output $y$. The auxiliary variable $\xi(t)$ is also called the *partial state* [6]

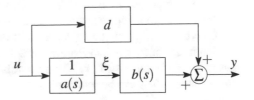

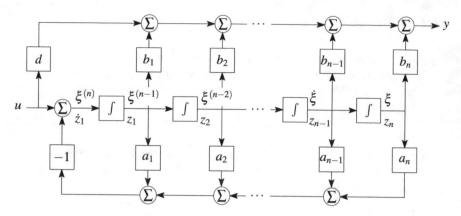

**Fig. 6.5** Block diagram for a system in reachable canonical form. The individual states of the system are represented by a chain of integrators whose input depends on the weighted values of the states. The output is given by an appropriate combination of the system input and other states

Taking the inverse Laplace transform of (6.20) and rearranging yields

$$\xi^{(n)} = -a_1 \xi^{(n-1)} - \cdots - a_{n-1}\dot{\xi} - a_n \xi + u(t), \tag{6.21}$$

where $\xi^{(i)}$ indicates the $i$th time derivative of $\xi$. Similarly, from Fig. 6.4, $Y(s) = b(s)\Xi(s) + dU(s)$ so that

$$Y(s) = \left( b_1 s^{n-1} + \cdots + b_{n-1}s + b_n \right) \Xi(s) + dU(s),$$

which has an inverse Laplace transform of

$$y(t) = b_1 \xi^{(n-1)} + \cdots + b_{n-1}\dot{\xi} + b_n \xi + du(t). \tag{6.22}$$

As is customary, let

$$z_1 := \xi^{(n-1)}, \quad z_2 := \xi^{(n-2)}, \quad \cdots \quad z_{(n-1)} := \dot{\xi}, \quad z_n := \xi, \tag{6.23}$$

so that (6.21) becomes

$$\dot{z}_1 = -a_1 z_1 - a_2 z_2 - \cdots - a_{n-1}z_{n-1} - a_n z_n + u(t). \tag{6.24}$$

A graphical representation of (6.21), (6.22) and (6.24) is shown in Fig. 6.5. Note that the lower order subscripts of the $z_i$ correspond to higher order derivatives of $\xi$ and that (6.18) is simply the vector form of equations (6.24) and (6.22).

The change of coordinates from $x$ to $z$ will be obtained by finding a transformation matrix $T_r$, such that $z = T_r x$. Using the relation $x = T_r^{-1}z$, it can be seen that the state equation (6.4) can be transformed to

$$T_r^{-1}\dot{z} = AT_r^{-1}z + Bu,$$

so that

$$\dot{z} = T_r AT_r^{-1}z + T_r Bu.$$

Thus, $A_r$, $B_r$ and $C_r$ in (6.18) are given by

$$A_r = T_r AT_r^{-1}, \quad B_r = T_r B \quad \text{and} \quad C_r = CT_r^{-1}.$$

The reachability matrix $W_r$ of the system expressed in the original state coordinates $x$ and the associated reachability matrix $\tilde{W}_r$ of the system expressed in the transformed state coordinates $z$ can be used to determine the transformation matrix $T_r$. The reachability matrix for the transformed system is

$$\tilde{W}_r = \begin{bmatrix} B_r & A_r B_r & A_r^2 B_r & \cdots & A_r^{n-1} B_r \end{bmatrix}.$$

By individually transforming each element of $\tilde{W}_r$ it can be seen that

$$\begin{aligned} B_r &= T_r B, \\ A_r B_r &= T_r AT_r^{-1} T_r B = T_r AB, \\ A_r^2 B_r &= (T_r AT_r^{-1})^2 T_r B = T_r AT_r^{-1} T_r AT_r^{-1} T_r B = T_r A^2 B, \\ &\vdots \\ A_r^{n-1} B_r &= T_r A^{n-1} B, \end{aligned}$$

and so the reachability matrix for the transformed system is

$$\tilde{W}_r = T_r \begin{bmatrix} B & AB & A^2 B & \cdots & A^{n-1} B \end{bmatrix} = T_r W_r. \tag{6.25}$$

Since $W_r$ is invertible, it can be used to obtain the transformation $T_r$ that takes the system into reachable canonical form,

$$T_r = \tilde{W}_r W_r^{-1}.$$

Note that the reachable canonical form is not always well-conditioned and must be used with care. For example, when analyzing high-order systems, small changes in the coefficients $a_i$ can give large changes in the eigenvalues.

**Example 6.3** A transfer function for the roll stabilization of an unmanned surface vehicle (USV) using a movable mass (Fig. 6.6) is given by

$$\frac{\Phi}{U} = \frac{K_0}{s^2 + 2\zeta\omega_n s + \omega_n^2}.$$

We seek the transformation $T_r$ that can be used to put the state equation into reachable canonical form. From the transfer function, it can be seen that

**Fig. 6.6** Roll stabilization of
a USV using a movable mass

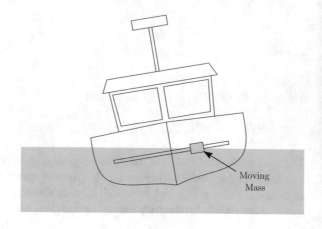

Moving
Mass

$$(s^2 + 2\zeta\omega_n s + \omega_n^2)\Phi = K_0 U, \tag{6.26}$$

which in turn gives the differential equation

$$(\ddot{\phi} + 2\zeta\omega_n\dot{\phi} + \omega_n^2\phi) = K_0 u.$$

Let $x = [x_1 \ x_2]^T$, where $x_1 = \phi$ and $x_2 = \dot{\phi}$, then the equation of the system can be written in state form $\dot{x} = Ax + Bu$ with

$$A = \begin{bmatrix} 0 & 1 \\ -\omega_n^2 & -2\zeta\omega_n \end{bmatrix} \quad \text{and} \quad B = \begin{bmatrix} 0 \\ K_0 \end{bmatrix}.$$

The corresponding reachability matrix is

$$W_r = \begin{bmatrix} 0 & K_0 \\ K_0 & -2\zeta\omega_n K_0 \end{bmatrix}$$

and its inverse is

$$W_r^{-1} = \frac{1}{K_0^2} \begin{bmatrix} 2\zeta\omega_n K_0 & K_0 \\ K_0 & 0 \end{bmatrix}.$$

The dynamics matrix and the control matrix in reachable canonical form are

$$A_r = \begin{bmatrix} -a_1 & -a_2 \\ 1 & 0 \end{bmatrix} \quad \text{and} \quad B_r = \begin{bmatrix} 1 \\ 0 \end{bmatrix},$$

respectively. Using (6.21), the coefficients $a_1$ and $a_2$ can be determined by examining the characteristic polynomial of the original system (6.17), giving $a_1 = 2\zeta\omega_n$ and $a_2 = \omega_n^2$.

$$\tilde{W}_r = [B_r \quad A_r B_r] = \begin{bmatrix} 1 & -2\zeta\omega_n \\ 0 & 1 \end{bmatrix}.$$

The resulting transformation matrix is

$$T_r = \tilde{W}_r W_r^{-1} = \frac{1}{K_0} \begin{bmatrix} 0 & 1 \\ 1 & 0 \end{bmatrix}$$

and the transformed coordinates are

$$z = T_r x = \frac{1}{K_0} \begin{bmatrix} x_2 \\ x_1 \end{bmatrix}.$$

$\square$

## 6.3 State Feedback

Here, state variable feedback will be used to obtain the roots of the closed loop system's characteristic equation, which provide a transient performance that meets the desired response. For simplicity, we will assume that the system to be controlled is described by a linear state model and has a single input and a single output (SISO).

The full system consists of the (linear) vehicle dynamics, the controller elements $K$ and $k_r$, the reference input $r$ and disturbances $w_d$ (Fig. 6.7). The goal is to regulate the output of the system $y$ such that it tracks the reference input in the presence of disturbances and also uncertainty in the vehicle dynamics.

The approach is based on the feedback of all the state variables, such that

$$\dot{x} = Ax + Bu, \quad y = Cx + Du, \tag{6.27}$$

where the disturbance $w_d$ is ignored for now and we assume that all components of the state vector are measured. For a linear feedback system, we have

$$u = -Kx + k_r r. \tag{6.28}$$

For the SISO case, when $x$ and $B$ are $n \times 1$ column vectors, $K$ will be a $1 \times n$ row vector.

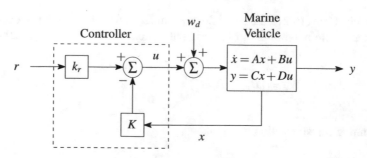

**Fig. 6.7** A feedback control system with state feedback. The controller uses the system state $x$ and the reference input $r$ to command the vehicle through its input $u$. In this diagram disturbances are modeled as an additive input $w_d$

When (6.28) is applied to (6.27), we obtain the following closed loop feedback system

$$\dot{x} = (A - BK)x + Bk_r r. \tag{6.29}$$

Here, the control problem is to solve for the feedback gain $K$ so that the closed loop system has the desired characteristic polynomial given by

$$p(s) = s^n + p_1 s^{n-1} + \cdots + p_{n-1}s + p_n. \tag{6.30}$$

Note that $\det[s\mathbf{1} - (A - BK)]$ corresponds to $p(s)$. When formulated in this way, the control problem is called the *eigenvalue assignment problem* or the *pole placement problem*. The gains $K$ and $k_r$ are used to design the dynamics of the closed loop system to satisfy our performance goals.

Note that $k_r$ does not affect the stability of the system (which is determined by the eigenvalues of $(A - BK)$, but does affect the steady state solution. In particular, the equilibrium point and steady state output for the closed loop system are given by

$$x_e = -(A - BK)^{-1} Bk_r r, \quad y_e = Cx_e + Du_e, \tag{6.31}$$

hence $k_r$ should be chosen such that $y_e = r$ (the desired output value). Since $k_r$ is a scalar, we can easily solve to show that if $D = 0$ (the most common case),

$$k_r = -\frac{1}{C(A - BK)^{-1}B}. \tag{6.32}$$

Notice that $k_r$ is exactly the inverse of the DC gain of the closed loop system (see Fig. 3.3).

## 6.3.1 Where Do I Place the Poles for State Feedback?

The state feedback approach provides more freedom in where the poles of the closed loop system can be placed during control design than either Root Locus Methods (Chap. 4) or Frequency Response Methods (Chap. 5). However, the same control design philosophy of only fixing the undesirable aspects of the open loop response still applies. More control effort is required to place the closed loop poles far away from the system's open loop poles, as well as to move the closed loop poles far away from the system zeros. Thus, if pole placement is conducted without regard for the positions of the open loop poles, the resulting controllers could require larger actuators and consume more power than an approach that only moves the poles as much as is needed (see Sect. 4.3.2).

For open loop systems that have a dominant pair of complex poles and behave similarly to a second order system with damping ratio $\zeta$ and natural frequency $\omega_n$, pole placement can be conducted using the rise time, overshoot and settling time criteria. The closed loop poles given by (6.30) can be selected (as with root locus and frequency response techniques) as a pair of dominant second order poles, with the real parts of any remaining poles selected to be sufficiently damped so that the closed loop system mimics a second order response with minimal control effort.

For higher order systems that cannot be readily approximated as a second order system, more complex pole placement techniques must be applied. Common approaches include selecting poles that minimize more complex performance measures to define a suitable closed loop characteristic equation (6.30) for the system, such as the Integral Square of the Error (ISE), the Integral of the Absolute magnitude of the Error (IAE), the Integral of Time multiplied by Absolute Error

$$\text{ITAE} := \int_0^T t|e(t)|\,dt, \tag{6.33}$$

and the Integral of Time multiplied by the Squared Error (ITSE) (see Sect. 3.5). The closed loop characteristic equations that minimize the ITAE performance measure (6.33) are given in Table 6.1.

Approaches that place the closed loop poles using either a dominant second order system model or the performance models described above do not explicitly account for the control effort, but instead require a control designer to keep the recommended design philosophy of minimally moving the locations of the open loop poles. An approach that allows one to explicitly include the control effort in placing the closed loop poles is Linear Quadratic Regulator (LQR) design, which is discussed in Sect. 6.3.5 below.

**Example 6.4** A transfer function for the simplified, linear one degree of freedom model that relates the output lateral deviation $y$ of a small unmanned surface vessel (USV) to input changes in its rudder angle $\delta_R$ is

**Table 6.1** Closed loop characteristic equations (6.30) that minimize the ITAE performance measure for a unit step input and a unit ramp input (after [4])

| Step | |
|---|---|
| $n$ | $p(s)$ |
| 1 | $s + \omega_n$ |
| 2 | $s^2 + 1.4\omega_n s + \omega_n^2$ |
| 3 | $s^3 + 1.75\omega_n s^2 + 2.15\omega_n^2 s + \omega_n^3$ |
| 4 | $s^4 + 2.1\omega_n s^3 + 3.4\omega_n^2 s^2 + 2.7\omega_n^3 s + \omega_n^4$ |
| 5 | $s^5 + 2.8\omega_n s^4 + 5.0\omega_n^2 s^3 + 5.5\omega_n^3 s^2 + 3.4\omega_n^4 s + \omega_n^5$ |
| 6 | $s^6 + 3.25\omega_n s^5 + 6.60\omega_n^2 s^4 + 8.60\omega_n^3 s^3 + 7.45\omega_n^4 s^2 + 3.95\omega_n^5 s + \omega_n^6$ |
| Ramp | |
| $n$ | $p(s)$ |
| 2 | $s^2 + 3.2\omega_n s + \omega_n^2$ |
| 3 | $s^3 + 1.75\omega_n s^2 + 3.25\omega_n^2 s + \omega_n^3$ |
| 4 | $s^4 + 2.41\omega_n s^3 + 4.93\omega_n^2 s^2 + 5.14\omega_n^3 s + \omega_n^4$ |
| 5 | $s^5 + 2.19\omega_n s^4 + 6.50\omega_n^2 s^3 + 6.30\omega_n^3 s^2 + 5.24\omega_n^4 s + \omega_n^5$ |

$$\frac{Y}{\Delta_R} = \frac{U/100}{s^2(s + 5/4)},$$

where $U = 1.5$ m/s is the constant surge speed of the USV. Here, we will design a controller by placing the closed loop poles to minimize the ITAE performance measure for a step response.

In order to provide a basis of comparison for the controller developed using the pole placement method and a controller designed using classical SISO control design techniques, a Proportional-Derivative (PD) controller of the form

$$D(s) = K\left(\frac{s}{b} + 1\right) \tag{6.34}$$

is designed using the Frequency Response Method. The design requirements are that the closed loop step response of the system has less than 30% overshoot and a settling time of less than 20 s. With $b = 0.1623$ rad/s and $K = 4.355$ the closed loop system has a step response with a settling time of 18.7 s, a rise time of 3.16 s and an overshoot of 29.7%. The natural frequency of the PD controlled system is $\omega_n = 0.4255$ rad/s.

Returning now to the design of a controller using pole placement, the differential equation corresponding to (6.4) is

$$\frac{d^3 y}{dt^3} + \frac{5}{4}\frac{d^2 y}{dt^2} = k_0 \delta_R,$$

where $k_0 := U/100$. The order of the system is $n = 3$ and so $x \in \mathbb{R}^3$. Let $u = \delta_R$, $x_1 = y$, $x_2 = \dot{y}$ and $x_3 = \ddot{y}$, then the differential equation can be written in state form $\dot{x} = Ax + Bu$, where

$$A = \begin{bmatrix} 0 & 1 & 0 \\ 0 & 0 & 1 \\ 0 & 0 & -\dfrac{5}{4} \end{bmatrix} \quad \text{and} \quad B = \begin{bmatrix} 0 \\ 0 \\ k_0 \end{bmatrix}.$$

The reachability matrix

$$W_r := \begin{bmatrix} B & AB & A^2B \end{bmatrix} = k_0 \begin{bmatrix} 0 & 0 & 1 \\ 0 & 1 & -\dfrac{5}{4} \\ 1 & -\dfrac{5}{4} & \dfrac{25}{16} \end{bmatrix}$$

has full rank, i.e. rank $(W_r) = 3$, so the system is reachable.

We seek to place the closed loop poles of the system so they correspond to the characteristic equation $p(s)$ for $n = 3$ in Table 6.1 that minimizes the ITAE performance measure,

$$p(s) = s^3 + 1.75\omega_n s^2 + 2.15\omega_n^2 s + \omega_n^3. \tag{6.35}$$

From (6.29), the polynomial $\det[s\mathbf{1} - (A - BK)]$, where $K = [k_1 \quad k_2 \quad k_3]$ is a row vector of control gains, gives the closed loop characteristic equation $p(s)$. Thus, the closed loop roots are computed using

$$A - BK = \begin{bmatrix} 0 & 1 & 0 \\ 0 & 0 & 1 \\ -k_0k_1 & -k_0k_2 & -\left(\dfrac{5}{4} + k_0k_3\right) \end{bmatrix}$$

so that

$$\det[s\mathbf{1} - (A - BK)] = \begin{vmatrix} s & -1 & 0 \\ 0 & s & -1 \\ k_0 k_1 & k_0 k_2 & s + \left(\dfrac{5}{4} + k_0 k_3\right) \end{vmatrix},$$

$$= s^3 + \left(\frac{5}{4} + k_0 k_3\right) s^2 + k_0 k_2 s + k_0 k_1.$$

(6.36)

Taking the natural frequency to be the same value as is used for the PD controller designed using frequency response methods above, $\omega_n = 0.4255$ rad/s, and equating (6.36) with (6.35), the controller gains $k_1$, $k_2$ and $k_3$ can be solved to give

$$k_1 = 5.136,$$
$$k_2 = 25.95,$$
$$k_3 = -33.69.$$

The reference gain is calculated using (6.32) to get $k_r = 5.136$.

The step response of the closed loop system has a settling time of $t_s = 17.72$ s, a rise time of 5.46 s and an overshoot of $M_p = 1.975\%$. The step response of the closed loop system designed using pole placement and that of the closed loop system designed with the PD frequency response method are plotted together in Fig. 6.8. While the rise time of the closed loop system designed using pole placement is longer than that of the frequency response designed system, it can be seen that the overshoot of the pole placement designed controller is significantly smaller and that the settling time is shorter. ☐

**Fig. 6.8** Comparison of step responses of USV lateral deviation with PD controller designed using frequency response $y_{fr}$ and using state space pole placement $y_{pp}$. The controller designed using pole placement provides a faster time response and a lower overshoot than the PD controller designed using frequency response methods

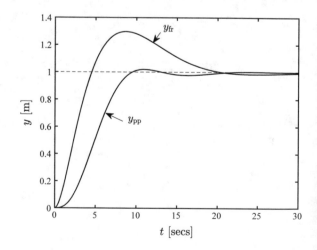

## 6.3.2   Reachable Canonical Form for State Feedback

When a system is in reachable canonical form, the parameters in the dynamics matrix are the coefficients of the characteristic polynomial, simplifying the process of selecting controller gains. Consider a system in reachable canonical form (6.18) with $d = 0$. Using (6.17), it can be seen that the open loop system has the characteristic polynomial

$$\det(s\mathbf{1} - A) = s^n + a_1 s^{n-1} + \cdots + a_{n-1}s + a_n.$$

Introducing the control law

$$u = -\tilde{K}z + k_r r = -\tilde{k}_1 z_1 - \tilde{k}_2 z_2 \cdots - \tilde{k}_n z_n + k_r r, \qquad (6.37)$$

the closed loop system becomes

$$\frac{dz}{dt} = \begin{bmatrix} -(a_1 + \tilde{k}_1) & -(a_2 + \tilde{k}_2) & -(a_3 + \tilde{k}_3) & \cdots & -(a_n + \tilde{k}_n) \\ 1 & 0 & 0 & \cdots & 0 \\ 0 & 1 & 0 & \cdots & 0 \\ \vdots & \ddots & \ddots & \ddots & \vdots \\ 0 & 0 & \cdots & 1 & 0 \end{bmatrix} z + \begin{bmatrix} k_r \\ 0 \\ 0 \\ \vdots \\ 0 \end{bmatrix} r,$$

$$y = [b_1 \ b_2 \ b_3 \ \cdots \ b_n]\, z.$$

$$(6.38)$$

The feedback parameters are combined with the elements of the first row of the matrix $\tilde{A}$, which corresponds to the parameters of the characteristic polynomial. The closed loop system thus has the characteristic polynomial

$$a_{CL}(s) = s^n + (a_1 + \tilde{k}_1)s^{n-1} + (a_2 + \tilde{k}_2)s^{n-2} + \cdots + (a_{n-1} + \tilde{k}_{n-1})s + a_n + \tilde{k}_n.$$

Requiring this polynomial to be equal to the desired closed loop polynomial (6.30), we find that the controller gains should be chosen as

$$\tilde{k}_1 = (p_1 - a_1), \quad \tilde{k}_2 = (p_2 - a_2), \quad \cdots \quad \tilde{k}_n = (p_n - a_n).$$

This feedback replaces the parameters $a_k$ in the system (6.18) by $p_k$. The feedback gain for a system in reachable canonical form is thus

$$\tilde{K} = [(p_1 - a_1) \ (p_2 - a_2) \ \cdots \ (p_n - a_n)]. \qquad (6.39)$$

To have DC gain equal to unity, the parameter $k_r$ should be chosen as

$$k_r = \frac{a_n + \tilde{k}_n}{b_n} = \frac{p_n}{b_n}. \qquad (6.40)$$

Notice that it is essential to know the precise values of parameters $a_n$ and $b_n$ in order to obtain the correct DC gain. The DC gain is thus obtained by precise calibration. This is very different from obtaining the correct steady state value by integral action.

### 6.3.3  Eigenvalue Assignment

Feedback can be used to design the dynamics of a system through the *assignment* of a set of desired closed loop eigenvalues. To solve the general case, we change coordinates so that the system is in reachable canonical form. Consider the system given by (6.27). We can change the coordinates with the linear transformation $z = T_r x$ so that the transformed system is in reachable canonical form (6.18). For such a system, the feedback is given by (6.37), where the coefficients are given by (6.39).

Transforming back to the original coordinates gives the feedback

$$u = -\tilde{K}z + k_r r = -\tilde{K}T_r x + k_r r = -Kx + k_r r, \tag{6.41}$$

where $K$ is given by

$$K = \tilde{K}T_r = [(p_1 - a_1)\ (p_2 - a_2)\ \cdots\ (p_n - a_n)]\,\tilde{W}_r W_r^{-1}. \tag{6.42}$$

This equation is known as *Ackermann's Formula*. It is implemented in the MATLAB functions *acker* and *place*. Note that *place* is preferable for systems of high order because it is numerically more well-conditioned. However, *place* cannot be used to position two or more poles at the same location while *acker* can.

Here, the $a_k$ are the coefficients of the characteristic polynomial of the matrix $A$, the reachability matrix $W_r$ is given by (6.14), the reference gain $k_r$ is given by (6.32) and

$$\tilde{W}_r = \begin{bmatrix} 1 & a_1 & a_2 & \cdots & a_{n-1} \\ 0 & 1 & a_1 & \cdots & a_{n-2} \\ \vdots & & \ddots & \ddots & \vdots \\ 0 & 0 & \cdots & 1 & a_1 \\ 0 & 0 & 0 & \cdots & 1 \end{bmatrix}^{-1}.$$

For simple problems, the eigenvalue assignment problem can be solved by introducing the elements $k_i$ of $K$ as unknown variables. We then compute the characteristic polynomial

$$a_{CL}(s) = \det(s\mathbf{1} - A + BK)$$

and equate coefficients of equal powers of $s$ to the coefficients of the desired characteristic polynomial (6.30). This gives a system of linear equations to determine the $k_i$. The equations can always be solved if the system is reachable.

The only formal requirement for eigenvalue assignment is that the system be reachable. In practice there are many other constraints because the selection of eigenvalues has a strong effect on the magnitude and rate of change of the control signal. Large eigenvalues will in general require large control signals as well as fast changes of the signals. The capabilities of the actuators, e.g. actuator saturation (Sect. 3.3.4) and actuator rate limits (Sect. 10.6), will therefore impose constraints on the possible location of closed loop eigenvalues.

### 6.3.4 State Space Integral Control

Generally, the state space controller design techniques presented thus far only produce proportional and derivative feedback. This may lead to closed loop systems exhibiting steady state error. Careful selection of the gain $k_r$ can be used to obtain the correct steady state response. However, owing to uncertainty of modeling real physical systems, it is usually more practical to use integral control to ensure that the steady state error is driven to zero.

The basic approach in state space control design is to create a new state variable (one that doesn't already appear in the vehicle dynamics) to compute the integral of the error $\dot{z} = y - r$, which is then used as a feedback term. The augmented state equation is

$$\frac{d}{dt}\begin{bmatrix} x \\ z \end{bmatrix} = \begin{bmatrix} Ax + Bu \\ y - r \end{bmatrix} = \begin{bmatrix} Ax + Bu \\ Cx - r \end{bmatrix}.$$

We search for a control such that

$$\lim_{t \to \infty} \dot{z} = 0 \quad \text{and} \quad \lim_{t \to \infty} y = r.$$

Let the control law be

$$u = -Kx - k_i z + k_r r,$$

where $K$ is the usual state feedback term, $k_i$ is the integral gain and $k_r$ is used to set the nominal input so that the desired steady state response is $y = r$. This type of control law is called a *dynamic compensator* because the integrator associated with the error signal has its own internal dynamics. The equilibrium point becomes

$$x_e = -(A - BK)^{-1} B(k_r r - k_i z_e), \quad Cx_e = r.$$

Here $z_e$ is not specified, but as long as the system is stable, automatically settles to a value such that $\dot{z} = y - r = 0$.

**Remark 6.1** Here, integral control was developed by defining a new state $z$ to be the integral of the error and by then adding another differential equation, which describes the time evolution of $z$, to the original state space system (6.4). The new system is then analyzed to determine control gains that satisfy a desired performance criteria. The process of adding equations describing the evolution of one or more additional states to a state equation and then identifying a controller or observer for the new, combined system is known as *integrator augmentation* and can also be used to design observers for unknown disturbances, include the effects of actuator rate limits and compute higher order derivatives. Integrator augmentation is also heavily used in the design of nonlinear backstepping controllers (see Chap. 10).

### 6.3.5   Linear Quadratic Regulators

The use of state feedback can be sometimes lead to PD or PID controllers that generate excessively large control forces. A way to mitigate this possibility is to design a controller that optimizes a cost function, rather than selecting the closed loop eigenvalues. This approach helps to ensure that the magnitudes of the control inputs are appropriately balanced. The design of a Linear Quadratic Regulator (LQR) controller involves minimizing a cost function $J$ subject to the constraint that the time-dependent state trajectories of the system $x(t)$ are governed by the state equation

$$\dot{x} = Ax + Bu,$$

where $x \in \mathbb{R}^n$ and $u \in \mathbb{R}^p$.

Insight into how LQR control design works can be obtained by first reviewing some of the basic principles behind the Method of Lagrange Multipliers [13]. We begin by first considering the geometric relation between a curve $C$ along which a function $L(x, y, z)$ varies and the gradient of the function $\nabla L$.

**Theorem 6.2** *Suppose that the function* $L(x, y, z) > 0$ *has continuous first order partial derivatives in a region that contains the differentiable curve*

$$C : x(t) = x(t)\hat{i} + y(t)\hat{j} + z(t)\hat{k}, \tag{6.43}$$

*where* $\hat{i}$, $\hat{j}$ *and* $\hat{k}$ *represent unit vectors defining an orthogonal coordinate system. If* $x^*$ *is a point on* $C$ *where* $L$ *has a local maximum or minimum relative to its values on* $C$, *then* $\nabla L$ *is perpendicular to* $C$ *at* $x^*$.

The above theorem can be proven by noting that the total derivative of $L$ with respect to $t$ on $C$ is

$$\frac{dL}{dt} = \frac{\partial L}{\partial x}\frac{dx}{dt} + \frac{\partial L}{\partial y}\frac{dy}{dt} + \frac{\partial L}{\partial z}\frac{dz}{dt},$$ (6.44)

$$= \nabla L \cdot v,$$

where $v = dx/dt = \dot{x}$ is the tangent vector to $C$. At any point $x^*$ on $C$ where $L$ has a maximum or minimum, $dL/dt = 0$ so that $\nabla L \cdot v = 0$.

Expanding upon this idea, consider the case when the curve $C$ (which can represent a state trajectory) lies along a level surface given by $g(x, y, z) = 0$, where the function $g(x, y, z)$ has continuous first order partial derivatives. At a point $x^*$ on $C$ where $L$ has a maximum or a minimum value, $\nabla L$ is perpendicular to the local tangent (velocity) vector, and hence perpendicular to $C$. At the same time, $\nabla g$ is also perpendicular to $C$ at $x^*$ because $\nabla g$ is perpendicular to the level surface $g(x, y, z) = 0$. Therefore, at $x^*$, $\nabla L$ is equal to a scalar multiple of $\nabla g$, i.e. $\nabla L = -\lambda \nabla g$. This provides the main idea behind the Method of Lagrange Multipliers.

### The Method of Lagrange Multipliers

Suppose that $L(x, y, z) > 0$ and $g(x, y, z)$ have continuous partial derivatives. To find the maximum and minimum values of $L$ subject to the constraint $g(x, y, z) = 0$, find the values of $x^* = (x^*, y^*, z^*)$ and $\lambda$ that simultaneously satisfy

$$\nabla L = -\lambda \nabla g \quad \text{and} \quad g(x, y, z) = 0.$$ (6.45)

**Example 6.5** (*Method of Lagrange Multipliers, adapted from* [13]) Find the minimum distance from the origin to the three dimensional surface

$$g(x, y, z) = x^2 - z^2 - 1 = 0.$$ (6.46)

This problem can be formulated as follows: Let $L(x, y, z) = x^2 + y^2 + z^2$. Minimize the function $L(x, y, z)$ subject to the constraint $g(x, y, z) = 0$. Before proceeding to the equations, consider the geometry of the problem shown in Fig. 6.9. Start with a sphere of radius squared equal to $L(x, y, z)$ centered at the origin and let it grow or shrink like a bubble until it just touches the two surfaces generated by the function $g(x, y, z) = 0$. At the two points of contact we expect the gradient of $L$ to be parallel to the gradient of $g$, so that $\nabla L = -\lambda \nabla g$, giving

$$2x\hat{i} + 2y\hat{j} + 2z\hat{k} = -\lambda(2x\hat{i} - 2z\hat{k}).$$ (6.47)

**Fig. 6.9** The surfaces
$L(x, y, z) = x^2 + y^2 + z^2$
(the spherical surface) and
$g(x, y, z) = x^2 - z^2 - 1 = 0$ (the hyperbolic surfaces).
At the two points of contact
between $L$ and $g$ we expect
$\nabla L$ to be parallel to $\nabla g$

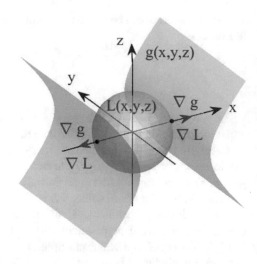

Equating the corresponding $\hat{i}$, $\hat{j}$, and $\hat{k}$ components of each side of the above equation gives

$$2x(1 + \lambda) = 0,$$
$$2y = 0, \qquad\qquad\qquad (6.48)$$
$$2z(1 - \lambda) = 0.$$

Since $x \neq 0$ at every point on the surface $g$, the solution to this system of equations is $\lambda = -1$, $y^* = 0$ and $z^* = 0$. When $y^* = 0$ and $z^* = 0$, the constraint equation $g(x, y, z) = x^2 - z^2 - 1 = 0$ requires that $x^* = \pm 1$. Thus, the points where $L$ has a minimum value on the constraint surface $g$ are $\boldsymbol{x}^* = (1, 0, 0)$ and $\boldsymbol{x}^* = (-1, 0, 0)$, and the minimum distance is $\sqrt{L(\boldsymbol{x}^*)} = 1$.                                              □

### The Method of Lagrange Multipliers, Alternate Formulation

An alternate formulation of the Method of Lagrange Multipliers can be obtained by modifying the function $L$ that we wish to maximize or minimize, by incorporated the constraints within its definition as

$$\tilde{L} := L + \lambda g. \qquad\qquad\qquad (6.49)$$

Now the conditions given by equation (6.45) for minimizing $L$ subject to the constraint $g(x, y, z) = 0$ become solving for $\boldsymbol{x}^*$ and $\lambda$, such that

$$\nabla \tilde{L} = 0. \qquad\qquad\qquad (6.50)$$

With this alternate formulation, minimizing $\tilde{L}$ is equivalent to the optimization given by

$$\min_{x} \left(L(x) + \lambda g(x)\right). \tag{6.51}$$

We are now ready to consider the *optimal control problem* of finding a control input $u$ that minimizes the cost function

$$J = \int_{t_0}^{t_f} L(x, u)\mathrm{d}t + V(x_f). \tag{6.52}$$

subject to the constraint

$$\dot{x} = f(x, u, t), \tag{6.53}$$

with the initial condition

$$x(t_0) = x_0,$$

where $L(x, u) > 0$, $x_f = x(t_f)$ and $V(x_f)$ is the *terminal cost*. Thus, the total cost $J$ is the sum of the terminal cost and the "integrated cost along the way".

To solve the optimal control problem, we will augment the cost function (6.52) with the constraint, as in (6.51), and use the alternate formulation of the Method of Lagrange Multipliers to find the $u$ that minimizes the augmented cost function. Using (6.53), let the constraint equation be $g(x, u, t) := f(x, u, t) - \dot{x} = 0$. Then the augmented cost function can be written as

$$\tilde{J} = \int_{t_0}^{t_f} \left(L + \lambda^T g\right) \mathrm{d}t + V(x_f) + v^T \psi(x_f),$$

$$= \int_{t_0}^{t_f} \left[L + \lambda^T \left(f(x, u, t) - \dot{x}\right)\right] \mathrm{d}t + V(x_f) + v^T \psi(x_f), \tag{6.54}$$

$$= \int_{t_0}^{t_f} \left\{\left[L + \lambda^T f(x, u, t)\right] - \lambda^T \dot{x}\right\} \mathrm{d}t + V(x_f) + v^T \psi(x_f),$$

where the Lagrange multipliers are now taken to be a time dependent vector $\lambda(t)$, known as the *costate vector* and $v^T \psi(x_f)$ represents the terminal constraints. The terminal constraints $\psi_i(x_f) = 0$ for $i = 1, \ldots, q$ can be used to specify the final value of the trajectory. When $q = n$, the final state is completely constrained and the value of $V(x_f)$ is fixed, such that $V(x_f)$ can be omitted from the augmented cost function. The expression in square brackets above is known as the *Hamiltonian*, defined as

$$H := L + \lambda^T f(x, u, t), \tag{6.55}$$

so that (6.54) is written as

$$\tilde{J} = \int_{t_0}^{t_f} \left(H - \lambda^T \dot{x}\right) dt + V(x_f) + v^T \psi(x_f). \tag{6.56}$$

Next, we linearize the augmented cost function near the optimal solution and note that when a solution is optimal, the difference between the augmented cost function and its linearization should tend to zero. Thus, taking $x(t) = x^*(t) + \delta x(t)$, $u(t) = u^*(t) + \delta u(t)$, $\lambda(t) = \lambda^*(t) + \delta \lambda(t)$ and $v = v^* + \delta v$, when the state trajectory and control input are optimal we have

$$\delta \tilde{J} = \lim_{\substack{\delta x \to 0 \\ \delta u \to 0 \\ \delta \lambda \to 0 \\ \delta v \to 0}} \left[\tilde{J}(x^* + \delta x, u^* + \delta u, \lambda^* + \delta \lambda, v^* + \delta v) - \tilde{J}(x^*, u^*, \lambda^*)\right] = 0.$$

$$\tag{6.57}$$

Computing the total variation of $\tilde{J}$ with respect to $x$, $u$ and $\lambda$ via application of the Chain Rule of calculus to (6.56) gives

$$\delta \tilde{J} = \int_{t_0}^{t_f} \left(\frac{\partial H}{\partial x} \delta x + \frac{\partial H}{\partial u} \delta u + \frac{\partial H}{\partial \lambda} \delta \lambda - \lambda^T \delta \dot{x} - \dot{x}^T \delta \lambda\right) dt$$

$$+ \frac{\partial V}{\partial x}(x_f) \delta x(t_f) + v^T \frac{\partial \psi}{\partial x} \delta x(t_f) + \delta v^T \psi(x_f),$$

$$\tag{6.58}$$

$$= \int_{t_0}^{t_f} \left[\frac{\partial H}{\partial x} \delta x + \frac{\partial H}{\partial u} \delta u + \left(\frac{\partial H}{\partial \lambda} - \dot{x}^T\right) \delta \lambda - \lambda^T \delta \dot{x}\right] dt$$

$$+ \frac{\partial V}{\partial x}(x_f) \delta x(t_f) + v^T \frac{\partial \psi}{\partial x} \delta x(t_f) + \delta v^T \psi(x_f).$$

Integration by parts can be used to simplify the term involving $-\lambda^T \delta \dot{x}$ in the preceding integral, as

$$-\int_{t_0}^{t_f} \lambda^T \delta \dot{x} dt = -\lambda^T \delta x \Big|_{t_0}^{t_f} + \int_{t_0}^{t_f} \dot{\lambda}^T \delta x$$

$$= -\lambda^T(t_f) \delta x(t_f) + \lambda^T(t_0) \delta x(t_0) + \int_{t_0}^{t_f} \dot{\lambda}^T \delta x. \tag{6.59}$$

Note that $\delta x(t_0) = 0$ because we specified that $x(t_0) = x_0$ in the problem formulation. Then, inserting (6.59) into (6.58) we get

$$\delta \tilde{J} = \int_{t_0}^{t_f} \left[\left(\frac{\partial H}{\partial x} + \dot{\lambda}^T\right) \delta x + \frac{\partial H}{\partial u} \delta u + \left(\frac{\partial H}{\partial \lambda} - \dot{x}^T\right) \delta \lambda\right] dt$$

$$+ \left[\frac{\partial V}{\partial x}(x_f) + v^T \frac{\partial \psi}{\partial x} - \lambda^T(t_f)\right] \delta x(t_f) + \delta v^T \psi(x_f). \tag{6.60}$$

The local conditions that make $\delta \tilde{J} = 0$ for any $\delta x(t)$, $\delta u(t)$, $\delta \lambda(t)$ and $\delta v$ are found by setting the terms in (6.60) that multiply $\delta x(t)$, $\delta u(t)$, $\delta \lambda(t)$ and $\delta v$ to zero. These conditions provide an optimal solution and are known as the *Pontryagin Maximum Principle* (PMP).

**Theorem 6.3** (Pontryagin Maximum Principle) *If $x^*(t)$, $u^*(t)$, $t_0 \le t \le t_f$ is the optimal state-control trajectory starting at $x(t_0) = x_0$ with terminal constraints $\psi_i(x_f) = 0$, where $x_f = x(t_f)$, then there exist a costate trajectory $\lambda(t)$ with*

$$\lambda^T(t_f) = \frac{\partial V}{\partial x}(x_f) + v^T \frac{\partial \psi}{\partial x}, \tag{6.61}$$

*and $v^*$, such that*

$$-\dot{\lambda}^T = \frac{\partial H}{\partial x}, \tag{6.62}$$

$$\dot{x}^T = \frac{\partial H}{\partial \lambda}, \tag{6.63}$$

*and*

$$H(x^*(t), u^*(t), \lambda^*(t)) \le H(x^*(t), u(t), \lambda^*(t)). \tag{6.64}$$

The PMP provides a means of solving optimal control problems by treating them as a differential equation with boundary conditions. By choosing the control law, one can then solve for a resulting trajectory that minimizes the cost. When $u \in \mathbb{R}^m$ and the Hamiltonian $H$ is differentiable, the optimal input must also satisfy

$$\frac{\partial H}{\partial u} = 0. \tag{6.65}$$

With knowledge of the PMP, we can now turn to the examining the *infinite horizon* (i.e. $t_f \to \infty$) LQR optimal control problem, which is given as

$$\dot{x} = Ax + Bu, \tag{6.66}$$

$$J = \frac{1}{2} \int_0^{\infty} \left( x^T Q_x x + u^T R_u u \right) dt, \tag{6.67}$$

where the weighting matrices $Q_x = Q_x^T \ge 0$, $Q_x \in \mathbb{R}^{n \times n}$ and $R_u = R_u^T > 0$, $R_u \in \mathbb{R}^{p \times p}$ are selected to minimize the cost function. The cost function (6.67) represents a trade-off between the length (magnitude) of the state vector $x$ and the cost of the control input $u$. This trade-off helps to balance the rate of convergence with the cost of the control.

To use the PMP to find the optimal control, we start by computing the Hamiltonian

$$H = \frac{1}{2}\left(x^T Q_x x + u^T R_u u\right) + \lambda^T (Ax + Bu),$$  (6.68)

so that the PMP can be applied. Then, from Theorem 6.3, we get the following conditions for optimality:

$$-\dot{\lambda} = \left(\frac{\partial H}{\partial x}\right)^T = Q_x x + A^T \lambda,$$  (6.69)

$$\dot{x} = \left(\frac{\partial H}{\partial \lambda}\right)^T = Ax + Bu,$$  (6.70)

and

$$\frac{\partial H}{\partial u} = u^T R_u + \lambda^T B = 0.$$  (6.71)

A common approach to solving the infinite horizon optimal LQR control problem is to assume that $\lambda(t) = Px$, where $P = P^T > 0$ is a matrix of constants. With this assumption, the time derivative of $\lambda$ is $\dot{\lambda}(t) = P\dot{x}$. Then using (6.66) and (6.69) gives

$$P(Ax + Bu) = -Q_x x - A^T Px.$$  (6.72)

Solving (6.71) for $u$ gives the control law

$$u = -R_u^{-1} B^T Px.$$  (6.73)

Substituting this into (6.72) and rearranging terms gives

$$\left(PA + A^T P - PBR_u^{-1}B^T P + Q_x\right)x = 0.$$  (6.74)

This latter equation is true for any $x$ if we can find a $P$, such that

$$PA + A^T P - PBR_u^{-1}B^T P + Q_x = 0.$$  (6.75)

Equation (6.75) is known as the *Algebraic Ricatti Equation*. If the system is reachable, a unique $P$ exists that makes the closed loop system stable.

The Matlab command lqr returns an optimal $u$ given $A$, $B$, $Q_x$, and $R_u$.

An important question in LQR control design is how to select $Q_x$ and $R_u$ in order to ensure a solution such that $Q_x \geq 0$ and $R_u \geq 0$. In general, there are also observability conditions on $Q_x$ that restrict the values of its elements. One can generally assume that $Q_x > 0$ to ensure that solutions to (6.75) exist [1].

A simple approach is to use diagonal weights, so that

$$Q_x = \begin{bmatrix} q_1 & 0 & \cdots & 0 \\ 0 & q_2 & & 0 \\ \vdots & & \ddots & \vdots \\ 0 & 0 & \cdots & q_n \end{bmatrix} \quad \text{and} \quad R_u = \begin{bmatrix} \rho_1 & 0 & \cdots & 0 \\ 0 & \rho_2 & & 0 \\ \vdots & & \ddots & \vdots \\ 0 & 0 & \cdots & \rho_p \end{bmatrix}.$$

With diagonal weighting, the individual elements of $Q_x$ and $R_u$ determine how much each state and input contribute to the overall cost $J$. Common strategies for choosing the weights include assigning larger values to the $q_i$ that correspond to states $x_i$ that should remain small and selecting the $\rho_i$ to penalize an input $u_i$ versus the states and other inputs.

## 6.4 Observability

When designing a control system there are many cases when it is not reasonable to assume that all states can be measured, or at least not always measured with a high degree of certainty. In such cases an *observer* can be used to estimate unmeasured states, or states with uncertain measurements, by using a mathematical model of the system combined with a noisy measurement, or with available measurements of other states (Fig. 6.10). The observability of a system provides a means of understanding whether the full system state can be determined from available measurements and the use of an observer. If a system is observable, there are no hidden dynamics. The state of the system can be determined via observation of the inputs and outputs over time. The observability of a system can be used to determine whether a set of available sensors is sufficient for controlling a system. In some sense, the observer can be thought of as a virtual sensor that estimates the states that can't be measured directly. The process of estimating states from multiple sensors using mathematical models is also called *sensor fusion*.

Consider the combined input-output system in (6.4) and (6.5),

$$\dot{x} = Ax + Bu, \quad y = Cx + Du, \tag{6.76}$$

where $x \in \mathbb{R}^n$ is the state, $u \in \mathbb{R}^p$ and $y \in \mathbb{R}^q$ is the measured output. As mentioned above, the measurement $y$ may be corrupted by noise $w_n$, but for now we will consider noise-free measurements.

**Definition 6.1** (*Observability*) The linear system (6.4) is observable if $\forall t_1 > 0$ it is possible to determine $x(t_1)$ using measurements of $y(t)$ and $u(t)$ on the interval $[0, t_1]$.

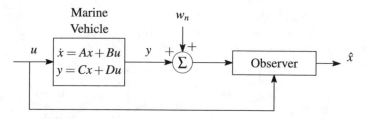

**Fig. 6.10** Block diagram for an observer. The observer uses the output measurement $y$ (possibly corrupted by noise $w_n$) and the control input $u$ to estimate the current state of the vehicle. The state estimate is denoted as $\hat{x}$

**Theorem 6.4** *The linear system in (6.76) is observable, if and only if, the the determinant of the observability matrix*

$$W_o := \begin{bmatrix} C \\ CA \\ \vdots \\ CA^{n-1} \end{bmatrix}, \tag{6.77}$$

*where $W_o \in \mathbb{R}^{n \times n}$ is invertible (i.e. has full rank n).*

### 6.4.1  Observable Canonical Form

A linear single input single output state space system is in the observable canonical form

$$\dot{z} = A_o z + B_o u, \\ y = C_o z + du, \tag{6.78}$$

if its dynamics are given by

$$\frac{dz}{dt} = \begin{bmatrix} -a_1 & 1 & 0 & \cdots & 0 \\ -a_2 & 0 & 1 & & 0 \\ \vdots & & & \ddots & 0 \\ -a_{n-1} & 0 & 0 & & 1 \\ -a_n & 0 & 0 & \cdots & 0 \end{bmatrix} z + \begin{bmatrix} b_1 \\ b_2 \\ \vdots \\ b_{n-1} \\ b_n \end{bmatrix} u, \tag{6.79}$$

$$y = \begin{bmatrix} 1 & 0 & 0 & \cdots & 0 \end{bmatrix} z + du.$$

The definition can be extended to multiple input systems, in which case $B_o$ is a matrix, rather than a vector.

**Fig. 6.11** Transfer function representation of the relation between the control input $u$ and the system output $y$

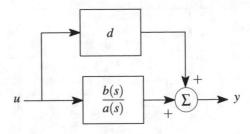

To see how the transformed coordinates $z$ are defined, consider the transfer function representation of a system with characteristic polynomial

$$a(s) = s^n + a_1 s^{n-1} + \cdots + a_{n-1} s + a_n \tag{6.80}$$

with control input $u$ and output $y$ as shown in Fig. 6.11.

Here, $d$ represents the direct term. From Fig. 6.11 it can be seen that

$$Y(s) = \frac{b(s)}{a(s)} U(s) + dU(s).$$

Rearranging terms gives

$$a(s)\,[Y(s) - dU(s)] = b(s)U(s).$$

If we define $Z_1(s) := Y(s) - dU(s)$, the corresponding differential equation is

$$z_1^{(n)} + a_1 z_1^{(n-1)} + \cdots + a_{n-1}\dot{z}_1 + a_n z_1 = b_1 u^{(n-1)} + \cdots + b_n u.$$

If the input $u$ and the output $y$, and hence $z_1 = y - du$, are available, the state $z_n$ can be computed. Define

$$\dot{z}_n := b_n u - a_n z_1$$
$$= z_1^{(n)} + a_1 z_1^{(n-1)} + \cdots + a_{n-1}\dot{z}_1 - b_1 u^{(n-1)} - \cdots - b_{n-1}\dot{u}$$

and integrate $\dot{z}_n$ to get

$$z_n = z_1^{(n-1)} + a_1 z_1^{(n-2)} + \cdots + a_{n-1} z_1 - b_1 u^{(n-2)} - \cdots - b_{n-1} u.$$

Define

$$\dot{z}_{n-1} := z_n + b_{n-1} u - a_{n-1} z_1$$

$$= z_1^{(n-1)} + a_1 z_1^{(n-2)} + \cdots + a_{n-2}\dot{z}_1 - b_1 u^{(n-2)} - \cdots - b_{n-2}\dot{u}$$

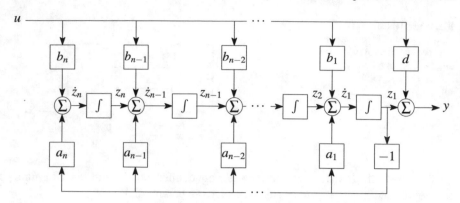

**Fig. 6.12** Block diagram of a system in observable canonical form. The states of the system are represented by individual integrators whose inputs are a weighted combination of the next integrator in the chain, the first state (rightmost integrator) and the system input. The output is a combination of the first state and the input

and integrate $\dot{z}_{n-1}$ to get

$$z_{n-1} = z_1^{(n-2)} + a_1 z_1^{(n-3)} + \cdots + a_{n-2} z_1 - b_1 u^{(n-3)} - \cdots - b_{n-2} u.$$

After repeating this process a total of $n - 1$ times we arrive at

$$\dot{z}_1 = z_2 + b_1 u - a_1 z_1.$$

Finally, after integrating $\dot{z}_1$ and adding $d\dot{u}$ we obtain at the output $y$,

$$y = z_1 + du.$$

A block diagram of this process is presented in Fig. 6.12. As in the case of the reachable canonical form (Sect. 6.2.1), it can be seen that the coefficients in the system description appear in the block diagram.

As is done when transforming a system into reachable canonical form (Sect. 6.2.1), a change of coordinates from $x$ to $z$ can be obtained by finding a transformation matrix $T_o$, such that $z = T_o x$. Using the relation $x = T_o^{-1} z$, it can be seen that the state equation (6.4) can be transformed to

$$T_o^{-1} \dot{z} = A T_o^{-1} z + B u,$$

so that

$$\dot{z} = T_o A T_o^{-1} z + T_o B u.$$

Thus, $A_o$, $B_o$ and $C_o$ in (6.78) are given by

$$A_o = T_o A T_o^{-1}, \quad B_o = T_o B \quad \text{and} \quad C_o = C T_o^{-1}.$$

The observability matrix $W_o$ of the system expressed in the original state coordinates $x$ and the associated observability matrix $\tilde{W}_o$ of the system expressed in the transformed state coordinates $z$ can be used to determine the transformation matrix $T_o$. The observability matrix for the transformed system is

$$\tilde{W}_o = \begin{bmatrix} C_o \\ C_o A_o \\ C_o A_o^2 \\ \vdots \\ C_o A_o^{n-1} \end{bmatrix}. \tag{6.81}$$

Using $A_o$ and $C_o$ in (6.79) to construct $\tilde{W}_o$ and then taking its inverse, it can be shown that the inverse of the observability matrix in reachable canonical form has a relatively simple structure given by

$$\tilde{W}_o^{-1} = \begin{bmatrix} 1 & 0 & 0 & \cdots & 0 \\ a_1 & 1 & 0 & \cdots & 0 \\ a_2 & a_1 & 1 & & 0 \\ \vdots & & & \ddots & 0 \\ a_{n-1} & a_{n-2} & a_{n-3} & \cdots & 1 \end{bmatrix}, \tag{6.82}$$

where the lower triangular part of the matrix consists of the coefficients $a_i$ of the open loop characteristic equation (6.80). We can use (6.82) to determine the observability transformation matrix $T_o$ by relating the observability matrix of system expressed in the observable canonical form $\tilde{W}_o^{-1}$ to the observability matrix of the original system $W_o$. By individually transforming each of the matrices comprising $\tilde{W}_o$ in (6.81), it can be seen that

$$\begin{aligned} C_o &= C T_o^{-1}, \\ C_o A_o &= C T_o^{-1} T_o A T_o^{-1} = C A T_o^{-1}, \\ C_o A_o^2 &= C T_o^{-1} \left( T_o A T_o^{-1} \right)^2 = C T_o^{-1} \left( T_o A T_o^{-1} \right) \left( T_o A T_o^{-1} \right) = C A^2 T_o^{-1}, \\ &\vdots \\ C_o A_o^{n-1} &= C T_o^{-1} \left( T_o A T_o^{-1} \right)^{n-1} = C A^{n-1} T_o^{-1}, \end{aligned}$$

and so the observability matrix for the transformed system is

$$\tilde{W}_o = \begin{bmatrix} C \\ CA \\ CA^2 \\ \vdots \\ CA^{n-1} \end{bmatrix} \quad T_o^{-1} = W_o T_o^{-1}. \tag{6.83}$$

When a system is observable $W_o$ is invertible, and can be used to obtain the transformation $T_o^{-1}$

$$T_o^{-1} = W_o^{-1} \tilde{W}_o. \tag{6.84}$$

Then the transformation that takes the system into reachable canonical form is

$$T_o = \tilde{W}_o^{-1} W_o. \tag{6.85}$$

As with the reachability matrix, the observability matrix may be poorly conditioned numerically.

## 6.5   State Estimation

An *observer* is a dynamical system, which uses a system's control input $u$ and measurements of its output $y$ to produce an estimate of its state $\hat{x} \in \mathbb{R}^n$. The observer is designed to drive the state estimate $\hat{x}$ towards its true value $x$, so that

$$\lim_{t \to \infty} \hat{x}(t) = x(t).$$

Consider the single input single output system in (6.27) and let $D = 0$ for simplicity. Define the observer dynamics as

$$\frac{d\hat{x}}{dt} = A\hat{x} + Bu + L(y - C\hat{x}). \tag{6.86}$$

Feedback from the measured output is provided by the term $L(y - C\hat{x})$, which is proportional to the difference between the observed output and the output predicted by the observer. Let $\tilde{x} := x - \hat{x}$ be the estimation error. It follows from (6.27) and (6.86) that

$$\frac{d\tilde{x}}{dt} = (A - LC)\tilde{x}.$$

If $L$ can be chosen in such a way that $(A - LC)$ has eigenvalues with negative real parts, the error $\tilde{x}$ will go to zero. The convergence rate is determined by an appropriate selection of the eigenvalues.

The problems of finding a state feedback controller and finding an observer are similar. The observer design problem is the *dual* of the state feedback design problem. State feedback design by eigenvalue assignment is mathematically equivalent to finding a matrix $K$ so that $(A - BK)$ has given eigenvalues. Designing an observer with prescribed eigenvalues is equivalent to finding a matrix $L$ so that $(A - LC)$ has given eigenvalues.

**Theorem 6.5** *Observer design by eigenvalue assignment: If the single input single output system (6.27) is observable, then the dynamical system (6.86) is a state observer when $L$ is chosen as*

$$L = T_o^{-1} \begin{bmatrix} p_1 - a_1 \\ p_2 - a_2 \\ \vdots \\ p_n - a_n \end{bmatrix}, \tag{6.87}$$

*where $T_o^{-1}$ is given by (6.84). The resulting observer error $\tilde{x} = x - \hat{x}$ is governed by a differential equation having the characteristic polynomial*

$$p(s) = s^n + p_1 s^{n-1} + \cdots + p_{n-1} s + p_n.$$

For simple low-order problems it is convenient to introduce the elements of the observer gain $L$ as unknown parameters and solve for the values required to give the desired characteristic polynomial. Owing to the duality between state feedback and observer design, the same rules (with proper substitutions of the control $B$ and the sensor matrices $C$) can be used to design observers.

Note that it is also possible to design *reduced order observers*, which estimate fewer than the $n$ states contained in the full state vector. However, it is generally better to implement a full state observer because in addition to estimating unmeasured states, a full state observer will also filter some of the noise present in the measured states.

## 6.5.1 Where Do I Place the Observer Poles?

In order to ensure that the state feedback poles dominate the dynamic response of the system, the observer poles must be chosen so that observer errors decay more rapidly than those of the state feedback controller. Thus, the observer poles are generally chosen to be 2 to 5 times faster than the controller poles [6]. As discussed in Sect. 6.3.1, selecting controller gains for a fast response increases the bandwidth of the controller, which in turn can lead to a large control effort. Similarly, selecting observer gains for a fast response will increase the bandwidth of the observer. However, this does not directly affect the control effort. Instead, it causes the observer to pass on more of the sensor noise to the controller. When the sensor noise is high, one could select observer poles to be 2 times faster than the controller poles, however, it is quite likely that the overall system response would be influenced by the location of the observer poles. Good estimator design requires one to balance a fast observer response with a sufficiently low bandwidth so that the sensor noise does not adversely affect actuator activity.

## 6.6   Separation Principle

As shown above, the problems of finding a state feedback controller and finding an observer are similar. However, the feedback gain $K$ depends only on the dynamics matrix $A$ and the control matrix $B$ and can be computed assuming that all of the states are known. At the same time, the observer gain $L$ depends only on the *open loop* dynamics matrix $A$ and the sensor matrix $C$. Thus, eigenvalue assignment for output feedback can be separated into an eigenvalue assignment problem for the feedback gain and a separate eigenvalue assignment problem for the observer. This idea is known as the *Separation Principle*. It permits state space controller design to be broken down into two steps. First, a feedback control law is designed, assuming that all state variables are available. Then an observer (also called an *estimator*) is designed to estimate the entire state using available output measurements.

To see this, consider the single input single output system (6.27). We wish to design a controller, which may depend upon several of the states, but for which only the single output measurement is available. Since all of the states are not measureable, we use the feedback law

$$u = -K\hat{x} + k_r r, \tag{6.88}$$

where $\hat{x}$ depends on the observer dynamics given by (6.86). To analyze the closed loop system, we can change coordinates by replacing the estimated state $\hat{x}$ with the estimation error $\tilde{x} = x - \hat{x}$. Subtracting (6.86) from (6.27), we can obtain an equation for the time evolution of the estimation error

$$\frac{d\tilde{x}}{dt} = A(x - \hat{x}) - LC(x - \hat{x}) = (A - LC)\tilde{x}.$$

Using the observer feedback control law (6.88) in (6.27) and using the relation $\tilde{x} = x - \hat{x}$ to eliminate $\hat{x}$ gives

$$\begin{aligned}
\frac{dx}{dt} &= Ax + Bu = Ax - BK\hat{x} + Bk_r r \\
&= Ax - BK(x - \tilde{x}) + Bk_r r \\
&= (A - BK)x + BK\tilde{x} + Bk_r r.
\end{aligned}$$

Then the combined state and error equations for full state feedback with an observer are

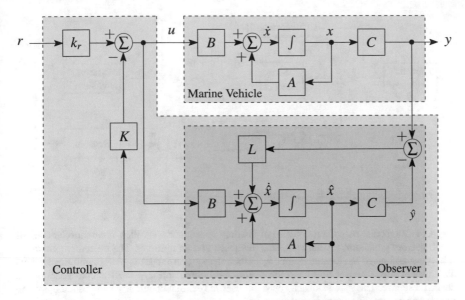

**Fig. 6.13** Block diagram of an observer-based control system. The observer uses the measured output $y$ and the input $u$ to construct an estimate of the state. This estimate is used by a state feedback controller to generate the corrective input. The controller consists of the observer and the state feedback

$$\frac{d}{dt}\begin{bmatrix} x \\ \tilde{x} \end{bmatrix} = \begin{bmatrix} Ax - BK & BK \\ 0 & A - LC \end{bmatrix}\begin{bmatrix} x \\ \tilde{x} \end{bmatrix} + \begin{bmatrix} Bk_r \\ 0 \end{bmatrix} r. \qquad (6.89)$$

Note that the observer error $\tilde{x}$ is not affected by the reference signal $r$. The characteristic polynomial of the closed loop system is

$$\lambda(s) = \det(s\mathbf{1} - A + BK)\det(s\mathbf{1} - A + LC),$$

which is a product of two terms—the characteristic polynomial of the closed loop system obtained with state feedback and the characteristic polynomial of the observer error.

A block diagram of the controller is shown in Fig. 6.13. The controller contains a model of the vehicle being controlled—this is known as the *Internal Model Principle*.

## 6.7 Two Degree of Freedom Controllers

Controller design often involves balancing multiple, sometimes conflicting, performance requirements, including robustness, stability, response time, overshoot and tracking. The use of a two degree of freedom controller can permit one to separate the problems of trajectory tracking and disturbance rejection by essentially creat-

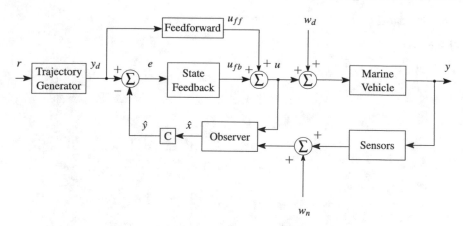

**Fig. 6.14** Block diagram of a two degree of freedom state space controller. The controller consists of a trajectory generator, feedforward, state feedback and an observer. The trajectory generation subsystem computes the desired output $y_d$. When the system model is perfect and there are no disturbances $w_d$ or sensor noise $w_n$, the feedforward command $u_{ff}$ itself should generate the desired response. The state feedback controller uses the estimated output $\hat{y}$ and desired output $y_d$ to compute a corrective control input $u_{fb}$

ing two closed loop transfer functions, each of which can be individually tuned. The controller is said to have two degrees of freedom since the responses to reference commands and disturbances is decoupled. The strategy consists of designing a feedback controller to satisfy stability and disturbance rejection requirements, and designing a second, feedforward controller according to a set of tracking performance requirements. An example of a two degree of freedom controller architecture is shown in Fig. 6.14. The controller consists of four parts—an observer, which computes estimates of the states based on a model and vehicle inputs and outputs; a state feedback controller; a feedforward controller; and a trajectory generator. The trajectory generator computes the desired states needed to achieve the reference command. When there is no modeling uncertainty or any disturbances, the *feedforward controller* generates a control signal $u_{ff}$ that should generate the desired response. When modeling errors or disturbances exist, the state feedback controller mitigates their effects by generating the signal $u_{fb}$ to counteract them.

## 6.8   Linear Disturbance Observer Based Control

Thus far, we have considered the design of state feedback controllers and observers in the absence of model uncertainty and exogenous disturbances. However, unmodeled dynamics (e.g. imprecise models of a vehicle's inertia tensor, drag, actuators, etc.) and external disturbances (such as wind, currents or waves) can generally affect the practical implementation of controllers and observers in a real marine system. While

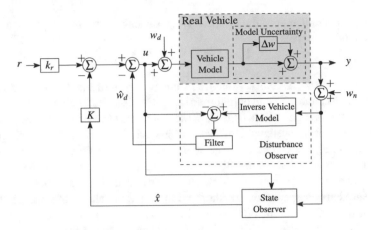

**Fig. 6.15** Conceptual design of a two DOF disturbance observer based controller

controllers and observer gains can be selected to attenuate the effects of disturbances, there may be conflicting control design requirements, such as stability, tracking, response time, etc. that make it difficult to also provide suitable disturbance rejection. The use of integral control (Sect. 6.3.4) can be used to drive the steady state error caused by constant disturbances to zero, but may also increase overshoot in tracking and negatively impact system stability. Controllers designed using optimal control techniques, such as the LQR controller discussed in Sect. 6.3.5, can be particularly sensitive to model uncertainty and disturbances [11]. As shown in Fig. 6.15, the basic idea in *disturbance observer based control* (DOBC) is to design an observer that estimates the total disturbance acting on a system, under the assumption that the total disturbance represents the combined effects of model uncertainty and exogenous disturbances, and to subtract the total disturbance estimate from the control input in order to mitigate the influence of the real disturbances.

Taking model uncertainty and exogenous disturbances to be lumped together the state equation (6.4) becomes

$$\dot{x} = Ax + B_u u + B_d w_d,$$

$$y = Cx,$$

(6.90)

where it is assumed that the direct term $D = 0$. When the influence of $w_d$ acts along the input channels of $u$, the disturbances are known as *matched disturbances*, and we take $B_d = B_u$. Alternatively, a more general matched disturbance can be modeled by taking $B_d = B_u \Gamma$, where $\Gamma$ is a transformation matrix that maps the disturbances and model uncertainties $w_d$ onto the input channels.

The time dependence of the lumped disturbances are assumed to obey a simple dynamic model, such as

$$\dot{\boldsymbol{\xi}} = A_\xi \boldsymbol{\xi}, \quad \boldsymbol{w}_d = \boldsymbol{C}_\xi \boldsymbol{\xi}, \tag{6.91}$$

where $\boldsymbol{\xi}$ is a state variable describing the time evolution of the disturbance, $A_\xi$ is a dynamics matrix and $\boldsymbol{C}_\xi$ is an output matrix. In principle, this model must be known *a priori* in the design of the DOBC, which is somewhat of a strict requirement for most engineering systems. However, in practice, it has been found that precise disturbance estimates can be obtained with very simple models [11]. For example, it is often possible to assume that disturbances, which vary slowly in time compared to the vehicle dynamics, can be modeled as being constant, i.e. $\dot{\boldsymbol{\xi}} = 0$. For example, the measured power spectra of nearshore wind over water shows that the wind tends to vary at time scales on the order of a minute, whereas the dynamic response time of unmanned surface vessels is on the order of about a second [10].

In addition, many environmental disturbances, such as winds and currents, are often modeled as first-order Markov processes [7, 9] of the form

$$T_b \dot{\boldsymbol{\xi}} = -\boldsymbol{\xi} + A_n \boldsymbol{b}_n(t), \tag{6.92}$$

where $\boldsymbol{\xi}$ is a vector of wind-/current-generated bias forces and moments, $T_b$ is a diagonal matrix of time constants, $\boldsymbol{b}_n(t)$ is a vector of zero-mean Gaussian white noise, and $A_n$ is a diagonal matrix that scales the amplitude of $\boldsymbol{b}_n$. These environmental models are similar to (6.91), but have a driving term governed by a random process. When this random process changes slowly in comparison to the dynamics of the vehicle, (6.91) can be a reasonable model for the disturbances. To see this, take $\boldsymbol{b}_n$ to be a constant, then (6.92) can be rewritten as

$$\dot{\boldsymbol{\xi}} = -T_b^{-1}\boldsymbol{\xi} + T_b^{-1}A_n \boldsymbol{b}_n. \tag{6.93}$$

Let $\boldsymbol{\xi} := \bar{\boldsymbol{\xi}} + \phi$, where $\phi$ is a constant to be computed. Then $\dot{\boldsymbol{\xi}} = \dot{\bar{\boldsymbol{\xi}}}$. Thus (6.93) gives

$$\dot{\boldsymbol{\xi}} = \dot{\bar{\boldsymbol{\xi}}} = -T_b^{-1}\boldsymbol{\xi} + T_b^{-1}A_n \boldsymbol{b}_n = -T_b^{-1}(\bar{\boldsymbol{\xi}} + \phi) + T_b^{-1}A_n \boldsymbol{b}_n.$$

If we take $\phi = A_n \boldsymbol{b}_n$, (6.93) can be rewritten as $\dot{\bar{\boldsymbol{\xi}}} = -T_b^{-1}\bar{\boldsymbol{\xi}}$, which has the same form as (6.91).

It is also possible to assume more complex time dependencies for the exogenous disturbances, such as sinusoidal variations, by using integrator augmentation of (6.92) to model higher order terms [11].

As shown in [2], both the state and disturbance can be simultaneously estimated by constructing a pair of observers, $\Sigma_1$ and $\Sigma_2$, with the form

$$\Sigma_1 : \begin{cases} \dot{\hat{x}} = A\hat{x} + B_u u + L_x \left(y - \hat{y}\right) + B_d \hat{w}_d, \\[2mm] \hat{y} = C\hat{x}, \end{cases} \tag{6.94}$$

$$\Sigma_2 : \begin{cases} \dot{\hat{\xi}} = A_\xi \hat{\xi} + L_d \left(y - \hat{y}\right), \\[2mm] \hat{w}_d = C_\xi \hat{\xi}, \end{cases} \tag{6.95}$$

where $\hat{x}$ is the state estimate, $L_x$ is the state observer gain, $\hat{w}_d$ is the disturbance estimate, $\hat{\xi}$ is the estimate of $\xi$, and $L_d$ is the disturbance observer gain. It can be shown that the estimate yielded by (6.94) and (6.95) can asymptotically track both the state of system (6.90) and the unknown disturbance governed by (6.91) if the observer gains $L_x$ and $L_d$ are chosen so that the observer error dynamics are stable [11]. The observer gains are determined using pole placement.

In the case of matched disturbances, the corresponding control input is then given by

$$u = -K\hat{x} - \Gamma \hat{w}_d, \tag{6.96}$$

where the control gain matrix is determined using pole placement to ensure that $(A - B_u K)$ has the desired closed loop roots. When disturbances are mismatched, the control input is given by

$$u = -K\hat{x} - K_d \hat{w}_d, \tag{6.97}$$

where

$$K_d = \left[C(A - B_u K)^{-1} B_u\right]^{-1} \left[C(A - B_u K)^{-1} B_d\right]. \tag{6.98}$$

## 6.9  Matlab for State Space Controller and Observer Design

Here, the use of Matlab for controller and observer design using state space methods will be demonstrated through an example.

**Example 6.6**  A Lagrangian float is a type of underwater robotic buoy often used for data collection in the open ocean (Fig. 6.16). They are typically designed to drift in ocean currents and so their horizontal motion is typically passive, while their vertical motion (depth) may be automatically controlled. Using precise depth control, a Lagrangian float can be used as a camera platform to provide high resolution seafloor imaging to evaluate the health of benthic habitats, such as coral reefs [12]. The linearized equation of motion for the depth of a Lagrangian float is given by

$$\ddot{z} = \frac{F_z}{m},$$

**Fig. 6.16** A Lagrangian
float designed for underwater
imaging

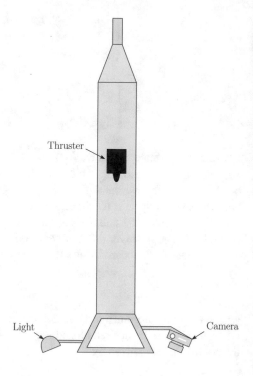

where $m$ is the mass of the float (including added mass), $z$ is float depth and $F_z$ is the
control force along the vertical direction, which could be developed using a thruster
or buoyancy-control mechanism. Let $u := F_z/m$ be the control input. The objective
is to design a control input $u$ that will permit the closed loop system to maintain a
desired depth. The state space representation of the system is

$$\frac{d}{dt}\begin{bmatrix} x_1 \\ x_2 \end{bmatrix} = \begin{bmatrix} 0 & 1 \\ 0 & 0 \end{bmatrix}\begin{bmatrix} x_1 \\ x_2 \end{bmatrix} + \begin{bmatrix} 0 \\ 1 \end{bmatrix} u,$$

where $x_1 = z$ represents a displacement from the desired depth and $x_2 = \dot{z}$ is the
vertical velocity. The output of the system is $y = Cx = [1\ 0]x$ and there is no direct
term, i.e. $D = 0$. The dynamics, control and sensor matrices are defined in Matlab
as follows.

```
A = [
    0    1
    0    0
    ];
```

```
B = [
     0
     1
     ];

C = [1  0];

D = 0;
```

First we explore the response of the open loop system. If the initial displacement of the drifter is non-zero, without any feedback to generate a correcting buoyancy force, the drifter should not move. The following code confirms this, as shown in Fig. 6.17.

```
%% Simulate open loop response
t = 0:0.01:50;
u = zeros(size(t));    % the buoyance force input is zero
x0 = [0.25 0];         % the initial position is set to 25 cm

sys = ss(A,B,C,D);   % construct the state space system

[y,t,x] = lsim(sys,u,t,x0); % compute the response
plot(t,y)                              % plot the response
xlabel('t [sec]')
ylabel('z [m]')
```

To get a fast response we select $\zeta = 1/\sqrt{2}$. We aim to have a settling time of about $t_s = 30$ seconds (2% criterion) and so set $\omega_n = 4/(t_s \zeta) = 0.1886$ rad/s. The

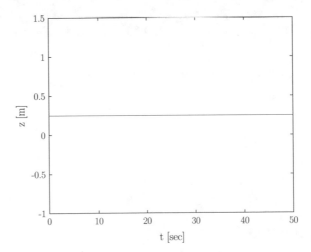

**Fig. 6.17** Open loop response of the Lagrangian float

closed loop system will be of second order, so we only need to specify $\zeta$ and $\omega_n$ for pole placement with the characteristic polynomial $p(s) = s^2 + 2\zeta\omega_n s + \omega_n^2$. The following code is used to verify that the system is reachable, specify the desired closed loop poles, perform pole placement and plot the response of the closed loop system to a non-zero initial displacement of the buoy. When running the code, it can be seen that the rank of the controllability (reachability) matrix is 2, and so the system is controllable. The response of the closed loop system to a non-zero initial condition is shown in Fig. 6.18.

```
%% Create and simulate closed loop system
% Construct controllability (reachability) matrix
co = ctrb(sys_ss);
% rank of reachability matrix
controllability = rank(co);

ts = 30;                                 % desired settling time
zeta = 1/sqrt(2);                % desired damping ratio
omegan = 4/ts/zeta;      % desired CL natural frequency

% Desired characteristic polynomial of CL system
p = roots([1 2*zeta*omegan omegan^2]);

% Perform pole placement
K = place(A,B,p);

% Construct the closed loop system
sys_cl = ss(A-B*K,B,C,0);
% Compute the reference input gain
kr = -1/(C*inv(A-B*K)*B);

% Simulate the closed loop system
[y,t,x] = lsim(sys_cl,kr*u,t,x0);
plot(t, y, t, zeros(size(y)),'--k')
xlabel('t [sec]')
ylabel('z [m]')
```

**Fig. 6.18** Closed loop response of the Lagrangian float

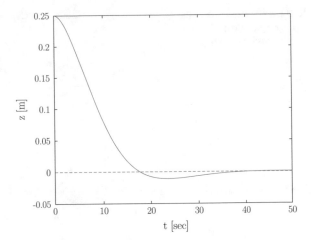

Next we design an observer for the system. We select the natural frequency of the observer to be 5 times larger than that of the closed loop system and choose $\zeta = 0.5$ as its damping ratio. The observer poles are then placed following the same procedure as was used to place the controller poles.

```
%% Observer design
zetaon = 0.5;
omegaon = 5*omegan;

% Characteristic equation of the observer
op = roots([1 2*zetaon*omegaon omegaon^2]);

L = place(A',C',op)';
```

Using the matrix of observer gains $L$ found above, we construct a compensator to control the system. From Fig. 6.19 it can be seen that the settling time of the compensated system is shorter than the system with state feedback only, although the overshoot is a little larger. The response could be further tuned by modification of the observer damping ratio.

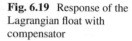

**Fig. 6.19** Response of the Lagrangian float with compensator

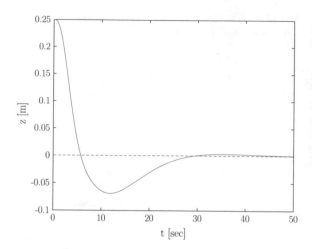

```
%% Construct the compensator matrices
At = [ A-B*K              B*K
       zeros(size(A))     A-L*C ];


Bt = [     B*kr
       zeros(size(B)) ];

Ct = [ C    zeros(size(C)) ];

% Construct CL compensated system
sys = ss(At,Bt,Ct,0);

% Simulate the response
% the same initial conditions are used for both
% state and observer
[y,t,x] = lsim(sys,zeros(size(t)),t,[x0 x0]);

% the third output of lsim is the state trajectory x
% with length(t) rows and as many columns as states

plot(t, y, t, zeros(size(y)),'--k')
xlabel('t [sec]')
ylabel('z [m]')
```

Lastly, we explore the time response of the observer state estimates. The following code is used to isolate the states and the state estimates from the output of the *lsim* command in the code section above. The results are plotted in Fig. 6.20 and demonstrate how the state estimates from the observer converge to the system states over time.

```
%% Observer convergence

% separate the state and state estimates from
% lsim output above
n = 2;
e = x(:,n+1:end);
x = x(:,1:n);
x_est = x - e;

% Explicitly name state variables to aid plotting
z = x(:,1);
z_dot = x(:,2);
z_est = x_est(:,1);
z_dot_est = x_est(:,2);

% Linewidth command used to make lines
% heavier for visibility
plot(t,z,'-r','LineWidth',1.5),
hold on
plot(t,z_est,':r','LineWidth',1.5);
plot(t,z_dot,'-b','LineWidth',1.5);
plot(t,z_dot_est,':b','LineWidth',1.5);
hold off

legend('z','z_{est}','zdot','zdot_{est}')
xlabel('t [sec]')
```

## Problems

**6.1** Consider vehicle dynamics given by

$$\dot{x} = \begin{bmatrix} 0 & 2 \\ 14 & -8 \end{bmatrix} x + \begin{bmatrix} 2 \\ 4 \end{bmatrix} u,$$

$$y = [2 \quad 6]x.$$

(a) Find the transfer function using matrix algebra.

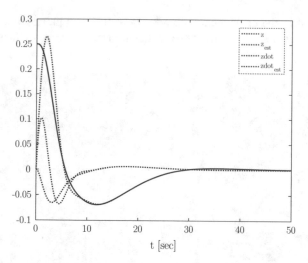

**Fig. 6.20** Convergence of the state estimates during the response of the Lagrangian float with compensator

(b) Draw the block diagram for the vehicle at the integrator level.
(c) Find the closed loop characteristic equation if the feedback is

   (i) $u = -[K_1 \quad K_2]x$,
   (ii) $u = -Ky$.

**6.2** Using the state variables indicated, give the state space representation for each of the systems shown in Fig. 6.21a–b. Give the corresponding transfer function for each system using block diagram reduction and matrix algebra.

**6.3** Consider the system shown in Fig. 6.22. Determine the relationship between $K$, $\tau$ and $\zeta$ that minimizes the ITAE criterion for a step input.

**6.4** For the transfer function

$$\frac{Y(s)}{U(s)} = \frac{5s + 10}{s^2 + 5s + 25},$$

(a) write the state equations in both reachable canonical and observable canonical form; and
(b) draw the block diagram at the integrator level.

**6.5** For the transfer function

$$\frac{Y(s)}{U(s)} = \frac{s^2 - 9}{s^2(s^2 - 4)}$$

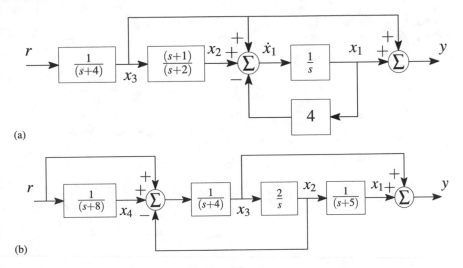

(a)

(b)

**Fig. 6.21 a** and **b** Block diagrams for Problem 6.2

**Fig. 6.22** Block diagram for Problem 6.3

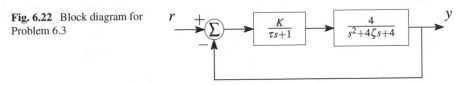

(a) write the state equations in both reachable canonical and observable canonical form; and

(b) draw the block diagram at the integrator level.

**6.6** A spar buoy is a long cylindrical buoy vertically positioned in the water column, which is designed to provide a stable platform for mounting scientific instrumentation with minimal wave-induced motion. As optical telescopes perform best when far away from the light-pollution of human-inhabited areas, one possible application of spar buoys is to provide a stable mount for astronomy using an optical telescope in the open ocean. Consider the spar-mounted telescope shown in Fig. 6.23. The system is designed to orient the heading angle of the telescope $\psi$ using a set of underwater thrusters, which provide a positioning torque $\tau_\psi$.

The equation of motion of the system is given by

$$I_z \ddot{\psi} = \tau_\psi.$$

If we take $x_1 = \psi$, $x_2 = \dot{\psi}$ and $u = \tau_\psi / I_z$, the equation of motion can be expressed in state-space form as

$$\frac{d}{dt}\begin{bmatrix} x_1 \\ x_2 \end{bmatrix} = \begin{bmatrix} 0 & 1 \\ 0 & 0 \end{bmatrix}\begin{bmatrix} x_1 \\ x_2 \end{bmatrix} + \begin{bmatrix} 0 \\ 1 \end{bmatrix} u(t).$$

**Fig. 6.23** A spar-mounted telescope

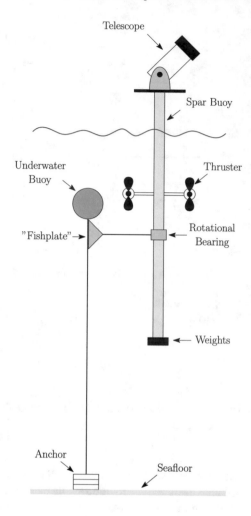

Let the LQR cost function $J$ in (6.67) be comprised of

$$Q_x = \begin{bmatrix} q^2 & 0 \\ 0 & 0 \end{bmatrix} \quad \text{and} \quad R_u = 1.$$

Take $P = P^T > 0$ to be

$$P = \begin{bmatrix} a & b \\ b & c \end{bmatrix}.$$

(a) Analytically solve the Algebraic Ricatti equation (6.75) for $P$ in terms of $q$.
(b) What is the resulting gain matrix $K$ in the state feedback control law $u = -Kx$ given by the LQR optimal control law (6.73)?

(c) The closed loop system is of second order. Find the characteristic equation of the closed loop system dynamics given by the eigenvalues of $A - BK$. Comparing this to the canonical form of a second order system (3.1), what are the natural frequency $\omega_n$ and damping ratio $\zeta$ of the closed loop system? Does this result agree with what is expected for a typical second order system?

**6.7** The rolling motion of a large unmanned surface vessel can be approximately described by the differential equation

$$\ddot{\phi} + \omega^2 \phi = u.$$

(a) Write the equations of motion in state space form.
(b) Design an observer that reconstructs both state variables using the measurements of the USV's roll rate $\dot{\phi}$. Assume the roll period of the USV is $T = 8$ s, where $\omega = 2\pi/T$ and place the observer poles at $s = -1.5 \pm 1.5j$ rad/s.
(c) Write the transfer function between measured value of $\dot{\phi}$ and $\hat{\phi}$ (the estimated value of $\phi$).
(d) Design a controller using state feedback, placing the poles at $s = -0.3 \pm 0.3j$ rad/s.
(e) Give the transfer function of the compensator.

**6.8** Consider the pitch stability of an AUV at constant forward speed. The transfer function of the system is

$$G(s) = \frac{1}{(s^2 - 9)}.$$

(a) Find $A_o$, $B_o$ and $C_o$ for this system when it is in observer canonical form (6.79).
(b) Is the system given by $A_o$ and $B_o$ reachable/controllable?
(c) Compute the state feedback matrix $K$ that places the closed loop poles at $s = -3 \pm 3j$.
(d) Is the system observable?
(e) Design an observer with observer poles placed at $s = -6 \pm 6j$.
(f) A PD controller has been designed to stabilize the pitch angle, giving a loop transfer function of

$$L(s) = \frac{k_{pd}(s + 4)}{s^2 - 9}.$$

It was found that the closed loop system with this PD controller is *conditionally stable* and that the gain must be $k_{pd} > 9/4$ for stability. Suppose we take $k_{pd} = 9/4$, such that $L(s)$ is right at the stability limit. Show that if we design a state feedback controller of the form $u = -Kx + k_r r$ to control the system $L(s)$, there is a feedback gain $K$ that makes the system unobservable.

**6.9** Consider an unmanned surface vehicle with dynamics $M\dot{u} + Du = \tau$, where $u := [u \ v \ r]^T$, $u$ is the surge speed, $v$ is the sway speed and $r$ is the yaw rate. The normalized inertia tensor, linear drag matrix and propeller force/moment input are

$$M = \begin{bmatrix} 1 & 0 & 0 \\ 0 & 1.9 & 0 \\ 0 & 0 & 2.2 \end{bmatrix}, \quad D = -\begin{bmatrix} 1/10 & 0 & 0 \\ 0 & 1/5 & 1/20 \\ 0 & 1/20 & 1/5 \end{bmatrix},$$

and

$$\tau = \begin{bmatrix} 1 & 1 \\ 0 & 0 \\ 1/2 & -1/2 \end{bmatrix} \begin{bmatrix} T_1 \\ T_2 \end{bmatrix},$$

respectively. This model assumes that the differential forces provided by two thrusters are used to propel and steer the vessel (see [8], for example).

(a) Find the dynamics matrix $A$ and the control matrix $B$ that result when the state equation is written in the form (6.4) as $\dot{u} = -M^{-1}Du + M^{-1}\tau$.

(b) Is the system reachable/controllable? Would the system be controllable if we set $D_{2,3} = D_{3,2} = 0$, why?

(c) Use pole placement to determine a state feedback matrix $K$ that will minimize the ITAE criterion for a step input with a closed loop natural frequency of $\omega_n = 1.25$ rad/s.

(d) Plot the closed loop response of $u$, $v$ and $r$ for the initial condition $u(0) = [0\,0\,0]^T$ when the system is commanded to follow a desired speed of $u_d(t) = [1(t)\,0\,0]^T$ for $0 \le t \le 30$ s.

**6.10** There has been an ever increasing interest in developing highly automated transportation systems, e.g. driverless trams and buses, which can essentially be thought of as mobile robots. A group of engineers is exploring the potential for a surface effect ship (SES) to be used as a fast ferry to carry cars across a bay. An SES is similar to hovercraft, but has rigid side hulls, and uses large fans to pressurize a cushion of air, which supports its weight (Fig. 6.24). Such a vessel can transport a large fraction of its own weight at a relatively high speed, making it an interesting candidate for this application. The total weight of the cars must be known in order to ensure safe operation of the SES. It is proposed that a DOBC system could be used to control the SES and also provide an estimate of the weight it is carrying. The dynamic equations describing the vertical displacement of the SES from its designed equilibrium position in the water $x_1$ is given by

$$\dot{x}_1 = x_2$$

$$\dot{x}_2 = -\omega_n^2 x_1 - 2\zeta\omega_n x_2 + \frac{u}{m_{ac}} + \frac{w_d}{m_{ac}},$$

$$y = x_1.$$

As shown in [5], typical numbers for a large SES are $\omega_n = 12.1$ rad/s, $\zeta = 0.08$ and $m_{ac} = 2.71$. The term $m_{ac}$ is related to the unloaded mass of the surface effect vessel, $M = 2.0 \times 10^5$ kg by the relation

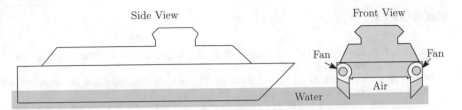

**Fig. 6.24** Outline of a surface effect ship. Fans are used to support part of the weight of the system by pressurizing a cushion of air underneath the vessel

$$\frac{M}{m_{ac}} = 7.37 \times 10^4.$$

(a) Take the scaled weight of the cars to be a third of the vessel's unloaded weight (i.e. $w_d = m_{ac}g/3$). Plot the open loop response of the system $y(t)$ for times $0 \le t \le 60$ s if the weight is added at time $t = 30$ s.

(b) Design a feedback controller so that the closed loop roots of the system have the characteristic equation

$$p(s) = s^2 + 2\zeta_p \omega_p s + \omega_p^2,$$

where $\zeta_p = 1/\sqrt{2}$ and $\omega_p = 1.2\omega_n$. Plot the closed loop response of the system $y(t)$ for the same test case examined in part (a).

(c) Design a state observer by using pole placement so that the closed loop roots of the observer's error system have the characteristic equation

$$p(s) = s^2 + 2\zeta_x \omega_x s + \omega_x^2,$$

where $\zeta_x = 0.85$ and $\omega_x = 5\omega_p$. The dynamics matrix and the output matrix for the disturbance are given by

$$A_\xi = \begin{bmatrix} 0 & 1 \\ -1 & -1 \end{bmatrix} \quad \text{and} \quad C_\xi = \begin{bmatrix} 1 & 0 \end{bmatrix},$$

respectively. Let the disturbance observer gain be $L_d = 1.5L_x$, where $L_x$ is the state observer gain. Plot the closed loop response of the system $y(t)$ to the test cases studied in parts (a) and (b) above. Also plot the estimated disturbance $w_d(t)$. Is it possible to use the disturbance observer to measure the total weight of the cars?

# References

1. Åström, K.J., Murray, R.M.: Feedback Systems: An Introduction for Scientists and Engineers. Princeton University Press (2010)
2. Chen, W.-H., Yang, J., Guo, L., Li, S.: Disturbance-observer-based control and related methods-an overview. IEEE Trans. Industr. Electron. **63**(2), 1083–1095 (2016)
3. Dhanak, M.R., Ananthakrishnan, P., Frankenfield, J., von Ellenrieder, K.: Seakeeping characteristics of a wave-adaptive modular unmanned surface vehicle. In: ASME 2013 32nd International Conference on Ocean, Offshore and Arctic Engineering, pp. V009T12A053–V009T12A053. American Society of Mechanical Engineers (2013)
4. Dorf, R.C., Bishop, R.H.: Modern Control Systems. Pearson (Addison-Wesley) (1998)
5. Faltinsen, O.M.: Hydrodynamics of High-speed Marine Vehicles. Cambridge University Press, Cambridge (2005)
6. Franklin, G.F., Powell, J.D., Emami-Naeini, A.: Feedback Control of Dynamic Systems, 3rd edn. Addison-Wesley, Reading (1994)
7. Jialu, D., Xin, H., Krstić, M., Sun, Y.: Robust dynamic positioning of ships with disturbances under input saturation. Automatica **73**, 207–214 (2016)
8. Klinger, W.B., Bertaska, I.R., von Ellenrieder, K.D., Dhanak, M.R.: Control of an unmanned surface vehicle with uncertain displacement and drag. IEEE J. Ocean. Eng. **42**(2), 458–476 (2017)
9. Refsnes, J., Sorensen, A.J., Pettersen, K.Y.: Output feedback control of an auv with experimental results. In: 2007 Mediterranean Conference on Control and Automation, pp. 1–8. IEEE (2007)
10. Sarda, E.I., Qu, H., Bertaska, I.R., von Ellenrieder, K.D.: Station-keeping control of an unmanned surface vehicle exposed to current and wind disturbances. Ocean Eng. **127**, 305–324 (2016)
11. Sariyildiz, E., Oboe, R., Ohnishi, K.: Disturbance observer-based robust control and its applications: 35th anniversary overview. IEEE Trans. Indust. Electron. (2019)
12. Snyder, W., Roman, C., Licht, S.: Hybrid actuation with complementary allocation for depth control of a lagrangian sea-floor imaging platform. J. Field Robot. **35**(3), 330–344 (2017)
13. Thomas, G.B. Jr., Finney, R.L.: Calculus and Analytic Geometry, 6th edn. Addison-Wesley (1985)

# Part II
# Nonlinear Methods

Marine environments are generally complex, unstructured and uncertain. The disturbances affecting a marine vehicle caused by wind, waves and currents are time varying and generally unpredictable.

Not only are the characteristics of the environment complex, but the dynamics of marine vehicles themselves is usually highly nonlinear. Consider how the wave drag of a surface vessel (also called the *residuary resistance $R_R$*) varies with its speed $U$ and length $L$, which is typically characterized using the Froude number $\text{Fr} := U/\sqrt{gL}$, as shown in the figure below. Even within a narrow range of operating conditions (e.g. $0 \leq \text{Fr} \leq 0.6$), the wave drag can vary substantially.

As the speed of a surface vessel increases, the hydrodynamic response of its hull changes substantially because the physical processes involved in generating the lift force that supports the vehicle's weight are Froude number dependent. The dynamics of the lift force transition from a response that can be characterized as:

a) *hydrostatic displacement*—generated mainly by the buoyancy arising from the weight of the water displaced by the underwater portion of the hull, i.e. Archimedes' Principle, (Fr $\leq 0.4$), to

b) *semi-displacement*—generated from a combination of buoyancy and hydrodynamic lift ($0.4 \leq \text{Fr} \leq 1 - 1.2$), to

c) *planing*—generated mainly by hydrodynamic lift ($1 - 1.2 \leq \text{Fr}$).

When a surface vessel operates across a range of Froude numbers the draft of the hull (sinkage) and its trim (pitch angle of the hull) can vary substantially, causing large changes in its drag characteristics.

Normally, the hydrodynamics of the main hull of a marine vessel and its actuators (fins, rudders and propellers) are characterized separately [3]. This can lead to uncertainty in the dynamic response of the vessel, for which correction factors are often used. At the same time nonlinear interactions (vortex shedding from lifting surfaces, such as rudders and fins, and complex changes in the flow around the hull caused by propellers and thrusters) between the actuator and main body of the vessel are almost always present. Such interactions may not always be captured well when fixed parameters are used in the maneuvering equations of motion.

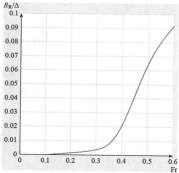

An example of how the residuary resistance (wave-making resistance) of a hull varies with Froude number. Here $R_R$ is normalized by the displacement $\Delta$ (weight) of the vessel. Reprinted from [4] with permission from Springer, ©(2016).

Further, complex, nonlinear hydrodynamic interactions can also arise when a vessel operates at/near the surface, where its damping depends on the encounter frequency (speed/direction-adjusted frequency) of incident waves [1, 2], near the seafloor, where pressure effects can change vessel dynamics, or when multiple vehicles operate near one another. For example, a significant component of the drag of a surface vessel operating in a seaway depends on the heights and frequencies of the waves it traverses. As the encounter frequency of the waves depends on the speed and orientation of the vessel, the drag can vary substantially depending on the relative angles between waves and vehicle (normally, control design is performed using the maneuvering equations, which assume calm flat water with no waves).

Finally, marine vessels are almost never operated in the configuration or at the operating conditions for which they were designed. Thus, the dynamics for which their control systems were originally developed may often no longer apply.

In short, the dynamics of marine vehicles are highly uncertain, both in terms of the model coefficients used (*parametric uncertainties*) and the equations that describe them (e.g. there are often unmodeled dynamics, called *unstructured uncertainties*). The dynamics are also generally nonlinear and time-varying. When coupled with the fact that the external environment imposes time varying and unpredictable disturbances on a vehicle, it becomes clear that the application of special robust nonlinear control techniques, which have been specifically developed to handle time varying nonlinear systems with uncertain models, can help to extend the range of safe operating conditions beyond what is possible using standard linear control techniques.

In the second part of the text, we will explore some of the basic tools of nonlinear control and their application to marine systems. The following topics are covered:

- Lyapunov and Lyapunov-like stability analysis techniques for nonlinear systems;
- feedback linearization;
- the control of underactuated systems;
- integrator backstepping and related methods, including nonlinear disturbance observers;

- adaptive control; and
- sliding model control, including higher order sliding mode observers and differentiators.

# References

1. Faltinsen, O.: Sea Loads on Ships and Offshore Structures, vol. 1. Cambridge University Press, Cambridge (1993)
2. Faltinsen, O.M.: Hydrodynamics of High-Speed Marine Vehicles. Cambridge University Press, Cambridge (2005)
3. Gillmer, T.C., Johnson, B.: Introduction to Naval Architecture. Naval Institute Press (1982)
4. von Ellenrieder, K.D., Dhanak, M.R.: Hydromechanics. In Dhanak, M.R., Xiros, N.I. (eds.) *Springer Handbook of Ocean Engineering*, pp. 127–176. Springer, Berlin (2016)

# Chapter 7
# Nonlinear Stability for Marine Vehicles

## 7.1 Introduction

Before a new controller for a marine vehicle is widely implemented in practice, it must
be conclusively shown to be stable. The dynamics of marine vehicles are generally
very nonlinear. The techniques required for the stability analysis of nonlinear systems
are quite different from those of linear systems, as some of the familiar tools, such as
Laplace and Fourier Transforms, cannot be easily applied. The analysis of controlled
nonlinear systems is most often performed using Lyapunov stability methods. We
will mostly focus on Lyapunov's Direct (Second) Method, which can be used to show
the boundedness of a closed loop system, even when the system has no equilibrium
points.

   Generally, nonlinear stability theory consists of three main components: (1) *definitions* of the different kinds of stability, which provide insight into the behavior
of a closed loop system; (2) the different *conditions* that a closed loop system must
satisfy in order to possess a certain type of stability; and (3) *criteria* that enable one
to check whether or not the required conditions hold, without having to explicitly
compute the solution of the differential equations describing the time evolution of
the closed loop system. The *conditions* and *criteria* required to establish the various
types of stability are often presented in the form of mathematical theorems.

   An engineering student encountering this approach of using a series of mathematical definitions and theorems to characterize the nonlinear stability of a system for
the first time may find it daunting. However, a significant advantage of the approach
is that such a series of mathematical statements can be used somewhat like a lookup
table, allowing one to fairly quickly isolate and characterize the stability properties
of a given nonlinear system by matching the criteria and conditions to the specific
problem at hand. To provide some guidance in this direction, a decision tree outlining
the basic process is provided in the chapter summary (Sect. 7.9).

© Springer Nature Switzerland AG 2021                                    267
K. D. von Ellenrieder, *Control of Marine Vehicles*, Springer Series on Naval
Architecture, Marine Engineering, Shipbuilding and Shipping 9,
https://doi.org/10.1007/978-3-030-75021-3_7

## 7.2  Stability of Time–Invariant Nonlinear Systems

Here we restrict our attention to time–invariant systems of the form

$$\dot{\mathbf{x}} = \mathbf{f}(\mathbf{x}, \mathbf{u}), \quad \mathbf{x} \in \mathbb{R}^n, \mathbf{u} \in \mathbb{R}^p$$
$$\mathbf{y} = \mathbf{h}(\mathbf{x}), \quad \mathbf{y} \in \mathbb{R}^q, \tag{7.1}$$

where $\mathbf{f}$ is a locally Lipschitz continuous function. Generally, the control input will be a function of the state so that $\mathbf{u} = \mathbf{u}(\mathbf{x})$ and one could also simply write $\dot{\mathbf{x}} = \mathbf{f}(\mathbf{x})$ to represent the dynamics of a closed loop system. In this text, the notation $\mathbf{f} : \mathbb{R}^n \to \mathbb{R}^n$ will be used to indicate that the vector function $\mathbf{f}$ maps an $n$-dimensional vector, i.e. $\mathbf{x} \in \mathbb{R}^n$, into another $n$-dimensional vector, in this case $\dot{\mathbf{x}} \in \mathbb{R}^n$.

**Definition 7.1** (*Lipschitz Continuous*) A function $\mathbf{f}(\mathbf{x})$ is defined to be *Lipschitz continuous* if for some constant $c > 0$,

$$\|\mathbf{f}(\mathbf{x}_2) - \mathbf{f}(\mathbf{x}_1)\| < c\|\mathbf{x}_2 - \mathbf{x}_1\|, \quad \forall \, \mathbf{x}_1, \mathbf{x}_2,$$

where $\| \bullet \|$ denotes the 2-norm (Euclidean norm) of a vector. A sufficient condition for a function to be Lipschitz continuous is that its Jacobian

$$\mathbf{A} = \frac{\partial \mathbf{f}}{\partial \mathbf{x}}$$

is uniformly bounded for all $\mathbf{x}$. When a function is *Lipschitz Continuous*, sometimes we simply say "the function is *Lipschitz*". The constant $c$ is referred to as the *Lipschitz constant*.

The existence and uniqueness of a solution to (7.2) is guaranteed when $\mathbf{f}(\mathbf{x}, \mathbf{u})$ is Lipschitz. The following sections summarize the basics of nonlinear stability. Excellent, detailed references on these topics, which focus on control theory, include [5–7].

### 7.2.1  Stability Definitions

Recall from Sect. 2.4 that equilibrium points correspond to fixed points, or stationary points, where $\dot{\mathbf{x}} = 0$.

**Definition 7.2** (*Equilibrium Point*) A state $\mathbf{x}_e \in \mathbb{R}^n$ is an *equilibrium point* of system (7.2), if $\mathbf{x}(t) = \mathbf{x}_e$ for all $t$.

Therefore, at an equilibrium point

$$\dot{\mathbf{x}}(t) = \mathbf{f}(\mathbf{x}_e) = 0.$$

**Fig. 7.1** A graphical illustration of Lyapunov stability. The dashed curve represents the trajectory of $\mathbf{x}(t)$ when $\tilde{\mathbf{x}}$ is a solution of (7.2)

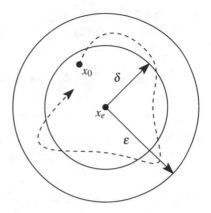

**Remark 7.1** A linear system, e.g. $\dot{\mathbf{x}} = \mathbf{A}\mathbf{x}$ has an isolated equilibrium point at $\mathbf{x} = 0$ if all of the eigenvalues of $\mathbf{A}$ are nonzero (i.e. $\det(\mathbf{A}) \neq 0$). Otherwise, a linear system will have a continuum of equilibrium points. On the other hand, a nonlinear system can have multiple isolated equilibrium points.

**Remark 7.2** For the purposes of conducting stability analysis for control design, it is often more convenient to identify a new variable $\tilde{\mathbf{x}} := (\mathbf{x} - \mathbf{x}_e)$ when $\mathbf{x}_e \neq 0$ and to rewrite the dynamics of the system (7.2) in the form

$$\dot{\tilde{\mathbf{x}}} = \mathbf{f}(\tilde{\mathbf{x}}), \quad \tilde{\mathbf{x}} \in \mathbb{R}^n \tag{7.2}$$

so that the equilibrium point is $\tilde{\mathbf{x}} = 0$, than it is to use the original system (7.2) and state variable $\mathbf{x}$. In this context the new state variable $\tilde{\mathbf{x}}$ is very often also simply referred to as the *state*. Alternatively, it is also sometimes called the *state error*, an *error surface*, or for tracking problems, i.e. when $\mathbf{x}_e = \mathbf{x}_e(t)$, as the *tracking error*.

**Definition 7.3** (*Stable*) Let the parameters $\delta$ and $\varepsilon$, where $\varepsilon > \delta > 0$, be the radii of two concentric $n$-dimensional "balls" (e.g. circles in two dimensions, spheres in three dimensions, etc.), where $n$ is the dimension of the state $\mathbf{x} \in \mathbb{R}^n$. In the sense of Lyapunov, a point $\mathbf{x}_e$ is a *stable equilibrium point* of the system (7.2), if for a given initial condition $\mathbf{x}_0 = \mathbf{x}(t_0)$, where $\|\mathbf{x}_0 - \mathbf{x}_e\| = \|\tilde{\mathbf{x}}_0\| < \delta$, the solution trajectories of $\mathbf{x}(t)$ remain in the region $\varepsilon$ for all $t > t_0$, i.e. such that $\|\mathbf{x}(t) - \mathbf{x}_e\| = \|\tilde{\mathbf{x}}\| < \varepsilon$, $\forall t > t_0$ (Fig. 7.1). In this case we will also simply say that the system (7.2) is *stable*.

**Definition 7.4** (*Unstable*) A system is *unstable* if it is not stable.

Note that stability is a property of equilibrium points. A given system may have both stable and unstable equilibrium points. Also note that the Lyapunov definition of the stability of an equilibrium point does not require the solution trajectories of the system to converge to $\mathbf{x}_e$.

**Definition 7.5** (*Asymptotically Stable*) The system (7.2) is *asymptotically stable* if, (1) it is stable and (2) the solution trajectories of the state $\mathbf{x}$ converge to $\mathbf{x}_e$ for initial conditions sufficiently close to $\mathbf{x}_e$, i.e. a $\delta$ can be chosen so that

$$\|\mathbf{x}_0 - \mathbf{x}_e\| = \|\tilde{\mathbf{x}}_0\| < \delta \Rightarrow \lim_{t \to \infty} \|\mathbf{x}(t) - \mathbf{x}_e\| = \lim_{t \to \infty} \|\tilde{\mathbf{x}}(t)\| = 0.$$

**Definition 7.6** (*Region of Attraction*) The set of all initial states $\mathbf{x}_0 = \mathbf{x}(t_0)$ from which the trajectories converge to an equilibrium point $\mathbf{x}_e$ is called the equilibrium point's *region of attraction*.

**Definition 7.7** (*Compact Set*) A set of real numbers $\mathcal{B}_\delta = \{\mathbf{x} \in \mathbb{R} : \|\mathbf{x}\| \leq \delta\}$ is a *compact set* if every sequence in $\mathcal{B}_\delta$ (such as a time dependent trajectory) has a sub-sequence that converges to an element, which is again contained in $\mathcal{B}_\delta$.

**Definition 7.8** (*Locally Attractive*) If the system is not necessarily stable, but has the property that all solutions with initial conditions $\mathbf{x}_0 = \mathbf{x}(t_0)$ that lie within some radius $\delta$ of the equilibrium point $\mathbf{x}_e$ converge to $\mathbf{x}_e$, then it is *locally attractive*. If $\|\mathbf{x}_0 - \mathbf{x}_e\| = \|\tilde{\mathbf{x}}_0\| > \delta$, the solution trajectories of $\mathbf{x}(t)$ might not converge to $\mathbf{x}_e$, or even diverge.

**Definition 7.9** (*Exponentially Stable*) An equilibrium point $\mathbf{x}_e$ is *exponentially stable* if there exist positive constants $\delta$, $k$ and $\lambda$ such that all solutions of (7.2) with $\|\mathbf{x}_0 - \mathbf{x}_e\| = \|\tilde{\mathbf{x}}_0\| \leq \delta$ satisfy the inequality

$$\|\tilde{\mathbf{x}}(t)\| \leq k \|\tilde{\mathbf{x}}_0\| e^{-\lambda t}, \quad \forall t \geq t_0. \tag{7.3}$$

The constant $\lambda$ in (7.3) is often referred to as the *convergence rate*.

**Remark 7.3** Exponential stability implies asymptotic stability, whereas asymptotic stability does not imply exponential stability.

**Definition 7.10** (*Globally Attractive*) Generally, the asymptotic stability and exponential stability of an equilibrium point are local properties of a system. However, when the region of attraction of an equilibrium point includes the entire space of $\mathbb{R}^n$, i.e. $\delta \to \infty$, the equilibrium point is *globally attractive*.

**Definition 7.11** (*Globally Asymptotically Stable*) An equilibrium point $\mathbf{x}_e$ is *globally asymptotically stable* (GAS) if it is stable and the state $\mathbf{x}(t)$ converges to $\mathbf{x}_e$ ($\lim_{t \to \infty} \|\tilde{\mathbf{x}}\| \to 0$) from any initial state $\mathbf{x}_0$.

**Definition 7.12** (*Globally Exponentially Stable*) An equilibrium point $\mathbf{x}_e$ is *globally exponentially stable* (GES) if the state $\mathbf{x}(t)$ converges exponentially to $\mathbf{x}_e$ ($\lim_{t \to \infty} \|\tilde{\mathbf{x}}\| \to 0$) from any initial state $\mathbf{x}_0$.

## 7.2.2  Lyapunov's Second (Direct) Method

Consider the nonlinear time-invariant system (7.2),

$$\dot{\tilde{\mathbf{x}}} = \mathbf{f}(\tilde{\mathbf{x}}), \quad (\mathbf{x} - \mathbf{x}_e) := \tilde{\mathbf{x}} \in \mathbb{R}^n$$

with a local equilibrium point $\mathbf{x}_e$, such that $\mathbf{f}(\tilde{\mathbf{x}} = 0) = 0$.

**Definition 7.13** (*Class* $\mathbb{C}^k$ *functions*) Let $k$ be a non-negative integer, i.e. $k \in \mathbb{Z}_{\geq 0}$. A function $V$ is a class $\mathbb{C}^k$ function if the derivatives $V', V'', \ldots, V^{(k)}$ exist and are continuous.

In the following definitions we will take $V : \mathbb{R}^n \to \mathbb{R}$ be a $\mathbb{C}^1$ function and explore the properties of $V$ within a neighborhood $D$ of $\mathbf{x}_e$ (Fig. 7.2).

**Definition 7.14** (*Positive Definite*) The function $V(\tilde{\mathbf{x}})$ is *positive definite* (PD) in $D$ if (a) $V(\tilde{\mathbf{x}} = 0) = 0$ and (b) $V(\tilde{\mathbf{x}}) > 0$ for all $\mathbf{x} \in D, \mathbf{x} \neq \mathbf{x}_e$.

**Definition 7.15** (*Negative Definite*) Similarly, the function $V(\tilde{\mathbf{x}})$ is *negative definite* (ND) in $D$ if (a) $V(\tilde{\mathbf{x}} = 0) = 0$ and (b) $V(\tilde{\mathbf{x}}) < 0$ for all $\mathbf{x} \in D, \mathbf{x} \neq \mathbf{x}_e$.

**Definition 7.16** (*Positive Semidefinite*) The function $V(\tilde{\mathbf{x}})$ is *positive semidefinite* (PSD) in $D$ if (a) $V(\tilde{\mathbf{x}} = 0) = 0$ and (b) $V(\tilde{\mathbf{x}}) \geq 0$ for all $\mathbf{x} \in D, \mathbf{x} \neq \mathbf{x}_e$. I.e. $V(\tilde{\mathbf{x}})$ can be zero at points other than $\mathbf{x} = \mathbf{x}_e$.

**Definition 7.17** (*Negative Semidefinite*) The function $V(\tilde{\mathbf{x}})$ is *negative semidefinite* (NSD) in $D$ if (a) $V(\tilde{\mathbf{x}} = 0) = 0$ and (b) $V(\tilde{\mathbf{x}}) \leq 0$ for all $\mathbf{x} \in D, \mathbf{x} \neq \mathbf{x}_e$. Here, $V(\tilde{\mathbf{x}})$ can be zero at points other than $\mathbf{x} = \mathbf{x}_e$.

**Definition 7.18** (*Radially Unbounded*) The function $V(\tilde{\mathbf{x}})$ is *radially unbounded* if

$$\lim_{\|\mathbf{x}\| \to \infty} V(\tilde{\mathbf{x}}) \to \infty.$$

**Fig. 7.2** A spherical neighborhood $D$ of $\mathbf{x}_e$

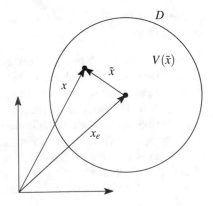

A *Lyapunov function* $V$, is an energy-like function that can be used to determine the stability of a system. If we can find a non-negative function that always decreases along trajectories of the system, we can conclude that the minimum of the function is a locally stable equilibrium point. Let $V(\tilde{\mathbf{x}})$ be radially unbounded and assume that there exist two class $\mathcal{K}_\infty$ functions $\alpha_1, \alpha_2$, such that $V$ satisfies

$$\alpha_1(\|\tilde{\mathbf{x}}\|) \leq V(\tilde{\mathbf{x}}) \leq \alpha_2(\|\tilde{\mathbf{x}}\|), \quad \forall \mathbf{x} \in D. \tag{7.4}$$

For time-invariant systems $V(\tilde{\mathbf{x}})$ is an implicit function of time $t$, such that the time derivative of $V$, $\dot{V}$ is given by the derivative of $V$ along the solution trajectories of (7.2), i.e.,

$$\frac{\mathrm{d}V}{\mathrm{d}t} = \dot{V}(\tilde{\mathbf{x}}) = \frac{\partial V}{\partial \tilde{\mathbf{x}}} \mathbf{f}(\tilde{\mathbf{x}}). \tag{7.5}$$

The main result of Lyapunov's stability theory is expressed in the following statement.

**Theorem 7.1** (Lyapunov Stability) *Suppose that there exists a* $\mathbb{C}^1$ *function* $V$ : $\mathbb{R}^n \to \mathbb{R}$ *which is positive definite in a neighborhood of an equilibrium point* $\mathbf{x}_e$,

$$V(\tilde{\mathbf{x}}) > 0, \forall \mathbf{x} \in D, \mathbf{x} \neq \mathbf{x}_e, \tag{7.6}$$

*whose time derivative along solutions of the system (7.2) is negative semidefinite, i.e.*

$$\dot{V}(\tilde{\mathbf{x}}) \leq 0, \quad \forall \mathbf{x} \in D, \tag{7.7}$$

*then the system (7.2) is stable. If the time derivative of* $V(\tilde{\mathbf{x}})$ *is negative definite,*

$$\dot{V}(\tilde{\mathbf{x}}) < 0, \quad \forall \mathbf{x} \in D, \mathbf{x} \neq \mathbf{x}_e, \tag{7.8}$$

*then (7.2) is asymptotically stable. If, in the latter case,* $V(\tilde{\mathbf{x}})$ *is also radially unbounded, then (7.2) is globally asymptotically stable.*

**Definition 7.19** (*Weak Lyapunov Function*) The function $V$ is a *weak Lyapunov function* if it satisfies (7.7).

**Definition 7.20** (*Lyapunov Function*) The function $V$ is a *Lyapunov function* if it satisfies (7.8).

Theorem 7.1 is valid when $V$ is merely continuous and not necessarily $\mathbb{C}^1$, provided that (7.7) and (7.8) are replaced by the conditions that $V$ is nonincreasing and strictly decreasing along nonzero solutions, respectively.

A graphical illustration of Theorem 7.1 is shown in Fig. 7.3. Assume (7.7) holds. Consider the ball around the equilibrium point of radius $\varepsilon > 0$. Pick a positive number $b < \min_{\|\tilde{\mathbf{x}}\|=\varepsilon} V(\tilde{\mathbf{x}})$. Let $\delta$ be the radius of some ball around $\mathbf{x}_e$, which is inside the set $\{\mathbf{x} : V(\tilde{\mathbf{x}}) \leq b\}$. Since $\dot{V}(\tilde{\mathbf{x}}) \leq 0$, $V$ is nonincreasing along the solution trajectories of $\tilde{\mathbf{x}}(t)$. Each solution starting in the smaller ball of radius $\delta$ satisfies $V(\tilde{\mathbf{x}}(t)) \leq b$, hence it remains inside the bigger ball of radius $\varepsilon$.

**Fig. 7.3** Lyapunov stability

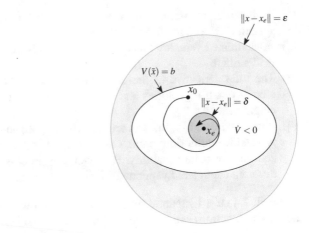

**Fig. 7.4** Geometric illustration of Lyapunov's Stability Theorem for time-invariant systems. When $\dot{V}(\tilde{\mathbf{x}}) < 0$, $V(\tilde{\mathbf{x}})$ decreases along the trajectory

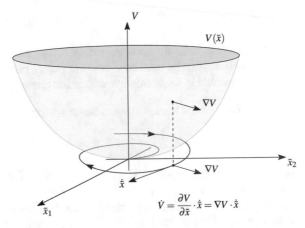

$$\dot{V} = \frac{\partial V}{\partial \tilde{x}} \cdot \dot{\tilde{x}} = \nabla V \cdot \dot{\tilde{x}}$$

From a geometric standpoint, the stronger condition that $\dot{V}(\tilde{\mathbf{x}}) < 0$ in (7.8) means that solution trajectories move to smaller and smaller values of $V$ (Fig. 7.4). If $\dot{V}(\tilde{\mathbf{x}})$ is negative definite, $\lim_{t \to \infty} \|\tilde{\mathbf{x}}\| \to 0$ so that

$$\lim_{t \to \infty} \mathbf{x}(t) \to \mathbf{x}_e.$$

**Definition 7.21** (*Candidate Lyapunov Functions*) In order to apply Theorem 7.1 we must first choose a suitable Lyapunov function. In many cases a suitable function is not known ahead of time and one may try one or more different possible functions. In such cases, the trial functions are often referred to as *Candidate Lyapunov Functions*.

Using Theorem 7.1 to examine the stability of a system in the form of (7.2) about an equilibrium point $\mathbf{x}_e$ is generally performed as follows:

(1) A change of coordinates from $\mathbf{x}(t)$ to $\tilde{\mathbf{x}} := (\mathbf{x}(t) - \mathbf{x}_e)$ is used to transform the system (7.2) into the form (7.2).
(2) A candidate Lyapunov function $V(\tilde{\mathbf{x}})$, which is PD in a neighborhood of $\tilde{\mathbf{x}} = 0$, is identified.
(3) The time derivative of the candidate Lyapunov function, $\dot{V}(\tilde{\mathbf{x}})$, is computed along system trajectories and checked to see if it is NSD or ND in the same neighborhood.
(4) If the candidate Lyapunov function satisfies (7.6) and, either (7.7) or (7.8), $\tilde{\mathbf{x}} = 0$ is a stable equilibrium point. Otherwise, no conclusion can be drawn about the equilibrium point's stability. However, the process can be repeated to see if a different candidate Lyapunov function will work.

**Remark 7.4** If a Lyapunov function $V(\tilde{\mathbf{x}})$ is known to be admitted by a system, additional functions can be generated using the relation

$$\bar{V}(\tilde{\mathbf{x}}) = \beta V^{\gamma}(\tilde{\mathbf{x}}), \quad \beta > 0, \gamma > 1. \tag{7.9}$$

Lyapunov functions are not always easy to find and are not unique. In most cases, energy functions can be used as a starting point. Fortunately, Lyapunov functions can be constructed in a systematic manner for linear systems of the form $\dot{\mathbf{x}} = \mathbf{A}\mathbf{x}$.

**Example 7.1** For the linear time–invariant system

$$\dot{\mathbf{x}} = \mathbf{A}\mathbf{x} \tag{7.10}$$

asymptotic stability, exponential stability, and their global versions are all equivalent and amount to the property that $\mathbf{A}$ is a *Hurwitz* matrix, i.e., all eigenvalues of $\mathbf{A}$ have negative real parts. To see this, consider the quadratic Lyapunov function of the form $V(\mathbf{x}) = \mathbf{x}^T \mathbf{P} \mathbf{x}$ where $\mathbf{P} \in \mathbb{R}^{n \times n}$ is a symmetric matrix $\mathbf{P} = \mathbf{P}^T$. The condition that $V$ be positive definite is equivalent to the condition that $\mathbf{P}$ be positive definite $\mathbf{x}^T \mathbf{P} \mathbf{x} > 0, \ \forall \mathbf{x} \neq 0$, which is written as $\mathbf{P} > 0$. A matrix is positive definite if and only if all of its eigenvalues are real and positive.

Computing the derivative of $V$ gives

$$\dot{V} = \frac{\partial V}{\partial \mathbf{x}} \frac{d\mathbf{x}}{dt} = \mathbf{x}^T \left( \mathbf{A}^T \mathbf{P} + \mathbf{P} \mathbf{A} \right) \mathbf{x} := -\mathbf{x}^T \mathbf{Q} \mathbf{x}. \tag{7.11}$$

Thus, the requirement that $\dot{V}$ be negative definite (for asymptotic stability) becomes a requirement that $\mathbf{Q}$ be positive definite. To find a Lyapunov function it is sufficient to choose a $\mathbf{Q} > 0$ and solve the *Lyapunov equation*

$$\mathbf{A}^T \mathbf{P} + \mathbf{P} \mathbf{A} = -\mathbf{Q}. \tag{7.12}$$

**Fig. 7.5**  A linear
mass-spring-damper system

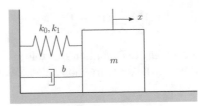

The explicit formula for **P** is

$$\mathbf{P} = \int_0^\infty e^{\mathbf{A}^T t} \mathbf{Q} e^{\mathbf{A}t} dt.$$

Indeed, we have

$$\mathbf{A}^T \mathbf{P} + \mathbf{P}\mathbf{A} = \int_0^\infty \frac{d}{dt} \left( e^{\mathbf{A}^T t} \mathbf{Q} e^{\mathbf{A}t} \right) dt = -\mathbf{Q}$$

if **A** is Hurwitz. Thus, it is always possible to find a quadratic Lyapunov function when all the eigenvalues of **A** have a negative real part.                                    □

However, for nonlinear systems, finding a suitable Lyapunov function using energy-like functions is not always straightforward.

**Example 7.2**  Consider the nonlinear mass-spring-damper system shown in Fig. 7.5. The equation of motion is given by

$$m\ddot{x} + b\dot{x}|\dot{x}| + k_0 x + k_1 x^3 = 0, \tag{7.13}$$

where $m$ is the mass, $b > 0$ is a nonlinear damping constant, $k_0 > 0$ is a linear spring constant and $k_1 > 0$ is a nonlinear spring constant.
  Define a state vector $\mathbf{x} := [x_1 \; x_2]^T$, where $x_1 = x$ and $x_2 = \dot{x}$. The equation of motion can be written as the following system of equations

$$\dot{x}_1 = x_2 \tag{7.14}$$

$$\dot{x}_2 = -\frac{b}{m} x_2 |x_2| - \frac{k_0}{m} x_1 - \frac{k_1}{m} x_1^3. \tag{7.15}$$

Physical intuition suggests that if the mass is positioned at some initial location $x_0$, where the spring is compressed or stretched, and then released, the mass will oscillate until the total energy of the system is dissipated. From Eq. (7.15) it can be seen that the only equilibrium point of the system is $\mathbf{x} = [0\;0]^T$. Thus, $\tilde{\mathbf{x}} = \mathbf{x}$. When the oscillations stop, the mass will be at rest at the equilibrium position $\mathbf{x}_e = 0$.
  Here, a Lyapunov function can be written using the total energy (kinetic and potential) of the mass spring-damper-system as

$$V(\mathbf{x}) = \frac{1}{2}mx_2^2 + \int_0^{x_1} (k_0\xi + k_1\xi^3)\,d\xi = \frac{1}{2}mx_2^2 + \frac{1}{2}k_0x_1^2 + \frac{1}{4}k_1x_1^4. \qquad (7.16)$$

According to Definition 7.14 this selection of $V(\mathbf{x})$ is positive definite and by Definition 7.18 it is radially unbounded. Taking the time derivative of (7.16) and using (7.15) gives

$$\dot{V} = x_2\,m\dot{x}_2 + k_0x_1\dot{x}_1 + k_1x_1^3\dot{x}_1,$$

$$= x_2\,m\dot{x}_2 + x_2(k_0x_1 + k_1x_1^3),$$

$$= x_2\left(-bx_2|x_2| - k_0x_1 - k_1x_1^3\right) + x_2\left(k_0x_1 + k_1x_1^3\right),$$

$$= -bx_2^2|x_2| < 0, \forall x_2 \neq 0.$$

Note that $\dot{V}(\mathbf{x})$ is not negative definite because $\dot{V}(\mathbf{x}) = 0$ when $x_2 = 0$, irrespective of the value of $x_1$. Thus, although there is only one possible equilibrium position in this problem, according to Theorem 7.1 the selected $V(\mathbf{x})$ can only be used to show that the system is stable, but cannot be used to classify the type of stability near the equilibrium point $\mathbf{x}_e = 0$. ☐

As can be seen in the previous example, physical reasoning about the total energy of a system can often be used to identify a suitable Lyapunov function. However, some trial and error is generally required. Another useful strategy for identifying a useful Lyapunov function involves constructing a quadratic form of the kind

$$V(\tilde{\mathbf{x}}) = \frac{1}{2}(\mathbf{x} - \mathbf{x}_e)^T\mathbf{P}(\mathbf{x} - \mathbf{x}_e) = \frac{1}{2}\tilde{\mathbf{x}}^T\mathbf{P}\tilde{\mathbf{x}}, \qquad (7.17)$$

where $\mathbf{P} \in \mathbb{R}^{n \times n}$ is a positive definite symmetric matrix with $\mathbf{P} = \mathbf{P}^T > 0$ and $\mathbf{x}^T\mathbf{P}\mathbf{x} > 0$ for any vector $\mathbf{x}$. To use this approach, one must find the elements of $\mathbf{P}$, $p_{ij}$ by simultaneously satisfying the following:

(1) Necessary and sufficient conditions for a symmetric matrix $\mathbf{P}$ to be positive definite are that $p_{11} > 0$ and $\det(\mathbf{P}) > 0$. These two conditions create a set of inequalities that must be satisfied by the $p_{ij}$.
(2) The second set of inequalities and equations results from taking the time derivative of (7.17) and selecting the $p_{ij}$ that ensure $\dot{V} < 0$ for any $\tilde{\mathbf{x}} \neq 0$.

This approach is illustrated in the following example.

**Example 7.3** Consider the closed loop control of the angular displacement $\theta$ of a pendulum comprised of a point mass $m$, a mass-less rod of length $l$ with a friction coefficient of $d$ at its pivot point, which is oscillating under the torque caused gravity and an electric motor mounted at the pivot point (Fig. 7.6). The equation of motion is

$$ml^2\ddot{\theta} + d\dot{\theta} + mgl\sin\theta = u. \qquad (7.18)$$

**Fig. 7.6** The motor-driven
pendulum

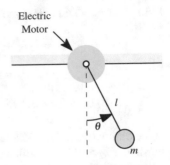

Let the state variables be $\mathbf{x} := [x_1 \ x_2]^T = [\theta \ \dot{\theta}]^T$. We will examine whether a control
input $u$ can be designed to stabilize the position of the pendulum at a desired angle
$\theta_d = x_{1d}$. The equations of motion of the pendulum in terms of the state variables
are

$$\dot{x}_1 = x_2,$$
$$\dot{x}_2 = -\frac{g}{l} \sin x_1 - \frac{d}{ml^2} x_2 + \frac{1}{ml^2} u. \tag{7.19}$$

Using feedback linearization, select the control input to be

$$u = -k_p \tilde{x}_1 - k_d \tilde{x}_2 + mgl \sin x_1, \tag{7.20}$$

where $\tilde{x}_1 := x_1 - x_{1d}$ is the position error, $\tilde{x}_2 := \dot{\tilde{x}}_1 = x_2$ is the velocity error, $k_p > 0$
is a proportional gain, $k_d > 0$ is a derivative gain, and the term $mgl \sin x_1$ linearizes
the closed loop system. With this control input, the equations of motion for the closed
loop system can be written as

$$\dot{\tilde{x}}_1 = \tilde{x}_2,$$
$$\dot{\tilde{x}}_2 = -\frac{k_p}{ml^2} \tilde{x}_1 - \frac{(k_d + d)}{ml^2} \tilde{x}_2, \tag{7.21}$$

which has an equilibrium point at the origin of the closed loop error system $\tilde{\mathbf{x}} = [\tilde{x}_1 \ \tilde{x}_2]^T = [0 \ 0]^T$.

Consider a candidate Lyapunov function composed of a term proportional to the
square of the position error $\tilde{x}_1^2$ and the kinetic energy (one half of the mass multiplied
by the square of the translational velocity),

$$V(\mathbf{x}) = \frac{1}{2}\tilde{x}_1^2 + \frac{1}{2}ml^2\tilde{x}_2^2. \tag{7.22}$$

Using $\dot{\tilde{x}}_2$ from (7.21), the time derivative of $V$ can be written as

$$\dot{V} = \tilde{x}_1\dot{\tilde{x}}_1 + ml^2\tilde{x}_2\dot{\tilde{x}}_2,$$

$$= \tilde{x}_1\tilde{x}_2 + ml^2\tilde{x}_2\left(-\frac{k_p}{ml^2}\tilde{x}_1 - \frac{(k_d + d)}{ml^2}\tilde{x}_2\right),$$

$$= -(k_p - 1)\tilde{x}_1\tilde{x}_2 - (k_d + d)\tilde{x}_2^2. \tag{7.23}$$

Since both terms in (7.23) include $\tilde{x}_2$ it is possible that $\dot{V} = 0$ when $\tilde{x}_1 \neq 0$, $\tilde{x}_2 = 0$. Further, if $\tilde{x}_2 > 0$ and $\tilde{x}_1 < 0$ it could be possible for the first term in (7.23) to be large enough that $\dot{V} > 0$. Unfortunately, the stability of the closed loop system cannot be proven with this candidate Lyapunov function.

Next, consider a different candidate Lyapunov function

$$V(\tilde{\mathbf{x}}) = \frac{1}{2}\tilde{\mathbf{x}}^T\mathbf{P}\tilde{\mathbf{x}}, \tag{7.24}$$

with

$$\mathbf{P} := \begin{bmatrix} p_{11} & p_{12} \\ p_{12} & p_{22} \end{bmatrix}, \tag{7.25}$$

such that

$$V(\tilde{\mathbf{x}}) = \frac{1}{2}\left[p_{11}\tilde{x}_1^2 + 2p_{12}\tilde{x}_1\tilde{x}_2 + p_{22}\tilde{x}_2^2\right]. \tag{7.26}$$

In order for $\mathbf{P}$ to be positive definite, we require that

$$p_{11} > 0 \tag{7.27}$$

$$\det(\mathbf{P}) = p_{11}p_{22} - p_{12}^2 > 0. \tag{7.28}$$

Let

$$a := \frac{k_p}{ml^2} \quad \text{and} \quad b := \frac{(k_d + d)}{ml^2} \tag{7.29}$$

in (7.21) above. Then taking the time derivative of $V$ and using (7.21) gives

$$\dot{V} = \frac{1}{2}\left[2p_{11}\tilde{x}_1\dot{\tilde{x}}_1 + 2p_{12}\dot{\tilde{x}}_1\tilde{x}_2 + 2p_{12}\tilde{x}_1\dot{\tilde{x}}_2 + 2p_{22}\tilde{x}_2\dot{\tilde{x}}_2\right],$$

$$= \left[p_{11}\tilde{x}_1\tilde{x}_2 + p_{12}\tilde{x}_2^2 + p_{12}\tilde{x}_1\dot{\tilde{x}}_2 + p_{22}\tilde{x}_2\dot{\tilde{x}}_2\right],$$

$$= \left[p_{11}\tilde{x}_1\tilde{x}_2 + p_{12}\tilde{x}_2^2 - (p_{12}\tilde{x}_1 + p_{22}\tilde{x}_2)(a\tilde{x}_1 + b\tilde{x}_2)\right], \tag{7.30}$$

$$= \left[-ap_{12}\tilde{x}_1^2 + (p_{11} - ap_{22} - bp_{12})\tilde{x}_1\tilde{x}_2 - (bp_{22} - p_{12})\tilde{x}_2^2\right].$$

In order to ensure that $\dot{V}$ is negative definite for every $\tilde{\mathbf{x}} \neq 0$, in addition to satisfying (7.28), the $p_{ij}$ must be selected to also satisfy

$$p_{12} > 0, \tag{7.31}$$

$$p_{11} - ap_{22} - bp_{12} = 0, \tag{7.32}$$

$$bp_{22} > p_{12}. \tag{7.33}$$

The system is underdetermined, so we can start by selecting a convenient value for one of the terms. Comparing (7.22) with (7.26), it can be seen that the term $p_{22}$ would have a nice physical meaning if we select

$$p_{22} = ml^2. \tag{7.34}$$

Then from (7.29) and (7.32) we have

$$p_{11} - k_p - bp_{12} = 0 \quad \Rightarrow \quad p_{11} = k_p + bp_{12}, \tag{7.35}$$

so that with our selection of $p_{22}$, (7.28) becomes

$$(k_p + bp_{12})ml^2 > p_{12}^2,$$

which, using (7.29), can be rearranged to get

$$p_{12}^2 - (k_d + d)p_{12} - k_p ml^2 < 0.$$

Solving this inequality as if though it were a quadratic equation gives

$$p_{12} < \frac{(k_d + d)}{2} \pm \frac{1}{2}\sqrt{(k_d + d)^2 + 4k_p ml^2},$$

$$< \frac{(k_d + d)}{2} \pm \frac{(k_d + d)}{2}\sqrt{1 + \frac{4k_p ml^2}{(k_d + d)^2}}.$$

To satisfy (7.31), we must take the positive term, giving

$$p_{12} < \frac{(k_d + d)}{2}\left[1 + \sqrt{1 + \frac{4k_p ml^2}{(k_d + d)^2}}\right].$$

This latter inequality can be satisfied by simply selecting $p_{12}$ to be

$$p_{12} = \frac{(k_d + d)}{2}. \tag{7.36}$$

Then from (7.35) we can now solve for $p_{11}$

$$p_{11} = k_p + \frac{(k_d + d)^2}{2ml^2}. \tag{7.37}$$

For the values of $p_{11}$, $p_{12}$ and $p_{22}$ selected, all of the inequalities are satisfied and from (7.26) the Lyapunov function becomes

$$V(\tilde{\mathbf{x}}) = \frac{1}{2} \left\{ \left[ k_p + \frac{(k_d + d)^2}{2ml^2} \right] \tilde{x}_1^2 + (k_d + d)\tilde{x}_1\tilde{x}_2 + ml^2\tilde{x}_2^2 \right\}, \tag{7.38}$$

and its time derivative is

$$\dot{V} = -\left[ \frac{k_p(k_d + d)}{2\,ml^2} \tilde{x}_1^2 + \frac{(k_d + d)}{2} \tilde{x}_2^2 \right] < 0, \forall \tilde{\mathbf{x}} \neq 0. \tag{7.39}$$

Note that the product $\tilde{x}_1\tilde{x}_2$ in (7.38) could be negative. Let the maximum value of $\tilde{x}_2$ along solution trajectories of the closed loop system be $\tilde{x}_{2M}$. Since $|\theta| = |\tilde{x}_1| < \pi$, we can ensure $V \geq 0$ by selecting $k_p$ and $k_d$ so that

$$\left[ k_p + \frac{(k_d + d)^2}{2ml^2} \right] \pi^2 + ml^2\tilde{x}_{2M}^2 > (k_d + d)\pi\tilde{x}_{2M}. \tag{7.40}$$

In this problem, the equilibrium point of the closed loop system $\tilde{\mathbf{x}} = [\tilde{x}_1 \ \tilde{x}_2]^T = [0 \ 0]^T$ is asymptotically stable. However, $V(\tilde{\mathbf{x}})$ is not radially unbounded. Instead, the maximum permissible magnitudes of $\tilde{x}_1$ and $\tilde{x}_2$ depend on the values of the control gains $k_p$ and $k_d$. This type of stability is known as *semiglobal asymptotic stability*.

The time response of the closed loop system for a pendulum of mass $m = 1$ kg, a length of $l = 1$ m a drag coefficient of $d = 0.1$ kg-m$^2$/s is shown in the Fig. 7.7. The

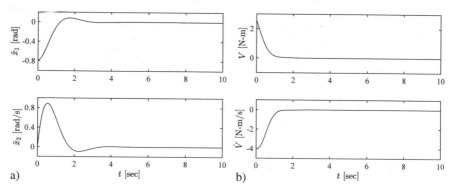

**Fig. 7.7**  Time response of the closed loop motor-driven pendulum: (a) position error $\tilde{x}_1$ and velocity error $\tilde{x}_2$; (b) Lyapunov function $V$ and time derivative of Lyapunov function $\dot{V}$. The pendulum starts at rest from an initial position of $\theta = 0$. The desired angle is $\theta_d = \pi/4$

desired angular position is $\theta_d = \pi/4 = 0.785$ rad and controller gains of $k_p = 5$ and $k_d = 2.5$ are used. It can be seen that the angular position is driven to within 2% of its final value in about 2.6 s. Both $V$ and $\dot{V}$ are also driven to zero. $\qquad\qquad$ □

## 7.3 Invariant Set Theorem

In many cases, it is preferable to know that the state of a marine vehicle can be controlled so that the closed loop control system approaches a desired equilibrium point, i.e. that the closed loop system is asymptotically stable. However, for nonlinear systems, it is sometimes difficult to find a positive definite function $V(\tilde{\mathbf{x}})$ whose derivative is strictly negative definite. When applying Lyapunov's Direct Method, one may find that the time derivative of the chosen Lyapunov function $\dot{V}(\tilde{\mathbf{x}})$ is only negative semi definite (7.7), rather than negative definite. When this happens, one can infer the stability of the closed loop system, but not the asymptotic stability of the system to $\tilde{\mathbf{x}} = 0$. In such a situation, the Invariant Set Theorem may allow one to analyze the stability of the system in more detail, without needing to identify a different Lyapunov function. Thus, the Invariant Set Theorem enables us to conclude the asymptotic stability of an equilibrium point under less restrictive conditions, which are easier to construct.

**Definition 7.22** (*Invariant Set*) A set of states, which form a subset $G \subseteq \mathbb{R}^n$ of the state space, is an *invariant set* of (7.2) if any trajectory starting from a point $\tilde{\mathbf{x}}_0 \in G$ always stays in $G$, i.e. if $\tilde{\mathbf{x}}(t) \in G$ for all $t \geq t_0$.

Recall that a state $\mathbf{x}_e \in \mathbb{R}^n$ is an equilibrium point, if $\mathbf{x}(t) = \mathbf{x}_e$ for all $t$. Thus, the notion of an invariant set is essentially a generalization of the concept of an equilibrium point. Examples of invariant sets include equilibrium points and the basin of attraction of an asymptotically stable equilibrium point.

The basic idea is that if $V(\tilde{\mathbf{x}}) > 0$ (PD) and $\dot{V}(\tilde{\mathbf{x}}) \leq 0$ (NSD) in a neighborhood of $\tilde{\mathbf{x}} = 0$, then if $V(\tilde{\mathbf{x}})$ approaches a limit value, then $\dot{V}(\tilde{\mathbf{x}}) \to 0$, at least under certain conditions.

**Theorem 7.2** (LaSalle's Local Invariant Set Theorem) *For time-invariant system (7.2), assume there exists a function $V(\tilde{\mathbf{x}}) \in \mathbb{C}^1$ such that*

1. *the region $\Omega_\alpha = \{\tilde{\mathbf{x}} \in \mathbb{R}^n : V(\tilde{\mathbf{x}}) \leq \alpha\}$ is bounded for some $\alpha > 0$, and*
2. *$\dot{V}(\tilde{\mathbf{x}}) \leq 0$ in $\alpha$*

*and define $P$, the set of points in $\Omega_\alpha$ where $\dot{V}(\tilde{\mathbf{x}}) = 0$, then any trajectory of the system that starts in $\Omega_\alpha$ asymptotically approaches $M$, the largest invariant set contained in $P$. A geometric interpretation of LaSalle's Invariant Set Theorem is illustrated in Fig. 7.8.*

An immediate consequence of the theorem is the following corollary.

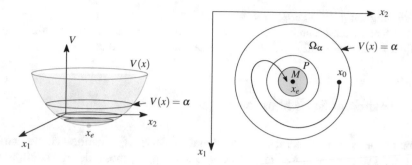

**Fig. 7.8** Geometric interpretation of Theorem 7.2 drawn in the original state coordinates of (7.2)

**Corollary 7.1** (Local Asymptotic Stability) *An equilibrium point of (7.2) is locally asymptotically stable if there exists a function $V(\tilde{\mathbf{x}}) \in C^1$ such that*

*(1)  $V(\tilde{\mathbf{x}})$ is PD in a set $D$ that contains $\tilde{\mathbf{x}} = 0$,*
*(2)  $\dot{V}(\tilde{\mathbf{x}})$ is NSD in the same set,*
*(3)  the largest invariant set $M$ in $P$ (the subset of $D$ where $\dot{V} = 0$) consists of $\tilde{\mathbf{x}} = 0$ only.*

*In addition, if the largest region defined by $V(\tilde{\mathbf{x}}) \leq \alpha$, $\alpha > 0$ and contained in $D$ is denoted as $\Omega_\alpha$, $\Omega_\alpha$ is an estimate of the basin of attraction of $\tilde{\mathbf{x}} = 0$.*

**Theorem 7.3** (LaSalle's Global Invariant Set Theorem) *For system (7.2), assume there exists a function $V(\tilde{\mathbf{x}}) \in \mathbb{C}^1$ such that*

*1.  $V(\tilde{\mathbf{x}})$ is radially unbounded, and*
*2.  $\dot{V}(\tilde{\mathbf{x}}) \leq 0$ in $\mathbb{R}^n$,*

*then any trajectory of the system asymptotically approaches the set $M$, the largest invariant set in $P$, the set of points of $\Omega_\alpha$ where $\dot{V} = 0$.*

**Remark 7.5** The radial unboundedness of $V(\tilde{\mathbf{x}})$ guarantees that any region $\Omega_\alpha = \{\tilde{\mathbf{x}} \in \mathbb{R}^n : V(\tilde{\mathbf{x}}) < \alpha\}$, $\alpha > 0$ is bounded.

There is also an additional corollary associated with Theorem 7.3.

**Corollary 7.2** (Global Asymptotic Stability) *An equilibrium point $\tilde{\mathbf{x}} = 0$ of (7.2) is globally asymptotically stable if there exists a function $V(\tilde{\mathbf{x}}) \in C^1$ such that*

*1.  $V(\tilde{\mathbf{x}})$ is PD in any neighborhood of $\tilde{\mathbf{x}} = 0$ and radially unbounded*
*2.  $\dot{V}(\tilde{\mathbf{x}})$ is NSD in any neighborhood of $\tilde{\mathbf{x}} = 0$*
*3.  the largest invariant set $M$ in $P$ (the subset of $D$ where $\dot{V} = 0$) consists of $\tilde{\mathbf{x}} = 0$ only.*

**Fig. 7.9** Use of a vectored thruster to control the pitch of an AUV

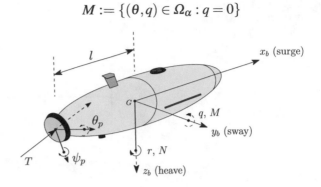

$$M := \{(\theta, q) \in \Omega_\alpha : q = 0\}$$

**Example 7.4** Let's re-examine the problem in Example 2.4 of using a vectored thruster to actively stabilize the pitch moment of an AUV (Fig. 7.9). Previously, we explored the stability of the linearized system, i.e. its linear stability, near a desired pitch angle of $\theta = 0$. Here, we will explore the nonlinear stability of the system. Adding a control input $u$ to (2.23), the closed loop equations of motion can be written in the simplified form

$$\dot{\theta} = q,$$
$$\dot{q} = c_0 \sin(2\theta) + u, \tag{7.41}$$

where $\theta$ is the pitch angle, $q := \dot{\theta}$ is the pitch rate and $c_0 > 0$ is a constant. When $u = 0$, the open loop system has an unstable equilibrium point at $\theta = q = 0$. A nonlinear controller of the form

$$u = -c_0 \sin(2\theta) - k_p\theta - k_d q, \tag{7.42}$$

where the constants $k_p > 0$ and $k_d > 0$ are control gains, can be used to stabilize the AUV at the origin $\tilde{\mathbf{x}} = [\theta \ q]^T = [0 \ 0]^T$. To explore the closed loop stability of the system, consider the following candidate Lyapunov function

$$V(\tilde{\mathbf{x}}) = \frac{1}{2}\theta^2 + \frac{1}{2}q^2.$$

Note that since $\theta = \theta + 2\pi k$, where $k \in \mathbb{Z}$ is an integer, this Lyapunov function is not radially unbounded. Thus, we will explore local stability for $|\theta| < \pi/2$.

The time derivative of $V(\mathbf{x})$ is

$$\dot{V} = \theta\dot{\theta} + q\dot{q}$$
$$= q\theta + q[c_0 \sin(2\theta) + u] \tag{7.43}$$
$$= -(k_p - 1)\theta q - k_d q^2.$$

**Fig. 7.10** Phase portrait of
the stabilized AUV pitch
angle

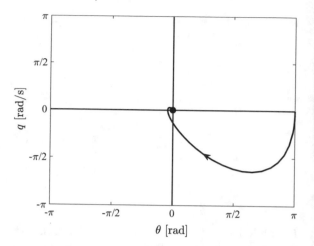

The second term of (7.43), $-k_d q^2$, is ND for any $q$. However, for $k_p > 1$ first term becomes positive when either $\theta < 0$ and $q > 0$, or $\theta > 0$ and $q < 0$. One could select $k_p = 1$, but it would then not be necessary for both $q = 0$ and $\theta = 0$ to obtain $\dot{V} = 0$. Therefore, we cannot conclude the asymptotic stability of the closed loop system to $\tilde{\mathbf{x}} = 0$ using Theorem 7.1 (recall the similar situation encountered in Example 7.2 above).

However, if we restrict our analysis to a small neighborhood of the origin $\Omega_\alpha$, $\alpha \ll \pi/2$, then we can define

$$M := \{(\theta, q) \in \Omega_\alpha : q = 0\}$$

and we can compute the largest invariant set inside $M$. For a trajectory to remain in this set we must have $q = 0$, $\forall t$ and hence $\dot{q} = 0$ as well. Using (7.41) and (7.42) we can write the closed loop dynamics of the system as

$$\begin{aligned} \dot{\theta} &= q \\ \dot{q} &= -k_p \theta - k_d q, \end{aligned} \tag{7.44}$$

so that $q(t) = 0$ and $\dot{q} = 0$ also implies both $\theta = 0$ and $\dot{\theta} = 0$. Hence the largest invariant set inside $M$ is $\tilde{\mathbf{x}} = [\theta \ q]^T = [0 \ 0]^T$ and we can use LaSalle's Local Invariant Set Theorem (Theorem 7.2) to conclude that the origin is locally asymptotically stable. A phase portrait of the closed loop system response is shown in Fig. 7.10 for $k_p = 2$, $k_d = 2$ and $\mathbf{x}_0 = [\pi \ 0]^T$. $\qquad \square$

## 7.4 Stability of Time–Varying Nonlinear Systems

Invariant set theorems only apply to time-invariant systems. Additional tools must be used to determine the stability of a nonlinear time-varying system, e.g. an adaptive control system, or a system with exogenous (externally-generated) time-varying disturbances.

Consider a time-varying differential equation of the form

$$\dot{\mathbf{x}} = \mathbf{f}(\mathbf{x}, t), \tag{7.45}$$

where $\mathbf{f} : \mathbb{R}^n \times \mathbb{R}_{\geq 0} \to \mathbb{R}^n$.

**Remark 7.6** As with nonlinear time-invariant systems, it is often convenient when performing stability analyses of time-varying nonlinear systems to transform the coordinates of the state variable $\tilde{\mathbf{x}} := (\mathbf{x} - \mathbf{x}_e)$ and to rewrite the dynamics of the system (7.45) in the form

$$\dot{\tilde{\mathbf{x}}} = \mathbf{f}(\tilde{\mathbf{x}}, t), \quad \tilde{\mathbf{x}} \in \mathbb{R}^n \tag{7.46}$$

so that the equilibrium point of the transformed system is $\tilde{\mathbf{x}} = 0$.

**Assumption 7.1** Let the origin $\tilde{\mathbf{x}} = 0$ be an equilibrium point of system (7.46) and $D \subseteq \mathbb{R}^n$ be a domain containing $\tilde{\mathbf{x}} = 0$. Assume that $\mathbf{f}(\tilde{\mathbf{x}}, t)$ is piecewise continuous in $t$ and locally Lipschitz in $\tilde{\mathbf{x}}$ (see Definition 7.1), for all $t \geq 0$ and $\tilde{\mathbf{x}} \in D$.

When analyzing the stability of time-varying systems, the solution $\mathbf{x}(t)$ is dependent on both time $t$ and on the initial condition $\mathbf{x}_0 = \mathbf{x}(t_0)$. Comparison functions can be used to formulate the stability definitions so that they hold uniformly in the initial time $t_0$ [5]. An additional advantage to using comparison functions is that they permit the stability definitions to be written in a compact way.

**Definition 7.23** (*Class $\mathcal{K}$ Functions*) A function $\alpha : [0, a) \to [0, \infty)$ is of class $\mathcal{K}$ if it is continuous, strictly increasing, and $\alpha(0) = 0$.

**Definition 7.24** (*Class $\mathcal{K}_\infty$ Functions*) If

$$\lim_{r \to \infty} \alpha(r) \to \infty, \tag{7.47}$$

then $\alpha$ is a class $\mathcal{K}_\infty$ function.

**Definition 7.25** (*Class $\mathcal{KL}$ Functions*) A continuous function $\beta : [0, a) \times [0, \infty) \to [0, \infty)$ is of class $\mathcal{KL}$ if $\beta(r, t)$ is of class $\mathcal{K}$ with respect to $r$ for each fixed $t \geq t_0$ and, if for each fixed $r \geq 0$, $\beta(r, t)$ decreases with respect to $t$ and

$$\lim_{t \to \infty} \beta(r, t) \to 0. \tag{7.48}$$

We will write $\alpha \in \mathcal{K}_\infty, \beta \in \mathcal{KL}$ to indicate that $\alpha$ is a class $\mathcal{K}_\infty$ function and $\beta$ is a class $\mathcal{KL}$ function, respectively. Note that the arguments of both functions $\alpha(r)$ and $\beta(r, t)$ are scalar.

Definitions for nonlinear time-varying systems, which are analogous to Definitions 7.3, 7.5, 7.11 and 7.12 above for nonlinear time-invariant systems, can be written using comparison functions.

**Definition 7.26** (*Uniformly Stable*) The equilibrium point $\tilde{\mathbf{x}} = 0$ of system (7.46) is *uniformly stable* (US) if there exists a class $\mathcal{K}$ function $\alpha$ and a constant $\delta > 0$, independent of the initial time $t_0$, such that

$$\|\tilde{\mathbf{x}}(t)\| \leq \alpha(\|\tilde{\mathbf{x}}_0\|), \quad \forall t \geq t_0 \geq 0, \quad \forall \|\tilde{\mathbf{x}}_0\| < \delta. \tag{7.49}$$

**Definition 7.27** (*Uniformly Asymptotically Stable*) The equilibrium point $\tilde{\mathbf{x}} = 0$ of system (7.46) is *uniformly asymptotically stable* (UAS) if there exists a class $\mathcal{KL}$ function $\beta$ and a constant $\delta > 0$, independent of the initial time $t_0$, such that

$$\|\tilde{\mathbf{x}}(t)\| \leq \beta(\|\tilde{\mathbf{x}}_0\|, t - t_0), \quad \forall t \geq t_0 \geq 0, \quad \forall \|\tilde{\mathbf{x}}_0\| < \delta. \tag{7.50}$$

**Definition 7.28** (*Uniformly Globally Asymptotically Stable*) The equilibrium point $\tilde{\mathbf{x}} = 0$ of system (7.46) is *uniformly globally asymptotically stable* (UGAS) if a class $\mathcal{KL}$ function $\beta$ exists such that inequality (7.50) holds for any initial state, i.e.

$$\|\tilde{\mathbf{x}}(t)\| \leq \beta(\|\tilde{\mathbf{x}}_0\|, t - t_0), \quad \forall t \geq t_0 \geq 0, \quad \forall \|\tilde{\mathbf{x}}_0\|. \tag{7.51}$$

**Definition 7.29** (*Exponentially Stable*) The equilibrium point $\tilde{\mathbf{x}} = 0$ of system (7.46) is *exponentially stable* (ES) if a class $\mathcal{KL}$ function $\beta$ with the form $\beta(r, s) = kre^{-\lambda s}$ exists for some constants $k > 0$, $\lambda > 0$ and $\delta > 0$, such that

$$\|\tilde{\mathbf{x}}(t)\| \leq k\|\tilde{\mathbf{x}}_0\|e^{-\lambda(t-t_0)}, \quad \forall t \geq t_0 \geq 0, \quad \forall \|\tilde{\mathbf{x}}_0\| < \delta. \tag{7.52}$$

**Definition 7.30** (*Globally Exponentially Stable*) The equilibrium point $\tilde{\mathbf{x}} = 0$ of system (7.46) is *globally exponentially stable* (GES) if inequality (7.52) holds for any initial state, i.e.

$$\|\tilde{\mathbf{x}}(t)\| \leq k\|\tilde{\mathbf{x}}_0\|e^{-\lambda(t-t_0)}, \quad \forall t \geq t_0 \geq 0, \quad \forall \|\tilde{\mathbf{x}}_0\|. \tag{7.53}$$

**Theorem 7.4** (Uniformly Stable) *Let* $\tilde{\mathbf{x}} = 0$ *be an equilibrium point of system (7.46) and* $V(t, \tilde{\mathbf{x}})$ *be a continuously differentiable function such that*

$$W_1(\tilde{\mathbf{x}}) \leq V(\tilde{\mathbf{x}}, t) \leq W_2(\tilde{\mathbf{x}})$$

*and*

$$\frac{\partial V}{\partial t} + \frac{\partial V}{\partial \tilde{\mathbf{x}}} \mathbf{f}(\tilde{\mathbf{x}}, t) \leq 0$$

*for all $t \geq 0$ and $\tilde{\mathbf{x}} \in D$, where $W_1(\tilde{\mathbf{x}})$ and $W_2(\tilde{\mathbf{x}})$ are continuous positive definite functions in D. Then, under Assumption 7.1, the equilibrium point $\tilde{\mathbf{x}} = 0$ is uniformly stable.*

The following definition will be subsequently used to characterize the stability of nonlinear time-varying systems.

**Definition 7.31**   *(Ball)* A closed ball of radius $\delta$ in $\mathbb{R}^n$ centered at $\tilde{\mathbf{x}} = 0$ is denoted by $\mathcal{B}_\delta$, i.e.

$$\mathcal{B}_\delta := \left\{ \tilde{\mathbf{x}} \in \mathbb{R}^n : \|\tilde{\mathbf{x}}\| \leq \delta \right\}.$$

**Theorem 7.5**   (Uniformly Asymptotically Stable) *Let $\tilde{\mathbf{x}} = 0$ be an equilibrium point of system (7.46) and $V(t, \tilde{\mathbf{x}})$ be a continuously differentiable function such that*

$$W_1(\tilde{\mathbf{x}}) \leq V(\tilde{\mathbf{x}}, t) \leq W_2(\tilde{\mathbf{x}})$$

*and*

$$\frac{\partial V}{\partial t} + \frac{\partial V}{\partial \tilde{\mathbf{x}}} \mathbf{f}(\tilde{\mathbf{x}}, t) \leq -W_3(\tilde{\mathbf{x}})$$

*for all $t \geq 0$ and $\tilde{\mathbf{x}} \in D$, where $W_1(\tilde{\mathbf{x}})$, $W_2(\tilde{\mathbf{x}})$ and $W_3(\tilde{\mathbf{x}})$ are continuous positive definite functions in D. Then, under Assumption 7.1, the equilibrium point $\tilde{\mathbf{x}} = 0$ is uniformly asymptotically stable.*

*Further, if there exist two positive constants $\delta$ and $c$, such that the ball $\mathcal{B}_\delta = \{\|\tilde{\mathbf{x}}_0\| \in \mathbb{R}^n : \|\tilde{\mathbf{x}}_0\| \leq \delta\} \subset D$ and $c < \min_{\|x\|=\delta} W_1(\tilde{\mathbf{x}})$, then every solution trajectory starting in $\{\tilde{\mathbf{x}} \in \mathcal{B}_\delta : W_2(\tilde{\mathbf{x}}) \leq c\}$ satisfies*

$$\|\tilde{\mathbf{x}}(t)\| \leq \beta(\|\tilde{\mathbf{x}}_0\|, t - t_0), \quad \forall t \geq t_0 \geq 0$$

*for some class $\mathcal{KL}$ function $\beta$.*

**Theorem 7.6**   (Uniformly Globally Asymptotically Stable) *If the assumptions of Theorem 7.5 above hold for any initial state $\tilde{\mathbf{x}}_0 \in \mathbb{R}^n$, $\|\tilde{\mathbf{x}}_0\| \to \infty$, the equilibrium point $\tilde{\mathbf{x}} = 0$ of system (7.46) is uniformly globally asymptotically stable (UGAS).*

**Theorem 7.7**   (Exponentially Stable) *If the assumptions of Theorem 7.5 above are satisfied with*

$$k_1 \|\tilde{\mathbf{x}}\|^p \leq V(\tilde{\mathbf{x}}, t) \leq k_2 \|\tilde{\mathbf{x}}\|^p$$

*and*

$$\frac{\partial V}{\partial t} + \frac{\partial V}{\partial \tilde{\mathbf{x}}} \mathbf{f}(\tilde{\mathbf{x}}, t) \leq -k_3 \|\tilde{\mathbf{x}}\|^p$$

*for all $t \geq 0$ and $\tilde{\mathbf{x}} \in D$, where $k_1$, $k_2$, $k_3$ and $p$ are positive constants. Then, the equilibrium point $\tilde{\mathbf{x}} = 0$ of system (7.46) is exponentially stable.*

**Theorem 7.8** (Globally Exponentially Stable)*] If the assumptions of Theorem 7.7 above hold for any initial state $\tilde{\mathbf{x}}_0 \in \mathbb{R}^n$, $\|\tilde{\mathbf{x}}_0\| \to \infty$, the equilibrium point $\tilde{\mathbf{x}} = 0$ of system (7.46) is globally exponentially stable.*

## 7.5 Input-to-State Stability

The development of automatic controllers for marine systems can be challenging because of the broad range of environmental conditions that they must operate within and the parametric uncertainty in dynamic models for them that arises because of changes in their configuration. Therefore, it is of important to extend stability concepts to include disturbance inputs. In the linear case, which can be represented by the system

$$\dot{\mathbf{x}} = \mathbf{A}\mathbf{x} + \mathbf{B}\mathbf{w}_d,$$

if the matrix $\mathbf{A}$ is Hurwitz, i.e., if the unforced system $\dot{\mathbf{x}} = \mathbf{A}\mathbf{x}$ is asymptotically stable, then bounded inputs $\mathbf{w}_d$ lead to bounded states, while inputs converging to zero produce states converging to zero. Now, consider a nonlinear system of the form

$$\dot{\mathbf{x}} = \mathbf{f}(\mathbf{x}, \mathbf{w}_d) \tag{7.54}$$

where $\mathbf{w}_d$ is a measurable, locally bounded disturbance input. In general, global asymptotic stability of the unforced system $\dot{\mathbf{x}} = \mathbf{f}(\mathbf{x}, 0)$ does not guarantee input-to-state properties of the kind mentioned above. For example, the scalar system

$$\dot{x} = -x + x w_d \tag{7.55}$$

has unbounded trajectories under the bounded input $w_d = 2$. This motivates the following important concept, introduced by Sontag and Wang [8].

The system (7.54) is called input-to-state stable (ISS) with respect to $\mathbf{w}_d$ if for some functions $\gamma \in \mathcal{K}_\infty$ and $\beta \in \mathcal{KL}$, for every initial state $\mathbf{x}_0$, and every input $\mathbf{w}_d$ the corresponding solution of (7.54) satisfies the inequality

$$\|\mathbf{x}(t)\| \leq \beta(\|\mathbf{x}_0\|, t) + \gamma(\|\mathbf{w}_d\|_{[0,t]}) \quad \forall t \geq 0 \tag{7.56}$$

where $\|\mathbf{w}_d\|_{[0,t]} := \text{ess sup}\{|\mathbf{w}_d(s)| : s \in [0, t]\}$ (supremum norm on [0, t] except for a set of measure zero). Since the system (7.54) is time-invariant, the same property results if we write

$$\|\mathbf{x}(t)\| \leq \beta(\|\mathbf{x}_0\|, t - t_0) + \gamma(\|\mathbf{w}_d\|_{[t_0,t]}) \quad \forall t \geq t_0 \geq 0.$$

The ISS property admits the following Lyapunov-like equivalent characterization: the system (7.54) is ISS if and only if there exists a positive definite radially unbounded $C^1$ function $V : \mathbb{R}^n \to \mathbb{R}$ such that for some class $\mathcal{K}_\infty$ functions $\alpha$ and $\chi$ we have

$$\frac{\partial V}{\partial \mathbf{x}}\mathbf{f}(\mathbf{x}, \mathbf{w}_d) \leq -\alpha(\|\mathbf{x}\|) + \chi(\|\mathbf{w}_d\|) \quad \forall \mathbf{x}, \mathbf{w}_d.$$

This is in turn equivalent to the following "gain margin" condition:

$$\|\mathbf{x}\| > \rho(\|\mathbf{w}_d\|) \quad \Rightarrow \quad \frac{\partial V}{\partial \mathbf{x}}\mathbf{f}(\mathbf{x}, \mathbf{w}_d) \leq -\bar{\alpha}(\|\mathbf{x}\|)$$

where $\bar{\alpha}, \rho \in \mathcal{K}_\infty$. Such functions $V$ are called *ISS-Lyapunov functions*.

The system (7.54) is *locally input-to-state stable* (locally ISS) if the bound (7.56) is valid for solutions with sufficiently small initial conditions and inputs, i.e., if there exists a $\delta > 0$ such that (7.56) is satisfied whenever $\|x_0\| \leq \delta$ and $\|u\|_{[0,t]} \leq \delta$. It turns out (local) asymptotic stability of the unforced system $\dot{x} = f(x, 0)$ implies local ISS.

For systems with outputs, it is natural to consider the following notion which is dual to ISS. A system

$$\begin{aligned} \dot{\mathbf{x}} &= \mathbf{f}(\mathbf{x}) \\ \mathbf{y} &= \mathbf{h}(\mathbf{x}) \end{aligned} \tag{7.57}$$

is called *output-to-state stable* if for some functions $\gamma \in \mathcal{K}_\infty$ and $\beta \in \mathcal{KL}$ and every initial state $\mathbf{x}_0$ the corresponding solution of (7.57) satisfies the inequality

$$\|\mathbf{x}(t)\| \leq \beta(\|\mathbf{x}_0\|, t) + \gamma(\|\mathbf{y}\|_{[0,t]})$$

as long as it is defined. While ISS is to be viewed as a generalization of stability, OSS can be thought of as a generalization of observability; it does indeed reduce to the standard observability property in the linear case. Given a system with both inputs and outputs

$$\begin{aligned} \dot{\mathbf{x}} &= \mathbf{f}(\mathbf{x}, \mathbf{w}_d) \\ \mathbf{y} &= \mathbf{h}(\mathbf{x}) \end{aligned} \tag{7.58}$$

one calls it *input/output-to-state stable* (IOSS) if for some functions $\gamma_1, \gamma_2 \in \mathcal{K}_\infty$ and $\beta \in \mathcal{KL}$, for every initial state $\mathbf{x}_0$, and every input $\mathbf{w}_d$ the corresponding solution of (7.58) satisfies the inequality

$$\|\mathbf{x}(t)\| \leq \beta(\|\mathbf{x}_0\|, t) + \gamma_1(\|\mathbf{w}_d\|_{[0,t]}) + \gamma_2(\|\mathbf{y}\|_{[0,t]})$$

as long as it exists.

## 7.6 Ultimate Boundedness

Even when an equilibrium point $\tilde{\mathbf{x}} = 0$ of system (7.46) does not exist, a Lyapunov-like stability analysis can be used to show whether or not system state trajectories $\tilde{\mathbf{x}}(t)$ remain bounded within a region of the state space. We begin with a motivating example, which illustrates the concept of *uniform ultimate boundedness* (UUB). When a system is uniformly ultimately bounded the solution trajectories of the the state do not necessarily approach an equilibrium point, but instead converge to, and remain, within some neighborhood of $\tilde{\mathbf{x}} = 0$ after a sufficiently long time.

**Example 7.5** Consider a surge speed-tracking controller for an unmanned surface vessel operating under the influence of time-varying exogenous disturbances (e.g. wind or waves). The equation of motion of the system is

$$m\dot{u} = -c_d u|u| - \tau + w_d(t), \tag{7.59}$$

where $m$ is the mass (including added mass) of the vessel, $u$ is the surge speed, $c_d$ is the drag coefficient, $\tau$ is the control input (thruster force), and $w_d(t)$ is the disturbance.

Assume that the magnitude of the disturbance can be upper-bounded by a known positive constant $w_{d0} > 0$, such that

$$|w_d(t)| < w_{d0}, \quad \forall t. \tag{7.60}$$

It is desired that the surge speed-controller track a time-dependent speed $u_d(t)$, which is assumed to be continuously differentiable.

Let the control input be

$$\tau = m\dot{u}_d + c_d u|u| - k_p \tilde{u}, \tag{7.61}$$

where $\tilde{u} := u - u_d$ is the speed tracking error and $k_p > 0$ is a control gain. Then the closed loop equation of motion is given by

$$m\dot{\tilde{u}} = -k_p \tilde{u} + w_d(t). \tag{7.62}$$

Using the concept of flow along a line (Sect. 2.3), it can be seen that when $w_d(t) = 0$ the closed loop system has an equilibrium point at $\tilde{u} = 0$. However, the equilibrium point at $\tilde{u} = 0$ no longer exists when $w_d(t) \neq 0$.

To investigate the stability, consider the Lyapunov function

$$V = \frac{1}{2}\tilde{u}^2, \tag{7.63}$$

which is radially unbounded. Then, taking the derivative along system trajectories using (7.62) gives

$$\begin{aligned}
\dot{V} = \tilde{u}\dot{\tilde{u}} &= \tilde{u}\left[-k_p\tilde{u} + w_d(t)\right], \\
&= -k_p\tilde{u}^2 + \tilde{u}w_d(t), \\
&\leq -k_p\tilde{u}^2 + \|\tilde{u}\|w_{d0}, \\
&\leq -\|\tilde{u}\|\left[k_p\|\tilde{u}\| - w_{d0}\right],
\end{aligned} \tag{7.64}$$

where the upper bound is obtained with (7.60).

It therefore follows that

$$\dot{V} < 0, \quad \text{for all } \|\tilde{u}\| > \delta := \frac{w_{d0}}{k_p}. \tag{7.65}$$

In other words, $\dot{V} < 0$ when $\tilde{u}$ is outside the compact set $\mathcal{B}_\delta = \{\tilde{u} \in \mathbb{R} : \|\tilde{u}\| \leq \delta\}$ (see Definition 7.7). This implies that all of the solution trajectories $\tilde{u}(t)$ of (7.62) that start outside of $\mathcal{B}_\delta$ will tend towards $\mathcal{B}_\delta$.

We can show that the system is *uniformly bounded*, meaning that the final (*ultimate*) bound of $\tilde{u}$ is independent of the value of $\tilde{u}$ at the initial time $t_0$. Let $0 < \delta < \Delta$, then all solution trajectories $\tilde{u}(t)$ of (7.62) that start in the set

$$\mathcal{B}_\Delta := \{\tilde{u} \in \mathbb{R} : \|\tilde{u}\| \leq \Delta\}$$

will remain within $\mathcal{B}_\Delta$ for all $t \geq t_0$, where $\mathcal{B}_\delta \subset \mathcal{B}_\Delta$, since $\dot{V} < 0, \forall \|\tilde{u}\| > \delta$, see (7.65).

An estimate for the value of the *ultimate bound* can also be determined. Let $\zeta$ be a positive constant defined so that $\delta < \zeta < \Delta$. Since $\delta < \|\tilde{u}\| < \Delta$, $V$ can be bounded as

$$\frac{1}{2}\delta^2 < \frac{1}{2}\tilde{u}^2 < \frac{1}{2}\Delta^2.$$

Inside the annular set $(\mathcal{B}_\Delta - \mathcal{B}_\delta)$, $\dot{V} < 0$ so that $V$ decreases monotonically in time until $|\tilde{u}(t)| \leq \zeta$. Denote the earliest time when $|\tilde{u}(t)| \leq \zeta$ as $t = (t_0 + T_\zeta)$, where $T_\zeta$ is the elapsed time, and define the compact set $\mathcal{B}_\zeta := \{\tilde{u} \in \mathbb{R} : \|\tilde{u}\| \leq \zeta, \delta < \zeta\}$. For any time $t \geq t_0 + T_\zeta$ the trajectory of $\tilde{u}(t)$ remains inside $\mathcal{B}_\zeta$ because $\dot{V} < 0$ outside $\mathcal{B}_\zeta$ and on its boundary. Thus, using (7.65), we can conclude that the state trajectories of the closed loop system are *uniformly ultimately bounded* (UUB) with the ultimate bound $|\tilde{u}(t_0 + T_\zeta)| = \zeta > w_{d0}/k_p$. Note that the size of the compact set $\mathcal{B}_\zeta$ can be reduced by increasing the control gain $k_p$.

A simulation of the closed loop USV surge speed control system (7.62) is shown in Fig. 7.12 using a model of the small vehicle shown in Fig. 7.11. The USV has an overall length of 1.7 m, a mass of $m = 15$ kg and is simulated operating in the presence of deep water waves with a wavelength of 1.5 m (wave period of about 1.0 s). The control gain is set to $k_p = 5$. The desired speed $u_d$ follows a step-like pattern in time. The wave-induced disturbance force has an amplitude of 4.5 N. In Fig. 7.12a it can be seen that the magnitude of the surge speed error decreases rapidly after the simulation starts, and oscillates about $\tilde{u} = 0$ with an amplitude of slightly less than

**Fig. 7.11**  A small lightweight USV. ©[2009] IEEE. Reprinted, with permission, from [3]

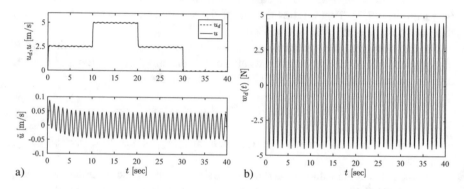

a)                                                              b)

**Fig. 7.12**  Time response of a surge speed tracking USV: (a) desired speed $u_d$, output speed $u$, and speed tracking error $\tilde{u}$, (b) wave disturbance forces $w_d(t)$

0.05 m/s. If we bound the anticipated magnitude of the disturbance as $w_{d0} = 4.5$ N, the UUB analysis above gives a very conservative estimate for the bounds of the error as

$$\|\tilde{u}\| > \frac{w_{d0}}{k_p} = 0.9 \text{ m/s}.$$

$\square$

**Definition 7.32** (*Uniformly Bounded*). The solution trajectories $\tilde{\mathbf{x}}(t)$ of (7.46) are *uniformly bounded* (UB) if there exist constants $\delta$ and $\varepsilon$, where $0 < \delta < \varepsilon$, and a function $\alpha(\delta) > 0$ such that

$$\|\tilde{\mathbf{x}}_0\| \leq \delta \Rightarrow \|\tilde{\mathbf{x}}(t)\| \leq \alpha(\delta), \quad \forall t \geq t_0 \geq 0. \tag{7.66}$$

Comparing the definition of UB to Definitions 7.3 and 7.49, it can be seen that while UB is similar, it is defined only with reference to the boundedness of solution trajectories, rather than with reference to the position of an equilibrium point.

**Definition 7.33** (*Uniformly Globally Bounded*). The solution trajectories $\tilde{\mathbf{x}}(t)$ of (7.46) are *uniformly globally bounded* (UGB) if (7.66) holds for $\delta \to \infty$.

**Definition 7.34** (*Uniformly Ultimately Bounded*) The solution trajectories $\tilde{\mathbf{x}}(t)$ of (7.46) are *uniformly ultimately bounded* (UUB), with ultimate bound $\zeta$, if there exist constants $\delta$, $\zeta$ and $\varepsilon$, where for every $0 < \delta < \varepsilon$, there exists an elapsed time $T = T(\delta, \zeta) \geq 0$ such that

$$\|\tilde{\mathbf{x}}_0\| < \delta \Rightarrow \|\tilde{\mathbf{x}}(t)\| \leq \zeta, \quad \forall t \geq t_0 + T. \tag{7.67}$$

**Definition 7.35** (*Uniformly Globally Ultimately Bounded*) The solution trajectories $\tilde{\mathbf{x}}(t)$ of (7.46) are *uniformly globally ultimately bounded* (UGUB) if (7.67) holds for $\delta \to \infty$.

In Definitions 7.32–7.35 above, the term *uniform* means that the ultimate bound $\zeta$ does not depend on the initial time $t_0$. The term *ultimate* indicates that boundedness holds after an elapsed time $T$. The constant $\varepsilon$ defines a neighborhood of the origin, which is independent of $t_0$, so that all trajectories starting within the neighborhood remain bounded. If $\varepsilon$ can be chosen arbitrarily large then the system then the UUB is global.

**Remark 7.7** Lyapunov stability Definitions 7.26–7.30 require that solution state trajectories $\tilde{\mathbf{x}}(t)$ remain arbitrarily close to the system equilibrium point $\tilde{\mathbf{x}} = 0$ by starting sufficiently close to it. In general, this requirement is too strong to achieve in practice, as real systems are usually affected by the presence of unknown disturbances. Further, the UUB bound $\zeta$ cannot be made arbitrarily small by starting closer to $\tilde{\mathbf{x}} = 0$. In practical systems, $\zeta$ depends on both the disturbances and system uncertainties.

To understand how a Lyapunov-like analysis can be used to study UUB, consider a continuously differentiable, positive definite function $V : \mathbb{R}^n \to \mathbb{R}$ and suppose that the sets

$$\Omega_c := \{\tilde{\mathbf{x}} \in \mathbb{R}^n : V(\tilde{\mathbf{x}}) \leq c\} \quad \text{and} \quad \Omega_\varepsilon := \{\tilde{\mathbf{x}} \in \mathbb{R}^n : V(\tilde{\mathbf{x}}) \leq \varepsilon\}$$

are compact and invariant from some $0 < \varepsilon < c$ (see Fig. 7.13). Let $\Lambda := \Omega_c - \Omega_\varepsilon = \{\tilde{\mathbf{x}} \in \mathbb{R}^n : \varepsilon \leq V(\tilde{\mathbf{x}}) \leq c\}$ and suppose that the time derivative along the solution trajectories $\tilde{\mathbf{x}}(t)$ of (7.46) is given by

$$\dot{V}(\tilde{\mathbf{x}}, t) = \frac{\partial V}{\partial \tilde{\mathbf{x}}} \mathbf{f}(\tilde{\mathbf{x}}) \leq -W_3(\tilde{\mathbf{x}}), \quad \forall \tilde{\mathbf{x}} \in \Lambda, \quad \forall t \geq 0,$$

**Fig. 7.13** Geometric
interpretation of uniform
ultimate boundedness

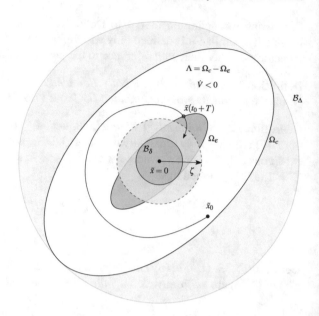

where $W_3(\tilde{\mathbf{x}})$ is a continuous positive definite function. Since $\dot{V}(\tilde{\mathbf{x}}, t) < 0$ for all
$\tilde{\mathbf{x}} \in \Lambda$ a trajectory $\tilde{\mathbf{x}}(t)$ starting in the set $\Lambda$ must move in the direction of decreasing
$V(\tilde{\mathbf{x}})$. Thus, $V(\tilde{\mathbf{x}})$ will decrease until the trajectory enters the set $\Omega_\varepsilon$, and once inside
$\Omega_\varepsilon$ the trajectory will remain inside $\Omega_\varepsilon$ for all future times.

Define

$$k := \min_{\tilde{\mathbf{x}} \in \Lambda} W_3(\tilde{\mathbf{x}}) > 0,$$

then

$$\dot{V}(\tilde{\mathbf{x}}, t) \leq -k, \quad \forall \tilde{\mathbf{x}} \in \Lambda, \quad \forall t \geq t_0 \geq 0,$$

so that

$$V(\tilde{\mathbf{x}}(t)) \leq V(\tilde{\mathbf{x}}_0) - k(t - t_0) = V_0 - k(t - t_0).$$

Thus, a state starting at the position $\tilde{\mathbf{x}}_0$ at time $t = t_0$ will arrive at the boundary of
the compact set $V = \Omega_\varepsilon$ after the elapsed time $T = (V_0 - \varepsilon)/k + t_0$.

In order to find an estimate of the ultimate bound, we define the balls (Definition 7.31)

$$\mathcal{B}_\delta := \left\{ \tilde{\mathbf{x}} \in \mathbb{R}^n : \|\tilde{\mathbf{x}}(t)\| \leq \delta \right\} \quad \text{and} \quad \mathcal{B}_\Delta := \left\{ \tilde{\mathbf{x}} \in \mathbb{R}^n : \|\tilde{\mathbf{x}}(t)\| \leq \Delta \right\},$$

such that the compact set $\Lambda$ occurs in a subset of the region of the state space where
$\delta \leq \|\tilde{\mathbf{x}}\| \leq \Delta$ (see Fig. 7.13) and

$$\dot{V}(\tilde{\mathbf{x}}, t) \leq -W_3(\tilde{\mathbf{x}}), \quad \forall \delta \leq \|\tilde{\mathbf{x}}\| \leq \Delta, \quad \forall t \geq t_0 \geq 0.$$

Let $\alpha_1$ and $\alpha_2$ be class $\mathcal{K}$ functions, such that

$$\alpha_1(\|\tilde{\mathbf{x}}\|) \leq V(\tilde{\mathbf{x}}) \leq \alpha_2(\|\tilde{\mathbf{x}}\|). \tag{7.68}$$

Referring to Fig. 7.13, it can be seen that $c = \alpha_2(\Delta)$ where the boundaries of $\mathcal{B}_\Delta$ and $\Omega_c$ touch. Thus, inside the compact set $\Omega_c$, $V(\tilde{\mathbf{x}}) \leq \alpha_2(\Delta)$, so that $\Omega_c \subseteq \mathcal{B}_\Delta$. Since the maximum value of $V(\tilde{\mathbf{x}})$ in the set $\Omega_\varepsilon$ occurs at its boundary, we have $\varepsilon = \alpha_2(\delta)$ where the surfaces of $\Omega_\varepsilon$ and $\mathcal{B}_\delta$ touch. However, inside of $\Omega_\varepsilon$, $\alpha_1(\|\tilde{\mathbf{x}}\|) \leq V(\tilde{\mathbf{x}})$, so that the relation $\alpha_1(\|\tilde{\mathbf{x}}\|) \leq \varepsilon = \alpha_2(\delta)$ holds. Taking the inverse of this latter inequality provides an estimate of the ultimate bound

$$\|\tilde{\mathbf{x}}\| \leq \alpha_1^{-1}[\alpha_2(\delta)]. \tag{7.69}$$

**Remark 7.8** The ultimate bound is independent of the initial state $\tilde{\mathbf{x}}_0 \in \Omega_c$ and is a class $\mathcal{K}$ function of $\delta$. As a consequence, the smaller the value of $\delta$, the smaller the ultimate bound.

**Theorem 7.9** (Uniformly Ultimately Bounded) *Suppose that*

$$\alpha_1(\|\tilde{\mathbf{x}}\|) \leq V(\tilde{\mathbf{x}}) \leq \alpha_2(\|\tilde{\mathbf{x}}\|),$$

$$\dot{V}(\tilde{\mathbf{x}}, t) = \frac{\partial V}{\partial \tilde{\mathbf{x}}}\mathbf{f}(\tilde{\mathbf{x}}) \leq -W_3(\tilde{\mathbf{x}}), \quad \forall \|\tilde{\mathbf{x}}\| \geq \delta > 0, \quad \forall t \geq 0,$$

*and* $\|\tilde{\mathbf{x}}\| \leq \Delta$, *where* $\alpha_1, \alpha_2 \in \mathcal{K}$, $W_3(\tilde{\mathbf{x}}) > 0$ *is continuous, and* $\delta < \alpha_2^{-1}[\alpha_1(\Delta)]$. *Then, for every initial state* $\tilde{\mathbf{x}}_0 \in \{\|\tilde{\mathbf{x}}\| \leq \alpha_2^{-1}[\alpha_1(\Delta)]\}$, *there exists an elapsed time* $T = T(\tilde{\mathbf{x}}_0, \delta) \geq 0$, *such that*

$$\|\tilde{\mathbf{x}}(t)\| \leq \alpha_1^{-1}[\alpha_2(\delta)], \quad \forall t \geq t_0 + T.$$

*If the conditions hold for* $\Delta \to \infty$ *and* $\alpha_1 \in \mathcal{K}_\infty$, *then the conclusions are valid for every initial state* $\tilde{\mathbf{x}}_0$ *and the system is uniformly globally ultimately bounded.*

## 7.7  Practical Stability

Thus far, we have mostly considered the stability of nonlinear time-varying systems when no external disturbances are acting on them and when there is no uncertainty in the dynamics of the system. However, most marine vehicles operate in uncertain environments and with inaccurate dynamic models. Thus, it is important to understand how disturbances and model uncertainty can affect the stability of the system. This is especially true for systems that operate near the upper limits of the capabilities of their actuators.

As discussed in Example 7.5, when systems operate in the presence of time-varying disturbances, an equilibrium point $\tilde{\mathbf{x}} = 0$ no longer exists. However, the solution trajectories of the state $\tilde{\mathbf{x}}(t)$ may be ultimately bounded within a finite region. In some cases, as in the example, the size of the bounded region can be decreased by tuning the control gains. This stability property is often referred to as *practical stability*.

When the region of attraction of the state trajectories of a system consists of the entire state space, the region is *globally attractive*, see Definitions 7.28, 7.30 and 7.35, for example. Instead, when the region of attraction of a closed loop system can be arbitrarily enlarged by tuning a set of control gains, the region is called *semiglobally attractive*.

In Sect. 7.3 it is seen that invariant sets can be thought of as a generalization of the concept of an equilibrium point. In a similar way, we will consider the stability of state trajectories towards closed *balls* (Definition 7.31) in state space as a similar generalization of the idea of an equilibrium point when characterizing the stability of nonlinear systems in the presence of disturbances. The following stability definitions, theorems and notation are developed in [1, 2, 4]. Additional details can be found in these references.

However, before proceeding, we will generalize the notion of uniform asymptotic stability (UAS) about an equilibrium point given in Definition 7.27 and Theorem 7.5 above to the UAS of a ball. To do this, define a bound for the shortest distance between a point $\tilde{\mathbf{z}}$ on the surface of the ball $\mathcal{B}_\delta$ and the point $\tilde{\mathbf{x}}$, as

$$\|\tilde{\mathbf{x}}\|_\delta := \inf_{\tilde{\mathbf{z}} \in \mathcal{B}_\delta} \|\tilde{\mathbf{x}} - \tilde{\mathbf{z}}\| \tag{7.70}$$

and let $\Delta > \delta > 0$ be two non-negative numbers.

**Definition 7.36** (*Uniform Asymptotic Stability of a Ball*) The ball $\mathcal{B}_\delta$ is *uniformly asymptotically stable* (UAS) on $\mathcal{B}_\Delta$ for the system (7.46), if and only if there exists a class $\mathcal{KL}$ function $\beta$, such that for all initial states $\tilde{\mathbf{x}}_0 \in \mathcal{B}_\Delta$ and all initial times $t_0 \geq 0$, the solution of (7.46) satisfies

$$\|\tilde{\mathbf{x}}(t)\|_\delta \leq \beta(\|\tilde{\mathbf{x}}_0\|, t - t_0), \quad \forall t \geq t_0. \tag{7.71}$$

**Definition 7.37** (*Uniform Global Asymptotic Stability of a Ball*) The ball $\mathcal{B}_\delta$ is *uniformly globally asymptotically stable* (UGAS) if (7.71) holds for any initial condition $\tilde{\mathbf{x}}_0$, i.e. $\Delta \to \infty$.

While the UAS and the UGAS of a ball imply the property of ultimate boundedness (with any $\zeta > \delta$ as the ultimate bound), they are stronger properties, as they guarantee that sufficiently small transients remain arbitrarily near $\mathcal{B}_\delta$. Thus, the definition of the stability of a system about a ball, is similar to the definition of stability about an equilibrium point, e.g. compare Definition 7.36 to Definition 7.5. Ultimate boundedness is really a notion of convergence, which does not necessary imply stability to perturbations.

Consider a nonlinear, time-varying system of the form

$$\dot{\tilde{x}} = f(\tilde{x}, t, \theta), \quad \tilde{x} \in \mathbb{R}^n, \quad t \geq 0, \tag{7.72}$$

where $\theta \in \mathbb{R}^m$ is a vector of constant parameters, and $f : \mathbb{R}^n \times \mathbb{R} \times \mathbb{R}^m$ is locally Lipschitz in $\tilde{x}$ and piecewise continuous in $t$. System (7.72) is representative of closed loop control systems, where $\theta$ would typically contain control gains, but could represent other parameters.

In the following definitions, let $\Theta \subset \mathbb{R}^m$ be a set of parameters.

**Definition 7.38** (*Uniform Global Practical Asymptotic Stability*) The system (7.72) is *uniformly globally practically asymptotically stable* (UGPAS) on $\Theta$, if for any $\delta > 0$ there exists a $\theta^\star(\delta) \in \Theta$, such that $\mathcal{B}_\delta$ is UGAS for system (7.72) when $\theta = \theta^\star$.

Thus, (7.72) is UGPAS if the ball $\mathcal{B}_\delta$, which is UGAS, can be arbitrarily diminished by a convenient choice of $\theta$.

**Definition 7.39** (*Uniform Semiglobal Practical Asymptotic Stability*) The system (7.72) is *uniformly semiglobally practically asymptotically stable* (USPAS) on $\Theta$, if for any $\Delta > \delta > 0$ there exists a $\theta^\star(\delta, \Delta) \in \Theta$, such that $\mathcal{B}_\delta$ is UAS on $\mathcal{B}_\Delta$ for system (7.72) when $\theta = \theta^\star$.

In this case, (7.72) is USPAS if the estimate of the domain of attraction $\mathcal{B}_\Delta$ and the ball $\mathcal{B}_\delta$, which is UAS, can be arbitrarily enlarged and diminished, respectively, by tuning the parameter $\theta$.

**Definition 7.40** (*Uniform Global Practical Exponential Stability*) System (7.72) is *uniformly globally practically exponentially stable* (UGPES) on $\Theta$, if for any $\delta > 0$ there exists a $\theta^\star(\delta) \in \Theta$, and positive constants $k(\delta)$ and $\gamma(\delta)$, such that for any initial state $\tilde{x}_0 \in \mathbb{R}^n$ and for any time $t \geq t_0 \geq 0$ the solution trajectories of (7.72) satisfy

$$\|\tilde{x}(t, \theta^\star)\| \leq \delta + k(\delta)\|\tilde{x}_0\| e^{-\gamma(\delta)(t-t_0)}, \quad \forall t \geq t_0.$$

**Definition 7.41** (*Uniform Semiglobal Practical Exponential Stability*) System (7.72) is *uniformly semiglobally practically exponentially stable* (USPES) on $\Theta$, if for any $\Delta > \delta > 0$ there exists a $\theta^\star(\delta, \Delta) \in \Theta$, and positive constants $k(\delta, \Delta)$ and $\gamma(\delta, \Delta)$, such that for any initial state $\tilde{x}_0 \in \mathcal{B}_\Delta$ and for any time $t \geq t_0 \geq 0$ the solution trajectories of (7.72) satisfy

$$\|\tilde{x}(t, \theta^\star)\| \leq \delta + k(\delta, \Delta)\|\tilde{x}_0\| e^{-\gamma(\delta, \Delta)(t-t_0)}, \quad \forall t \geq t_0.$$

**Theorem 7.10** (UGPAS) *Suppose that given any* $\delta > 0$ *there exist a parameter* $\theta^\star(\delta) \in \Theta$, *a continuously differentiable function* $V : \mathbb{R}^n \times \mathbb{R}_{\geq 0} \to \mathbb{R}_{\geq 0}$, *and three class* $\mathcal{K}_\infty$ *functions* $\alpha_1$, $\alpha_2$ *and* $\alpha_3$, *such that for all* $\tilde{\mathbf{x}} \in \mathbb{R}^n \backslash \mathcal{B}_\delta$ *(i.e. all points* $\tilde{\mathbf{x}}$ *that are not inside* $\mathcal{B}_\delta$*) and for all* $t \geq 0$

$$\alpha_1(\|\tilde{\mathbf{x}}\|) \leq V(\tilde{\mathbf{x}}, t) \leq \alpha_2(\|\tilde{\mathbf{x}}\|),$$

$$\frac{\partial V}{\partial t} + \frac{\partial V}{\partial \tilde{\mathbf{x}}} \mathbf{f}(\tilde{\mathbf{x}}, t) \leq -\alpha_3(\|\tilde{\mathbf{x}}\|),$$

*and*

$$\lim_{\delta \to 0} \alpha_1^{-1} [\alpha_2(\delta)] = 0. \tag{7.73}$$

*Then, (7.72) is UGPAS on the parameter set* $\Theta$.

The first two conditions of Theorem 7.10 can often be verified using a Lyapunov function that establishes the UGAS of the system (Theorem 7.6) when the disturbances and uncertainties are assumed to be zero [1]. The Lyapunov function may depend on the tuning parameter $\theta$, and so on the radius $\delta$. Hence, (7.73) is required to links the bounds on the Lyapunov function.

**Theorem 7.11** (USPAS) *Suppose that given any* $\Delta > \delta > 0$ *there exist a parameter* $\theta^\star(\delta, \Delta) \in \Theta$, *a continuously differentiable function* $V : \mathbb{R}^n \times \mathbb{R}_{\geq 0} \to \mathbb{R}_{\geq 0}$, *and three class* $\mathcal{K}_\infty$ *functions* $\alpha_1$, $\alpha_2$ *and* $\alpha_3$, *such that for all* $\delta < \|\tilde{\mathbf{x}}\| < \Delta$ *and for all* $t \geq 0$

$$\alpha_1(\|\tilde{\mathbf{x}}\|) \leq V(\tilde{\mathbf{x}}, t) \leq \alpha_2(\|\tilde{\mathbf{x}}\|),$$

*and*

$$\frac{\partial V}{\partial t} + \frac{\partial V}{\partial \tilde{\mathbf{x}}} \mathbf{f}(\tilde{\mathbf{x}}, t) \leq -\alpha_3(\|\tilde{\mathbf{x}}\|).$$

*Assume further that for any constants* $0 < \delta^\star < \Delta^\star$, *there exist* $0 < \delta < \Delta$ *such that*

$$\alpha_1^{-1} [\alpha_2(\delta)] \leq \delta^\star \quad \text{and} \quad \alpha_2^{-1} [\alpha_1(\Delta)] \geq \Delta^\star. \tag{7.74}$$

*Then, (7.72) is USPAS on the parameter set* $\Theta$.

Since the Lyapunov function $V(\tilde{\mathbf{x}}, t)$ is not required to be the same for all $\delta$ and all $\Delta$, the conditions in (7.74) must be imposed to ensure that the estimate of the domain of attraction $\mathcal{B}_\Delta$ and the set $\mathcal{B}_\delta$, which is UAS, can be arbitrarily enlarged and diminished, respectively [2].

**Theorem 7.12** (UGPES) *Suppose that for any* $\delta > 0$ *there exist a parameter* $\theta^\star(\delta) \in \Theta$, *a continuously differentiable function* $V : \mathbb{R}^n \times \mathbb{R}_{\geq 0} \to \mathbb{R}_{\geq 0}$, *and positive constants* $k_1(\delta)$, $k_2(\delta)$ *and* $k_3(\delta)$, *such that for all* $\tilde{\mathbf{x}} \in \mathbb{R}^n \backslash \mathcal{B}_\delta$ *(i.e. all points* $\tilde{\mathbf{x}}$ *that are not inside* $\mathcal{B}_\delta$*) and for all* $t \geq 0$

$$k_1(\delta)\|\tilde{\mathbf{x}}\|^p \leq V(\tilde{\mathbf{x}}, t) \leq k_2(\delta)\|\tilde{\mathbf{x}}\|^p,$$

$$\frac{\partial V}{\partial t} + \frac{\partial V}{\partial \tilde{\mathbf{x}}}\mathbf{f}(\tilde{\mathbf{x}}, t) \leq -k_3(\delta)\|\tilde{\mathbf{x}}\|^p,$$

*where $p > 0$ is a constant and*

$$\lim_{\delta \to 0} \frac{k_2(\delta)\delta^p}{k_1(\delta)} = 0. \tag{7.75}$$

*Then, (7.72) is UGPES on the parameter set $\Theta$.*

**Theorem 7.13** (USPES) *Suppose that for any $\Delta > \delta > 0$ there exist a parameter $\theta^\star(\delta, \Delta) \in \Theta$, a continuously differentiable function $V : \mathbb{R}^n \times \mathbb{R}_{\geq 0} \to \mathbb{R}_{\geq 0}$, and positive constants $k_1(\delta, \Delta)$, $k_2(\delta, \Delta)$ and $k_3(\delta, \Delta)$, such that for all $\tilde{\mathbf{x}} \in \mathcal{B}_\Delta \backslash \mathcal{B}_\delta$ (i.e. all points $\tilde{\mathbf{x}}$ in $\mathcal{B}_\Delta$ that are not inside $\mathcal{B}_\delta$) and for all $t \geq 0$*

$$k_1(\delta, \Delta)\|\tilde{\mathbf{x}}\|^p \leq V(\tilde{\mathbf{x}}, t) \leq k_2(\delta, \Delta)\|\tilde{\mathbf{x}}\|^p,$$

$$\frac{\partial V}{\partial t} + \frac{\partial V}{\partial \tilde{\mathbf{x}}}\mathbf{f}(\tilde{\mathbf{x}}, t) \leq -k_3(\delta, \Delta)\|\tilde{\mathbf{x}}\|^p,$$

*where $p > 0$ is a constant. Assume further that for any constants $0 < \delta^\star < \Delta^\star$, there exist $0 < \delta < \Delta$ such that*

$$\frac{k_2(\delta, \Delta)\delta^p}{k_1(\delta, \Delta)} \leq \delta^\star \quad \text{and} \quad \frac{k_1(\delta, \Delta)\Delta^p}{k_2(\delta, \Delta)} \geq \Delta^\star. \tag{7.76}$$

*Then, (7.72) is USPES on the parameter set $\Theta$.*

The following inequalities can be very useful for simplifying the stability analysis of nonlinear systems.

**Theorem 7.14** (Young's Inequality) *Assume that $a \in \mathbb{R}$, $b \in \mathbb{R}$ and $\varepsilon > 0$. Then*

$$ab \leq \frac{a^2}{2\varepsilon} + \frac{\varepsilon b^2}{2}. \tag{7.77}$$

*This can also be expressed in vector form as*

$$\|\mathbf{a}^T\mathbf{b}\| \leq \frac{\|\mathbf{a}\|^2}{2\varepsilon} + \frac{\varepsilon\|\mathbf{b}\|^2}{2}, \quad \forall \mathbf{a}, \mathbf{b} \in \mathbb{R}^n. \tag{7.78}$$

**Theorem 7.15** (Triangle Inequality) *Assume that $\|\mathbf{a}\|$ and $\|\mathbf{b}\|$ are two vectors. Then*

$$\|\mathbf{a} + \mathbf{b}\| \leq \|\mathbf{a}\| + \|\mathbf{b}\|, \quad \forall \mathbf{a}, \mathbf{b} \in \mathbb{R}^n. \tag{7.79}$$

**Theorem 7.16** (Cauchy–Schwarz Inequality) *Assume that* $\|\mathbf{a}\|$ *and* $\|\mathbf{b}\|$ *are two vectors. Then*

$$\|\mathbf{a}^T\mathbf{b}\| \leq \|\mathbf{a}\|\|\mathbf{b}\|, \quad \forall \mathbf{a}, \mathbf{b} \in \mathbb{R}^n. \tag{7.80}$$

**Example 7.6** Let us return to Example 7.5 where we considered the surge speed-tracking control of an unmanned surface vessel operating under the influence of a time-varying exogenous disturbance. It can be seen that the term $-k_p\tilde{u}^2$ in (7.64) is ND. How to handle the second term may be unclear, as it depends on the unknown disturbance. Fortunately, even though we don't know the value of $w_d(t)$ at any given time, we can use the known upper bound on the magnitude of the disturbance $|w_d(t)| < w_{d0}$, $\forall t$ to find an upper bound for $\dot{V}$.

Applying Young's Inequality (Theorem 7.14) with $\varepsilon = 1$ to (7.64) and using $w_d(t)^2 < w_{d0}^2$ gives

$$\begin{aligned}\dot{V} &\leq -k_p\tilde{u}^2 + \frac{\tilde{u}^2}{2} + \frac{w_d(t)^2}{2}, \\ &\leq -\frac{(2k_p - 1)}{2}\tilde{u}^2 + \frac{w_{d0}^2}{2}.\end{aligned} \tag{7.81}$$

From (7.63) we have

$$-\frac{(2k_p - 1)}{2}\tilde{u}^2 = -(2k_p - 1)V \tag{7.82}$$

so that

$$\dot{V} \leq -(2k_p - 1)V + \frac{w_{d0}^2}{2}. \tag{7.83}$$

For $k_p > 1/2$, $V$ will decrease until $\dot{V} = 0$. Thus, $\dot{V}$ is NSD because we can have $\dot{V} = 0$ even though $\tilde{u} \neq 0$. Since $w_d = w_d(t)$ the system is not time-invariant, so LaSalle's Invariant Set Theorem (Theorem 7.3) cannot be used to analyze its stability.

However, note that integrating (7.83) gives

$$V \leq \left[V_0 - \frac{w_{d0}^2}{2(2k_p - 1)}\right]e^{-(2k_p-1)(t-t_0)} + \frac{w_{d0}^2}{2(2k_p - 1)}, \tag{7.84}$$

where $V_0$ is the value of $V$ at the initial time $t = t_0$. Thus, it can be seen that the value of $V$ decreases to its equilibrium value (when $\dot{V} = 0$) exponentially in time.

Using (7.63), we can solve for the bounds of $\|\tilde{u}\|$ from the bounds of $V$ to get

$$\|\tilde{u}\| \leq \sqrt{\left[\tilde{u}_0^2 - \frac{w_{d0}^2}{(2k_p - 1)}\right]e^{-(2k_p-1)(t-t_0)} + \frac{w_{d0}^2}{(2k_p - 1)}}. \tag{7.85}$$

This can be simplified into the form of a USPES system (Definition 7.41), using the Triangle Inequality (Theorem 7.15), to get

$$\|\tilde{u}\| \leq \sqrt{\tilde{u}_0^2 - \frac{w_{d0}^2}{(2k_p - 1)}}\, e^{-(k_p - 1/2)(t - t_0)} + \frac{w_{d0}}{\sqrt{2k_p - 1}},$$

$$\leq \frac{w_{d0}}{\sqrt{2k_p - 1}} + \|\tilde{u}_0\| \sqrt{1 - \frac{w_{d0}^2}{\tilde{u}_0^2(2k_p - 1)}}\, e^{-(k_p - 1/2)(t - t_0)}, \qquad (7.86)$$

$$\leq \delta + k(\delta, \Delta)\|\tilde{u}_0\| e^{-\gamma(t - t_0)},$$

where $\delta := w_{d0}/\sqrt{2k_p - 1}$, $\Delta := \|\tilde{u}_0\|$,

$$k(\delta, \Delta) := \sqrt{1 - \frac{w_{d0}^2}{\tilde{u}_0^2(2k_p - 1)}} = \sqrt{1 - \frac{\delta^2}{\Delta^2}},$$

and $\gamma(\delta, \Delta) := (k_p - 1/2)$.

Note that with (7.86) it can be seen that

$$\lim_{t \to \infty} = \|\tilde{u}\| = \frac{w_{d0}}{\sqrt{2k_p - 1}}.$$

Thus, the size of the bounded region into which the trajectories of $\tilde{u}(t)$ converge in time can be decreased by increasing the control gain $k_p$. As discussed above, this type of stability is known as *practical stability*.

In order to actually prove that the system is USPES, we need to confirm that it satisfies the three conditions given in Theorem 7.13.

Note that simultaneously satisfying both the first and third conditions of Theorem 7.13 effectively provides an upper and lower bounds for $V$. These bounds depend on the relative sizes of $\delta$ and $\Delta$. Using the numerical values provided in Example 7.5 above, we have $w_{d0} = 4.5$ N, $k_p = 5$ (this gain would effectively have units of kg/s) and $\|\tilde{u}_0\| = 2.5$ m/s (the maximum initial surge speed error occurring in the problem), so that $\delta = 1.5$ m/s and $\Delta = 2.5$ m/s.

With $\Delta^* > \delta^* > 0$ the third condition of Theorem 7.13 can also be expressed as $k_1 \Delta / \delta > k_2 > 0$. Combining this with the first condition of Theorem 7.13 we can also write this as $k_1 \Delta / \delta > k_2 > k_1$, which places an upper and lower bounds on the range of $k_2$ with respect to $k_1$. For this example, the numerical range of possible values for $k_1$ and $k_2$ is $5k_1/3 > k_2 > k_1$ and $k_1 < 1/2$ (because of the first condition and the fact that $V = \|\tilde{u}\|^2/2$). Thus, by selecting $k_1 = 1/4$ and $k_2 = 3/8$ both the first and third conditions of Theorem 7.13 are satisfied.

In order to show that the closed loop system is USPES, it only remains to prove that the second condition of Theorem 7.13 is also satisfied. Using (7.84) in (7.83) gives

$$\dot{V} \le -(2k_p - 1)\left[V_0 - \frac{w_{d0}^2}{2(2k_p - 1)}\right]e^{-(2k_p-1)(t-t_0)},$$

$$\le -\frac{(2k_p - 1)}{2}\left[\tilde{u}_0^2 - \frac{w_{d0}^2}{(2k_p - 1)}\right]e^{-(2k_p-1)(t-t_0)} + \frac{w_{d0}^2}{2},$$

$$\le -\frac{(2k_p - 1)}{2}\left\{\left[\tilde{u}_0^2 - \frac{w_{d0}^2}{(2k_p - 1)}\right]e^{-(2k_p-1)(t-t_0)} + \frac{w_{d0}^2}{(2k_p - 1)}\right\}.$$

Comparing this result with (7.85), it can be seen that

$$\dot{V} \le -k_3\|\tilde{u}\|^2,$$

where $k_3 := (k_p - 1/2)$ so that the second condition of Theorem 7.13 is also satisfied.

Thus, as the closed loop system has been shown to satisfy all three conditions of Theorem 7.13, it is guaranteed to be uniformly semiglobally exponentially stable.□

## 7.8  Barbalat's Lemma

When using Lyapunov Theory to solve adaptive control problems, $dV/dt$ is often only negative semidefinite and additional conditions must be imposed on the system, in such cases the following lemma can be useful.

**Lemma 7.1** (Barbalat's Lemma) *For a time-invariant nonlinear system $\dot{\mathbf{x}} = \mathbf{f}(\mathbf{x}, t)$. Consider a function $V(\mathbf{x}, t) \in \mathbb{C}^1$. If*

*(1)  $V(\mathbf{x}, t)$ is lower bounded,*
*(2)  $\dot{V}(\mathbf{x}, t) \le 0$, and*
*(3)  $\dot{V}(\mathbf{x}, t)$ is uniformly continuous,*

*then $\dot{V}(\mathbf{x}, t) \le 0$ converges to zero along the trajectories of the system.*

**Remark 7.9** The third condition of Barbalat's Lemma is often replaced by the stronger condition $\ddot{V}(\mathbf{x}, t)$ is bounded. In general, Barbalat's Lemma relaxes some conditions, e.g. $V$ is not required to be PD.

**Example 7.7** (*Stability of MIMO adaptive control for marine systems*) As will be shown in Sect. 11.4, the Lyapunov function for a MIMO marine system using adaptive control via feedback linearization will have the form

$$V = \frac{1}{2}\tilde{\boldsymbol{\eta}}^T \mathbf{K}_p \tilde{\boldsymbol{\eta}} + \frac{1}{2}\mathbf{s}^T \mathbf{M}_\eta(\boldsymbol{\eta})\mathbf{s} + \frac{1}{2}\tilde{\boldsymbol{\theta}}^T \boldsymbol{\Gamma}^{-1}\tilde{\boldsymbol{\theta}},$$

where $\tilde{\boldsymbol{\theta}} := (\hat{\boldsymbol{\theta}} - \boldsymbol{\theta})$ is the parameter estimation error, $\mathbf{M}_\eta(\boldsymbol{\eta}) = \mathbf{M}_\eta^T(\boldsymbol{\eta}) > 0$ is the inertia tensor in the NED frame, $\mathbf{K}_p > 0$ is a diagonal control design matrix, $\boldsymbol{\Gamma} = \boldsymbol{\Gamma}^T > 0$ is a weighting matrix, and both $\tilde{\boldsymbol{\eta}} := (\boldsymbol{\eta} - \boldsymbol{\eta}_d)$ and $\mathbf{s} := \dot{\tilde{\boldsymbol{\eta}}} + \Lambda\tilde{\boldsymbol{\eta}}$ are measures of the tracking error. The parameter $\Lambda > 0$ in $s$ is also a diagonal control design matrix.

Here we show that in order for the closed loop system to be stable, the conditions of Lemma 7.1 must be satisfied.

The time derivative of $V$ is

$$\dot{V} = -\tilde{\eta}^T \mathbf{K}_p \Lambda \tilde{\eta} - \mathbf{s}^T \left[ \mathbf{D}_\eta(\mathbf{v}, \eta) + \mathbf{K}_d \right] \mathbf{s} + \tilde{\theta}^T \Gamma^{-1} \left[ \dot{\tilde{\theta}} + \Gamma \Phi^T \mathbf{J}^{-1}(\eta) \mathbf{s} \right], \quad (7.87)$$

where $\mathbf{D}_\eta(\mathbf{v}, \eta) = \mathbf{D}_\eta^T(\mathbf{v}, \eta) > 0$ is the drag (dissipation) tensor in the NED frame and $\mathbf{K}_d > 0$ is a diagonal control design matrix. The first two terms in the above equation are negative definite for all $\mathbf{s} \neq 0$ and for all $\tilde{\eta} \neq 0$. To obtain $\dot{V} \leq 0$, the *parameter update law* is selected as

$$\dot{\tilde{\theta}} = -\Gamma \Phi^T \mathbf{J}^{-1}(\eta) \mathbf{s}.$$

However, it can be seen that $\dot{V}$ is still only negative semidefinite, not negative definite, because it does not contain any terms that are negative definite in $\tilde{\theta}$. As shown in [9], using the parameter update law, Eq. (7.87) can be written in the form

$$\dot{V} = -\tilde{\eta}^T \mathbf{K}_p \Lambda \tilde{\eta} - \mathbf{s}^T \left[ \mathbf{D}_\eta(\mathbf{v}, \eta) + \mathbf{K}_d \right] \mathbf{s} = -\mathbf{e}^T \mathbf{Q} \mathbf{e}, \quad (7.88)$$

where $\mathbf{Q} = \mathbf{Q}^T$ and $\mathbf{e} = [\tilde{\eta} \ \dot{\tilde{\eta}}]^T$ is a vector of tracking errors. Since $\dot{V} \leq 0$, $V$ is always less than or equal to its initial value $V_0$, so that $\tilde{\eta}, \dot{\tilde{\eta}}, \mathbf{s}$ and $\tilde{\theta}$ are necessarily bounded, if $V_0$ is finite.

If $\dot{V}$ is continuous in time, i.e. if $\ddot{V}$ is bounded in time (see Remark 7.9), then $\dot{V}$ will be integrable. If this condition is met, (7.88) can be integrated to obtain

$$V - V_0 = -\int_0^t \mathbf{e}^T \mathbf{Q} \mathbf{e} \, dt < \infty, \quad (7.89)$$

which implies that $\mathbf{e}$ is a square integrable function and that $\mathbf{e} \to 0$ as $t \to \infty$ [9]. Note that, the condition that $\dot{V}$ is continuous requires $\dot{\tilde{\eta}}$ to also be continuous (it is a component of $\mathbf{e}$). In turn, this requires both $\ddot{\eta}$ and $\ddot{\eta}_d$ to be bounded. The equations of motion (1.58)–(1.59) can be used to verify that $\ddot{\eta}$ is bounded and restricting the permissible reference trajectories $\ddot{\eta}_d$ will ensure that $\dot{\tilde{\eta}}$ is continuous. Provided these latter conditions are satisfied, the velocity error also converges to zero $\dot{\tilde{\eta}} \to 0$. $\quad \Box$

## 7.9  Summary

A decision tree for analyzing the stability of a nonlinear closed loop system is shown
in Fig. 7.14. Firstly, one should ask if the system is time-varying (i.e. is it affected
by time-varying exogenous disturbances or model parameters) or not? In both cases,
one can also try to identify a suitable candidate Lyapunov function $V$ for the closed
loop system at this stage (Definition 7.21).

If the system is time-invariant (i.e. autonomous), one should compute $\dot{V}$ and
determine if it is negative definite (ND) or negative semidefinite (NSD). If the system
is ND, one can proceed to analyze it using Lyapunov's Second (Direct) Method
(Sect. 7.2.2). If NSD, one can either try a different CLF to see if it might be ND,
such that Lyapunov's Second Method can be used, or continue the stability analysis
using LaSalle's Invariant Set Theorem (Sect. 7.3) with the original CLF.

If the nonlinear closed loop system is time-varying, one would first ask whether
the system is affected by time-varying model parameters (e.g. an adaptive control

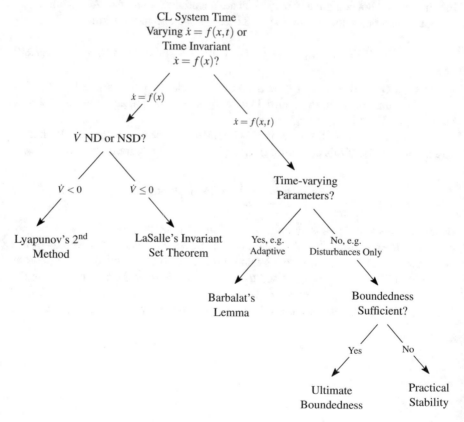

**Fig. 7.14**  Decision tree for analyzing the stability of a nonlinear closed loop system

system) or time-varying disturbances. In the former case, one would then apply Barbalat's Lemma (Sect. 7.8) to explore system stability. Otherwise, one can either explore the Ultimate Boundedness of the system or its Practical Stability. In some cases, it may be sufficient for the control system designer to simply know that the output of the closed loop system converges to some bounded value, so that identifying the Ultimate Boundedness of the controlled system is enough (Sect. 7.6). However, there may also be situations in which it is better to more precisely characterize the performance of the controlled system in order to know, for example, whether the system is exponentially stable or simply asymptotically stable and to have a sense of the ranges of the initial conditions and final output values over which stability is guaranteed. In such situations, the practical stability characteristics of the system will need to be determined (Sect. 7.7).

Note that, as with linear stability analysis techniques, the nonlinear stability analysis techniques presented here permit one to determine the stability of a closed loop system without having to explicitly solve the differential equations governing the system.

## Problems

**7.1** Consider the second–order system

$$\dot{x}_1 = -ax_1,$$
$$\dot{x}_2 = -bx_1 - cx_2,$$

where $a, b, c > 0$. Investigate whether the functions

$$V_1(x) = \frac{1}{2}x_1^2 + \frac{1}{2}x_2^2, \quad V_2(x) = \frac{1}{2}x_1^2 + \frac{1}{2}\left(x_2 + \frac{b}{c-a}x_1\right)^2$$

are Lyapunov functions for the system and give any conditions that must hold.

**7.2** Consider the heave motion of a spar buoy with the dynamics

$$m\ddot{z} + c\dot{z} + kz = 0.$$

A natural candidate for a Lyapunov function is the total energy of the system, given by

$$V = \frac{1}{2}m\dot{w}^2 + \frac{1}{2}kz^2,$$

where $z$ is the vertical displacement of the buoy from its equilibrium position and $w = \dot{z}$ is the velocity in the heave direction. Use LaSalle's Invariance Principle to show that the system is asymptotically stable.

**7.3** As shown in Example 2.4, the pitch angle $\theta$ of an AUV is governed by an equation of the form

$$\ddot{\theta} = \frac{c}{2}\sin(2\theta),$$

where $c > 0$ is a physical constant based on a model of the AUV. The use of a nonlinear control input $u = -k(V_0 - V)\dot{\theta}\cos^2(2\theta)$ is proposed to control the pitch, where

$$V(\theta, \dot{\theta}) = \frac{c}{4}\cos(2\theta) - 1 + \frac{1}{2}\dot{\theta}^2$$

is a Lyapunov function and $V_0$ is an estimate for the initial value of $V$. In practice, one might take $V_0$ to be an upper bound for the largest value of $V$.

With this controller, the closed loop equation of motion is

$$\ddot{\theta} = \frac{c}{2}\sin(2\theta) - k(V_0 - V)\dot{\theta}\cos^2(2\theta).$$

(a) Show that the controller stabilizes the pitch angle of the AUV to $\theta = 0$.
(b) Show that the stability analysis breaks down if the dynamics of the AUV are uncertain, i.e. if

$$V(\theta, \dot{\theta}) = \frac{c^*}{4}\cos(2\theta) - 1 + \frac{1}{2}\dot{\theta}^2,$$

where $c^* \neq c$.

**7.4** In Example 7.5 the closed loop equation of motion of surge speed-tracking controller for an unmanned surface vessel has the form

$$\dot{\tilde{u}} = -k\tilde{u} + w_d(t),$$

where, $\tilde{u}$ is the surge speed error, $k$ is constant and $w_d(t)$ is a non vanishing time-varying disturbance.

(a) Let $\beta_d > 0$ be an upper bound for the magnitude of the disturbance, i.e. take $\beta_d = \sup_t w_d(t)$. Show that the solution for the magnitude of $\tilde{u}(t)$ can be bounded as

$$|\tilde{u}(t)| \leq \left|\tilde{u}(0) - \frac{\beta_d}{k}\right| e^{-kt} - \frac{\beta_d}{k}.$$

(b) For what range of $k$ is the system stable?
(c) When $k$ is properly selected, $\tilde{u}(t)$ will exponentially converge to a ball around the origin $\tilde{u} = 0$. What is the radius of this ball?

**7.5** Consider the problem of station keeping in a strong following current, as shown in Fig. 7.15. Take the speed of the current along the longitudinal axis of the vehicle to be $u_c$. The equation of motion for the surge speed of the vehicle is given by

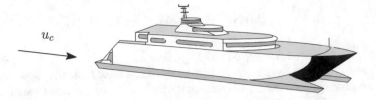

**Fig. 7.15** Station keeping a vessel in a strong following current

$$m\dot{u} = -c(u - u_c)|u - u_c| + \tau,$$

where $u$ is the surge speed, $m$ is the mass (including added mass) of the vessel, $c$ is the drag coefficient, and $\tau$ is a control input.

(a) Let $\tau = c(u - u_c)|u - u_c| - ku$, where $k > 0$ is a constant control gain. Show that the closed loop system is UGES.

(b) Let $\tau = -ku$.

    (i) Explain why the closed loop system is semiglobally attractive (see Sect. 7.7).

    (ii) Show that the domain of attraction is the set

$$\mathcal{B}_\Delta = \left\{ u < u_c + \frac{k}{2c^2}\left[ 1 + \sqrt{\frac{4u_c c^2}{k} + 1} \right] \right\}.$$

    (iii) Show that the ball $\mathcal{B}_\delta$ is given by the set

$$\mathcal{B}_\delta = \left\{ u \leq \frac{c^2(u - u_c)^2}{k} \right\}.$$

    (iv) Explain how $\mathcal{B}_\Delta$ and $\mathcal{B}_\delta$ vary with $k$.

    (v) Numerically integrate the equation of motion for times $0 \leq t \leq 300$ s with $m = 50 \times 10^3$ kg, $c = 15$ kg/m, $u_c = 3$ m/s, and $k = 10^3$ for various values of the initial velocity $u(0) = u_0$. Using the simulations, show that when $u_0 \leq 9.49$ m/s the system is stable and that when $u_0 > 9.49$ m/s the system becomes unstable.

**7.6** The kinetic energy of a marine vehicle $V_k$ is often included as one of the terms in a candidate Lyapunov functions used for control design. Let

$$V_k = \frac{1}{2}\mathbf{v}^T \mathbf{M} \mathbf{v},$$

where, in body-fixed coordinates, $\mathbf{M}$ is the inertial tensor and $\mathbf{v}$ is the velocity.

(a) Show that the upper and lower bounds for $V_k$ can be written as

$$\frac{1}{2}\lambda_{\min}(\mathbf{M})\|\mathbf{v}\|^2 \leq \frac{1}{2}\mathbf{v}^T\mathbf{M}\mathbf{v} \leq \frac{1}{2}\lambda_{\max}(\mathbf{M})\|\mathbf{v}\|^2.$$

Recall from Sect. 1.5.2 that $\mathbf{M} = \mathbf{M}^T > 0$.

(b) When performing Lyapunov stability analysis one must typically show that $\dot{V} < 0$, which involves showing that each of the individual terms in the time derivative of the candidate Lyapunov function are negative definite. Use the inequality above to find an upper bound for $-\|\mathbf{v}\|^2$ (note that this is a negative value).

**7.7** A proportional derivative controller is proposed to control the speed of a marine vehicle. The dynamics of the vehicle are governed by (1.59),

$$\mathbf{M}\dot{\mathbf{v}} + \mathbf{C}(\mathbf{v})\mathbf{v} + \mathbf{D}(\mathbf{v})\mathbf{v} + \mathbf{g}(\boldsymbol{\eta}) = \boldsymbol{\tau}.$$

Take the control input to be

$$\boldsymbol{\tau} = \mathbf{u} + \mathbf{C}(\mathbf{v})\mathbf{v} + \mathbf{D}(\mathbf{v})\mathbf{v} + \mathbf{g}(\boldsymbol{\eta}),$$

where

$$\mathbf{u} = \mathbf{M}\dot{\mathbf{v}}_d - \mathbf{K}_p\tilde{\mathbf{v}} - \mathbf{K}_d\dot{\tilde{\mathbf{v}}}.$$

The variable terms in the control input are the desired velocity $\mathbf{v}_d = \mathbf{v}_d(t)$ and the velocity error $\tilde{\mathbf{v}} := \mathbf{v} - \mathbf{v}_d$. The remaining terms, $\mathbf{K}_p > 0$ and $\mathbf{K}_d > 0$, are diagonal proportional gain and derivative gain matrices, respectively.

Consider the candidate Lyapunov function

$$V = \frac{1}{2}\tilde{\mathbf{v}}^T\mathbf{M}\tilde{\mathbf{v}} + \frac{1}{2}\tilde{\mathbf{v}}^T\mathbf{K}_d\tilde{\mathbf{v}}.$$

(a) Show that $V$ can be upper and lower bounded as

$$\frac{1}{2}\lambda_{\min}(\mathbf{M} + \mathbf{K}_d)\|\tilde{\mathbf{v}}\|^2 \leq \frac{1}{2}\tilde{\mathbf{v}}^T(\mathbf{M} + \mathbf{K}_d)\tilde{\mathbf{v}} \leq \frac{1}{2}\lambda_{\max}(\mathbf{M} + \mathbf{K}_d)\|\tilde{\mathbf{v}}\|^2.$$

(b) Show that $\dot{V} < 0$ for all $\tilde{\mathbf{v}} \neq 0$.

(c) Using the bounds from part (a), show that $\dot{V} \leq -2\mu V$, where

$$\mu = \frac{\lambda_{\min}(\mathbf{K}_p)}{\lambda_{\max}(\mathbf{M} + \mathbf{K}_d)}.$$

(d) Now suppose that an exogenous disturbance acts upon the system, so that the equation of motion becomes

$$\mathbf{M}\dot{\mathbf{v}} + \mathbf{C}(\mathbf{v})\mathbf{v} + \mathbf{w}_d(\mathbf{v})\mathbf{v} + \mathbf{g}(\boldsymbol{\eta}) = \boldsymbol{\tau} + \mathbf{w}_d(t).$$

Show that the time derivative of the Lyapunov function is now given by

$$\dot{V} = -\tilde{\mathbf{v}}^T \mathbf{K}_p \tilde{\mathbf{v}} + \tilde{\mathbf{v}}^T \mathbf{w}_d.$$

(e) Use Young's Inequality (Theorem 7.77) and the bounds from part (a) to show that

$$\dot{V} \leq -2\mu^* V + \frac{1}{2}\|\mathbf{w}_d\|^2,$$

where

$$\mu^* = \frac{\lambda_{\min}(\mathbf{K}_p) - \dfrac{1}{2}}{\lambda_{\max}(\mathbf{M} + \mathbf{K}_d)}.$$

   (i) From part (c), what is the minimum value of $\mathbf{K}_p$ required for stability when $\mathbf{w}_d = 0$?
   (ii) What is the minimum value of $\mathbf{K}_p$ required for stability when $\mathbf{w}_d \neq 0$?

(f) Let the disturbance be upper bounded by a positive constant, so that $C_d \leq \|\mathbf{w}_d\|^2/2$. Then the inequality in part (e) can be written as

$$\dot{V} \leq -2\mu^* V + C_d.$$

Integrate both sides of this inequality to show that it implies

$$V \leq \left(V_0 - \frac{C_d}{2\mu^*}\right) e^{-2\mu^* t} + \frac{C_d}{2\mu^*},$$

where $V_0$ is the initial value of $V$.

(g) Lastly, show that the result from part f) implies that $\|\tilde{\mathbf{v}}\|$ is uniformly ultimately bounded. Give the ultimate bound and describe its dependence on $\mathbf{K}_p$ and $\mathbf{K}_d$. What type of stability is expected?

**7.8** An adaptive surge speed controller is designed to permit an AUV to make environmental measurements in a kelp forest. Here, we will use Barbalat's Lemma (Sect. 7.8) to show that we expect the controller to drive the surge speed error to zero.

As the AUV moves through the underwater forest, it brushes past the kelp leaves giving it an uncertain drag coefficient, which we will model as being time dependent and bounded. The equation of motion for the forward speed of the AUV is

$$m\dot{u} = \tau - cu^2,$$

where $u$ is the surge speed, $c = c(t)$ is the unknown drag coefficient, $m$ is the mass (including added mass) of the AUV, and $\tau$ is the control input. Let the control input be

$$\tau = m\dot{u}_d - mk\tilde{u} + \hat{c}u^2,$$

where $\dot{u}_d$ is the time derivative of the desired velocity, $\tilde{u} = u - u_d$ is the surge speed error, $k > 0$ is a constant proportional controller gain, and $\hat{c}$ is an estimate of the drag coefficient $c$. Let the error of the drag estimate be $\tilde{c} := \hat{c} - c$. Since the estimated drag $\hat{c}$ is computed using a fast computer inside the AUV, it is assumed that it can change in time much more rapidly than the true (unknown) drag coefficient $c$, which depends on the speed of the vehicle and the density of the trees in the kelp forest. Thus, take the time derivative of the error of the drag estimate to be

$$\dot{\tilde{c}} = \dot{\hat{c}} - \dot{c},$$

$$\approx \dot{\hat{c}}.$$

It is also assumed that we can accurately measure $u$, e.g. using a precise data velocimetry logger (DVL).

Let the update law for the $\hat{c}$ be given by

$$\dot{\hat{c}} = -\frac{u^2}{\gamma m}\tilde{u},$$

where $\gamma > 0$ is a constant.

Consider the candidate Lyapunov function

$$V = \frac{1}{2}\tilde{u}^2 + \frac{1}{2}\gamma\tilde{c}^2.$$

(a)  Show that

$$\dot{V} = -k\tilde{u}^2.$$

(b)  As the value of $\dot{V}$ computed in part (a) above is always less than zero, why is it not sufficient to show that $\dot{V}$ is negative definite?

(c)  Compute $\ddot{V}$ and explain how Barbalat's Lemma can be applied to conclude that $\tilde{u} \to 0$ for $t \to 0$.

**7.9**  The equations of motion for a marine vehicle, Eqs. (1.58) and (1.59), can be expressed as

$$\dot{\boldsymbol{\eta}} = \mathbf{J}(\boldsymbol{\eta})\mathbf{v},$$

$$\mathbf{M}\dot{\mathbf{v}} = \mathbf{N}(\mathbf{v}, \boldsymbol{\eta}) + \boldsymbol{\tau},$$

where $\mathbf{N}(\mathbf{v}, \boldsymbol{\eta}) = -\mathbf{C}(\mathbf{v})\mathbf{v} - \mathbf{D}(\mathbf{v})\mathbf{v} - \mathbf{g}(\boldsymbol{\eta})$ and it is assumed that no exogenous disturbances act on the system. Let us investigate the stability of this system when a backstepping control input is used.

(a)  Define the pose tracking error to be

$$\tilde{\boldsymbol{\eta}} := \boldsymbol{\eta} - \boldsymbol{\eta}_d, \quad \tilde{\boldsymbol{\eta}} \in \mathbb{R}^n,$$

and the velocity tracking error to be

$$\tilde{\mathbf{v}} := \mathbf{v} - \boldsymbol{\alpha}_1, \quad \tilde{\mathbf{v}} \in \mathbb{R}^n,$$

where $\boldsymbol{\alpha}_1 \in \mathbb{R}^n$ is a backstepping stabilizing function, which shall be defined shortly. Show that the time derivatives of $\tilde{\boldsymbol{\eta}}$ and $\tilde{\mathbf{v}}$ are

$$\dot{\tilde{\boldsymbol{\eta}}} = \mathbf{J}\tilde{\mathbf{v}} + \mathbf{J}\boldsymbol{\alpha}_1 - \dot{\boldsymbol{\eta}}_d,$$
$$\mathbf{M}\dot{\tilde{\mathbf{v}}} = \mathbf{N} + \boldsymbol{\tau} - \mathbf{M}\dot{\boldsymbol{\alpha}}_1.$$

(b) Consider the candidate Lyapunov function

$$V = \frac{1}{2}\tilde{\boldsymbol{\eta}}^T\tilde{\boldsymbol{\eta}} + \frac{1}{2}\tilde{\mathbf{v}}^T\mathbf{M}\tilde{\mathbf{v}}.$$

Show that its time derivative can be written as

$$\dot{V} = \tilde{\boldsymbol{\eta}}^T\left[\mathbf{J}\boldsymbol{\alpha}_1 - \dot{\boldsymbol{\eta}}_d\right] + \tilde{\mathbf{v}}^T\left[\mathbf{J}^T\tilde{\boldsymbol{\eta}} + \mathbf{N} - \mathbf{M}\dot{\boldsymbol{\alpha}}_1 + \boldsymbol{\tau}\right].$$

(c) Show that selecting the backstepping stabilizing function to be

$$\boldsymbol{\alpha}_1 = -\mathbf{J}^{-1}\left[\mathbf{K}_p\tilde{\boldsymbol{\eta}} - \dot{\boldsymbol{\eta}}_d\right],$$

and the control input to be

$$\boldsymbol{\tau} = -\mathbf{K}_d\tilde{\mathbf{v}} - \mathbf{J}^T\tilde{\boldsymbol{\eta}} + \mathbf{M}\dot{\boldsymbol{\alpha}}_1 - \mathbf{N}(\mathbf{v}, \boldsymbol{\eta}), \quad (7.90)$$

where $\mathbf{K}_p = \mathbf{K}_p^T > 0$ and $\mathbf{K}_d = \mathbf{K}_d^T > 0$ are diagonal control design matrices, renders $\dot{V} < 0$ for any $\tilde{\boldsymbol{\eta}}, \tilde{\mathbf{v}} \neq 0$.

(d) What type of stability does this system have?

**7.10** When using the inverse dynamics method of feedback linearization to design a proportional derivative (PD) control input for marine systems the resulting control input is

$$\boldsymbol{\tau} = \mathbf{M}\mathbf{J}^{-1}(\boldsymbol{\eta})\left\{\mathbf{a}_n - \frac{d\left[\mathbf{J}(\boldsymbol{\eta})\right]}{dt}\mathbf{v}\right\} + \mathbf{C}(\mathbf{v})\mathbf{v} + \mathbf{D}(\mathbf{v})\mathbf{v} + \mathbf{g}(\boldsymbol{\eta}),$$

where

$$\mathbf{a}_n = \ddot{\boldsymbol{\eta}}_d - \mathbf{K}_p\tilde{\boldsymbol{\eta}} - \mathbf{K}_d\dot{\tilde{\boldsymbol{\eta}}},$$

$\boldsymbol{\eta}_d = \boldsymbol{\eta}_d(t)$ is the desired pose of the vehicle, $\tilde{\boldsymbol{\eta}} := \boldsymbol{\eta} - \boldsymbol{\eta}_d$ is the pose tracking error, and $\mathbf{K}_p = \mathbf{K}_p^T > 0$ and $\mathbf{K}_d = \mathbf{K}_d^T > 0$ are diagonal proportional and derivative gain matrices, respectively.

Let us examine the stability of this PD controller.

Recall that the equations of motion (1.58)–(1.59) for a marine vehicle in the NED frame are

$$\dot{\eta} = \mathbf{J}(\eta)\mathbf{v},$$

$$\mathbf{M}\dot{\mathbf{v}} = -\mathbf{C}(\mathbf{v})\mathbf{v} - \mathbf{D}(\mathbf{v})\mathbf{v} - \mathbf{g}(\eta) + \tau.$$

(a) Show that the following equation can be obtained from the kinematic part of the equations of motion above,

$$\dot{\mathbf{v}} = \mathbf{J}^{-1}(\eta) \left\{ \ddot{\eta} - \frac{d\,[\mathbf{J}(\eta)]}{dt}\mathbf{v} \right\}.$$

(b) Substitute the result from part (a) and the control law into the dynamics portion of the equations of motion and prove that the closed loop system satisfies the equation

$$\ddot{\tilde{\eta}} + \mathbf{K}_d\dot{\tilde{\eta}} + \mathbf{K}_p\tilde{\eta} = 0.$$

(c) Consider the candidate Lyapunov function

$$V = \frac{1}{2}\dot{\tilde{\eta}}^T\dot{\tilde{\eta}} + \frac{1}{2}\tilde{\eta}^T\tilde{\eta}.$$

Use the equation of the closed loop system found in part (b) and Young's Inequality to show that

$$\dot{V} \le -\left[\lambda_{\min}(\mathbf{K}_d) - \frac{1}{2}\right]\|\dot{\tilde{\eta}}\|^2 - \left[\lambda_{\min}(\mathbf{K}_p) - \frac{1}{2}\right]\|\tilde{\eta}\|^2,$$

where $\lambda_{\min}(\mathbf{K}_p)$ and $\lambda_{\min}(\mathbf{K}_d)$ are the minimum eigenvalues of $\mathbf{K}_p$ and $\mathbf{K}_d$, respectively.

What are the minimum permissible values of $\lambda_{\min}(\mathbf{K}_p)$ and $\lambda_{\min}(\mathbf{K}_d)$?

(d) Let

$$\mu = \min\left\{\lambda_{\min}(\mathbf{K}_d) - \frac{1}{2}, \ \lambda_{\min}(\mathbf{K}_p) - \frac{1}{2}\right\}.$$

Show that $\dot{V} \le -2\mu V$.

(e) Let $V_0$ be the value of $V$ at time $t = 0$ (the initial condition). Integrate the inequality $\dot{V} \le -2\mu V$ to show that $V$ decays exponentially in time. What does this imply about the time dependence of $\tilde{\eta}$ and $\dot{\tilde{\eta}}$?

(f) What type of stability does this system have?

# References

1. Chaillet, A., Loría, A.: Uniform global practical asymptotic stability for time-varying cascaded systems. Eur. J. Control **12**(6), 595–605 (2006)
2. Chaillet, A., Loría, A.: Uniform semiglobal practical asymptotic stability for non-autonomous cascaded systems and applications. Automatica **44**(2), 337–347 (2008)
3. Furfaro, T.C., Dusek, J.E., von Ellenrieder, K.D.: Design, construction, and initial testing of an autonomous surface vehicle for riverine and coastal reconnaissance. In: OCEANS 2009, pp. 1–6. IEEE (2009)
4. Grotli, E.I., Chaillet, A., Gravdahl, J.T.: Output control of spacecraft in leader follower formation. In: 2008 47th IEEE Conference on Decision and Control, pp. 1030–1035. IEEE (2008)
5. Khalil, H.K.: Nonlinear Systems, 3rd edn. Prentice Hall, Englewood Cliffs (2002)
6. Liberzon, D.: Switching in Systems and Control. Springer Science & Business Media, Berlin (2012)
7. Slotine, J.-J.E., Li, W.: Applied Nonlinear Control. Prentice-Hall, Englewood Cliffs (1991)
8. Sontag, E.D., Wang, Y.: On characterizations of the input-to-state stability property. Syst. Control Lett. **24**(5), 351–359 (1995)
9. Spong, M.W., Hutchinson, S., Vidyasagar, M.: Robot Modeling and Control. Wiley, Hoboken (2006)

# Chapter 8
# Feedback Linearization

## 8.1 Introduction

Feedback linearization can be used to transform the nonlinear dynamics of a vehicle into a linear system upon which conventional linear control techniques, such as pole placement (Chap. 6), can be applied. Provided that the full state of a system can be measured, and that a general observability condition holds, it may be possible to identify nonlinear transformations that leave the transformed system linear. A linear controller is then designed for the transformed model and the control input signal from the linear controller is transformed back into a nonlinear signal before being passed to the actuators/plant. In this way, knowledge of the nonlinearities in the system are built into the controller. Feedback linearization is very different from simply linearizing a nonlinear system about one or more operating points, using a Taylor's series expansion for example (see Sect. 2.4.1), and then designing a controller for the linearized system. Instead, feedback linearization is accomplished by an exact state transformation and feedback.

Feedback linearization can be categorized as being either *input-state linearization* or *input-output linearization*. Input-state linearization involves the generation of a linear differential relation between the state $\mathbf{x}$ and a new input $v$. Similarly, input-output linearization involves the generation of a linear differential relation between the output $y$ and a new input $v$. However, whereas input-state linearization generally results in a complete linearization between $\mathbf{x}$ and $v$, input-output linearization may result in a partial linearization of the system that does not include all of the closed loop dynamics. Input-output linearization may render one or more of the states unobservable. These unobservable states are known as *internal dynamics*. Thus, the successful implementation of input-output linearization requires that the internal dynamics of the system are stable.

Before exploring more generalized approaches to feedback linearization, we consider the use of the inverse dynamics approach, as it is fairly intuitive and demonstrates the main features of feedback linearization.

© Springer Nature Switzerland AG 2021
K. D. von Ellenrieder, *Control of Marine Vehicles*, Springer Series on Naval
Architecture, Marine Engineering, Shipbuilding and Shipping 9,
https://doi.org/10.1007/978-3-030-75021-3_8

## 8.2  Inverse Dynamics

Inverse dynamics is a special case of input-output feedback linearization. A detailed discussion of input-output linearization techniques is presented in Sect. 8.6 below. For marine vehicles, inverse dynamics can be separated into velocity control in the body-fixed frame and position and attitude (pose) control in the NED frame [4].

### 8.2.1  Body-Fixed Frame Inverse Dynamics

Consider the kinetic equation of motion for a marine vehicle (1.59),

$$\mathbf{M}\dot{\mathbf{v}} + \mathbf{C}(\mathbf{v})\mathbf{v} + \mathbf{D}(\mathbf{v})\mathbf{v} + \mathbf{g}(\boldsymbol{\eta}) = \boldsymbol{\tau}.$$

A control input

$$\boldsymbol{\tau} = \mathbf{f}(\boldsymbol{\eta}, \mathbf{v}, t)$$

that linearizes the closed loop system is sought. By inspection, it can be seen that if a control input of the form

$$\boldsymbol{\tau} = \mathbf{M}\mathbf{a}_b + \mathbf{C}(\mathbf{v})\mathbf{v} + \mathbf{D}(\mathbf{v})\mathbf{v} + \mathbf{g}(\boldsymbol{\eta})$$

can be found, the equation of motion reduces to

$$\dot{\mathbf{v}} = \mathbf{a}_b,$$

where $\mathbf{a}_b$ can be thought of as the commanded acceleration of the vehicle in the body-fixed frame. This new system is linear. Further, if the input $\mathbf{a}_b$ is designed so that each of its components are decoupled, i.e. so that $a_{bi}$ is only a function of $\eta_i$ and $v_i$ for $i = 1, \ldots, n$, the closed loop system is also decoupled. A simple approach is to design a linear proportional integral derivative control law of the form

$$\mathbf{a}_b = \dot{\mathbf{v}}_d - \mathbf{K}_p \tilde{\mathbf{v}} - \mathbf{K}_i \int_0^t \tilde{\mathbf{v}}(\tau) \mathrm{d}\tau - \mathbf{K}_d \dot{\tilde{\mathbf{v}}},$$

where $\mathbf{v}_d = \mathbf{v}_d(t)$ is the desired velocity, $\tilde{\mathbf{v}} := \mathbf{v} - \mathbf{v}_d$ is the velocity error, and $\mathbf{K}_p > 0$, $\mathbf{K}_i > 0$ and $\mathbf{K}_d > 0$ are diagonal proportional, integral and derivative gain matrices, respectively. Note that the first term of $\mathbf{a}_b$, $\dot{\mathbf{v}}_d$, functions as a feedforward term for the desired acceleration.

## 8.2.2  NED Frame Inverse Dynamics

Next, consider the full equations of motion (1.58)–(1.59) for a marine vehicle

$$\dot{\eta} = \mathbf{J}(\eta)\mathbf{v},$$
$$\mathbf{M}\dot{\mathbf{v}} = -\mathbf{C}(\mathbf{v})\mathbf{v} - \mathbf{D}(\mathbf{v})\mathbf{v} - \mathbf{g}(\eta) + \boldsymbol{\tau}.$$

Position and orientation trajectory tracking can be accomplished by commanding an acceleration $\mathbf{a}_n$ (measured with respect to an Earth-fixed inertial reference frame) of the form

$$\ddot{\eta} = \mathbf{a}_n$$

that linearizes the closed loop system. Taking the time derivative of the first term in the vehicle equations of motion gives

$$\ddot{\eta} = \frac{\mathrm{d}\,[\mathbf{J}(\eta)]}{\mathrm{d}t}\mathbf{v} + \mathbf{J}(\eta)\dot{\mathbf{v}}.$$

Solving for $\dot{\mathbf{v}}$ gives

$$\dot{\mathbf{v}} = \mathbf{J}^{-1}(\eta)\left\{\ddot{\eta} - \frac{\mathrm{d}\,[\mathbf{J}(\eta)]}{\mathrm{d}t}\mathbf{v}\right\}.$$

Substituting this into the equations of motion yields

$$\mathbf{M}\mathbf{J}^{-1}(\eta)\left\{\ddot{\eta} - \frac{\mathrm{d}\,[\mathbf{J}(\eta)]}{\mathrm{d}t}\mathbf{v}\right\} = -\mathbf{C}(\mathbf{v})\mathbf{v} - \mathbf{D}(\mathbf{v})\mathbf{v} - \mathbf{g}(\eta) + \boldsymbol{\tau}.$$

If $\boldsymbol{\tau}$ is selected as

$$\boldsymbol{\tau} = \mathbf{M}\mathbf{J}^{-1}(\eta)\left\{\mathbf{a}_n - \frac{\mathrm{d}\,[\mathbf{J}(\eta)]}{\mathrm{d}t}\mathbf{v}\right\} + \mathbf{C}(\mathbf{v})\mathbf{v} + \mathbf{D}(\mathbf{v})\mathbf{v} + \mathbf{g}(\eta)$$

the closed loop system becomes

$$\ddot{\eta} = \mathbf{a}_n.$$

As above, a linear proportional integral derivative control law can now be designed as

$$\mathbf{a}_n = \ddot{\eta}_d - \mathbf{K}_p\tilde{\eta} - \mathbf{K}_i\int_0^t \tilde{\eta}(\sigma)\mathrm{d}\sigma - \mathbf{K}_d\dot{\tilde{\eta}},$$

where $\sigma$ is used as an integration variable, $\eta_d = \eta_d(t)$ is the desired position and orientation of the vehicle, $\tilde{\eta} := \eta - \eta_d$ is the pose error, and $\mathbf{K}_p > 0$, $\mathbf{K}_i > 0$ and $\mathbf{K}_d > 0$ are diagonal proportional, integral and derivative gain matrices, respectively. As above, the first term of $\mathbf{a}_n$, $\ddot{\eta}_d$, functions as a feedforward term. Note that the transformation matrix $\mathbf{J}(\boldsymbol{\eta})$ must be nonsingular to use this approach, which limits the magnitude of the pitch angle of the vehicle to the range $-\pi/2 < \theta < \pi/2$.

## 8.3  Fundamental Concepts in Feedback Linearization

The use of inverse dynamics to design a nonlinear controller, is a special case of feedback linearization. It can be seen that the use of inverse dynamics essentially splits the controller into two parts, an inner loop that exactly linearizes the nonlinear system, and an outer loop, which can be designed using linear techniques according to tracking or disturbance rejection performance requirements (Fig. 8.1). The more general process of feedback linearization works in the same way, except that the outer-loop process of linearizing the nonlinear system may also involve a transformation of the state variable into a new set of coordinates. While inverse dynamics can be sufficient for many problems involving the control of marine vessels, the use of the more general form of feedback linearization may be needed when solving marine vehicle control problems that must take actuator dynamics or underactuation into account.

### 8.3.1  Use of a Linearizing Control Law

The following example illustrates the fundamental idea of how a linearizing control law can be developed.

**Example 8.1**  Tugboat pushing a cargo ship
Consider the situation shown in Fig. 8.2 in which a tugboat is used to guide the path of a cargo ship coming into port. Modeling the hull of the cargo ship as being symmetrical about the vertical centerline plane and ignoring any air or wave drag,

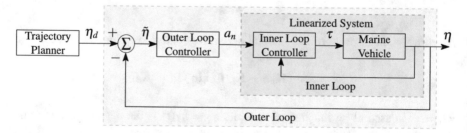

**Fig. 8.1**  Block diagram of a controller designed using feedback linearization. The system consists of an inner loop, which uses $\mathbf{a}_n$, as well as the states $\mathbf{v}$ and $\boldsymbol{\eta}$, to compute a control input $\boldsymbol{\tau}$ that compensates for nonlinearities in the plant model. The outer-loop consists of a trajectory tracking linear controller, which can be designed based on the decoupled linear plant model

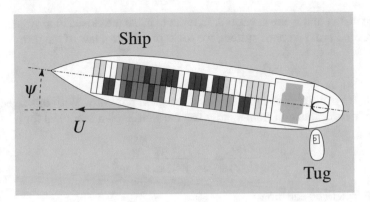

**Fig. 8.2** Tugboat aligning a cargo ship

we will take the yaw response of the ship to result from a combination of the Munk moment (see Example 2.4) and the moment applied by the tugboat $Tl\cos(\psi)$, where $T$ is the thrust of the tug and $l\cos(\psi)$ is a moment arm. Modifying (2.22), we can write the equation for the yaw angle of the ship as

$$(I_z - N_{\dot{r}})\ddot{\psi} = -U^2\sin(\psi)\cos(\psi)(Y_{\dot{v}} - X_{\dot{u}}) - Tl\cos(\psi).$$

Here $I_z$ and $N_{\dot{r}}$ are the rigid-body mass and added mass moments of inertia about the yaw axis of the ship. The yaw angle is $\psi$, $r = \dot{\psi}$ is the yaw rate, and $Y_{\dot{v}}$ and $X_{\dot{u}}$ are the added mass in the sway ($y_b$) and surge ($x_b$) directions, respectively. Owing to the geometry of the ship $X_{\dot{u}} > Y_{\dot{v}}$ so that $-U^2\sin(\psi)\cos(\psi)(Y_{\dot{v}} - X_{\dot{u}})$ is a destabilizing term. The control aim is to regulate the thrust of the tugboat to drive $\psi \to 0$. Let $x_1 := \psi$ and $x_2 := r$, after normalizing by $(I_z - N_{\dot{r}})$ we approximate the system equations as

$$\begin{aligned}
\dot{x}_1 &= x_2 \\
\dot{x}_2 &= \sin(x_1)\cos(x_1) + u\cos(x_1) \\
y &= x_1
\end{aligned} \qquad (8.1)$$

Introduce the transformed control signal

$$v(t) = \sin(x_1)\cos(x_1) + u(t)\cos(x_1).$$

This gives the linear system

$$\dot{\mathbf{x}} = \begin{bmatrix} 0 & 1 \\ 0 & 0 \end{bmatrix}\mathbf{x} + \begin{bmatrix} 0 \\ 1 \end{bmatrix}v.$$

Assume that $x_1$ and $x_2$ are measured. In order to make the closed loop system behave like a second order damped system, we look for a control law of the form

$$v(t) = -\omega_n^2 x_1 - 2\zeta\omega_n x_2 + \omega_n^2 u_c(t), \tag{8.2}$$

where $u_c(t)$ is a linear control law to be designed presently. With this selection of $v(t)$ the transfer function from $u_c$ to $y$ has the well-known canonical form

$$\frac{Y(s)}{U_c(s)} = \frac{\omega_n^2}{s^2 + 2\zeta\omega_n s + \omega_n^2}$$

with characteristic equation

$$s^2 + 2\zeta\omega_n s + \omega_n^2 = 0 \tag{8.3}$$

(see Sect. 3.1). Transformation back to the original system gives

$$u = \frac{1}{\cos(x_1)}\left[-\omega_n^2 x_1 - 2\zeta\omega_n x_2 + \omega_n^2 u_c(t) - \sin(x_1)\cos(x_1)\right],$$

which is highly nonlinear. To provide a basis of comparison for the performance of this control law, compare with the fixed-gain control law developed by linearizing (8.1) about $x_1 = \psi = 0$ and solving for a controller of the form

$$u = -k_1 x_1 - k_2 x_2 + k_3 u_c(t), \tag{8.4}$$

where the gains $k_1 = (1 + \omega_n^2)$, $k_2 = 2\zeta\omega_n$ and $k_3 = \omega_n^2$ are selected so that the characteristic equation of the linearized closed loop system is the same as (8.3).

The performance of the controller designed via feedback linearization and the fixed gain controller for $\zeta = 1/\sqrt{2}$, $\omega_n = 1$ and with $u_c$ selected as a PD controller $u_c = -k_p x_1 - k_d x_2$, where $k_p = 1.5$ and $k_d = 1$ is shown in Fig. 8.3. □

### 8.3.2  Coordinate Transformations for Feedback Linearization

In Example 8.1 it can be seen that feedback linearization can be readily applied to systems of the form

$$\begin{aligned}
\dot{x}_1 &= x_2 \\
\dot{x}_2 &= f_1(x_1, x_2, u) \\
y &= x_1.
\end{aligned} \tag{8.5}$$

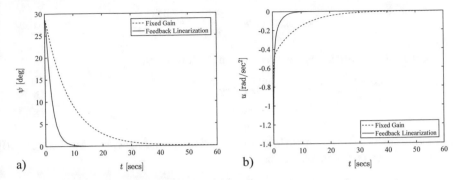

**Fig. 8.3** Tugboat aligning a cargo ship: **a** time response of the closed loop systems; **b** control inputs of the fixed gain controller and the controller developed using feedback linearization. The initial value of the yaw angle is 0.5 rads, i.e. 28.6°

Let us now ask, whether it can be extended to nonlinear systems of the more general form

$$\begin{aligned}
\dot{x}_1 &= f_1(x_1, x_2) \\
\dot{x}_2 &= f_2(x_1, x_2, u) \\
y &= x_1.
\end{aligned} \tag{8.6}$$

As in Example 8.1, we would like to identify a linearizing control law that converts the response of the system into a second order linear system with a transfer function $G(s)$ from input $u_c$ to output $y = x_1 = z_1$ with the canonical form (Sect. 3.1)

$$G(s) = \frac{\omega_n^2}{s^2 + 2\zeta\omega_n s + \omega_n^2}. \tag{8.7}$$

Consider the set of transformed coordinates $z_1$ and $z_2$, given by

$$\begin{aligned}
z_1 &:= x_1, \\
z_2 &:= \dot{x}_1 = f_1(x_1, x_2).
\end{aligned}$$

Taking the time derivative of the transformed coordinates gives

$$\begin{aligned}
\dot{z}_1 &:= z_2, \\
\dot{z}_2 &:= \frac{\partial f_1}{\partial x_1}\dot{x}_1 + \frac{\partial f_1}{\partial x_2}\dot{x}_2 = \frac{\partial f_1}{\partial x_1}f_1 + \frac{\partial f_1}{\partial x_2}f_2.
\end{aligned} \tag{8.8}$$

Examining the right hand side of (8.8), let us define a new control signal $v$ as

$$v = F(x_1, x_2, u) = \frac{\partial f_1}{\partial x_1} f_1 + \frac{\partial f_1}{\partial x_2} f_2. \tag{8.9}$$

Thus, these transformations have resulted in the linear system

$$\dot{z}_1 = z_2, \tag{8.10}$$
$$\dot{z}_2 = v. \tag{8.11}$$

Now the transformed system can be stabilized using simple linear feedback. Using (8.10) and (8.11), it can be seen that selecting

$$v = \omega_n^2 (u_c - z_1) - 2\zeta \omega_n z_2 \tag{8.12}$$

gives the desired closed loop system $\ddot{z}_1 + 2\zeta \omega_n \dot{z}_1 + \omega_n^2 z_1 = \omega_n^2 u_c$, which has the desired transfer function (8.7).

All that remains is to transform the system back to its original variables. It follows from (8.9) and (8.12) that

$$F(x_1, x_2, u) = \frac{\partial f_1}{\partial x_1} f_1 + \frac{\partial f_1}{\partial x_2} f_2 = \omega_n^2 (u_c - x_1) - 2\zeta \omega_n f_1(x_1, x_2). \tag{8.13}$$

This equation can be solved for $u$ to find the desired feedback, as long as $\partial F / \partial u \neq 0$.

Thus, it can be seen that if feedback linearization cannot be applied to the original system, it may be possible to transform the state variables so that feedback linearization can then be applied to the system when expressed in terms of the transformed state variables. The fundamental concepts introduced here in Sect. 8.3 are generalized and further elaborated upon in the remainder of this chapter.

## 8.4   Structural Properties of Feedback-Linearizable Systems

Controller design via feedback linearization requires one to first cancel the nonlinearities in a nonlinear system. Here we explore whether or not the dynamic system must possess a specific form that generally permits such cancellations to be performed.

First, let us consider a nonlinear system with the form

$$\dot{\mathbf{x}} = \mathbf{f}(\mathbf{x}) + \mathbf{G}(\mathbf{x})\mathbf{u}, \tag{8.14}$$

and ask if there is a nonlinear feedback control input

$$\mathbf{u} = \alpha(\mathbf{x}) + \beta(\mathbf{x})\mathbf{v} \tag{8.15}$$

that can be used to transform (8.14) into a linear system of the form

$$\dot{\mathbf{x}} = \mathbf{A}\mathbf{x} + \mathbf{B}\mathbf{v}. \tag{8.16}$$

As shown in [5, 6, 9], in order to cancel the nonlinear term $\boldsymbol{\alpha}(\mathbf{x})$ by subtraction, the control input $\mathbf{u}$ and $\boldsymbol{\alpha}(\mathbf{x})$ must appear as a sum $\mathbf{u} + \boldsymbol{\alpha}(\mathbf{x})$. To cancel the nonlinear term $\boldsymbol{\gamma}(\mathbf{x})$ by division, the control $\mathbf{u}$ and the nonlinearity $\boldsymbol{\gamma}(\mathbf{x})$ must be multiplied by $\boldsymbol{\gamma}(\mathbf{x})\mathbf{u}$. If the matrix $\boldsymbol{\gamma}(\mathbf{x})$ is nonsingular in the domain of interest, then it can be canceled by $\mathbf{u} = \boldsymbol{\beta}(\mathbf{x})\mathbf{v}$, where $\boldsymbol{\beta}(\mathbf{x}) := \boldsymbol{\gamma}^{-1}(\mathbf{x})$ is the inverse of the matrix $\boldsymbol{\gamma}(\mathbf{x})$. Therefore, the use of feedback to convert a nonlinear state equation into a reachable linear state equation by canceling nonlinearities requires the nonlinear state equation to have the structure

$$\dot{\mathbf{x}} = \mathbf{A}\mathbf{x} + \mathbf{B}\boldsymbol{\gamma}(\mathbf{x})\left[\mathbf{u} - \boldsymbol{\alpha}(\mathbf{x})\right], \tag{8.17}$$

where the pair $(\mathbf{A}, \mathbf{B})$ is reachable (Sect. 6.2). It can be seen that substituting (8.15) into (8.17) yields (8.16), as desired. In addition, state feedback can be used to design the linearized input $\mathbf{v} = -\mathbf{K}\mathbf{x}$ using pole placement (see Sect. 6.3), so that the closed loop system is stable and satisfies a set of desired performance requirements. The overall nonlinear stabilizing state feedback control is

$$\mathbf{u} = \boldsymbol{\alpha}(\mathbf{x}) - \boldsymbol{\beta}(\mathbf{x})\mathbf{K}\mathbf{x}.$$

Next, suppose that the nonlinear state equation (8.14) does not have the nice, desired structure given by (8.17). Note that the state model of a system is not unique, but depends on the choice of state variables. If (8.14) does not have the structure of (8.17) for one set of state variables, it might for a different set. Therefore, it may still be possible to linearize the system using feedback by first transforming the state variables into a set that provides a state equation with the needed structure. To accomplish this, we seek a change of state variables

$$\mathbf{z} = \mathbf{T}(\mathbf{x}) \tag{8.18}$$

that converts (8.14) to a linear reachable system of the form

$$\dot{\mathbf{z}} = \mathbf{A}\mathbf{z} + \mathbf{B}\mathbf{v}. \tag{8.19}$$

**Example 8.2** Consider a surge speed control system for a AUV. As with most real actuators, an AUV propeller is only capable of generating a finite amount of thrust (actuator saturation). Similarly, when the commanded thrust value suddenly changes, the propeller cannot generate the new thrust instantaneously. Instead, there is a time period over which the thrust generated continuously increases or decreases until the desired thrust is attained (rate limit). For the purposes of this example, the drag of the AUV will be neglected. Note that the drag can be included through the use of a suitable transformation of variables, see Problem 8.5. The dynamic response of an AUV driven by a propeller with saturation and rate limits can be modeled as

$$m\frac{dv}{dt} = u_{max} \tanh\left(\frac{u}{u_{max}}\right),$$
$$T_d\frac{du}{dt} = -u + u_c$$

where $m$ is the mass (and added mass) of the AUV, $v$ is the surge speed, $u_{max}$ is the maximum thrust that can be generated by the propeller, $T_d$ is the time constant of the propeller rate limit, $u$ is the state of the propeller and $u_c$ is the commanded thrust.

We would like to use feedback linearization to design a controller for the AUV. Before doing so, let us rescale the terms in these equations so that they can be written in a more concise form. Multiply both sides of the first equation by $T_d$ and divide the second equation by $u_{max}$ to get

$$T_d\frac{dv}{dt} = \frac{T_d u_{max}}{m} \tanh\left(\frac{u}{u_{max}}\right),$$
$$T_d\frac{d}{dt}\left(\frac{u}{u_{max}}\right) = -\frac{u}{u_{max}} + \frac{u_c}{u_{max}}.$$

Let $\tilde{t} := t/T_d$, $a := T_d u_{max}/m$, $\tilde{u} := u/u_{max}$ and $\tilde{u}_c := u/u_{max}$. Then the equations become

$$\frac{dv}{d\tilde{t}} = a \tanh(\tilde{u}),$$
$$\frac{du}{d\tilde{t}} = -\tilde{u} + \tilde{u}_c,$$
$$y = v.$$

where the output $y$ is the desired speed. Since the nonlinear term $\tilde{u}$ appears in the first equation, it is not possible to select a control input $\tilde{u}_c$ that will linearize the equation for $v$. Therefore, in order to use feedback linearization, we will need to identify a change of coordinates that enable us to then select a linearizing control input.

Take $z_1 = v$ and look for a $z_2$ defined so that $z_2 = \dot{z}_1$ and that enables us to find a linear set of state equations

$$\dot{z}_1 = z_2,$$
$$\dot{z}_2 = v_c.$$

By inspection, it can be seen that if we take $z_2 = \dot{z}_1 = a \tanh \tilde{u}$ the transformed state equation will have the desired form. Now, using $z_1 = v$, it is possible to select a simple linear controller, such as a PID controller

$$v_c = \ddot{v}_d - k_p(v - v_d) - k_d(\dot{v} - \dot{v}_d) - k_i \int_0^t (v - v_d) d\tilde{t} \tag{8.20}$$

for the inner control loop, where $v_d$ is the desired surge speed and $k_p > 0$, $k_d > 0$ and $k_i > 0$ are proportional, integral and derivative control gains.

With our selection of $z_2 = a \tanh \tilde{u}$ the transformed state equation becomes

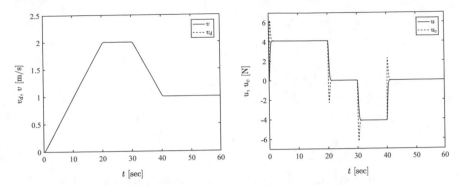

**Fig. 8.4** Use of the control law (8.23) to enable an AUV to track a desired velocity. **a** Desired velocity $v_d$ and output velocity $v$; **b** control input $u_c$ and actuator state $u$

$$\dot{z}_1 = z_2,$$
$$\dot{z}_2 = a\frac{\partial \tanh \tilde{u}}{\partial \tilde{u}} \cdot \frac{d\tilde{u}}{d\tilde{t}} = \frac{a}{\cosh^2 \tilde{u}}(-\tilde{u} + \tilde{u}_c) = v_c. \tag{8.21}$$

Therefore, the final control input is given by

$$\tilde{u}_c = \frac{\cosh^2 \tilde{u}}{a} \cdot v_c + \tilde{u}. \tag{8.22}$$

In terms of the original variables, this can be written as

$$\begin{aligned}
u_c &= \frac{u_{\max}}{a}\cosh^2\left(\frac{u}{u_{\max}}\right) \cdot v_c + u, \\
&= m\cosh^2\left(\frac{u}{u_{\max}}\right)\left[T_d\frac{d^2 v_d}{dt^2} - \frac{k_p}{T_d}e - k_d\frac{de}{dt} - \frac{k_i}{T_d^2}\int_0^t e\,dt\right] + u.
\end{aligned} \tag{8.23}$$

where $e := (v - v_d)$ is the velocity tracking error.

The performance of the controller can be seen in Fig. 8.4, where the desired velocity $v_d$, velocity output $v$, actuator state $u$ and control input $u_c$ are plotted for an AUV with a mass of $m = 39$ kg, maximum thrust of $u_{\max} = 10$ N, an actuator time constant of $T_d = 0.5$ s and control gains of $k_p = 5$, $k_d = 0$ and $k_i = 0$. The velocity output tracks the desired velocity with very little error, even without the need for derivative or integral control. As can be seen from (8.20), in order for the linearized control law to be continuous the second derivative of the desired velocity must also be continuous ($\ddot{v}_d$). An additional implementation issue with the controller is that the desired velocity profile must take the saturation limit of the propeller into account. If the commanded acceleration is too large, the controller will not perform well and may become unstable. □

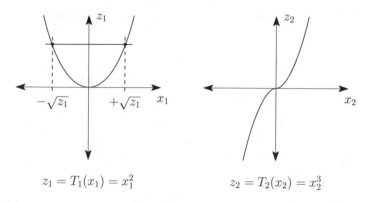

$$z_1 = T_1(x_1) = x_1^2 \qquad\qquad z_2 = T_2(x_2) = x_2^3$$

**Fig. 8.5** The transformations $z_1 = T_1(x_1) = x_1^2$ and $z_2 = T_2(x_2) = x_2^3$, for $x_1, x_2 \in \mathbb{R}$

As can be seen in Example 8.2 above, if the nonlinear state equation describing a system does not have the structure given by (8.17), it may still be possible to linearize the system using feedback by finding a change of state variables that converts the system into a linear reachable form. In general, the change of variables $\mathbf{z} = \mathbf{T}(\mathbf{x})$ used to transform the state equation from $\mathbf{x}$-coordinates to $\mathbf{z}$-coordinates, must have an inverse $\mathbf{x} = \mathbf{T}^{-1}(\mathbf{z})$. In addition, both $\mathbf{T}(\cdot)$ and $\mathbf{T}^{-1}(\cdot)$ must be continuously differentiable because the derivatives of $\mathbf{z}$ and $\mathbf{x}$ should be continuous. Functions with these properties (invertible and differentiable) are known as *diffeomorphisms*. Here, it will be assumed that $\mathbf{T}(\cdot)$ and its inverse are infinitely differentiable.

**Definition 8.1** (*Diffeomorphism*) A function $\mathbf{z} = \mathbf{T}(\mathbf{x})$, where $\mathbf{T} : \mathbb{R}^n \to \mathbb{R}^n$, is a diffeomorphism if it is smooth (continuously differentiable), one-to-one, and its inverse $\mathbf{x} = \mathbf{T}^{-1}(\mathbf{z})$ exists and is smooth.

**Example 8.3** Consider the transformations $z_1 = T_1(x_1) = x_1^2$ and $z_2 = T_2(x_2) = x_2^3$, for $x_1, x_2 \in \mathbb{R}$. Both $T_1$ and $T_2$ are continuously differentiable. As can be seen from Fig. 8.5, the inverse of $T_2$ is unique. However, the inverse of $T_1$ is not unique (one-to-one) because there are two possible values of $x_1$ for each value of $z_1$. Therefore, $T_2$ is a diffeomorphism, but $T_1$ is not.                                                       □

## 8.4.1  Manifolds, Lie Derivatives, Lie Brackets and Vector Fields

Concepts from differential geometry are often used in the analysis of nonlinear control systems. Recall from Sect. 2.3 that it is possible to graphically determine the stability of a one dimensional system (one state variable) by examining the flow of solution trajectories near the system's critical points (also called stationary points or equilibrium points). Similarly, as shown in Sect. 2.4, the stability of a two dimensional system can be explored by linearizing the system about its critical points and examining the trajectories (flows) of solutions in the space of its state variables (the phase plane). These ideas can be generalized to systems with $n$ state variables.

From a geometric perspective, the differential equation describing a dynamical system, e.g.

$$\dot{\mathbf{x}} = \mathbf{f}(\mathbf{x}),$$

can be viewed as representing the flow of a vector field along a space defined using the state variables of the system. In particular, the space in which the flow of the system occurs is known as a *differentiable manifold*, which we denote here as $M$. Such a system is deterministic, as all future states $\mathbf{x}(\mathbf{x}_0, t)$ are completely determined by the time $t$ and initial state $\mathbf{x}_0$.

In order to control the dynamics of the system we must consider a family of vector fields. Thus, a control system evolving on $M$ is a family of vector fields $\mathbf{f}(\mathbf{x}, \mathbf{u})$ on $M$, parameterized by the controls $\mathbf{u}$, i.e.

$$\dot{\mathbf{x}} = \mathbf{f}(\mathbf{x}, \mathbf{u}), \quad \mathbf{x} \in M, \mathbf{u} \subset \mathbb{R}^m.$$

Here, we restrict our attention to nonlinear control systems of the form

$$\begin{aligned} \dot{\mathbf{x}} &= \mathbf{f}(\mathbf{x}) + \mathbf{g}_1(\mathbf{x})u_1 + \cdots + \mathbf{g}_m(\mathbf{x})u_m, \\ &= \mathbf{f}(\mathbf{x}) + \mathbf{G}(\mathbf{x}) \cdot \mathbf{u}, \\ y_i &= h_i(\mathbf{x}), \quad 1 \le i \le p, \end{aligned} \tag{8.24}$$

where $\mathbf{f}(\mathbf{x})$ is known as the *drift vector field*, which specifies the dynamics of the system in the absence of a control input, and the vectors $\mathbf{g}_i(\mathbf{x}), i = \{1, \ldots, m\}$ are the *control vector fields*.

Here, the manifold $M$ is a subset of $\mathbb{R}^n$ ($M \subseteq \mathbb{R}^n$) and can be defined by the zero set of the smooth vector valued function $h : \mathbb{R}^n \to \mathbb{R}^p$, for $p < n$,

$$h_1(x_1, \ldots, x_n) = 0$$
$$\vdots \tag{8.25}$$
$$h_p(x_1, \ldots, x_n) = 0.$$

The functions $\mathbf{f}(\mathbf{x}), \mathbf{g}_1(\mathbf{x}), \ldots, \mathbf{g}_m(\mathbf{x})$ are smooth vector fields on $M$, where we define $\mathbf{G}(\mathbf{x}) := [\mathbf{g}_1(\mathbf{x}), \ldots, \mathbf{g}_m(\mathbf{x})]$ and $\mathbf{u} = [u_1, \ldots, u_m]^T$. Since the closed loop system takes an input $\mathbf{u} \in \mathbb{R}^m$ to produce an output $\mathbf{y} \in \mathbb{R}^p$, we say that $M$ is a topological space, which is locally diffeomorphic to the Euclidean space $\mathbb{R}^m$.

Next, we further explore the notions of vector fields and manifolds. If the total differentials $dh_1, \ldots, dh_p$ of (8.25) are linearly independent at each point, the dimension of the manifold is $m = (n - p)$.

**Definition 8.2** (*Smooth Vector Field*) A smooth vector field on a manifold $M$ is an infinitely differentiable function $\mathbf{f} : M \to T_x M$ (smooth mapping), which assigns to each point $\mathbf{x} \in M$ a tangent vector $\mathbf{f}(\mathbf{x}) \in T_x M$. A smooth vector field can be represented as a column vector

$$\mathbf{f}(\mathbf{x}) = \begin{bmatrix} f_1(\mathbf{x}) \\ \vdots \\ f_m(\mathbf{x}) \end{bmatrix}.$$

**Definition 8.3** (*Tangent Space*) A tangent space is an $m$-dimensional vector space specifying the set of possible velocities (directional derivatives) at $\mathbf{x}$. A tangent space $T_x M$ can be attached at each point $\mathbf{x} \in M$.

The concepts of a cotangent space and a covector field are also useful for the analysis of nonlinear control systems.

**Definition 8.4** (*Cotangent Space*) The cotangent space $T_x^* M$ is the dual space of the tangent space. It is an $m$-dimensional vector space specifying the set of possible differentials of functions at $\mathbf{x}$. Mathematically, $T_x^* M$ is the space of all linear functionals on $T_x M$, that is, the space of functions from $T_x M$ to $\mathbb{R}$.

**Definition 8.5** (*Smooth Covector Field*) A smooth covector field is a function $\mathbf{w} : M \rightarrow T_x^* M$, which is infinitely differentiable, represented as a row vector,

$$\mathbf{w}(\mathbf{x}) = [w_1(\mathbf{x}) \quad w_2(\mathbf{x}) \quad \cdots \quad w_m(\mathbf{x})].$$

**Example 8.4** (*Tangent Space and Covector Field on a Parabolic Surface*) Consider the parabolic surface in $\mathbb{R}^3$ given by

$$h(x, y, z) = z - x^2 - y^2 = 0. \tag{8.26}$$

The surface is a two-dimensional submanifold of $\mathbb{R}^3$, which we denote as $M$ (Fig. 8.6). At each point $\mathbf{x}_0 \in M$ on this surface we can attach a tangent space $T_{x_0} M$. For the paraboloid, each tangent space is a plane. As $h = 0$ is a constant on the surface and the gradient of the surface lies in the direction of maximum change of $h$, the gradient is perpendicular to the surface and its tangent plane at each $\mathbf{x}_0$. Here, the gradient will be used to specify the tangent plane at a point $\mathbf{x}_0$, as well as a pair of tangent vectors that span the tangent plane. The gradient of $h$ is

$$\begin{aligned} \nabla h &= \frac{\partial h}{\partial x}\hat{i} + \frac{\partial h}{\partial y}\hat{j} + \frac{\partial h}{\partial z}\hat{k}, \\ &= -2x\hat{i} - 2y\hat{j} + \hat{k}, \end{aligned} \tag{8.27}$$

where $\hat{i}$, $\hat{j}$ and $\hat{k}$ are unit vectors in the $x$, $y$ and $z$ directions, respectively. Note that the gradient is a covector field of $h(\mathbf{x})$ (see [5]). Thus, a smooth covector field on $M$ is given by the row vector

**Fig. 8.6** A parabolic manifold in $\mathbb{R}^3$ defined by the surface $h(x, y, z) = z - x^2 - y^2 = 0$. The tangent space at the point $\mathbf{x}_0$, $T_{\mathbf{x}_0} M$ is a plane spanned by the vectors $\mathbf{v}_1$ and $\mathbf{v}_2$. The gradient (Jacobian) of $h$ at $\mathbf{x}_0$ is perpendicular to the surface and tangent plane

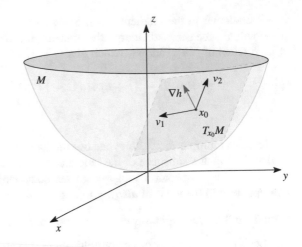

$$\mathbf{w} = [-2x \quad -2y \quad 1]. \tag{8.28}$$

Let $\mathbf{x}$ be a point in the tangent plane. Then the vector from $\mathbf{x}_0$ to $\mathbf{x}$, $\mathbf{dx} := \mathbf{x} - \mathbf{x}_0$ lies within the tangent plane. Owing to the fact that the gradient and tangent plane are perpendicular at $\mathbf{x}_0$, $\nabla h|_{\mathbf{x}_0} \cdot \mathbf{dx} = 0$.

Let the point $\mathbf{x}_0$ to be located at $\mathbf{x}_0 = (x_0, y_0, z_0) = (1, 1, 2)$. The tangent plane $T_{\mathbf{x}_0} M$ is therefore defined by the equation

$$
\begin{aligned}
dh|_{\mathbf{x}_0} &= \nabla h|_{\mathbf{x}_0} \cdot \mathbf{dx}|_{\mathbf{x}_0} \\
&= \left[ \frac{\partial h}{\partial x}\hat{i} + \frac{\partial h}{\partial y}\hat{j} + \frac{\partial h}{\partial z}\hat{k} \right]_{\mathbf{x}_0} \cdot \left[ (x - x_0)\hat{i} + (y - y_0)\hat{j} + (z - z_0)\hat{k} \right] \\
&= \left.\frac{\partial h}{\partial x}\right|_{\mathbf{x}_0}(x - x_0) + \left.\frac{\partial h}{\partial y}\right|_{\mathbf{x}_0}(y - y_0) + \left.\frac{\partial h}{\partial z}\right|_{\mathbf{x}_0}(z - z_0) \\
&= -2(x - 1) - 2(y - 1) + (z - 2) \\
&= -2x - 2y + z + 2 = 0,
\end{aligned} \tag{8.29}
$$

which can be rewritten as

$$z = 2(x + y). \tag{8.30}$$

At $\mathbf{x}_0$ the value of the gradient is

$$\nabla h|_{\mathbf{x}_0} = -2\hat{i} - 2\hat{j} + \hat{k}. \tag{8.31}$$

We can identify two tangent vectors at $\mathbf{x}_0$, which span the tangent plane, by noting that they must both be perpendicular to the gradient at that point. By inspection of (8.31) it can be seen that the vector

$$\mathbf{v}_1 = 1\hat{i} - 1\hat{j} + 0\hat{k} = \hat{i} - \hat{j} \tag{8.32}$$

is perpendicular to the gradient at $\mathbf{x}_0$. Since the other vector spanning the tangent plane will also be perpendicular to the gradient, it can be found by taking the vector cross product of $\nabla h|_{\mathbf{x}_0}$ and $\mathbf{v}_1$, which gives

$$\mathbf{v}_2 = (-2\hat{i} - 2\hat{j} + \hat{k}) \times (1\hat{i} - 1\hat{j} + 0\hat{k}) = \hat{i} + \hat{j} + 4\hat{k}. \tag{8.33}$$

$\square$

Note that multiple vector fields can be simultaneously defined on a given manifold so that they span a subspace of the tangent space at each point. Similarly, multiple covector fields may span a subspace of the cotangent space at each point. These concepts lead to the ideas of *distributions* and *codistributions*.

**Definition 8.6** (*Distributions and Codistributions*)

(1) Let $\mathbf{X}_1(\mathbf{x}), \ldots, \mathbf{X}_k(\mathbf{x})$ be vector fields on $M$ that are linearly independent at each point. A distribution $\Delta$ is the linear span (at each $\mathbf{x} \in M$)

$$\Delta = \mathrm{span}\,\{\mathbf{X}_1(\mathbf{x}), \ldots, \mathbf{X}_k(\mathbf{x})\}. \tag{8.34}$$

(2) Likewise, let $w_1(\mathbf{x}), \ldots, w_k(\mathbf{x})$ be covector fields on $M$, which are linearly independent at each point. The codistribution $\Omega$ is defined as the linear span (at each $\mathbf{x} \in M$)

$$\Omega = \mathrm{span}\,\{w_1(\mathbf{x}), \ldots, w_k(\mathbf{x})\}. \tag{8.35}$$

Thus, a distribution assigns a vector space $\Delta(\mathbf{x})$ to each point $\mathbf{x} \in M$, where $\Delta(\mathbf{x})$ is a $k$-dimensional subspace of the $m$-dimensional tangent space $T_x M$. Similarly, a codistribution defines a $k$-dimensional subspace $\Omega(\mathbf{x})$ at each $\mathbf{x}$ of the $m$-dimensional cotangent space $T_x^* M$.

The Lie bracket and Lie derivative are defined next. Both of these mathematical operations are often used when analyzing nonlinear systems, especially for trajectory planning or path following control problems.

**Definition 8.7** (*Lie Derivative*) Let $\mathbf{f} : \mathbb{R}^n \to \mathbb{R}^n$ be a vector field on $\mathbb{R}^n$ and let $h : \mathbb{R}^n \to \mathbb{R}$ be a scalar function. The *Lie derivative* of $h$ with respect to $\mathbf{f}$, denoted $L_f h$, is defined as

$$L_f h = \nabla h \cdot \mathbf{f}(\mathbf{x}) = \frac{\partial h}{\partial \mathbf{x}} \cdot \mathbf{f}(\mathbf{x}) = \sum_{i=1}^{n} \frac{\partial h}{\partial x_i} f_i(\mathbf{x}). \tag{8.36}$$

**Remark 8.1** The Lie derivative is simply the gradient of $h$ in the direction of $\mathbf{f}$. When the state equation of a system is given by $\dot{\mathbf{x}} = \mathbf{f}(\mathbf{x})$, the Lie derivative $L_f h$ represents the gradient of $h$ along the trajectories of $\dot{\mathbf{x}}$ in the system's state space (see Sect. 2.4).

Denote the Lie derivative of $L_f h$ with respect to $\mathbf{f}$ as $L_f^2 h$, that is,

$$L_f^2 h = L_f(L_f h). \tag{8.37}$$

In general, we can recursively define $L_f^k h$ as

$$L_f^k h = L_f(L_f^{k-1} h) \quad \text{for } k = 1, \ldots, n \tag{8.38}$$

with $L_f^0 h = h$.

**Definition 8.8** (*Lie Bracket*) Let $\mathbf{f}$ and $\mathbf{g}$ be two vector fields on $\mathbb{R}^n$. The *Lie bracket* of $\mathbf{f}$ and $\mathbf{g}$, denoted by $[\mathbf{f}, \mathbf{g}]$, is a vector field defined by

$$[\mathbf{f}, \mathbf{g}] = \nabla \mathbf{g} \cdot \mathbf{f} - \nabla \mathbf{f} \cdot \mathbf{g} = \frac{\partial \mathbf{g}}{\partial \mathbf{x}} \cdot \mathbf{f} - \frac{\partial \mathbf{f}}{\partial \mathbf{x}} \cdot \mathbf{g}, \tag{8.39}$$

where $\nabla \mathbf{g}$ and $\nabla \mathbf{f}$ are $n \times n$ Jacobian matrices with their $ij^{th}$ entries being

$$(\nabla \mathbf{g})_{ij} = \frac{\partial g_i}{\partial x_j} \quad \text{and} \quad (\nabla \mathbf{f})_{ij} = \frac{\partial f_i}{\partial x_j},$$

respectively.

We also denote $[\mathbf{f}, \mathbf{g}]$ as $\mathrm{ad}_f(\mathbf{g})$ and define $\mathrm{ad}_f^k(\mathbf{g})$ inductively by

$$\mathrm{ad}_f^k(\mathbf{g}) = [\mathbf{f}, \mathrm{ad}_f^{k-1}(\mathbf{g})], \tag{8.40}$$

where $\mathrm{ad}_f^0(\mathbf{g}) = \mathbf{g}$.

The following lemma relates the Lie bracket to the Lie derivative and is crucial to the subsequent development.

**Lemma 8.1** *Let $h : \mathbb{R}^n \to \mathbb{R}$ be a scalar function and $\mathbf{f}$ and $\mathbf{g}$ be vector fields on $\mathbb{R}^n$. Then we have the following identity*

$$L_{[f,g]} h = L_f L_g h - L_g L_f h. \tag{8.41}$$

**Example 8.5** (*Geometric Interpretation of the Lie Bracket*) Recall from Chap. 2 that the dynamics of a system can be modeled using a differential equation and that the solution trajectories of this differential equation can be represented as a flow in the phase space of a system (Sects. 2.3–2.4). As discussed above, here the space in which the flow of the system occurs is the differentiable manifold $M$. Let the smooth

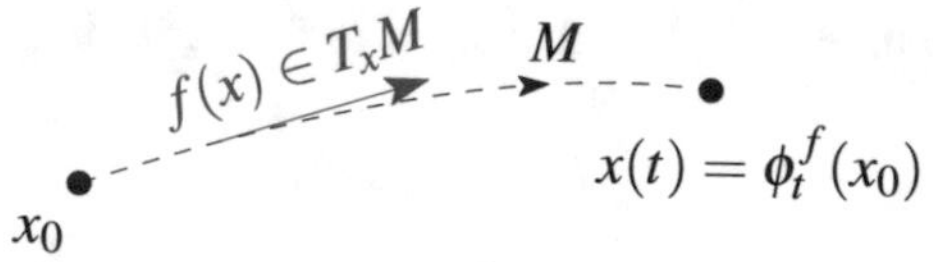

**Fig. 8.7** The flow $\phi_t^f(\mathbf{x}_0)$ (dashed line) is a solution trajectory of the equation $\dot{\mathbf{x}} = \mathbf{f}(\mathbf{x})$ on the manifold $M$ (which corresponds to the solution trajectory itself in this case). The flow maps points $\mathbf{x} \in M$ back into $M$. The vector $\mathbf{f}(\mathbf{x})$ is in the tangent space of $M$, $\mathbf{f}(\mathbf{x}) \in T_x M$

vector field $\mathbf{f}(\mathbf{x})$ be a smooth mapping $\mathbf{f} : \mathbb{R}^n \rightarrow T_x(\mathbb{R}^n)$ that assigns a tangent vector $\mathbf{f}(\mathbf{x}) \in T_x(\mathbb{R}^n)$ to each point $\mathbf{x} \in \mathbb{R}^n$. If the differential equation describing a system is given as

$$\dot{\mathbf{x}} = \mathbf{f}(\mathbf{x}), \tag{8.42}$$

for an initial state $\mathbf{x}_0$, the flow $\phi_t^f(\mathbf{x}_0)$ of $\mathbf{f}(\mathbf{x})$ can be considered to be a mapping that associates the solution of (8.42) to each point $\mathbf{x} \in M$ at time $t$, or

$$\frac{d}{dt}\left[\phi_t^f(\mathbf{x}_0)\right] = \mathbf{f}\left(\phi_t^f(\mathbf{x}_0)\right), \tag{8.43}$$

see Fig. 8.7.

The set of mappings $\left\{\phi_t^f\right\}$ is a one parameter group of diffeomorphisms (Definition 8.1) under the operation of composition,

$$\phi_{t_1}^f \circ \phi_{t_2}^f = \phi_{t_1+t_2}^f. \tag{8.44}$$

In general, the composition of the flows related to two different vector fields, e.g. $\mathbf{f}$ and $\mathbf{g}$, is not commutative, such that

$$\phi_t^f \circ \phi_t^g \neq \phi_t^g \circ \phi_t^f. \tag{8.45}$$

However, two vector fields commute if their Lie bracket (Definition 8.8) is zero, $[\mathbf{f}, \mathbf{g}] = 0$. We can obtain a geometric understanding of the Lie bracket by considering the following differential equation

$$\dot{\mathbf{x}} = \mathbf{f}(\mathbf{x})u_1 + \mathbf{g}(\mathbf{x})u_2. \tag{8.46}$$

If the two inputs $u_1$ and $u_2$ are never active at the same instant, the solution of (8.46) is obtained by composing the flows relative to $\mathbf{f}$ and $\mathbf{g}$. Consider the flows generated by (8.46) for the input sequence

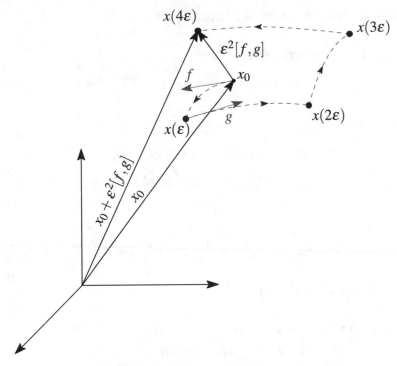

**Fig. 8.8** The flow (dashed line) associated with the input sequence (8.47). For short time segments $\varepsilon$, infinitesimal motion occurs in the directions spanned by the vector field $\mathbf{f}$ and $\mathbf{g}$, as well as in the direction of their Lie bracket $[\mathbf{f}, \mathbf{g}]$

$$u(t) = \begin{cases} u_1(t) = +1, & u_2(t) = 0, & 0 < t \le \varepsilon, \\ u_1(t) = 0, & u_2(t) = +1, & \varepsilon < t \le 2\varepsilon, \\ u_1(t) = -1, & u_2(t) = 0, & 2\varepsilon < t \le 3\varepsilon, \\ u_1(t) = 0, & u_2(t) = -1, & 3\varepsilon < t \le 4\varepsilon, \end{cases} \qquad (8.47)$$

where $\varepsilon$ is an infinitesimal interval of time.

The composition of the flows can be represented as

$$\mathbf{x}(4\varepsilon) = \phi_\varepsilon^{-g} \circ \phi_\varepsilon^{-f} \circ \phi_\varepsilon^{g} \circ \phi_\varepsilon^{f} (\mathbf{x}_0) \qquad (8.48)$$

and can be approximated by computing $\mathbf{x}(\varepsilon)$ as a series expansion about $\mathbf{x}_0 = \mathbf{x}(0)$ along $\mathbf{f}$, $\mathbf{x}(2\varepsilon)$ as a series expansion about $\mathbf{x}(\varepsilon)$ along $\mathbf{f}$, and so on (Fig. 8.8).

Approximating the flow corresponding to the first step of the input sequence gives

$$\begin{aligned} \mathbf{x}(\varepsilon) &= \mathbf{x}_0 + \dot{\mathbf{x}}\varepsilon + \frac{1}{2}\ddot{\mathbf{x}}\varepsilon^2 + O(\varepsilon^3), \\ &= \mathbf{x}_0 + \varepsilon\mathbf{f}(\mathbf{x}_0) + \frac{1}{2}\frac{\partial\mathbf{f}}{\partial\mathbf{x}}\mathbf{f}\varepsilon^2 + O(\varepsilon^3). \end{aligned} \qquad (8.49)$$

Then, the position attained after the second input is given by

$$
\begin{aligned}
\mathbf{x}(2\varepsilon) &= \mathbf{x}(\varepsilon) + \varepsilon\mathbf{g}(\mathbf{x}(\varepsilon)) + \frac{1}{2}\frac{\partial\mathbf{g}}{\partial\mathbf{x}}\mathbf{g}\varepsilon^2 + O(\varepsilon^3), \\
&= \mathbf{x}_0 + \varepsilon\mathbf{f}(\mathbf{x}_0) + \varepsilon\mathbf{g}(\mathbf{x}(\varepsilon)) + \frac{1}{2}\left(\frac{\partial\mathbf{f}}{\partial\mathbf{x}}\mathbf{f} + \frac{\partial\mathbf{g}}{\partial\mathbf{x}}\mathbf{g}\right)\varepsilon^2 + O(\varepsilon^3).
\end{aligned}
\tag{8.50}
$$

Expanding $\varepsilon\mathbf{g}(\mathbf{x}(\varepsilon))$ gives

$$
\begin{aligned}
\varepsilon\mathbf{g}(\mathbf{x}(\varepsilon)) &= \varepsilon\mathbf{g}(\mathbf{x}_0 + \varepsilon\mathbf{f}(\mathbf{x}_0) + O(\varepsilon^2)), \\
&= \varepsilon\mathbf{g}(\mathbf{x}_0) + \varepsilon^2\frac{\partial\mathbf{g}}{\partial\mathbf{x}}\mathbf{f}(\mathbf{x}_0) + O(\varepsilon^3),
\end{aligned}
\tag{8.51}
$$

so that

$$
\mathbf{x}(2\varepsilon) = \mathbf{x}_0 + \varepsilon[\mathbf{f}(\mathbf{x}_0) + \mathbf{g}(\mathbf{x}_0)] + \varepsilon^2\frac{\partial\mathbf{g}}{\partial\mathbf{x}}\mathbf{f}(\mathbf{x}_0) + \frac{\varepsilon^2}{2}\left(\frac{\partial\mathbf{f}}{\partial\mathbf{x}}\mathbf{f} + \frac{\partial\mathbf{g}}{\partial\mathbf{x}}\mathbf{g}\right) + O(\varepsilon^3).
\tag{8.52}
$$

Next, the position after the third input is

$$
\begin{aligned}
\mathbf{x}(3\varepsilon) &= \mathbf{x}(2\varepsilon) - \varepsilon\mathbf{f}(\mathbf{x}(2\varepsilon)) + \frac{1}{2}\frac{\partial\mathbf{f}}{\partial\mathbf{x}}\mathbf{f}\varepsilon^2 + O(\varepsilon^3), \\
&= \mathbf{x}_0 - \varepsilon\mathbf{f}(\mathbf{x}(2\varepsilon)) + \varepsilon[\mathbf{f}(\mathbf{x}_0) + \mathbf{g}(\mathbf{x}_0)] \\
&\quad + \varepsilon^2\left(\frac{\partial\mathbf{g}}{\partial\mathbf{x}}\mathbf{f}(\mathbf{x}_0) + \frac{\partial\mathbf{f}}{\partial\mathbf{x}}\mathbf{f} + \frac{1}{2}\frac{\partial\mathbf{g}}{\partial\mathbf{x}}\mathbf{g}\right) + O(\varepsilon^3).
\end{aligned}
\tag{8.53}
$$

Expanding $\varepsilon\mathbf{f}(\mathbf{x}(2\varepsilon))$ gives

$$
\begin{aligned}
-\varepsilon\mathbf{f}(\mathbf{x}(2\varepsilon)) &= -\varepsilon\mathbf{f}(\mathbf{x}_0 + \varepsilon[\mathbf{f}(\mathbf{x}_0) + \mathbf{g}(\mathbf{x}_0)] + O(\varepsilon^2)), \\
&= -\varepsilon\mathbf{f}(\mathbf{x}_0) - \varepsilon^2\frac{\partial\mathbf{f}}{\partial\mathbf{x}}\left[\mathbf{f}(\mathbf{x}_0) + \mathbf{g}(\mathbf{x}_0)\right] + O(\varepsilon^3),
\end{aligned}
\tag{8.54}
$$

so that

$$
\begin{aligned}
\mathbf{x}(3\varepsilon) &= \mathbf{x}_0 + \varepsilon\mathbf{g}(\mathbf{x}_0) \\
&\quad + \varepsilon^2\left(\frac{\partial\mathbf{g}}{\partial\mathbf{x}}\mathbf{f}(\mathbf{x}_0) - \frac{\partial\mathbf{f}}{\partial\mathbf{x}}\mathbf{g}(\mathbf{x}_0) + \frac{1}{2}\frac{\partial\mathbf{g}}{\partial\mathbf{x}}\mathbf{g}\right) + O(\varepsilon^3)
\end{aligned}
\tag{8.55}
$$

Lastly, after the fourth input the flow arrives at the position

$$
\begin{aligned}
\mathbf{x}(4\varepsilon) &= \mathbf{x}(3\varepsilon) - \varepsilon\mathbf{g}(\mathbf{x}(3\varepsilon)) + \frac{1}{2}\frac{\partial\mathbf{g}}{\partial\mathbf{x}}\mathbf{g}\varepsilon^2 + O(\varepsilon^3), \\
&= \mathbf{x}_0 + \varepsilon\mathbf{g}(\mathbf{x}_0) - \varepsilon\mathbf{g}\left(\mathbf{x}_0 + \varepsilon\mathbf{g}(\mathbf{x}_0) + O(\varepsilon^2)\right) \\
&\quad + \varepsilon^2\left(\frac{\partial\mathbf{g}}{\partial\mathbf{x}}\mathbf{f}(\mathbf{x}_0) - \frac{\partial\mathbf{f}}{\partial\mathbf{x}}\mathbf{g}(\mathbf{x}_0) + \frac{\partial\mathbf{g}}{\partial\mathbf{x}}\mathbf{g}\right) + O(\varepsilon^3)
\end{aligned}
\tag{8.56}
$$

Expanding $\varepsilon\mathbf{g}(\mathbf{x}(3\varepsilon))$ gives

$$-\varepsilon g(x(3\varepsilon)) = -\varepsilon g(x_0 + \varepsilon g(x_0) + O(\varepsilon^2)),$$
$$= -\varepsilon g(x_0) - \varepsilon^2 \frac{\partial g}{\partial x} g(x_0) + O(\varepsilon^3), \tag{8.57}$$

so that

$$x(4\varepsilon) = x_0 + \varepsilon^2 \left[ \frac{\partial g}{\partial x} f(x_0) - \frac{\partial f}{\partial x} g(x_0) \right] + O(\varepsilon^3),$$
$$= x_0 + \varepsilon^2 [f, g] + O(\varepsilon^3). \tag{8.58}$$

Thus, it can be seen from the preceding result that when two vector fields commute so that $[f, g] = 0$, $x(4\varepsilon) = x_0$ to order $O(\varepsilon^3)$. From a geometrical point of view, this means that no net flow results from the input sequence (8.47). As shown in Fig. 8.8, the flow returns to the starting point $x_0$ if $[f, g] = 0$.

One can also state more broadly that the computation leading to (8.58) shows that infinitesimal motion is possible not only in the directions spanned by the vector fields $f$ and $g$, but also in the directions of their Lie brackets [3]. The same is also true for higher order Lie brackets, e.g. $[f_1, [f_2, f_3]]$.

As discussed in [2, 3, 7], this seemingly innocuous result has profound implications for the reachability/controllability of nonlinear systems (see Sect. 6.2 for an introduction to reachability). For the uncontrolled system in (8.42) the solution trajectories of the system can only reach points along the manifold determined by the tangent space of $f(x)$. However, for the system in (8.46) the manifold along which solution trajectories can exist consists not only of the space spanned by the tangent vectors $f(x)$ and $g(x)$, but the larger space spanned by $f(x)$, $g(x)$, and the Lie bracket $[f, g]$. This implies that by a suitable selection of control inputs a larger region of the state space is reachable. This idea is known is known as *Chow's Theorem* (Theorem 9.1), which roughly stated, says that if we can move in every direction of the state space using Lie bracket motions (possibly of higher order than one), then the system is controllable [7]. The Frobenius Theorem, presented in Sect. 8.4.2 below, provides a basis for determining the reachability of a nonlinear control system, by establishing a property known as *integrability*.                                                  □

### 8.4.2   Frobenius Theorem

The Frobenius Theorem provides a means of understanding whether a system of first order partial differential equations is solvable, without the need to actually solve the partial differential equations. The theorem is presented here, as it is generally instrumental in determining whether nonlinear systems can be decomposed into reachable and unreachable parts, as well as into observable and unobservable parts [5].

Consider an unknown function $z = \phi(x, y)$, which defines the height of a surface above the $x$-$y$ plane. The surface is given by the mapping $\Phi : \mathbb{R}^2 \to \mathbb{R}^3$ defined by

$$\Phi = (x, y, \phi(x, y)). \tag{8.59}$$

The rates of change in the height of the surface with respect to the $x$ and $y$ directions are given by

$$\frac{\partial z}{\partial x} = \frac{\partial \phi}{\partial x} = f(x, y, z) = f(x, y, \phi(x, y)), \tag{8.60}$$

$$\frac{\partial z}{\partial y} = \frac{\partial \phi}{\partial y} = g(x, y, z) = g(x, y, \phi(x, y)), \tag{8.61}$$

respectively. Assume that $f(x, y, \phi(x, y))$ and $g(x, y, \phi(x, y))$ are known functions. Is it possible to solve the system of partial differential equations (8.60)–(8.61) to find $\phi(x, y)$?

At each point $(x, y)$ the tangent plane to the surface is spanned by two vectors, which can be found by taking the partial derivatives of $\Phi$ (8.59) in the $x$ and $y$ directions

$$\begin{aligned} \mathbf{X}_1 &= [1 \quad 0 \quad f(x, y, \phi(x, y))]^T, \\ \mathbf{X}_2 &= [0 \quad 1 \quad g(x, y, \phi(x, y))]^T. \end{aligned} \tag{8.62}$$

Note that the vector fields $\mathbf{X}_1$ and $\mathbf{X}_2$ are linearly independent and span a two dimensional subspace (plane) at each point. Mathematically, $\mathbf{X}_1$ and $\mathbf{X}_2$ are completely specified by Eqs. (8.60) and (8.61). Therefore, the problem of solving (8.60) and (8.61) for $\phi(x, y)$ is geometrically equivalent to finding a surface in $\mathbb{R}^3$, whose tangent space at each point is spanned by the vector fields $\mathbf{X}_1$ and $\mathbf{X}_2$ [10]. Such a surface, is an *integral manifold* of (8.60) and (8.61). When such an integral manifold exists, the set of vector fields (8.62), or equivalently, the system of partial differential equations (8.60)–(8.61), is completely integrable.

Alternatively, let the function

$$h(x, y, z) := z - \phi(x, y) = 0 \tag{8.63}$$

specify the surface. Using (8.36) and (8.60)–(8.62), it can be seen that

$$\begin{aligned} L_{\mathbf{X}_1} h &= \begin{bmatrix} -\dfrac{\partial \phi}{\partial x} & -\dfrac{\partial \phi}{\partial y} & 1 \end{bmatrix} \cdot \mathbf{X}_1 = -\dfrac{\partial \phi}{\partial x} + f(x, y, \phi(x, y)) = 0, \\ L_{\mathbf{X}_2} h &= \begin{bmatrix} -\dfrac{\partial \phi}{\partial x} & -\dfrac{\partial \phi}{\partial y} & 1 \end{bmatrix} \cdot \mathbf{X}_2 = -\dfrac{\partial \phi}{\partial y} + g(x, y, \phi(x, y)) = 0. \end{aligned} \tag{8.64}$$

Suppose a scalar function $h(x, y, z)$ can be found that satisfies (8.64) and that we can solve (8.63) for $z$. Then it can be shown that $\phi$ satisfies (8.60)–(8.61) [10]. Therefore, the complete integrability of the vector fields $\{\mathbf{X}_1, \mathbf{X}_2\}$ is equivalent to the existence of $h$ satisfying (8.64). Which leads to the following concepts.

**Definition 8.9** (*Integrable*) A distribution $\Delta = \text{span}\{\mathbf{X}_1, \dots, \mathbf{X}_m\}$ on $\mathbb{R}^n$ is completely integrable if and only if there are $n - m$ linearly independent functions $h_1, \dots, h_{n-m}$ satisfying the system of partial differential equations

$$L_{\mathbf{X}_i} h_j = 0 \quad \text{for } 1 \le i \le m, \quad 1 \le j \le n - m. \tag{8.65}$$

Another important concept is the notion of involutivity as defined next.

**Definition 8.10**   (*Involutive*) A distribution $\Delta = \mathrm{span}\{\mathbf{X}_1, \ldots, \mathbf{X}_m\}$ is involutive if and only if there are scalar functions $\alpha_{ijk} : \mathbb{R}^n \to \mathbb{R}$ such that

$$[\mathbf{X}_i, \mathbf{X}_j] = \sum_{k=1}^{m} \alpha_{ijk} \mathbf{X}_k \quad \forall i, j, k. \tag{8.66}$$

Involutivity simply means that if one forms the Lie bracket of any pair of vector fields in $\Delta$ then the resulting vector field can be expressed as a linear combination of the original vector fields $\mathbf{X}_1, \ldots, \mathbf{X}_m$. An involutive distribution is thus closed under the operation of taking Lie brackets. Note that the coefficients in this linear combination are allowed to be smooth functions on $\mathbb{R}^n$.

The Frobenius Theorem, stated next, gives the conditions for the existence of a solution to the system of partial differential equations (8.65).

**Theorem 8.1**   (Frobenius) *Suppose the dimension of the distribution $\Delta$ is constant. Then, $\Delta$ is completely integrable, if and only if, it is involutive.*

As mentioned above, the importance of the Frobenius theorem is that it allows one to determine whether or not a given distribution is integrable, and hence provides information about the reachability/controllability of an associated dynamical system (e.g. a linear or nonlinear controlled system), without having to actually solve the related partial differential equations. The involutivity condition can, in principle, be computed from the given vector fields alone.

**Remark 8.2**   The distribution generated by a single vector field is always involutive and, therefore integrable.

As will be seen in the following section, feedback controller design techniques for nonlinear systems often involve a change of coordinates of the state variables. When such a transformation of variables is a diffeomorphism (Definition 8.1), the integrability/reachability properties of the system are preserved.

## 8.5   Input-State Linearization

When all of the states of a nonlinear system are observable and reachable (controllable), it may be possible to use input-state feedback linearization to identify a linearizing feedback control input that transforms the state equation of the nonlinear closed loop system into a reachable linear system.

**Definition 8.11**   (*Input-State Linearization*) A nonlinear system

$$\dot{\mathbf{x}} = \mathbf{f}(\mathbf{x}) + \mathbf{G}(\mathbf{x})\mathbf{u} \tag{8.67}$$

where $\mathbf{f} : D \to \mathbb{R}^n$ and $\mathbf{G} : D \to \mathbb{R}^{n \times p}$ are sufficiently smooth on a domain $D \subset \mathbb{R}^n$, is input-state linearizable if there exists a diffeomorphism $\mathbf{T} : D \to \mathbb{R}^n$, such that $D_z = \mathbf{T}(D)$ contains the origin, and the change of variables $\mathbf{z} = \mathbf{T}(\mathbf{x})$ together with the linearizing feedback control input

$$\mathbf{u} = \boldsymbol{\alpha}(\mathbf{x}) + \boldsymbol{\beta}(\mathbf{x})\mathbf{v} \tag{8.68}$$

transforms the system (8.67) into the controllable linear system

$$\dot{\mathbf{z}} = \mathbf{A}\mathbf{z} + \mathbf{B}\mathbf{v}. \tag{8.69}$$

The feedback linearization is global if the region $D$ is all of $\mathbb{R}^n$.

The idea of feedback linearization is easiest to understand in the context of single-input systems, where (8.67) becomes

$$\dot{\mathbf{x}} = \mathbf{f}(\mathbf{x}) + \mathbf{g}(\mathbf{x})u \tag{8.70}$$

and (8.69) becomes

$$\dot{\mathbf{z}} = \mathbf{A}\mathbf{z} + \mathbf{b}v. \tag{8.71}$$

Without loss of generality, we can assume that the linear transformed state has the *linear companion form*

$$\mathbf{A} = \begin{bmatrix} 0 & 1 & 0 & \cdots & 0 \\ 0 & 0 & 1 & & 0 \\ \vdots & & & \ddots & 0 \\ 0 & 0 & 0 & & 1 \\ 0 & 0 & 0 & \cdots & 0 \end{bmatrix}, \qquad \mathbf{b} = \begin{bmatrix} 0 \\ 0 \\ \vdots \\ 0 \\ 1 \end{bmatrix}. \tag{8.72}$$

As discussed in [9], any representation of a linear controllable system can be transformed into the companion form (8.72) by using additional linear transformations on the state variable $\mathbf{x}$ and control input $u$.

Let us explore what the necessary and sufficient conditions on $f$ and $g$ in (8.70) are for the existence of the linearizing transformation $\mathbf{T}(\mathbf{x})$. Taking the time derivative of the transformation

$$\mathbf{z} = \mathbf{T}(\mathbf{x}) = \begin{bmatrix} z_1(\mathbf{x}) \\ z_2(\mathbf{x}) \\ \vdots \\ z_{n-1}(\mathbf{x}) \\ z_n(\mathbf{x}) \end{bmatrix} \tag{8.73}$$

gives

$$\dot{\mathbf{z}} = \nabla \mathbf{T} \cdot \dot{\mathbf{x}} = \frac{\partial \mathbf{T}}{\partial \mathbf{x}} \dot{\mathbf{x}}, \tag{8.74}$$

where $\nabla\mathbf{T}$ is the Jacobian of $\mathbf{T}(\mathbf{x})$. Using (8.70) and (8.71), (8.74) can be rewritten as

$$\dot{\mathbf{z}} = \nabla\mathbf{T} \cdot \left[\mathbf{f}(\mathbf{x}) + \mathbf{g}(\mathbf{x})u\right] = \mathbf{A}\mathbf{z} + \mathbf{b}v. \tag{8.75}$$

Note that

$$\nabla\mathbf{T} \cdot \left[\mathbf{f}(\mathbf{x}) + \mathbf{g}(\mathbf{x})u\right] =
\begin{bmatrix}
\dfrac{\partial z_1}{\partial x_1} & \dfrac{\partial z_1}{\partial x_2} & \cdots & \dfrac{\partial z_1}{\partial x_n} \\[2mm]
\dfrac{\partial z_2}{\partial x_1} & \dfrac{\partial z_2}{\partial x_2} & \cdots & \dfrac{\partial z_2}{\partial x_n} \\[2mm]
\vdots & \vdots & \ddots & \vdots \\[2mm]
\dfrac{\partial z_{n-1}}{\partial x_1} & \dfrac{\partial z_{n-1}}{\partial x_2} & \cdots & \dfrac{\partial z_{n-1}}{\partial x_n} \\[2mm]
\dfrac{\partial z_n}{\partial x_1} & \dfrac{\partial z_n}{\partial x_2} & \cdots & \dfrac{\partial z_n}{\partial x_n}
\end{bmatrix}
\cdot \left[\mathbf{f}(\mathbf{x}) + \mathbf{g}(\mathbf{x})u\right]$$

$$=
\begin{bmatrix}
\dfrac{\partial z_1}{\partial x_1}(f_1 + g_1 u) + \cdots + \dfrac{\partial z_1}{\partial x_n}(f_n + g_n u) \\[3mm]
\dfrac{\partial z_2}{\partial x_1}(f_1 + g_1 u) + \cdots + \dfrac{\partial z_2}{\partial x_n}(f_n + g_n u) \\[3mm]
\vdots \\[3mm]
\dfrac{\partial z_{n-1}}{\partial x_1}(f_1 + g_1 u) + \cdots + \dfrac{\partial z_{n-1}}{\partial x_n}(f_n + g_n u) \\[3mm]
\dfrac{\partial z_n}{\partial x_1}(f_1 + g_1 u) + \cdots + \dfrac{\partial z_n}{\partial x_n}(f_n + g_n u)
\end{bmatrix}. \tag{8.76}$$

Using (8.36) in Definition 8.7, (8.76) can be rewritten as

$$\nabla\mathbf{T} \cdot \left[\mathbf{f}(\mathbf{x}) + \mathbf{g}(\mathbf{x})u\right] =
\begin{bmatrix}
\sum\limits_{i=1}^{n} \dfrac{\partial z_1}{\partial x_i} f_i(x) + \sum\limits_{i=1}^{n} \dfrac{\partial z_1}{\partial x_i} g_i(x) u \\[3mm]
\sum\limits_{i=1}^{n} \dfrac{\partial z_2}{\partial x_i} f_i(x) + \sum\limits_{i=1}^{n} \dfrac{\partial z_2}{\partial x_i} g_i(x) u \\[3mm]
\vdots \\[3mm]
\sum\limits_{i=1}^{n} \dfrac{\partial z_{n-1}}{\partial x_i} f_i(x) + \sum\limits_{i=1}^{n} \dfrac{\partial z_{n-1}}{\partial x_i} g_i(x) u \\[3mm]
\sum\limits_{i=1}^{n} \dfrac{\partial z_n}{\partial x_i} f_i(x) + \sum\limits_{i=1}^{n} \dfrac{\partial z_n}{\partial x_i} g_i(x) u.
\end{bmatrix} \tag{8.77}$$

$$=
\begin{bmatrix}
L_f z_1 + L_g z_1 u \\
L_f z_2 + L_g z_2 u \\
\vdots \\
L_f z_{n-1} + L_g z_{n-1} u \\
L_f z_n + L_g z_n u.
\end{bmatrix}$$

With (8.71), (8.72) and (8.75), (8.77) can be rewritten in component form as

$$
\begin{bmatrix} L_f z_1 + L_g z_1 u \\ L_f z_2 + L_g z_2 u \\ \vdots \\ L_f z_{n-1} + L_g z_{n-1} u \\ L_f z_n + L_g z_n u \end{bmatrix} = \begin{bmatrix} 0 & 1 & 0 & \cdots & 0 \\ 0 & 0 & 1 & & 0 \\ \vdots & & & \ddots & 0 \\ 0 & 0 & 0 & & 1 \\ 0 & 0 & 0 & \cdots & 0 \end{bmatrix} \begin{bmatrix} z_1 \\ z_2 \\ \vdots \\ z_{n-1} \\ z_n \end{bmatrix} + \begin{bmatrix} 0 \\ 0 \\ \vdots \\ 0 \\ 1 \end{bmatrix} v = \begin{bmatrix} z_2 \\ z_3 \\ \vdots \\ z_n \\ v \end{bmatrix}.
$$

Assume that the transformations $z_1, \ldots, z_n$ are independent of $u$, but that $v$ depends on $u$. This requires that

$$
\begin{aligned}
L_g z_1 = L_g z_2 = \cdots = L_g z_{n-1} = 0 \\
L_g z_n \neq 0.
\end{aligned}
\tag{8.78}
$$

This leads to the system of partial differential equations

$$
L_f z_i = z_{i+1}, \quad i = 1, \ldots, n-1
\tag{8.79}
$$

and

$$
L_f z_n + L_g z_n u = v.
\tag{8.80}
$$

Using Lemma 8.1 together with (8.78) we can derive a system of partial differential equations in terms of $z_1$ alone as follows. Using $h = z_1$ in Lemma 8.1 we have

$$
L_{[f,g]} z_1 = L_f L_g z_1 - L_g L_f z_1 = 0 - L_g z_2 = 0.
\tag{8.81}
$$

Therefore,

$$
L_{[f,g]} z_1 = 0.
\tag{8.82}
$$

By proceeding inductively it can be shown that

$$
L_{\text{ad}_f^k g} z_1 = 0, \quad k = 0, \ldots, n-2
\tag{8.83}
$$

$$
L_{\text{ad}_f^{n-1} g} z_1 \neq 0.
\tag{8.84}
$$

If we can find $z_1$ satisfying the system of partial differential equations (8.83), then $z_2, \ldots, z_n$ are found inductively from (8.79) and the control input $u$ is found from (8.80) as

$$
u = \frac{1}{L_g z_n} \left( v - L_f z_n \right).
\tag{8.85}
$$

We have thus reduced the problem to solving the system given by (8.83) for $z_1$. When does such a solution exist?

First note that the vector fields $g, \text{ad}_f(g), \ldots, \text{ad}_f^{n-1}(g)$ must be linearly independent. If not, that is, if for some index $i$

$$\text{ad}^i_f(g) = \sum_{k=0}^{i-1} \alpha_k \text{ad}^k_f(g) \tag{8.86}$$

then $\text{ad}^{n-1}_f(g)$ would be a linear combination of $g, \text{ad}_f(g), \ldots, \text{ad}^{n-1}_f(g)$ and (8.84) could not hold. Now, by the Frobenius Theorem (Theorem 8.1), (8.83) has a solution if and only if the distribution $\Delta = \text{span}\{g, \text{ad}_f(g), \ldots, \text{ad}^{n-2}_f(g)\}$ is involutive. Putting this together leads to the following theorem.

**Theorem 8.2** (Feedback Linearizable) *The single-input nonlinear system*

$$\dot{\mathbf{x}} = \mathbf{f}(\mathbf{x}) + \mathbf{g}(\mathbf{x})u \tag{8.87}$$

*is feedback linearizable if and only if there exists an open region D containing the origin in $\mathbb{R}^n$ in which the following conditions hold:*

*(1) The vector fields $\{g, \text{ad}_f(g), \ldots, \text{ad}^{n-1}_f(g)\}$ are linearly independent in D.*
*(2) The distribution $\Delta = \text{span}\{g, \text{ad}_f(g), \ldots, \text{ad}^{n-2}_f(g)\}$ is involutive in D.*

Recall from Remark 7.2 that if the equilibrium point of a system $\mathbf{x}_e$ is not located at the origin $\mathbf{x}_e \neq 0$, it is common practice to perform a change of coordinates $\tilde{\mathbf{x}} = (\mathbf{x} - \mathbf{x}_e)$, so that the equilibrium point of the system when expressed in the new coordinates is located at the origin $\tilde{\mathbf{x}} = 0$. Such a change of coordinates is implicitly assumed in Theorem 8.2. Thus, the requirement that the region $D$ includes the origin is important, since stabilization of the system will drive the states to the origin. The control input may become unbounded if either the first or second condition of Theorem 8.2 does not hold when the state variables are zero.

It is also important to note that a controller designed using the input-state approach may not be well behaved if either condition 1 or condition 2 breaks down at points within the state space of the system.

**Example 8.6** Consider a nonlinear system of the form

$$\begin{aligned} \dot{x}_1 &= f_1(x_2), \\ \dot{x}_2 &= -x_1^2 + u, \end{aligned} \tag{8.88}$$

where $f_1(0) = 0$. Here, $\mathbf{x} = [x_1 \quad x_2]^T$,

$$\mathbf{f}(\mathbf{x}) = \begin{bmatrix} f_1(x_2) \\ -x_1^2 \end{bmatrix} \quad \text{and} \quad \mathbf{g} = \begin{bmatrix} 0 \\ 1 \end{bmatrix}. \tag{8.89}$$

To check whether this system is feedback linearizeable, compute $\text{ad}_f(g)$ and then verify that conditions 1 and 2 above hold.

$$\text{ad}_f(g) = [\mathbf{f}, \mathbf{g}] = \underbrace{\nabla \mathbf{g} \cdot \mathbf{f}}_{=0} - \nabla \mathbf{f} \cdot \mathbf{g} = -\nabla \mathbf{f} \cdot \mathbf{g}$$

$$= -\begin{bmatrix} 0 & \dfrac{\partial f_1}{\partial x_2} \\ -2x_1 & 0 \end{bmatrix} \begin{bmatrix} 0 \\ 1 \end{bmatrix} = -\begin{bmatrix} \dfrac{\partial f_1}{\partial x_2} \\ 0 \end{bmatrix}. \tag{8.90}$$

It can be seen that $\mathbf{g}$ and $\text{ad}_f(g)$ are linearly independent by assembling a matrix of the vectors as

$$\begin{bmatrix} -\dfrac{\partial f_1}{\partial x_2} & 0 \\ 0 & 1 \end{bmatrix}. \tag{8.91}$$

Since this matrix has rank 2, its columns are linearly independent and condition 1 holds, as long as $\partial f_1/\partial x_2 \neq 0$. To check the involutivity of $\mathbf{g}$ and $\text{ad}_f(g)$, construct the Lie bracket

$$[\mathbf{g}, \text{ad}_f(g)] = \nabla \text{ad}_f(g) \cdot \mathbf{g} - \underbrace{\nabla \mathbf{g} \cdot \text{ad}_f(g)}_{=0} = \nabla \text{ad}_f(g) \cdot \mathbf{g},$$

$$= \begin{bmatrix} 0 & -\dfrac{\partial^2 f_1}{\partial x_2^2} \\ 0 & 0 \end{bmatrix} \begin{bmatrix} 0 \\ 1 \end{bmatrix} = \begin{bmatrix} -\dfrac{\partial^2 f_1}{\partial x_2^2} \\ 0 \end{bmatrix} = \dfrac{\dfrac{\partial^2 f_1}{\partial x_2^2}}{\dfrac{\partial f_1}{\partial x_2}} \text{ad}_f(g). \tag{8.92}$$

Since $[\mathbf{g}, \text{ad}_f(g)] = \alpha(\mathbf{x})\text{ad}_f(g)$, condition 2 holds and the set of vectors $\{\mathbf{g}, \text{ad}_f(g)\}$ is involutive. Therefore, it should be possible to control this system using input-state feedback linearization. We can select any $f_1(x_2)$ and find an associated transformation for feedback linearization.

(1)  Let $f_1(x_2) = e^{x_2} - 1$. In this case $\partial f_1/\partial x_2 = e_2^x \neq 0$ for any $x_2$. In this case, using (8.91) and (8.92), it can be seen that the linear independence and involutivity conditions hold globally. A transformation $z_1$ that satisfies (8.78) is sought, so that

$$\nabla z_1 \cdot \mathbf{g} = 0 \Rightarrow \begin{bmatrix} \dfrac{\partial z_1}{\partial x_1} & \dfrac{\partial z_1}{\partial x_2} \end{bmatrix} \begin{bmatrix} 0 \\ 1 \end{bmatrix} = \dfrac{\partial z_1}{\partial x_2} = 0. \tag{8.93}$$

Selecting $z_1 = z_1(x_1) = x_1$ will satisfy this condition. Then, the transformed state variable $z_2$ is found from (8.79)

$$z_2 = L_f z_1 = \begin{bmatrix} \dfrac{\partial z_1}{\partial x_1} & \dfrac{\partial z_1}{\partial x_2} \end{bmatrix} \begin{bmatrix} e^{x_2} - 1 \\ -x_1^2 \end{bmatrix} = \left( e^{x_2} - 1 \right). \tag{8.94}$$

Then, using (8.88) the transformed system becomes

$$\dot{z}_1 = z_2$$
$$\dot{z}_2 = \dfrac{d}{dt}\left( e^{x_2} - 1 \right) = e^{x_2}\dot{x}_2 = e^{x_2}\left( -x_1^2 + u \right) = v, \tag{8.95}$$

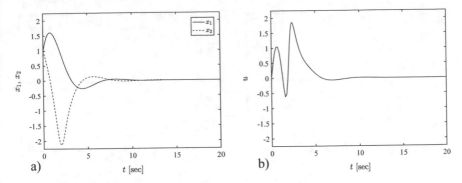

**Fig. 8.9** Response of control law (8.96) for $k_1 = 1$, $k_2 = 1$ with initial conditions $x_1(0) = 1$ and $x_2(0) = 1$

where $v$ is a linear control law in $z_1$, $z_2$, which can be taken as a proportional derivative controller $v = -k_1 z_1 - k_2 z_2$, for example. The control law $u$ is then given by

$$u = x_1^2 - \frac{(k_1 z_1 + k_2 z_2)}{e^{x_2}} = z_1^2 - \frac{(k_1 z_1 + k_2 z_2)}{z_2 + 1}. \tag{8.96}$$

The response of the controller with gains $k_1 = 1$, $k_2 = 1$ for the initial conditions $x_1(0) = 1$ and $x_2(0) = 1$ can be seen in Fig. 8.9. The controller stabilizes the states to the origin $(x_1, x_2) = (0, 0)$.

(2) Next, explore the use of input-state feedback linearization controller design when $f_1(x_2) = \sin(x_2)$ in (8.88). Since $\partial f_1/\partial x_2 = \cos(x_2) = 0$ when $x_2 = \pm\pi/2$, $x_2$ must always lie within the range $|x_2| < \pi/2$ to ensure that the control input is well-behaved. For the purposes of this example, we accomplish this through the appropriate selection of the initial conditions.

As above, selecting $z_1 = z_1(x_1) = x_1$ satisfies the condition

$$\nabla z_1 \cdot \mathbf{g} = 0 \Rightarrow \begin{bmatrix} \dfrac{\partial z_1}{\partial x_1} & \dfrac{\partial z_1}{\partial x_2} \end{bmatrix} \begin{bmatrix} 0 \\ 1 \end{bmatrix} = \frac{\partial z_1}{\partial x_2} = 0. \tag{8.97}$$

Then, the transformed state variable $z_2$ is found from (8.79)

$$z_2 = L_f z_1 = \begin{bmatrix} \dfrac{\partial z_1}{\partial x_1} & \dfrac{\partial z_1}{\partial x_2} \end{bmatrix} \begin{bmatrix} \sin(x_2) \\ -x_1^2 \end{bmatrix} = \sin(x_2). \tag{8.98}$$

Then, using (8.88) the transformed system becomes

$$\begin{aligned}
\dot{z}_1 &= z_2 \\
\dot{z}_2 &= \frac{d}{dt}\sin(x_2) = \cos(x_2)\dot{x}_2 = \cos(x_2)\left(-x_1^2 + u\right) = v,
\end{aligned} \tag{8.99}$$

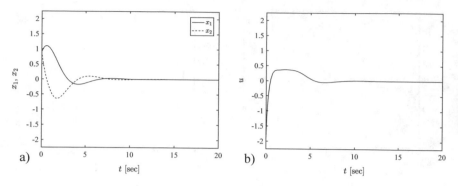

**Fig. 8.10** Response of control law (8.96) for $k_1 = 1$, $k_2 = 1$ with initial conditions $x_1(0) = 1$ and $x_2(0) = 1$

where $v$ is a linear control law in $z_1, z_2$. As above, let $v = -k_1 z_1 - k_2 z_2$. The control law $u$ is then given by

$$u = x_1^2 - \frac{(k_1 z_1 + k_2 z_2)}{\cos(x_2)} = z_1^2 - \frac{(k_1 z_1 + k_2 z_2)}{\cos\left[\sin^{-1}(z_2)\right]}. \qquad (8.100)$$

To ensure that the controller is well-behaved the initial conditions are selected so that at time $t = 0$, $|x_2(0)| < \pi/2$. Therefore, the initial states are taken as $x_1(0) = 0.9$ and $x_2(0) = 0.9$. The response of the controller with gains $k_1 = 1$ and $k_2 = 1$ can be seen in Fig. 8.10. The controller stabilizes the states to the origin $(x_1, x_2) = (0, 0)$.

□

---

### Guidelines for SISO Control Design via Input-State Feedback Linearization

1. Write the equations of motion in the form (8.67), i.e.

$$\dot{\mathbf{x}} = \mathbf{f}(\mathbf{x}) + \mathbf{g}(\mathbf{x})u.$$

2. Verify that the system satisfies the two conditions of Theorem 8.2 and that feedback linearization can be used. If these conditions hold, proceed to the next step. Otherwise, the system is not input-state feedback linearizeable and the use of other control techniques should be explored.
3. Find the transformation $\mathbf{z} = \mathbf{T}(\mathbf{x})$ and linearizing feedback control input $u$ that transforms the nonlinear system $\dot{\mathbf{x}} = \mathbf{f}(\mathbf{x}) + \mathbf{g}(\mathbf{x})u$ into a controllable linear system of the form (8.69),

$$\dot{\mathbf{z}} = \mathbf{A}\mathbf{z} + \mathbf{B}v.$$

(a) Take the output of the transformed system to be $h = z_1$ and identify a suitable function $z_1$, which satisfies the system of equations

$$L_{\mathrm{ad}_f^k \mathbf{g}} z_1 = 0, \quad k = 0, \ldots, n-2$$

$$L_{\mathrm{ad}_f^{n-1} \mathbf{g}} z_1 \neq 0,$$

where $\mathrm{ad}_f^0(\mathbf{g}) = \mathbf{g}$,

$$\mathrm{ad}_f(\mathbf{g}) = [\mathbf{f}, \mathbf{g}] = \nabla \mathbf{g} \cdot \mathbf{f} - \nabla \mathbf{f} \cdot \mathbf{g} = \frac{\partial \mathbf{g}}{\partial \mathbf{x}} \cdot \mathbf{f} - \frac{\partial \mathbf{f}}{\partial \mathbf{x}} \cdot \mathbf{g},$$

and

$$\mathrm{ad}_f^k(\mathbf{g}) = [\mathbf{f}, \mathrm{ad}_f^{k-1}(\mathbf{g})].$$

In principle, there are many possible solutions for $z_1$. In practice, it typically suffices to take one of the original state variables $x_i$, $i = 1, \ldots, n$ to be $z_1$, or if necessary, to use a simple additive combination of two or more of the original state variables.

(b) After a suitable $z_1$ is identified, the remaining transformed state variables $z_2, \ldots, z_n$ can be found inductively using (8.79),

$$z_{i+1} = L_f z_i = \nabla z_i \cdot \mathbf{f}, \quad i = 1, \ldots, n-1.$$

(c) Finally, the control input $u$ is found from (8.80) as

$$u = \frac{1}{L_g z_n} \left( v - L_f z_n \right),$$

where $L_g z_n = \nabla z_n \cdot \mathbf{g}$ and $L_f z_n = \nabla z_n \cdot \mathbf{f}$.

## 8.6 Input-Output Linearization

When certain output variables are of interest, as in tracking control problems, the state model is described by state and output equations. The system outputs are generally functions of the states. In vector form this is expressed as $\mathbf{y} = \mathbf{h}(\mathbf{x}) \in \mathbb{R}^m$, where $\mathbf{x} \in \mathbb{R}^n$. When $n > m$ some states are not observable or are not reachable (controllable). This is a situation which is commonly encountered in practice, for example in the control of underactuated vessels (see Chap. 9). In such a situation, linearizing the input-state equations may not also linearize the input-output map. However, it may still be possible to use feedback linearization to design a controller, provided that the unobservable/unreachable states are stable. The dynamics of the unobservable/unreachable states is known as the *zero dynamics*.

**Example 8.7** Consider the system

$$\dot{x}_1 = a \sin x_2$$
$$\dot{x}_2 = -x_1^2 + u$$

has an output $y = x_2$, then the change of variables and state feedback control

$$z_1 = x_1, \quad z_2 = a \sin x_2, \quad \text{and} \quad u = x_1^2 + \frac{1}{a \cos x_2} v$$

yield

$$\dot{z}_1 = z_2$$
$$\dot{z}_2 = v$$
$$y = \sin^{-1}\left(\frac{z_2}{a}\right)$$

While the state equation is linear, solving a tracking control problem for $y$ is still complicated by the nonlinearity of the output equation. Inspection of both the state and output equations in the $x$-coordinates shows that, if we use the state feedback control $u = x_1^2 + v$, we can linearize the input-output map from $u$ to $y$, which will be described by the linear model

$$\dot{x}_2 = v$$
$$y = x_2.$$

We can now proceed to solve the tracking control problem using linear control theory. This discussion shows that sometimes it is more beneficial to linearize the input-output map even at the expense of leaving part of the state equation nonlinear. In this case, the system is said to be input-output linearizable. One catch about input-output linearization is that the linearized input-output map may not account for all the dynamics of the system. In the foregoing example, the full system is described by

$$\dot{x}_1 = a \sin x_2$$
$$\dot{x}_2 = v$$
$$y = x_2$$

Note that the state variable $x_1$ is not connected to the output $y$. In other words, the linearizing feedback control has made $x_1$ unobservable from $y$. When we design tracking control, we should make sure that the variable $x_1$ is well behaved; that is, stable or bounded in some sense. A naive control design that uses only the linear input-output map may result in an ever-growing signal $x_1(t)$. For example, suppose we design a linear control to stabilize the output $y$ at a constant value $r$. Then, $x_1(t) =$

$x_1(0) + ta \sin r$ and, for $\sin r \neq 0$, $x_1(t)$ will grow unbounded. This internal stability issue will be addressed using the concepts of *relative degree* and *zero dynamics*. $\Box$

### 8.6.1 Relative Degree

The relative degree of a system is the lowest order derivative of the system output $y$ that explicitly depends on the control input $u$. Consider the single-input-single-output system

$$\dot{x} = f(x) + g(x)u$$
$$y = h(x)$$

$(8.101)$

where $f : D \to \mathbb{R}^n$, $g : D \to \mathbb{R}^n$, and $h$ are sufficiently smooth vector fields in a domain $D \subset \mathbb{R}^n$. The time derivative of the output $\dot{y}$ is given by

$$\dot{y} = \frac{dh}{dx}[f(x) + g(x)u] := L_f h(x) + L_g h(x)u.$$

If $L_g h(x) = 0$, then $\dot{y} = L_f h(x)$, which does not explicitly depend on $u$. If we calculate the second derivative of $y$, we obtain

$$\ddot{y} = y^{(2)} = \frac{\partial(L_f h)}{\partial x}[f(x) + g(x)u] = L_f^2 h(x) + L_g L_f h(x)u,$$

where the parenthesized superscript $(i)$ in $y^{(i)}$, for example, is used to represent the $i^{\text{th}}$-order time derivative of $y$. Once again, if $L_g L_f h(x) = 0$, then $y^{(2)} = L_f^2 h(x)$, which is not explicitly dependent on $u$. Repeating this process, we see that if $h(x)$ satisfies

$$L_g L_f^{i-1} h(x) = 0, \quad i = 1, 2, \ldots, r-1; \quad L_g L_f^{r-1} h(x) \neq 0$$

$u$ does not explicitly appear in the equations for $y, \dot{y}, \ldots, y^{(r-1)}$, but does finally appear with a nonzero coefficient in the equation for $y^{(r)}$,

$$y^{(r)} = L_f^r h(x) + L_g L_f^{r-1} h(x)u.$$

$(8.102)$

The foregoing equation shows that the system is input-output linearizable, since the state feedback control

$$u = \frac{1}{L_g L_f^{r-1} h(x)} \left[-L_f^r h(x) + v\right]$$

$(8.103)$

reduces the input-output map to

$$y^{(r)} = v,$$

$(8.104)$

which is a chain of $r$ integrators. In this case, the integer $r$ is called the *relative degree* of the system, according to the following definition:

**Definition 8.12** (*Relative Degree*) The nonlinear system (8.101) has relative degree $r$, $1 \leq r \leq n$, in a region $D_0 \subset D$ if

$$L_g L_f^{i-1} h(x) = 0, \quad i = 1, 2, \ldots, r - 1; \quad L_g L_f^{r-1} h(x) \neq 0$$

for all $x \in D_0$.

**Remark 8.3** When $r = n$, where $n$ is the order of the system, input-output linearization is equivalent to input-state linearization.

**Remark 8.4** (*Undefined relative degree*) Note that we require $L_g L_f^{r-1} h(x) \neq 0$ in order for the feedback control law (8.103) to be bounded. If an operating point $x_0$ exists where $L_g L_f^{r-1} h(x_0) = 0$, but $L_g L_f^{r-1} h(x) \neq 0$ at other points $x$ arbitrarily close to $x_0$, the relative degree of the system is *undefined* at $x_0$ [9]. In such cases it may be possible to find a change of coordinates in which the transformed system is stable. Here, we will only consider systems with a relative degree that can be defined.

**Example 8.8** Consider the controlled Van der Pol equation

$$\dot{x}_1 = x_2,$$
$$\dot{x}_2 = -x_1 + \varepsilon(1 - x_1^2)x_2 + u, \quad \varepsilon > 0,$$

with output $y = x_1$. Calculating the derivatives of the output, we obtain

$$\dot{y} = \dot{x}_1 = x_2,$$
$$\ddot{y} = \dot{x}_2 = -x_1 + \varepsilon(1 - x_1^2)x_2 + u, \quad \varepsilon > 0,$$

Hence, the system has relative degree two in $\mathbb{R}^2$. For the output $y = x_2$,

$$\dot{y} = -x_1 + \varepsilon(1 - x_1^2)x_2 + u$$

and the system has relative degree one in $\mathbb{R}^2$. For the output $y = x_1 + x_2^2$,

$$\dot{y} = x_2 + 2x_2[-x_1 + \varepsilon(1 - x_1^2)x_2 + u]$$

and the system has relative degree one in $D_0 = \{x \in \mathbb{R}^2 | x_2 \neq 0\}$. □

## 8.6.2 The Normal Form and Zero Dynamics

The nonlinear system (8.101) can be transformed into a *normal form* using the output $y$ and its time derivatives to order $(r - 1)$ when the relative degree of the system is $r < n$. Define the vector $\mu$ as

$$\mu := [\mu_1 \quad \mu_2 \quad \cdots \quad \mu_r]^T = [y \quad \dot{y} \quad \cdots \quad y^{(r-1)}]^T. \tag{8.105}$$

Then the normal form of (8.101) is given by

$$\dot{\mu} = \begin{bmatrix} \mu_2 \\ \vdots \\ \mu_r \\ a(\mu, \psi) + b(\mu, \psi)u \end{bmatrix}, \tag{8.106}$$

$$\dot{\psi} = w(\mu, \psi), \tag{8.107}$$

where the output is given by $y = \mu_1$ and, from (8.102), $a(\mu, \psi) = L_f^r h(x)$ and $b(\mu, \psi) = L_g L_f^{r-1} h(x)$. Note that $\dot{\mu}$ in (8.106) can be considered to be another representation of (8.102) and that the control input $u$ does not appear in (8.107).

As shown in Sect. 8.6.1 above, the control input (8.103) reduces the system (8.101) to a linear input-output relation of the form $y^{(r)} = v$ so that it is straightforward to design the input $v$ so that the output $y$ behaves as desired. However, such a control design can only account for the states of the system occurring in the first $r$ derivatives of the output. The dynamics of the system corresponding to the $(n - r)$ states which are unaccounted for are known as the *internal dynamics* and correspond to the $(n - r)$ equations represented by (8.107). In order for the closed loop system to be stable, the internal dynamics must remain bounded.

Generally, the internal dynamics depend on the output states $\mu$. However, the stability of the internal dynamics can be assessed by studying the stability of the system when the control input is selected as $u_0(x)$ so that $\mu = 0$. From (8.103), it can be seen that

$$u_0(\mathbf{x}) = -\frac{L_f^r h(x)}{L_g L_f^{r-1} h(x)}. \tag{8.108}$$

**Definition 8.13** (*Zero Dynamics*)  Assuming that $\mu(0) = 0$ and the control input is given by (8.108), the zero dynamics of (8.101) are given by

$$\dot{\mu} = 0, \tag{8.109}$$

$$\dot{\psi} = w(0, \psi), \tag{8.110}$$

From (8.106), it can be seen that the control input (8.108) can also be represented as a function of the internal states

$$u_0(\psi) = -\frac{a(0, \psi)}{b(0, \psi)}.$$

A linear system with stable zero dynamics is known as a *minimum phase* system [6, 9]. Conversely, a linear system with a zero in the right half of the $s$-plane (i.e. a zero with a positive real part) is known as a *nonminimum phase* system (see Sect. 3.1.3). The definition of minimum phase can be extended to nonlinear systems.

**Definition 8.14** (*Minimum Phase Systems*)  If the zero dynamics of (8.101) are asymptotically stable, the system is known as an asymptotically minimum phase system.

**Example 8.9**  Consider the controlled Van der Pol equation

$$\dot{x}_1 = x_2$$
$$\dot{x}_2 = -x_1 + \varepsilon(1 - x_1^2)x_2 + u$$
$$y = x_2$$

The system has relative degree one in $\mathbb{R}^2$. Taking $\mu = y$ and $\psi = x_1$, we see that the system is already in the normal form. The zero dynamics are given by $\dot{x}_1 = 0$, which does not have an asymptotically stable equilibrium point. Hence, the system is not minimum phase.                                                    □

To design a controller based on input-output linearization one first generally designs the controller based on the linear input-output relation and then checks whether or not the internal dynamics are stable.

### 8.6.3  Stabilization

To design a controller that will stabilize (8.101), let $v$ in (8.104) be

$$v = -k_{r-1}y^{(r-1)} - \cdots - k_1\dot{y} - k_0y, \tag{8.111}$$

where the control gains $k_i > 0$ are selected using pole placement such that the roots of the characteristic polynomial corresponding to the linearized input-output map

$$a(s) = s^r + k_{r-1}s^{(r-1)} + \cdots + k_1 s + k_0 \tag{8.112}$$

satisfy desired performance criteria (here $s$ is the Laplace transform variable). From (8.103), the corresponding control input $u$ is

$$u = \frac{1}{L_g L_f^{r-1} h(x)} \left[ -L_f^r h(x) - k_{r-1} y^{(r-1)} - \cdots - k_1 \dot{y} - k_0 y \right] \tag{8.113}$$

When the zero-dynamics of (8.101) are locally asymptotically stable, (8.113) yields a locally asymptotically stable closed loop system.

By applying Lyapunov control design techniques, globally asymptotically stable controller can be developed. To do so $\mu$ is treated as an input to the internal dynamics and $\psi$ as an output. The control design proceeds in two steps:

(1) A control law $\mu_0 = \mu_0(\psi)$ and an associated Lyapunov function $V_0$ that stabilizes the internal dynamics is identified. Since the reduced dynamics are of lower order, finding $\mu_0$ and $V_0$ is generally simpler that finding a control input and Lyapunov function for the original system (8.101).
(2) Identify a candidate Lyapunov function $V$ for the original system (8.101) as a modified version of $V_0$ and choose $v$ so that $V$ is a Lyapunov function for all of the closed loop dynamics.

Stabilization with this approach can often be used recursively on more complex systems, as well as on nonminimum phase systems.

### 8.6.4  Tracking

The simple stabilization controller developed with pole placement above can be extended for use on trajectory tracking controllers. Let $y_d(t)$ be a desired trajectory and define

$$\mu_d := [y_d \ \dot{y}_d \ \cdots \ y_d^{(r-1)}]^T, \tag{8.114}$$

where $\mu_d$ is smooth and bounded. Define the tracking errors as

$$\tilde{\mu}(t) = \mu(t) - \mu_d(t), \tag{8.115}$$

with $\mu(0) = \mu_d(0)$. Assuming that a solution to the equation $\dot{\psi}_d = w(\mu_d, \psi_d)$, where $\psi_d(0) = 0$, exists, is bounded and uniformly asymptotically stable, the tracking errors exponentially converge to zero when the control input is selected as

$$u = \frac{1}{L_g L_f^{r-1} \mu_1} \left[ -L_f^r \mu_1 + y_d^{(r)} - k_{r-1} \tilde{\mu}_r - \cdots - k_1 \tilde{\mu}_2 - k_0 \tilde{\mu}_1 \right], \tag{8.116}$$

**Fig. 8.11** The flow of a current of speed $U$ past a section of an underwater cable of cylindrical cross section. Vortex shedding gives rise to an oscillating lift force along the $y$ axis

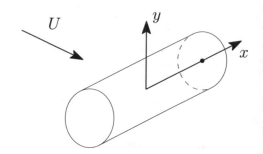

where the control gains $k_i > 0$ are selected such that the roots of the characteristic polynomial (8.112) are stable. Note that the control law (8.116) cannot be applied to nonminimum phase systems.

**Example 8.10** (*Matlab for Feedback Linearization*) Computation of the Lie brackets, distributions and state transformations using the methods of Sects. 8.5–8.6 can become quite involved. Use of the symbolic toolbox in Matlab can simplify the process. This is illustrated with an example.

Consider the vortex-shedding-induced motion of an underwater cable in a current Fig. 8.11. When the Reynolds number of the flow past a cable of circular cross section is greater than about 40, a series of vortices is shed from the cable in a flow structure known as a von Karman vortex street. The vortices are shed with an alternating direction of rotation, e.g. counter-clockwise, clockwise, counter-clockwise, clockwise, ..., counter-clockwise, clockwise. As each vortex is shed from the cable, it induces a change in the sign of the circulation of the flow around the cable which causes a "lift force" perpendicular to the direction of the current. Thus, the direction of the lift force also oscillates, causing a transverse motion of the cable, which is often referred to as *strumming*.

Consider the following (highly) simplified version of the cable strumming dynamics proposed by [8],

$$\ddot{q} = (1 - 25q^2)\dot{q} - q + \dot{y},$$
$$\ddot{y} = -(1 + 2\zeta\omega_n)\dot{y} - \omega_n^2 y + 25q, \tag{8.117}$$

where $y$ is the transverse displacement of the cable (see Fig. 8.11), $\zeta$ is a damping ratio, $\omega_n$ is the natural frequency of the motion of the cable in the transverse direction, and $q$ is related to the hydrodynamically generated oscillating lift force.

The first sub equation of (8.117) is a forced Van der Pol equation. The term $(1 - 25q^2)\dot{q}$ represents the effects of the vortex shedding and leads to a sustained limit cycle oscillation. When the lift force is small, $(1 - 25q^2)\dot{q} > 0$, so that energy is added to the system from the flow increasing $q$. If $q$ is too large, $(1 - 25q^2)\dot{q} < 0$, so that energy is extracted from the cable's transverse motion by the flow (in the form of energy extracted via viscous vortex shedding). Note that the tension in the cable serves as a restoring force, which pulls the cable back towards its equilibrium position.

Many underwater cables are designed with passive devices to help reduce strumming (see [1], for example). Let us explore whether it might be possible to mitigate strumming using feedback linearization, from a purely control engineering perspective. Questions of how to design and physically implement such a system are complex. Here, we will only explore whether a hypothetical control input exists. Add a control input $u$ to the dynamic equation for the transverse displacement of the cable to get

$$\ddot{q} = (1 - 25q^2)\dot{q} - q + \dot{y},$$
$$\ddot{y} = -(1 + 2\zeta\omega_n)\dot{y} - \omega_n^2 y + 25q + u. \tag{8.118}$$

First, we will check whether or not the system is feedback linearizable by testing the two criteria in Theorem 8.2. Equation (8.118) can be written in the form of (8.67) by taking

$$\mathbf{x} = [x_1 \ \ x_2 \ \ x_3 \ \ x_4]^T = [q \ \ \dot{q} \ \ y \ \ \dot{y}]^T \in \mathbb{R}^n, \tag{8.119}$$

where $n = 4$,

$$\mathbf{f(x)} = \begin{bmatrix} x_2 \\ (1 - 25x_1^2)x_2 - x_1 + x_4 \\ x_4 \\ -\omega_n^2 x_3 - (1 + 2\zeta\omega_n)x_4 + 25x_1 \end{bmatrix}, \quad \text{and} \quad \mathbf{g} = \begin{bmatrix} 0 \\ 0 \\ 0 \\ 1 \end{bmatrix} \tag{8.120}$$

so that

$$\dot{\mathbf{x}} = \mathbf{f(x)} + \mathbf{g}u. \tag{8.121}$$

In Matlab, we start by declaring symbolic variables for $x_1$, $x_2$, $x_3$, $x_4$, $\mathbf{f}$ and $\mathbf{g}$ and then defining $\mathbf{f}$ and $\mathbf{g}$.

```
syms x1 x2 x3 x4 f g zeta omega_n

f = [
    x2
    (1-25*x1^2)*x2 - x1 + x4
    x4
    -omega_n^2*x3 - (1 + 2*zeta*omega_n)*x4 + 25*x1
    ];

g = [0 0 0 1]';
```

In order to check the first condition of Theorem 8.2, i.e. that the vector fields $\{g, \text{ad}_f(g), \ldots, \text{ad}_f^{n-1}(g)\}$ are linearly independent, we can construct a matrix using the vector fields as columns with $n = 4$. If the matrix is full rank the columns are linearly independent, and thus, so are the vector fields. Recall that $\text{ad}_f^i(g)$ can be obtained from the Lie bracket of $\mathbf{f}$ and $\mathrm{ad}_f^{i-1}(g)$ for all $i$, so that $\mathrm{ad}_f(g) = [\mathbf{f}, \mathbf{g}]$, $\mathrm{ad}_f^2(g) = [\mathbf{f}, [\mathbf{f}, \mathbf{g}]]$ and $\mathrm{ad}_f^2(g) = [\mathbf{f}, [\mathbf{f}, [\mathbf{f}, \mathbf{g}]]]$. We will use this recursive property of $\mathrm{ad}_f^i(g)$ and the Jacobian-based definition of the Lie bracket in (8.39) in Matlab. Note that for this example $\mathbf{g}$ is constant so that its Jacobian is $\nabla \mathbf{g} = 0$. Using these ideas, the first condition of Theorem 8.2 can be checked in Matlab with the following code.

```
syms J adj1 adj2 adj3

J = jacobian(f,[x1 x2 x3 x4]);

adj1 = -J*g;
adj2 = -J*adj1;
adj3 = -J*adj2;

rankJ = rank([g adj1 adj2 adj3]);
```

Executing the code gives

$$
J = \nabla \mathbf{f} = \begin{bmatrix}
0 & 1 & 0 & 0 \\
-(50x_1x_2 + 1) & (1 - 25x_1^2) & 0 & 1 \\
0 & 0 & 0 & 1 \\
25 & 0 & -\omega_n^2 & -(2\omega_n\zeta + 1)
\end{bmatrix},
$$

$$
\mathrm{adj1} = \mathrm{ad}_f(g) = \begin{bmatrix} 0 \\ -1 \\ -1 \\ 2\omega_n\zeta + 1 \end{bmatrix}, \quad
\mathrm{adj2} = \mathrm{ad}_f^2(g) = \begin{bmatrix} 1 \\ -25x_1^2 - 2\omega_n\zeta \\ -2\omega_n\zeta - 1 \\ (2\omega_n\zeta + 1)^2 - \omega_n^2 \end{bmatrix},
$$

$$
\mathrm{adj3} = \mathrm{ad}_f^3(g) = \begin{bmatrix}
25x_1^2 + 2\omega_n\zeta \\
50x_1x_2 - (2\omega_n\zeta + 1)^2 + \omega_n^2 - (25x_1^2 - 1)(25x_1^2 + 2\omega_n\zeta) + 1 \\
\omega_n^2 - (2\omega_n\zeta + 1)^2 \\
\left[(2\omega_n\zeta + 1)^2 - \omega_n^2\right](2\omega_n\zeta + 1) - \omega_n^2(2\omega_n\zeta + 1) - 25
\end{bmatrix}
$$

and

```
rankJ = rank([g adj1 adj2 adj3]) = 4.
```

Since the rank of the matrix is 4, the vector fields $\{g, \mathrm{ad}_f(g), \ldots, \mathrm{ad}_f^{n-1}(g)\}$ are linearly independent so that the first condition of Theorem 8.2 is satisfied.

Next, we check whether or not the second condition of Theorem 8.2 holds, i.e. whether or not the distribution $\Delta = \mathrm{span}\{g, ad_f(g), ad_f^2(g)\}$ is involutive. To do this, according to Definition 8.10, we can take the Lie brackets of any vector field with the other two and check whether or not the result can be expressed as a linear combination of the original vector fields $\mathbf{g}$, $ad_f(g)$, and $ad_f^2(g)$. In practice this is often not too difficult as the result is often zero or one of the original vector fields, which implies that they are involutive. To check this in Matlab, since $\nabla \mathbf{g} = 0$, it is simplest to take the Lie brackets of $ad_f(g)$ and $ad_f^2(g)$ with $\mathbf{g}$, i.e. $[ad_f(g), \mathbf{g}]$ and $[ad_f^2(g), \mathbf{g}]$. This is implemented in Matlab with the following code.

```
lb1 = -jacobian(adj1,[x1 x2 x3 x4])*g;

lb2 = -jacobian(adj2,[x1 x2 x3 x4])*g;
```

Executing the code gives `lb1 = 0` and `lb2 = 0`, so that $\Delta$ is involutive and the second condition of Theorem 8.2 also holds. As both conditions of Theorem 8.2 are satisfied, the system is feedback linearizeable.

Next, we can use (8.79) to solve for the transformations (8.73) that transform the original system (8.121) into the form (8.71). As we would like to mitigate the transverse oscillations of the cable, we will take $z_1 = y = x_3$ and solve for the $z_2, z_3$ and $z_4$ recursively. This can be accomplished using the following Matlab code.

```
syms z1 z2 z3 z4

z1 = x3;
z2 = jacobian(z1,[x1 x2 x3 x4])*f;
z3 = jacobian(z2,[x1 x2 x3 x4])*f;
z4 = jacobian(z3,[x1 x2 x3 x4])*f;
```

The resulting transformations are given by

$$z_1 = x_3 = y, \quad z_2 = x_4 = \dot{y},$$

$$z_3 = 25x_1 - x_4(2\omega_n\zeta + 1) - \omega_n^2 x_3,$$

and

$$z_4 = 25x_2 + (2\omega_n\zeta + 1)\left[x_4(2\omega_n\zeta + 1) - 25x_1 + \omega_n^2 x_3\right] - \omega_n^2 x_4.$$

Now, the feedback linearized control input can be found using (8.85), using the following segment of Matlab code

```
syms unum uden nu
```

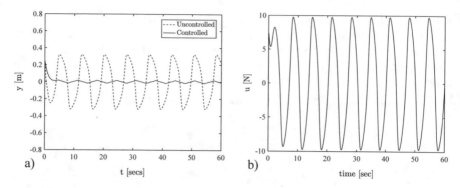

**Fig. 8.12** The controlled and uncontrolled strumming of an underwater cable: **a** transverse displacement of the cable, and **b** the control input

```
unum = nu - jacobian(z4,[x1 x2 x3 x4])*f;
uden = jacobian(z4,[x1 x2 x3 x4])*g;
```

to get

$$\text{unum} = v + 25x_1 - 25x_4 + \left[(2\omega_n\zeta + 1)^2 - \omega_n^2\right]\left[x_4(2\omega_n\zeta + 1) - 25x_1 + \omega_n^2 x_3\right]$$
$$+ 25x_2(25x_1^2 - 1) + x_2(50\omega_n\zeta + 25) - \omega_n^2 x_4(2\omega_n\zeta + 1),$$

and

$$\text{uden} = (2\omega_n\zeta + 1)^2 - \omega_n^2.$$

Then the control input using feedback linearization is obtained with $u = \text{unum}/\text{uden}$, where $v$ is a linear control input in terms of the transformed variables. For example, one could take $v = -k_p z_1 - k_d z_2$, where $k_p > 0$ and $k_d > 0$ are control gains.

As a case study, consider the flow-induced transverse motion of a cable with $\zeta = 0.707$ and $\omega_n = 2\pi$ rad/s in (8.118) with initial conditions $y(0) = 0.25$ m, and $q(0) = \dot{q}(0) = \dot{y}(0) = 0$ at $t = 0$. The uncontrolled and controlled response of the cable is shown in Fig. 8.12a. The controller gains were manually tuned to $k_p = 5000$ and $k_d = 5000$. It can be seen that the controller is able to substantially reduce the amplitude of the oscillations. The corresponding control input is shown in Fig. 8.12b.                                                                                     □

## Problems

**8.1** Consider the single-input single-output system

$$\dot{x} = f(x) + g(x)u,$$
$$y = h(x).$$

If $L_g h = 0$ and $L_g L_f h = 0$, compute the third derivative of the output with respect to time $y^{(3)}$ using Lie derivative notation, as in Sect. 8.6.1.

**8.2** Consider the moving mass roll stabilization system discussed in Example 6.3. The linearized equation of motion is

$$\ddot{\phi} + 2\zeta\omega_n\dot{\phi} + \omega_n^2 = \tau,$$

where $\phi$ is the roll angle of the vessel, $\zeta$ is the damping ratio, $\omega_n$ is the natural roll frequency and $\tau$ is a control input. Let $x_1 = \phi$ and $x_2 = \dot{\phi}$.

(a) Repeat the linear approach presented in Example 6.3.

  (i) Write the equation of motion in the state space form $\dot{x} = Ax + Bu$, where $x = [x_1 \ x_2]^T$ and $u = \tau$.
  (ii) Compute the reachability matrix $W_r$.

(b) Let us analyze this linear system using the nonlinear techniques presented in this chapter, to see how the linear and nonlinear analyses techniques are correlated.

  (i) Write the equation of motion in the vector form $\dot{x} = f(x) + g(x)u$, where $x = [x_1 \ x_2]^T$ and $u = \tau$.
  (ii) Compute the set of vector fields $\{g, \text{ad}_f(g)\}$. Are the vector fields linearly independent?
  (iii) Is the distribution $\Delta = \{g, \text{ad}_f(g)\}$ involutive?
  (iv) How does $\Delta$ compare with $W_r$?

(c) Examine the relative degree of the system.

  (i) Say we measure the roll angle. Let $y = x_1$. What is the relative degree of the system? If there are any internal dynamics, show whether or not they are stable.
  (ii) Say we can only measure the roll rate, i.e. $y = x_2$. What is the relative degree of the system? If there are any internal dynamics, show whether or not they are stable.

**8.3** Consider the longitudinal motion of an underwater vehicle. The equation of motion is

$$m\dot{u} = -c|u|u + \tau,$$

where $m$ is the mass of the vehicle, $c$ is the drag coefficient, $u$ is the surge speed and $\tau$ is the control input (thrust force generated by a propeller). Let us explore control of the vehicle's position $x_1$.

(a) Let $x_2 = \dot{x}_1 = u$ and $\mathbf{x} = [x_1 \ x_2]^T$. Write the equation of motion in the form
$\dot{\mathbf{x}} = \mathbf{f}(\mathbf{x}) + \mathbf{g}(\mathbf{x})\tau$.
(b) Compute the distribution $\Delta = \{\mathbf{g}, \mathrm{ad}_f(\mathbf{g})\}$. Is the system reachable?
(c) Assume we can measure the position of the vehicle. Take $y = x_1$. What is the relative degree of the system?
(d) Find a transformation $\mathbf{z} = T(\mathbf{x})$ that makes the system input-state linearizable.
(e) Solve for the associated linearizing control input of the form

$$\tau = \frac{1}{L_g z_n}\left(v - L_f z_n\right). \tag{8.122}$$

(f) Replace $\tau$ in the equation of motion $m\dot{u} = -c|u|u + \tau$ to show the linearized closed loop equation.

**8.4** Reconsider the longitudinal motion of an underwater vehicle examined in Exercise 8.3. The equation of motion is

$$m\dot{u} = -c|u|u + \tau,$$

where $m$ is the mass of the vehicle, $c$ is the drag coefficient, $u$ is the surge speed and $\tau$ is the control input (thrust force generated by a propeller). Let $x_2 = \dot{x}_1 = u$ and $\mathbf{x} = [x_1 \ x_2]^T$.

(a) Use the inverse dynamics approach (Sect. 8.2.2) to find a linearizing PID control input $\tau$ to track a desired time dependent position $x_d(t)$.
(b) Take $m = 40\,\mathrm{kg}$ and $c = 1.25\,\mathrm{kg/m}$. Create a numerical simulation using your result from part (a) to follow the trajectory

$$x_d(t) = \begin{cases} 5\,\mathrm{m}, & 0 < t \le 20\,\mathrm{s}, \\ 15\,\mathrm{m}, & 20 < t \le 60\,\mathrm{s}, \\ 0\,\mathrm{m}, & 60 < t \le 120\,\mathrm{s}. \end{cases}$$

Assume that the vehicle starts from rest, i.e. $x_1(0) = x_2(0) = 0$. Experiment with different proportional, derivative and integral gains to see how the response varies.
(c) Repeat part (b) when the model of the system is uncertain. I.e., let the value of the drag coefficient in the control input $c_\tau$ be slightly different from what it is in the system model $c$. How do the simulations vary when $c_\tau = \{0.5c, 0.75c, 1.25c, 1.5c\}$?

**8.5** The dynamic response of an AUV driven by a propeller with saturation and rate limits can be modeled as

$$m\frac{dv}{dt} = X_{v|v|}v|v| + u_{max}\tanh\left(\frac{u}{u_{max}}\right),$$
$$T_d\frac{du}{dt} = -u + u_c$$

where $m$ is the mass (and added mass) of the AUV, $v$ is the surge speed, $X_{v|v|} < 0$ is a drag coefficient, $u_{max}$ is the maximum thrust that can be generated by the propeller, $T_d$ is the time constant of the propeller rate limit, $u$ is the state of the propeller, and $u_c$ is the commanded thrust.

We would like to use feedback linearization to design a controller for the AUV. Before doing so, let us rescale the terms in these equations so that they can be written in a more concise form. Multiply both sides of the first equation by $T_d$ and divide the second equation by $u_{max}$ to get

$$T_d \frac{dv}{dt} = \frac{T_d X_{v|v|}}{m} v|v| + \frac{T_d u_{max}}{m} \tanh\left(\frac{u}{u_{max}}\right),$$

$$T_d \frac{d}{dt}\left(\frac{u}{u_{max}}\right) = -\frac{u}{u_{max}} + \frac{u_c}{u_{max}}.$$

Let $\tilde{t} := t/T_d$, $a := -T_d X_{v|v|}/m$, $b := T_d u_{max}/m$, $\tilde{u} := u/u_{max}$ and $\tilde{u}_c := u/u_{max}$. Then the equations become

$$\frac{dv}{d\tilde{t}} = -av|v| + b \tanh(\tilde{u}),$$

$$\frac{d\tilde{u}}{d\tilde{t}} = -\tilde{u} + \tilde{u}_c$$

Taking $x_1 := v$ and $x_2 := \tilde{u}$, the equations can be written as

$$\dot{x}_1 = -ax_1|x_1| + b \tanh x_2,$$
$$\dot{x}_2 = -x_2 + \tilde{u}_c,$$
$$y = x_1,$$

where the output $y$ is the desired speed. Since the nonlinear term $\tanh x_2$ appears in the first equation, it is not possible to select a control input $\tilde{u}_c$ that will linearize the equation for $x_1$. Therefore, in order to use feedback linearization, we will need to identify a change of coordinates that enable us to find a linearizing control input.

(a) What is the relative degree of the system?
(b) Write the equations of motion in the form $\dot{x} = f(x) + g(x)\tilde{u}_c$.
(c) Show that the system satisfies the two conditions of Theorem 8.2, so that it is feedback linearizable.
(d) Identify suitable transformed variables $z_1$ and $z_2$ for which the system can be represented in the linear form

$$\dot{z}_1 = z_2,$$
$$\dot{z}_2 = v.$$

(e) Let us construct a speed tracking controller, as discussed in Sect. 8.6.4. Take the desired speed to be $x_{1d}(t) = v_d(t)$. Then the acceleration and second derivative of the speed are given by $\dot{x}_{1d}$ and $\ddot{x}_{1d}$, respectively. Then, the transformed system can be represented as

$$\dot{\tilde{z}}_1 = \tilde{z}_2,$$
$$\dot{\tilde{z}}_2 = v.$$

Let $v = -k_1\tilde{z}_1 - k_2\tilde{z}_2$. Show that the resulting feedback linearizing tracking control input is

$$\tilde{u}_c = x_2 + \frac{\cosh^2(x_2)}{b}\{2a|x_1|[-ax_1|x_1| + b\tanh(x_2)] + \ddot{x}_{1d} - k_1\tilde{z}_1 - k_2\tilde{z}_2\}.$$

**8.6** Consider the fully-actuated control of an unmanned surface vessel. As usual, the equations of motion can be represented as

$$\dot{\eta} = \mathbf{J}(\eta)\mathbf{v},$$
$$\mathbf{M}\dot{\mathbf{v}} = -\mathbf{C}(\mathbf{v})\mathbf{v} - \mathbf{D}(\mathbf{v})\mathbf{v} + \boldsymbol{\tau},$$

where $\eta = [x\ y\ \psi]^T, \mathbf{v} = [u\ v\ r]^T, \mathbf{M} = \mathrm{diag}\{m_{11}, m_{22}, m_{33}\}, \mathbf{D} = \mathrm{diag}\{d_{11}, d_{22}, d_{33}\},$

$$\mathbf{J}(\eta) = \begin{bmatrix} \cos\psi & -\sin\psi & 0 \\ \sin\psi & \cos\psi & 0 \\ 0 & 0 & 1 \end{bmatrix},$$

$$\mathbf{C}(\mathbf{v}) = \begin{bmatrix} 0 & 0 & -m_{22}v \\ 0 & 0 & m_{11}u \\ m_{22}v & -m_{11}u & 0 \end{bmatrix},$$

and the control inputs are $\boldsymbol{\tau} = [\tau_x\ \tau_y\ \tau_\psi]^T$. Note that the expression $\mathrm{diag}\{a, b, c\}$ is used to represent a $3 \times 3$ diagonal matrix in which the only nonzero entries are the diagonal terms $a, b, c$, i.e.

$$\mathrm{diag}\{a, b, c\} := \begin{bmatrix} a & 0 & 0 \\ 0 & b & 0 \\ 0 & 0 & c \end{bmatrix}.$$

Rewrite the equations of motion in the form

$$\dot{\mathbf{x}} = \mathbf{f}(\mathbf{x}) + \mathbf{G}(\mathbf{x})\boldsymbol{\tau},$$
$$= \mathbf{f}(\mathbf{x}) + \mathbf{g}_1(\mathbf{x})\tau_x + \mathbf{g}_2(\mathbf{x})\tau_y + \mathbf{g}_3(\mathbf{x})\tau_\psi,$$

where

$$\mathbf{x} = \begin{bmatrix} \eta \\ \mathbf{v} \end{bmatrix}.$$

**8.7** (*Adapted from* [9]) A large electric motor is used to control the rapid positioning of an underwater arm. Assuming the arm is neutrally buoyant, the equations of motion for the system can be written as

$$I\ddot{q} = -I\alpha_1\dot{q}|\dot{q}| - I\alpha_0\dot{q} + \tau,$$
$$\dot{\tau} = -\lambda_2\lambda\tau + \lambda u,$$

where $q(t)$ is the motor shaft angle, $\tau$ is the torque on the motor shaft, $u$ is the control input (motor current), $I$ is the inertia of the motor shaft and propeller, $\alpha_1 > 0$ is a nonlinear drag coefficient, $\alpha_0 > 0$ is a linear drag coefficient associated with the back-emf generated in the electric motor, $\lambda$ is a scaling coefficient and $\lambda_1$ is the first-order time constant of the motor torque.

Perform input to state feedback linearization and design a controller for the system.

(a) Define $[x_1 \ x_2 \ x_3]^T = [q \ \dot{q} \ \tau]^T$ and rewrite the equations of motion in the vector form

$$\dot{\mathbf{x}} = \mathbf{f}(\mathbf{x}) + \mathbf{g}(\mathbf{x})u.$$

(b) Confirm that the controllability and involutivity conditions given in Theorem 8.2 are satisfied.
(c) Identify the first state $z_1$ using (8.78).
(d) Compute the state transformation and the input transformation that transform the equations of motion into a linear controllable system of the form (8.69).
(e) Select a linear control law for the transformed input $v$ and simulate the motion of the system using Matlab/Simulink. Tune the controller to obtain satisfactory performance.

**8.8** Let us use feedback linearization to reexamine the problem of speed control of a CTD cable reel, which was studied in Exercise 3.7 using a linear proportional controller. A tachometer is used as a feedback sensor for measuring the cable speed. Let the radius of the reel $r$ be 0.3 m when full and 0.25 m when empty. The rate of change of the radius is

$$\frac{dr}{dt} = -\frac{d_c^2\omega}{2\pi w},$$

where $d_c$ is the diameter of the cable, $w$ is the width of the reel and $\omega$ is the angular velocity of the reel.

The relation between reel angular velocity and the input torque from the controller $\tau$ is

$$I\frac{d\omega}{dt} = \tau.$$

The inertia $I$ changes when the cable is wound or unwound. The equation $I = 50r^4 - 0.1$ can be used to account for the changing inertia.

Rather than writing out all of the terms involved, the analysis can be simplified by expressing the dynamics of the system in the following form

$$\dot{r} = -a\omega,$$
$$\dot{\omega} = i(r)\tau,$$
$$y = r\omega,$$

where the output $y$ is the profiling speed of the CTD, the constant $a := d_c^2/(2\pi w)$, and the function $i(r) := 1/I = 1/(50r^4 - 0.1)$.

(a) Write the dynamics of the system in the form $\dot{\mathbf{x}} = \mathbf{f}(\mathbf{x}) + \mathbf{g}(\mathbf{x})\tau$.
(b) Compute the relative degree of the system to show whether input–state feedback linearization or input–output linearization must be used.
(c) Show that the system satisfies the two necessary conditions for feedback linearization of Theorem 8.2.
(d) Determine the transformed state variables $\mathbf{z}$ of the corresponding linearized system $\dot{\mathbf{z}} = \mathbf{A}\mathbf{z} + \mathbf{b}v$.
(e) Determine the linearizing feedback $\tau$ corresponding to a proportional-derivative controller for the linearized system, $v = -k_1 z_1 - k_2 z_2$.
(f) CTDs are often designed to be used with profiling speeds in the range $0.5 \le |v| \le 2$ m/s, with speeds of about 1 m/s generally being the best compromise between data quality and profile resolution. Take the desired cable speed to be $v_d = 1$ m/s. Simulate this system and obtain the step response of the speed over 100 s when the cable is let out. Assume $w = 0.5$ m, $d_c = 0.01$ m, and $r = 0.3$ m at $t = 0$ s. Tune $k_1$ and $k_2$ to obtain the best response.

**8.9** In Exercise 8.8, the portion of the CTD cable off of the reel was not accounted for. Let us assume that the cable weight can be modeled as being linearly proportional to the extended length of cable. This will require us to model the dynamics of the cable as

$$\dot{r} = -a\omega,$$
$$\dot{l} = r\omega,$$
$$\dot{\omega} = i(r)\tau + i(r)\rho gl,$$
$$y = r\omega,$$

where $\rho$ is the mass of the cable per unit length (e.g. kg/m) and $g$ is the acceleration of gravity. Take all other terms to be defined as in Exercise 8.8.

(a) Write the dynamics of the system in the form $\dot{\mathbf{x}} = \mathbf{f}(\mathbf{x}) + \mathbf{g}(\mathbf{x})\tau$.
(b) Using Matlab, as demonstrated in Example 8.10, show that the system no longer satisfies the two necessary conditions for feedback linearization of Theorem 8.2. Note that the computation of the Lie bracket shown in the example will need to be modified slightly, as the vector $\mathbf{g}(\mathbf{x})$ is not constant here.

**8.10** Let us reexamine the use of a tug to guide the path of a cargo ship, as discussed in Example 8.1. The equations of motion are given by

$$\dot{x}_1 = x_2,$$
$$\dot{x}_2 = \sin(x_1)\cos(x_1) + u\cos(x_1),$$
$$y = x_1,$$

where $x_1 := \psi$ and $x_2 := r$ are the yaw angle and yaw rate of the cargo ship. The control objective is to stabilize the system at $\mathbf{x} = [\psi \ r]^T = 0$.

(a) Write the dynamics of the system in the form $\dot{\mathbf{x}} = \mathbf{f}(\mathbf{x}) + \mathbf{g}(\mathbf{x})\tau$.

(b) Compute the relative degree of the system to show whether input–state feedback linearization or input–output linearization must be used.

(c) Show that the system satisfies the two necessary conditions for feedback linearization of Theorem 8.2.

(d) Determine the transformed state variables $\mathbf{z}$ of the corresponding linearized system $\dot{\mathbf{z}} = \mathbf{A}\mathbf{z} + \mathbf{b}v$.

(e) Determine the linearizing feedback $u$ corresponding to a proportional-derivative controller for the linearized system, $v = -k_1 z_1 - k_2 z_2$. How does this control input compare with the one found in Example 8.1?

# References

1. Blevins, R.D.: Flow-Induced Vibration (1977)
2. Brockett, R.W.: Nonlinear systems and differential geometry. Proc. IEEE **64**(1), 61–72 (1976)
3. De Luca, A., Oriolo, G.: Modelling and control of nonholonomic mechanical systems. In: Kinematics and Dynamics of Multi-body Systems, pp. 277–342. Springer, Berlin (1995)
4. Fossen, T.I.: Handbook of Marine Craft Hydrodynamics and Motion Control. Wiley, Hoboken (2011)
5. Isidori, A.: Nonlinear Control Systems. Springer Science & Business Media, Berlin (2013)
6. Khalil, H.K.: Nonlinear Systems, 3rd edn. Prentice Hall, Englewood Cliffs (2002)
7. Murray, R.M., Sastry, S.S.: Nonholonomic motion planning: steering using sinusoids. IEEE Trans. Autom. Control **38**(5), 700–716 (1993)
8. Skop, R.A., Balasubramanian, S.: A new twist on an old model for vortex-excited vibrations. J. Fluids Struct. **11**(4), 395–412 (1997)
9. Slotine, J.-J.E., Li, W.: Applied Nonlinear Control. Prentice-Hall, Englewood Cliffs (1991)
10. Spong, M.W., Hutchinson, S., Vidyasagar, M.: Robot Modeling and Control. Wiley, Hoboken (2006)

# Chapter 9
# Control of Underactuated Marine Vehicles

## 9.1 Introduction

When designing motion control systems for marine vehicles, it is important to distinguish between under-actuated vehicles and fully-actuated vehicles. Underactuated systems cannot be arbitrarily moved from some initial pose to some final pose because they cannot generate control forces/moments along every degree of freedom (missing actuators), or because actuator magnitude constraints, or rate limits, restrict their ability to accelerate in a certain direction. It is generally easier to control a fully-actuated vehicle, while under-actuation puts limitations on the control objectives that can be satisfied. Unfortunately, full-actuation is often impractical, because of considerations involving cost, weight or energy consumption, and so most marine vehicles are under-actuated. Here, we will consider three types of underactuated control problems: stabilization, path following and trajectory tracking.

The *stabilization* problem involves controlling a vehicle such that it moves from some initial pose $\eta_0$ to some final pose $\eta_f$, and the intermediate poses that the vehicle must pass through when traveling between $\eta_0$ and $\eta_f$ are unspecified. Station keeping a marine vehicle, so that it maintains a fixed heading and position, is an example of stabilization. Consider the problem of moving a surface vessel from $\eta_0$ to $\eta_f$, as shown in Fig. 9.1. One can imagine that if the vehicle is fully-actuated and has the ability to generate control forces in every desired direction, it will be able to smoothly move from $\eta_0$ to $\eta_f$ passing through any desired intermediate poses along the way. However, if the vehicle is underactuated, such that it can only generate a yaw moment and a surge force, a more complex strategy will need to be employed to go from $\eta_0$ to $\eta_f$. On the one hand, one might be able to plan the motion such that smooth surge control forces and smooth yaw moment control forces can be used. For example, a vehicle employing two thrusters transversely spaced at its stern can use differential thrust to simultaneously produce a yaw moment and surge force. With such a thruster configuration, a constant yaw moment could be applied to turn the vehicle clockwise, while at the same time a surge force is independently produced to first reverse, slow to

© Springer Nature Switzerland AG 2021
K. D. von Ellenrieder, *Control of Marine Vehicles*, Springer Series on Naval
Architecture, Marine Engineering, Shipbuilding and Shipping 9,
https://doi.org/10.1007/978-3-030-75021-3_9

**Fig. 9.1** Stabilization from
an initial pose
$\eta_0 = [x_0 \ \ y_0 \ \ \psi_0]^T$ to a final
pose $\eta_f = [x_f \ \ y_f \ \ \psi_f]^T$

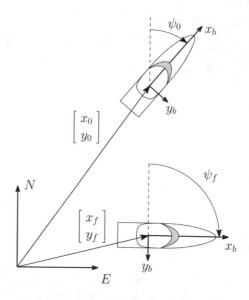

a stop and then smoothly accelerate forward to arrive at $\eta_f$. With this strategy, the use of smooth control inputs would require careful motion planning. A second possible strategy is to apply a series of time-dependent, discontinuous control forces to drive the vehicle from $\eta_0$ to $\eta_f$, perhaps moving the vehicle forwards and backwards several times before arriving at $\eta_f$. In fact, these are generally the two strategies most often used to solve the stabilization problem: (1) smooth control inputs with careful motion planning, or (2) time dependent, discontinuous control inputs.

The *path following* control problem requires a marine vehicle to follow a predefined path in an inertial coordinate system, without any time-based restrictions on the system's motion along the path. A path may consist of a series of straight-line segments connected by GPS waypoints, or may be defined by a smooth curve, such as a circle parameterized by a fixed distance to a GPS waypoint, for example. Consider the path following control of a surface vehicle, as shown in Fig. 9.2.

The vehicle starts at the position $\eta_0$ and it is desired that it to move along the line segment defined by the points WP1 and WP2. If the vehicle is fully-actuated, it can reach the path and start to follow it in any arbitrary fashion. For example, it can move sideways, along its sway axis, until its longitudinal axis is coincident with the line segment and then start to move forward. An underactuated vehicle must approach and follow the path in a different way. A common strategy for the path following control of underactuated vehicles is to formulate the control objective so that it can still be achieved when there are fewer actuators than degrees of freedom along which the vehicle can move. This often involves moving forward at a constant surge speed and steering so that the path of the vehicle tangentially approaches the path to be followed, somewhat like a driver might steer a car to follow lane markings when changing lanes. An approach to path following for surface vessels, known as Line-of-Sight Guidance will be presented in Sect. 9.6.

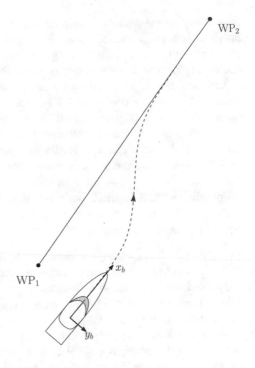

**Fig. 9.2** An underactuated vehicle approaching and following a path defined by two waypoints, WP1 and WP2

In *trajectory tracking control* a marine vehicle must track a desired, time varying pose $\eta_d(t)$. If a fully-actuated vehicle does not have any actuator constraints, it can track any arbitrary time-dependent trajectory. However, even when a vehicle possesses actuators that can produce forces in any desired direction, when the forces required to maintain the speed or acceleration required to catch up to and track a rapidly varying pose exceed the capabilities of the actuators, the vehicle is considered to be underactuated. Here, we will focus on the case when a marine vehicle is missing an actuator along one of the desired directions of motion and explore the use of its kinematics and dynamics to generate a dynamically feasible trajectory, which can then be tracked using a standard control approach (Sect. 9.7).

While the mathematical implementation of the strategies for stabilization via motion planning, path following control and trajectory tracking control to be presented in this chapter may appear quite different, a common thread runs through them. With each strategy, the control objectives are formulated taking the vehicle dynamics and kinematics into account, as well as the fact that the number of directions in which the actuators can move the vehicle is less than the number of directions in which the vehicle can be moved. This common thread is analogous to the control design philosophy presented in Sect. 4.3.2, where it is remarked that aiming to only fix the undesireable properties of a system's open loop response is preferable arbitrarily picking the locations of the closed loop system's poles without regard for the locations of the system's open loop poles, because it results in the use of smaller control efforts, which use less power and require smaller actuators. The analogous philosophy for the design of controllers for underactuated systems is that, whenever possible, it is better to incorporate the natural open loop dynamics of an underactuated system into the control design, rather than trying to overcome them with a controller, because doing so requires fewer actuators and a lower control effort.

Lastly, it should be noted that a broad range of powerful control design techniques exists for the control of fully-actuated systems, including optimal control, robust control and adaptive control. The use of these techniques is possible because the structure of the dynamic equations, which govern fully-actuated vehicles, possesses special mathematical properties that facilitate the control design, such as feedback linearizeability, passivity, matching conditions, and linear parameterizeability. The application of one or more of these special mathematical properties is often not possible when a system is underactuated. In addition, undesirable mathematical characteristics, such as higher relative degree and nonminimum phase behavior are typically present in underactuated systems.

**Example 9.1** As an example of a situation in which feedback linearization cannot be used because a system is underactuated, consider the design of a controller using body-fixed frame inverse dynamics, as shown in Sect. 8.2.1. Let us assume that the control input vector $\tau \in \mathbb{R}^n$ is produced on a marine vehicle using a combination of thrusters along $r$ independent directions. If we assume that the forces produced by these thrusters can be represented by a vector $\mathbf{u} \in \mathbb{R}^r$, we can write the control input as $\tau = \mathbf{B}\mathbf{u}$, where $\mathbf{B} \in \mathbb{R}^n \times \mathbb{R}^r$ is a matrix specifying the configuration of the $r$ thrusters. Then, the kinetic equation of motion of the system (1.59) becomes

$$\mathbf{M}\dot{\mathbf{v}} + \mathbf{C}(\mathbf{v})\mathbf{v} + \mathbf{D}(\mathbf{v})\mathbf{v} + \mathbf{g}(\eta) = \mathbf{B}\mathbf{u}. \tag{9.1}$$

Using a similar approach to that applied in Sect. 8.2.1, we seek the thruster control input forces

$$\mathbf{u} = \mathbf{u}(\eta, \mathbf{v}, t)$$

that linearize the closed loop system by taking

$$\mathbf{u} = \mathbf{B}^\dagger \left[ \mathbf{M}\mathbf{a}_b + \mathbf{C}(\mathbf{v})\mathbf{v} + \mathbf{D}(\mathbf{v})\mathbf{v} + \mathbf{g}(\eta) \right], \tag{9.2}$$

where

$$\mathbf{B}^\dagger := \mathbf{B}^T (\mathbf{B}\mathbf{B}^T)^{-1} \tag{9.3}$$

is the *Moore-Penrose pseudo-inverse* of $\mathbf{B}$. A pseudo-inverse matrix is similar to an inverse matrix, but can be also defined for non-square matrices. Then, using (9.2), (9.1) can be reduced to

$$\dot{\mathbf{v}} = \mathbf{a}_b,$$

where $\mathbf{a}_b$ can be thought of as the commanded acceleration of the vehicle in the body-fixed frame.

However, here we have a potential problem. If

$$\text{rank} \, [\mathbf{B}] < \dim [\mathbf{v}]$$

the pseudo-inverse of the control input matrix $\mathbf{B}$ does not exist and use of the standard inverse dynamics feedback linearization approach cannot be used to design a controller. This can happen, for example, if the vehicle has a degree of freedom along which any combination of the available thrusters cannot produce a control force. $\square$

## 9.2 The Terminology of Underactuated Vehicles

In order to analyze the motion and control of underactuated vehicles, knowledge of the following basic notation and terminology is needed.

**Definition 9.1** (*Configuration Space*) The configuration of a marine vehicle specifies the location of every point on the vehicle. The $n$-dimensional *configuration space* is the set of all configurations, i.e the set of all possible positions and orientations that a vehicle can have, possibly subject to external constraints.

**Definition 9.2** (*Degrees of Freedom—DOF*) A marine vehicle is said to have $n$ *degrees of freedom* if its configuration can be minimally specified by $n$ parameters. Therefore, the number of DOF is the same as the number of dimensions of the configuration space. The set of DOF is the set of independent displacements and rotations that completely specify the displaced position and orientation of the vehicle. A rigid body, such as a marine vehicle, that can freely move in three dimensions has six DOF: three translational DOF (linear displacements) and three rotational DOF (angular displacements).

When simulating the motion of a vehicle with 6 DOF, a system of 12 first-order, ordinary differential equations are needed—6 for the kinematic relations (1.58) and 6 for the kinetic equations of motion (1.59). The *order of a system* of the system of equations required to model a vehicle's motion is two times the number of its degrees of freedom $n$.

**Definition 9.3** (*Number of Independent Control Inputs*) The *number of independent control inputs* $r$ is the number of *independently controlled directions* in which a vehicle's actuators can generate forces/moments.

**Definition 9.4** (*Underactuated Marine Vehicles*) A marine vehicle is *underactuated* if it has fewer control inputs than generalized coordinates ($r < n$).

**Definition 9.5** (*Fully Actuated Marine Vehicles*) A marine vehicle is *fully actuated* if the number of control inputs is equal to, or greater than, the number of generalized coordinates ($r \geq n$).

**Definition 9.6** (*Workspace*) An underactuated vehicle can only produce independent control forces in $r < n$ directions. Therefore, when developing a controller, it makes sense to explore whether or not a space of fewer dimensions $m < n$ might exist in which the vehicle can be suitably controlled. Following [10], we define the workspace as the reduced space of dimension $m < n$ in which the control objective is defined.

In the parlance of feedback linearization, the processes of stabilization, path following control and trajectory tracking generally involve input–output linearization (Sect. 8.6). For these processes, the controlled outputs are the pose or velocity of a vehicle. Underactuated vehicles have more states in their configuration space $n$ than independent control inputs $r$. Because of this, some vehicle states are uncontrollable (unreachable). Thus, while it is possible to design a motion control system for an underactuated marine vehicle when its workspace is fully-actuated $m = r$, one must ensure that the zero dynamics of the closed loop system are stable when the dimension of the configuration space is reduced to that of the workspace. The uncontrolled equations of motion will appear as $k$ dynamic constraints that must have bounded solutions in order to prevent the system from becoming unstable.

## 9.3   Motion Constraints

In general, the constraints affecting a system can arise from both input and state constraints. Examples of dynamic constraints include missing actuators, but could also be caused by the magnitude or rate limitations of those actuators which are present. Constraints can also arise because of a physical barrier in the environment, e.g. a ship is generally constrained to move along the free surface of the water.

Constraints, which depend on both vehicle state and inputs can be expressed in the form

$$h(\boldsymbol{\eta}, \mathbf{u}, t) \geq 0. \tag{9.4}$$

Often the constraints are separated into those that depend only on the input (e.g. actuator constraints, also called *input constraints*) $h(\mathbf{u}) \geq 0$ and those that depend on the vehicle pose $h(\boldsymbol{\eta}) \geq 0$, which are known as *state constraints*. As $\boldsymbol{\eta}$ is an $n$-dimensional vector, if $k$ geometric constraints exist,

$$h_i(\boldsymbol{\eta}) \geq 0, \quad i = 1, \ldots, k \tag{9.5}$$

the possible motions of the vehicle are restricted to an $(n - k)$-dimensional submanifold (space). Thus, the state constraints reduce the dimensionality of the system's available state space.

When the constraints have the form

$$h_i(\eta) = 0, \quad i = 1, \ldots, k < n, \tag{9.6}$$

they are known as *holonomic constraints*.

System constraints that depend on both the pose and its first time derivative are *first order constraints*

$$h_i(\eta, \dot{\eta}) = 0, \quad i = 1, \ldots, k < n, \tag{9.7}$$

and are also called *kinematic constraints*, or *velocity constraints*. First order kinematic constraints limit the possible motions of a vehicle by restricting the set of velocities $\dot{\eta}$ that can be obtained in a given configuration. These constraints are usually encountered in the form

$$\mathbf{A}^T(\eta)\dot{\eta} = 0. \tag{9.8}$$

Holonomic constraints of the form (9.6) imply kinematic constraints of the form

$$\nabla h \cdot \dot{\eta} = 0. \tag{9.9}$$

However, the converse is not true. Kinematic constraints of the form (9.8) cannot always be integrated to obtain constraints of the form (9.6). When this is true, the constraints (and the system) are *nonholonomic*.

Note that (9.7) is a first order constraint. However, in many underactuated systems, including marine systems, the nonholonomic (non integrable) constraints are usually of second order and involve the acceleration of the system. They can be represented in the form

$$h_i(\eta, \dot{\eta}, \ddot{\eta}) = 0, \quad i = 1, \ldots, k < n. \tag{9.10}$$

Nonholonomic constraints limit the mobility of a system in a completely different way from holonomic constraints. A nonholonomic constraint does not restrain the possible configurations of the system, but rather how those configurations can be reached. While nonholonomic constraints confine the velocity or acceleration of a system to an $m = (n - k)$ dimensional subspace (the workspace), the entire $n$ dimensional configuration space of the system can still be reached. Instead, each holonomic constraint reduces the number of degrees of freedom of a system by one, so that the motion of a holonomic system with $k$ constraints is constrained to an $(n - k)$ dimensional subset of the full $n$ dimensional configuration space [8].

## 9.4   The Dynamics of Underactuated Marine Vehicles

As shown in [30], the equations of motion of an underactuated marine vehicle can be expressed in the form

$$\mathbf{M}_\eta(\boldsymbol{\eta})\ddot{\boldsymbol{\eta}} = -\mathbf{C}_\eta(\mathbf{v}, \boldsymbol{\eta})\dot{\boldsymbol{\eta}} - \mathbf{D}_\eta(\mathbf{v}, \boldsymbol{\eta})\dot{\boldsymbol{\eta}} - \mathbf{g}_\eta(\boldsymbol{\eta}) + \mathbf{B}_\eta(\boldsymbol{\eta})\mathbf{u}, \tag{9.11}$$

where $\mathbf{M}_\eta(\boldsymbol{\eta})$, $\mathbf{C}_\eta(\mathbf{v}, \boldsymbol{\eta})$, $\mathbf{D}_\eta(\mathbf{v}, \boldsymbol{\eta})$ and $\mathbf{g}_\eta(\boldsymbol{\eta})$ are the inertia tensor, Coriolis/centripetal force tensor, drag tensor and hydrostatic restoring forces, respectively, represented in an Earth-fixed coordinate system. The tensor $\mathbf{B}_\eta(\boldsymbol{\eta}) \in \mathbb{R}^n \times \mathbb{R}^r$ specifies the configuration of the $r < n$ actuator input forces on the vehicle. The relationships between $\mathbf{M}$, $\mathbf{C}(\mathbf{v})$, $\mathbf{D}(\mathbf{v})$ and $\mathbf{g}(\boldsymbol{\eta})$ in the body-fixed coordinate system and $\mathbf{M}_\eta(\boldsymbol{\eta})$, $\mathbf{C}_\eta(\mathbf{v}, \boldsymbol{\eta})$, $\mathbf{D}_\eta(\mathbf{v}, \boldsymbol{\eta})$ and $\mathbf{g}_\eta(\boldsymbol{\eta})$ in the Earth-fixed coordinate system are shown in Table 9.1 [10], where the matrix $\mathbf{B}$ is the *control input configuration matrix* in body-fixed coordinates.

Taking the time derivative of the kinematic equation (1.58) gives

$$\ddot{\boldsymbol{\eta}} = \dot{\mathbf{J}}(\boldsymbol{\eta})\mathbf{v} + \mathbf{J}(\boldsymbol{\eta})\dot{\mathbf{v}}. \tag{9.12}$$

Using this latter result, the first term in (9.11) can be rewritten as

$$\begin{aligned} \mathbf{M}_\eta(\boldsymbol{\eta})\ddot{\boldsymbol{\eta}} &= \mathbf{M}_\eta(\boldsymbol{\eta}) \left[ \dot{\mathbf{J}}(\boldsymbol{\eta})\mathbf{v} + \mathbf{J}(\boldsymbol{\eta})\dot{\mathbf{v}} \right], \\ &= \mathbf{J}^{-T}(\boldsymbol{\eta})\mathbf{M}\mathbf{J}^{-1}(\boldsymbol{\eta}) \left[ \dot{\mathbf{J}}(\boldsymbol{\eta})\mathbf{v} + \mathbf{J}(\boldsymbol{\eta})\dot{\mathbf{v}} \right], \\ &= \mathbf{J}^{-T}(\boldsymbol{\eta})\mathbf{M}\mathbf{J}^{-1}(\boldsymbol{\eta})\dot{\mathbf{J}}(\boldsymbol{\eta})\mathbf{v} + \mathbf{J}^{-T}(\boldsymbol{\eta})\mathbf{M}\dot{\mathbf{v}}, \end{aligned} \tag{9.13}$$

where the definition of $\mathbf{M}_\eta(\boldsymbol{\eta})$ in Table 9.1 has been applied. Using (9.13) and the relations in Table 9.1, (9.11) can be expressed as

$$\begin{aligned} \mathbf{J}^{-T}\mathbf{M}\mathbf{J}^{-1}\dot{\mathbf{J}}\mathbf{v} + \mathbf{J}^{-T}\mathbf{M}\dot{\mathbf{v}} = &-\mathbf{J}^{-T}[\mathbf{C}(\mathbf{v}) - \mathbf{M}\mathbf{J}^{-1}\dot{\mathbf{J}}]\mathbf{J}^{-1}\mathbf{J}\mathbf{v} \\ &-\mathbf{J}^{-T}\mathbf{D}(\mathbf{v})\mathbf{J}^{-1}\mathbf{J}\mathbf{v} - \mathbf{J}^{-T}\mathbf{g}(\boldsymbol{\eta}) + \mathbf{J}^{-T}\mathbf{B}\mathbf{u}, \end{aligned} \tag{9.14}$$

**Table 9.1** Relations between the inertia tensor, Coriolis/centripetal acceleration tensor, drag tensor and hydrostatic force vector in Earth-fixed and body-fixed coordinate systems

$$\mathbf{M}_\eta(\boldsymbol{\eta}) := \mathbf{J}^{-T}(\boldsymbol{\eta})\mathbf{M}\mathbf{J}^{-1}(\boldsymbol{\eta})$$

$$\mathbf{C}_\eta(\mathbf{v}, \boldsymbol{\eta}) := \mathbf{J}^{-T}(\boldsymbol{\eta})[\mathbf{C}(\mathbf{v}) - \mathbf{M}\mathbf{J}^{-1}(\boldsymbol{\eta})\dot{\mathbf{J}}(\boldsymbol{\eta})]\mathbf{J}^{-1}(\boldsymbol{\eta})$$

$$\mathbf{D}_\eta(\mathbf{v}, \boldsymbol{\eta}) := \mathbf{J}^{-T}(\boldsymbol{\eta})\mathbf{D}(\mathbf{v})\mathbf{J}^{-1}(\boldsymbol{\eta})$$

$$\mathbf{g}_\eta(\boldsymbol{\eta}) := \mathbf{J}^{-T}(\boldsymbol{\eta})\mathbf{g}(\boldsymbol{\eta})$$

$$\mathbf{B}_\eta(\boldsymbol{\eta}) := \mathbf{J}^{-T}(\boldsymbol{\eta})\mathbf{B}$$

where the explicit functional dependence of $\mathbf{J}(\eta)$ and $\mathbf{B}(\eta)$ upon $\eta$ is omitted for brevity. Eliminating the term $\mathbf{J}^{-T}\mathbf{M}\mathbf{J}^{-1}\dot{\mathbf{J}}\mathbf{v}$ from both the left and right hand sides of (9.14) and multiplying the remaining terms by $\mathbf{J}^T$ gives

$$\mathbf{M}\dot{\mathbf{v}} = -\mathbf{C}(\mathbf{v})\mathbf{v} - \mathbf{D}(\mathbf{v})\mathbf{v} - \mathbf{g}(\eta) + \mathbf{B}\mathbf{u}. \tag{9.15}$$

The $k = (n - m)$ components of the unactuated dynamics in (9.15) can be used to define a set of $k$ constraints on the accelerations. As shown by [30], for marine vehicles these constraints are non integrable second order constraints.

Multiplying all of the terms in this latter expression by $\mathbf{M}^{-1}$ we get the acceleration in the body fixed coordinate system

$$\dot{\mathbf{v}} = -\mathbf{M}^{-1}\left[\mathbf{C}(\mathbf{v})\mathbf{v} + \mathbf{D}(\mathbf{v})\mathbf{v} + \mathbf{g}(\eta) - \mathbf{B}\mathbf{u}\right]. \tag{9.16}$$

If we define a new state vector

$$\mathbf{x} := \begin{bmatrix} \eta \\ \mathbf{v} \end{bmatrix} \in \mathbb{R}^{2n}, \tag{9.17}$$

the following system of equations (of order $2n$) can be attained using the kinematic equations (1.58) and (9.16)

$$\dot{\mathbf{x}} = \begin{bmatrix} \mathbf{J}(\eta)\mathbf{v} \\ -\mathbf{M}^{-1}\left[\mathbf{C}(\mathbf{v})\mathbf{v} + \mathbf{D}(\mathbf{v})\mathbf{v} + \mathbf{g}(\eta)\right] \end{bmatrix} + \begin{bmatrix} \mathbf{0}_{n \times n} \\ \mathbf{M}^{-1} \end{bmatrix} \mathbf{B}\mathbf{u}. \tag{9.18}$$

Equation (9.18) can be written variously as

$$\begin{aligned} \dot{\mathbf{x}} &= \mathbf{f}(\mathbf{x}) + \mathbf{G}\mathbf{B}\mathbf{u}, \\ &= \mathbf{f}(\mathbf{x}) + \mathbf{G}\boldsymbol{\tau}, \\ &= \mathbf{f}(\mathbf{x}) + \sum_{i=1}^{r} \mathbf{g}_i \tau_i, \end{aligned} \tag{9.19}$$

where

$$\mathbf{f}(\mathbf{x}) = \begin{bmatrix} \mathbf{J}(\eta)\mathbf{v} \\ -\mathbf{M}^{-1}\left[\mathbf{C}(\mathbf{v})\mathbf{v} + \mathbf{D}(\mathbf{v})\mathbf{v} + \mathbf{g}(\eta)\right] \end{bmatrix} \in \mathbb{R}^{2n}, \quad \mathbf{G}(\mathbf{x}) = \begin{bmatrix} \mathbf{0} \\ \mathbf{M}^{-1} \end{bmatrix}, \tag{9.20}$$

the vectors $\mathbf{g}_i \in \mathbb{R}^{2n}$ are the columns of the matrix $\mathbf{G} \in \mathbb{R}^{2n \times n}$, and the $\tau_i$ are the individual components of the control input vector $\boldsymbol{\tau} \in \mathbb{R}^r$.

As discussed at the end of Sect. 9.3, when a vehicle is nonholonomic it should be possible to reach any desired configuration in its $n$ dimensional configuration space, even though the velocities or accelerations are confined to an $m < n$ workspace. Here, we will assume that the vehicle is fully-actuated in its workspace, so that $r = m$. Think of the control inputs $\tau_i$ in (9.19) as a set of steering commands. The central question is, whether or not it is possible to reach any desired configuration by using the inputs to suitably steer the vehicle.

### 9.4.1   The Dynamics of Underactuated Surface Vessels

Recall from Sect. 1.5.1 that in three DOF, the kinematic equations (1.58), reduce to

$$\dot{\eta} = \mathbf{R}(\psi)\mathbf{v}, \qquad (9.21)$$

where $\psi$ is the heading angle of the vehicle, $\mathbf{R}(\psi)$ is the transformation matrix from the body-fixed system to the a North-East-Down (NED) inertial coordinate system, which is given by

$$\mathbf{R}(\psi) := \begin{bmatrix} \cos\psi & -\sin\psi & 0 \\ \sin\psi & \cos\psi & 0 \\ 0 & 0 & 1 \end{bmatrix} \in SO(3), \qquad (9.22)$$

and

$$\eta := \begin{bmatrix} x_n \\ y_n \\ \psi \end{bmatrix} \in \mathbb{R}^2 \times \mathcal{S}, \quad \text{and} \quad \mathbf{v} := \begin{bmatrix} u \\ v \\ r \end{bmatrix} \in \mathbb{R}^3 \qquad (9.23)$$

are the position and orientation (pose) vector and velocity vector (in body-fixed coordinates), respectively (see Fig. 9.3). The variables appearing in (9.23) include position Northward $x_n$, position Eastward $y_n$, surge speed $u$, sway speed $v$ and yaw rate $r$.

Consider the kinetic equation of motion (1.59), which is reproduced here

$$\mathbf{M}\dot{\mathbf{v}} + \mathbf{C}(\mathbf{v})\mathbf{v} + \mathbf{D}(\mathbf{v})\mathbf{v} = \tau. \qquad (9.24)$$

Assuming that the origin of the body-fixed coordinate system is the located at the center of mass of the vessel and the vehicle is moving through still water,

$$\mathbf{M} := \begin{bmatrix} m - X_{\dot{u}} & 0 & 0 \\ 0 & m - Y_{\dot{v}} & -Y_{\dot{r}} \\ 0 & -N_{\dot{v}} & I_z - N_{\dot{r}} \end{bmatrix} = \begin{bmatrix} m_{11} & 0 & 0 \\ 0 & m_{22} & m_{23} \\ 0 & m_{32} & m_{33} \end{bmatrix}, \qquad (9.25)$$

**Fig. 9.3**  3 DOF
maneuvering coordinate
system definitions

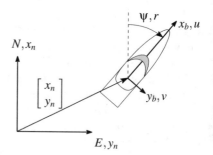

$$\mathbf{C}(\mathbf{v}) := \begin{bmatrix} 0 & 0 & -m_{22}v - \dfrac{(m_{23}+m_{32})}{2}r \\ 0 & 0 & m_{11}u \\ m_{22}v + \dfrac{(m_{23}+m_{32})}{2}r & -m_{11}u & 0 \end{bmatrix}, \quad (9.26)$$

$$\mathbf{D}_{nl}(\mathbf{v}) := -\begin{bmatrix} X_{u|u|}|u| & 0 & 0 \\ 0 & Y_{v|v|}|v| + Y_{v|r|}|r| & Y_{r|v|}|v| + Y_{r|r|}|r| \\ 0 & N_{v|v|}|v| + N_{v|r|}|r| & N_{r|v|}|v| + N_{r|r|}|r| \end{bmatrix}, \quad (9.27)$$

$$\mathbf{D}_l := -\begin{bmatrix} X_u & 0 & 0 \\ 0 & Y_v & Y_r \\ 0 & N_v & N_r \end{bmatrix}. \quad (9.28)$$

where $m$ is the mass of the vehicle, $I_z$ is the mass moment of inertia about the $z_b$ axis of the vehicle, and $X_{u|u|} < 0$ is the drag coefficient along the longitudinal axis of the vehicle.

Here, the Society of Naval Architects and Marine Engineers (SNAME) [27] nomenclature is used to represent the hydrodynamic coefficients (added mass and drag), which give rise to the forces and moments on the vessel in the body-fixed frame. The $X$ coefficients give rise to forces in the surge direction, the $Y$ coefficients give rise to forces in the sway direction and the $N$ coefficients give rise to moments about the yaw axis, as indicated by the vector of forces and moments (1.9). The subscript(s) on each coefficient correspond to the velocities of the vessel in the body-fixed frame, as indicated by (1.8), and denote(s) the motion which gives rise to the corresponding force/moment. For example, the coefficient $X_{\dot{u}}$ characterizes the surge force arising from acceleration in the surge direction (i.e. it is an *added mass* term).

The total drag $\mathbf{D}(\mathbf{v})\mathbf{v}$ is obtained by combining the nonlinear drag $\mathbf{D}_{nl}(\mathbf{v})\mathbf{v}$ with the linear drag $\mathbf{D}_l\mathbf{v}$, as

$$\mathbf{D}(\mathbf{v})\mathbf{v} = [\mathbf{D}_{nl}(\mathbf{v}) + \mathbf{D}_l]\,\mathbf{v} = \begin{bmatrix} d_x \\ d_y \\ d_\psi \end{bmatrix}. \quad (9.29)$$

Then, expanding terms, the equations of motion can be rewritten component-wise as

$$m_{11}\dot{u} - m_{22}vr - \left(\dfrac{m_{23}+m_{32}}{2}\right)r^2 + d_x = \tau_x, \quad (9.30)$$

$$m_{22}\dot{v} + m_{23}\dot{r} + m_{11}ur + d_y = 0, \quad (9.31)$$

$$m_{32}\dot{v} + m_{33}\dot{r} + (m_{22} - m_{11})uv + \left(\dfrac{m_{23}+m_{32}}{2}\right)ur + d_\psi = \tau_\psi. \quad (9.32)$$

Let $\mathbf{x} := [x_n \ y_n \ \psi \ u \ v \ r]^T$. We can write the combined kinematic and dynamic equations in the form (9.19) by first solving for $\dot{u}$, $\dot{v}$ and $\dot{r}$ from (9.30)–(9.32) and then assembling the resulting equations into vector form. From (9.30) we have

$$
\begin{aligned}
\dot{u} &= \frac{1}{m_{11}} \left[ m_{22} vr + \left( \frac{m_{23} + m_{32}}{2} \right) r^2 - d_x \right] + \frac{\tau_x}{m_{11}}. \\
&= f_x + \frac{\tau_x}{m_{11}}.
\end{aligned}
\tag{9.33}
$$

Dividing (9.31) by $m_{22}$ and (9.32) by $m_{33}$, they can be written as

$$
\begin{aligned}
\dot{v} + \frac{m_{23}}{m_{22}} \dot{r} &= -\frac{1}{m_{22}} \left[ m_{11} ur + d_y \right], \\
&= f'_y,
\end{aligned}
\tag{9.34}
$$

and

$$
\begin{aligned}
\frac{m_{32}}{m_{33}} \dot{v} + \dot{r} &= -\frac{1}{m_{33}} \left[ (m_{22} - m_{11}) uv + \left( \frac{m_{23} + m_{32}}{2} \right) ur + d_\psi \right] + \frac{\tau_\psi}{m_{33}}, \\
&= f'_\psi + \frac{\tau_\psi}{m_{33}},
\end{aligned}
\tag{9.35}
$$

respectively.

These latter two equations can be decoupled by multiplying (9.34) by $m_{23}/m_{22}$, subtracting the result from (9.35) and solving for $\dot{v}$. Similarly, one can find the decoupled differential equation for $\dot{r}$ by multiplying (9.35) by $m_{32}/m_{33}$, subtracting the result from (9.34) and solving for $\dot{r}$. The decoupled equations are

$$
\begin{aligned}
\dot{v} &= a_\psi \left[ f'_y - \frac{m_{23}}{m_{22}} f'_\psi \right] + a_y \frac{\tau_\psi}{m_{33}}, \\
&= f_y + a_y \frac{\tau_\psi}{m_{33}},
\end{aligned}
\tag{9.36}
$$

and

$$
\begin{aligned}
\dot{r} &= a_\psi \left[ f'_\psi - \frac{m_{32}}{m_{33}} f'_y \right] + a_\psi \frac{\tau_\psi}{m_{33}}, \\
&= f_\psi + a_\psi \frac{\tau_\psi}{m_{33}},
\end{aligned}
\tag{9.37}
$$

where

$$
a_\psi = \frac{m_{22} m_{33}}{m_{22} m_{33} - m_{23} m_{32}} \quad \text{and} \quad a_y = -\frac{m_{23}}{m_{22}} a_\psi.
\tag{9.38}
$$

Then, from (9.21)–(9.23), (9.33) and (9.36)–(9.38) we have

$$
\dot{\mathbf{x}} =
\begin{bmatrix}
u \cos \psi - v \sin \psi \\
u \sin \psi + v \cos \psi \\
r \\
f_x \\
f_y \\
f_\psi
\end{bmatrix}
+
\begin{bmatrix}
0 \\
0 \\
0 \\
\dfrac{1}{m_{11}} \\
0 \\
0
\end{bmatrix}
\tau_x
+
\begin{bmatrix}
0 \\
0 \\
0 \\
0 \\
\dfrac{a_y}{m_{33}} \\
\dfrac{a_\psi}{m_{33}}
\end{bmatrix}
\tau_\psi,
\qquad (9.39)
$$

which has the form

$$
\dot{\mathbf{x}} = \mathbf{f}(\mathbf{x}) + \mathbf{g}_1 \tau_x + \mathbf{g}_2 \tau_\psi.
$$

We will first explore the problem of stabilizing nonholonomic systems, as it provides the underlying rationale for the commonly used strategy of separating the control of underactuated vehicles into a motion planning problem and a feedback control problem, and then proceed to investigate the path following control and trajectory tracking of underactuated surface vessels.

## 9.5 Stabilization of Nonholonomic Vehicles

The control of an underactuated vehicle with kinematics or dynamics that satisfy nonholonomic constraints is particularly interesting because, in principle, the vehicle can be controlled from any given initial position and orientation to any desired final position and orientation, within its entire configuration space, while using a reduced number of actuators. This can be beneficial from a vehicle design standpoint, as cost, weight and energetic requirements can be lower than those of a fully-actuated system. Unfortunately, motion planning and control of an underactuated nonholonomic system are also generally more difficult than they are for a fully-actuated system.

When all of the constraints on a system are nonholonomic, it may be possible to use feedback linearization to find a feedback law can be used to drive the system to a Lyapunov stable equilibrium point. Recall from Definition 8.11 that a nonlinear system of the form

$$
\dot{\mathbf{x}} = \mathbf{f}(\mathbf{x}) + \mathbf{G}(\mathbf{x})\boldsymbol{\tau}
$$

is input-state feedback linearizable if there exists a diffeomorphic transformation $\mathbf{z} = \mathbf{T}(\mathbf{x})$ and a linearizing feedback control input

$$
\boldsymbol{\tau} = \boldsymbol{\alpha}(\mathbf{x}) + \boldsymbol{\beta}(\mathbf{x})\mathbf{v},
$$

which convert the system into the controllable linear system of the form

$$\dot{z} = Az + Bv.$$

Note that in the above equation, $B$ is the control input matrix of the transformed linear system and not the control input configuration matrix. A similar approach can also be used to design controllers for underactuated holonomic systems. However, the resulting input-state feedback linearization will be nonsmooth and non regular, and the associated state transformation $z = T(x)$ will be discontinuous. In fact, as will be shown here, a common approach for the stabilization of underactuated systems involves the following three steps:

(1) Verify that the system is controllable to any desired configuration by confirming that it is *locally accessible* and *small time locally controllable*.
(2) If the system is controllable, find a transformation that maps its nonlinear kinematic and dynamic equations of motion into a so-called *chained form*.
(3) Use the chained form of the underactuated systems dynamics to either:

   (a) find a kinematically feasible (for systems with first order constraints) or dynamically feasible (systems with second order constraints) trajectory that can be followed using a smooth open loop control input (or a two degree of freedom control architecture), or
   (b) use a discontinuous or time-varying control law to stabilize the system.

Before delving into the details, it is noted that the implementation of stabilization techniques for underactuated marine vehicles is not widely used in practice. However, as the development of these methods is a very active area of research (particularly in robotics and unmanned systems), they are presented here to help provide an understanding of the key considerations important for their future use.

### 9.5.1 The Controllability of Nonlinear Systems

Recall from Sect. 6.2 that the reachability of a system is its ability to reach an arbitrary desired state $x(t)$ in a transient fashion through suitable selection of a control inputs $\tau(t)$. This means that a closed loop system is reachable if, given any two points $x_0$ and $x_f$, there is a finite time $t_f$ and a control input $\tau$ such that $x_f = x(t_f, x_0, \tau)$. Define the reachable set $\mathcal{R}(x_0, t \le t_f)$ as the set of all points $x_f$, such that there exists an input $\tau(t), 0 \le t \le t_f$ that steers the system from $x(0) = x_0$ to $x(t_f) = x_f$ (see Fig. 6.3) in the finite time $t_f$.

As a general set of criteria do not exist for testing this intuitive definition of controllability, existing methods of establishing the reachability of a system rely on characterizing the structure of the system of equations in (9.19). To do this, the concept of a reachable set introduced in Sect. 6.2 is expanded slightly. As in [8], define $V$ to be a neighborhood near $x_0$ that trajectories starting at $x_0$ pass through

**Fig. 9.4** The reachable set $\mathscr{R}_{t_f}^V(\mathbf{x}_0)$

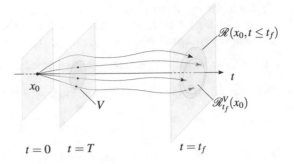

after a time $T$ and $\mathscr{R}^V(\mathbf{x}_0, t_f)$ be the set of states reachable at time $t_f$ from $\mathbf{x}_0$ with trajectories that pass though $V$ (see Fig. 9.4), so that

$$\mathscr{R}_{t_f}^V(\mathbf{x}_0) = \bigcup_{T \leq t_f} \mathscr{R}^V(\mathbf{x}_0, T).$$

**Definition 9.7** (*Locally Accessible*) A control system represented by (9.19) is considered *locally accessible* from $\mathbf{x}_0$, if for all neighborhoods[1] $V$ of $\mathbf{x}_0$ and all $t_f$, the reachable set $\mathscr{R}_{t_f}^V(\mathbf{x}_0)$ contains a non-empty open set[2] $\Omega$.

**Definition 9.8** (*Small-Time Locally Controllable*) A control system represented by (9.19) is considered *small-time locally controllable* from $\mathbf{x}_0$, if for all neighborhoods $V$ of $\mathbf{x}_0$ and $t_f$, the set $\mathscr{R}_{t_f}^V(\mathbf{x}_0)$ contains a non-empty neighborhood of $\mathbf{x}_0$.

If a system is locally accessible, or small-time locally controllable for any $\mathbf{x}_0 \in \mathbb{R}^n$, the system is globally controllable. To test the controllability of underactuated systems we will make extensive use of the concept of a distribution. Recall from Definition 8.6, that the distribution of a set of vector fields is its span.

**Definition 9.9** (*Accessibility Distribution*) The *accessibility distribution* $\Delta_C$ of the nonlinear system (9.19) is the involutive closure (see Definition 8.10) of the distribution associated with the vector fields $\mathbf{f}, \mathbf{g}_1, \ldots, \mathbf{g}_m$, where the vectors $\mathbf{g}_i$ are the columns of $\mathbf{G}$ in (9.19).

The accessibility distribution can be iteratively computed as follows:

(1) Compute $\Delta_1 = \text{span}\{\mathbf{f}, \mathbf{g}_1, \ldots, \mathbf{g}_m\}$.

---

[1] A *neighborhood* $V$ of a point $\mathbf{x}_0$ is a set of points containing $\mathbf{x}_0$ where one can move some amount in any direction away from $\mathbf{x}_0$ without leaving the set.

[2] One can think of an *open set* as a collection of points, like a sphere in three dimensions, about a given point that does not include a boundary. E.g. the set of points $x^2 + y^2 + z^2 < 4$ is an open set about the origin $x = y = z = 0$. It is bounded by the sphere $\sqrt{x^2 + y^2 + z^2} = 2$, which is a *closed set*.

(2) Then, for $i \geq 2$, compute $\Delta_i = \Delta_{i-1} + \text{span}\{[\mathbf{g}, \mathbf{v}] | \mathbf{g} \in \Delta_1, \mathbf{v} \in \Delta_{i-1}\}$, where $[\mathbf{g}, \mathbf{v}]$ is the Lie bracket of $\mathbf{g}$ and $\mathbf{v}$ (Definition 8.8).

(3) Repeat step 2 until $\Delta_{\kappa+1} = \Delta_\kappa$, giving $\Delta_C = \Delta_\kappa$.

Since $m \leq n$, step 2 only needs to be performed at most $n - m$ times.

**Theorem 9.1** (Chow's Theorem) *The system* (9.19) *is locally accessible from* $\mathbf{x}_0$ *if*

$$\dim \Delta_C(\mathbf{x}_0) = n. \tag{9.40}$$

*If* (9.40) *holds for all* $\mathbf{x}_0 \in \mathbb{R}^n$, *the closed loop system* (9.19) *is locally accessible.*

Chow's Theorem provides a sufficient condition for controllability when $\mathbf{f}(\mathbf{x}) = 0$ so that (9.19) represents a *driftless* control system[3], as well as when the drift term is present $\mathbf{f}(\mathbf{x}) \neq 0$ and

$$\mathbf{f}(\mathbf{x}) \in \text{span}\{\mathbf{g}_1(\mathbf{x}), \ldots, \mathbf{g}_m(\mathbf{x})\}, \quad \forall \mathbf{x} \in \mathbb{R}^n.$$

Additionally, when $\mathbf{f}(\mathbf{x}) = 0$ and (9.19) is driftless, if

$$\begin{aligned} \dot{\mathbf{x}} &= \mathbf{G}(\mathbf{x})\boldsymbol{\tau}, \\ &= \sum_{i=1}^{m} \mathbf{g}_i(\mathbf{x})\tau_i, \quad m = n - k, \end{aligned} \tag{9.41}$$

is controllable, so is its dynamic extension

$$\begin{aligned} \dot{\mathbf{x}} &= \sum_{i=1}^{m} \mathbf{g}_i(\mathbf{x})v_i, \quad m = n - k, \\ \dot{v}_i &= \tau_i, \quad\quad\quad i = 1, \ldots, m. \end{aligned} \tag{9.42}$$

**Remark 9.1** In the linear case, (9.19) can be represented in the form

$$\dot{\mathbf{x}} = \mathbf{A}\mathbf{x} + \mathbf{B}\mathbf{u},$$

and the accessibility rank condition (9.40) reduces to the reachability rank condition (6.15) for the reachability matrix presented in Theorem 6.1

$$\text{rank}(\mathbf{W}_r) = \text{rank}\begin{bmatrix} \mathbf{B} & \mathbf{A}\mathbf{B} & \mathbf{A}^2\mathbf{B} & \cdots & \mathbf{A}^{n-1}\mathbf{B} \end{bmatrix} = n.$$

---

[3] See Sect. 8.4.1 for brief definitions of the drift and control terms in a closed loop system.

**Example 9.2** (*USV with first order constraints*) Consider a simplified model of an unmanned surface vehicle with an $n = 3$ dimensional configuration space in surge, sway and yaw. Assume the body-fixed sway velocity is zero, $v = 0$, which means that the number of constraints is $k = 1$ and the control inputs must be defined in an $m = n - k = 2$ dimensional workspace. Such an assumption may roughly hold when the speed of the vehicle is very low. In the configuration space, the kinematic equations (9.50) reduce to (9.21)–(9.23). Taking $v = 0$, we have

$$\dot{\eta} = \begin{bmatrix} \cos\psi & -\sin\psi & 0 \\ \sin\psi & \cos\psi & 0 \\ 0 & 0 & 1 \end{bmatrix} \begin{bmatrix} u \\ 0 \\ r \end{bmatrix} = \begin{bmatrix} \cos\psi & 0 \\ \sin\psi & 0 \\ 0 & 1 \end{bmatrix} \begin{bmatrix} u \\ r \end{bmatrix}, \qquad (9.43)$$

where $\eta = [x \ y \ \psi]^T$.

Writing (9.43) in the form of the summation shown in (9.50), we have

$$\dot{\eta} = \begin{bmatrix} 0 \\ 0 \\ 1 \end{bmatrix} r + \begin{bmatrix} \cos\psi \\ \sin\psi \\ 0 \end{bmatrix} u, \qquad (9.44)$$
$$= \mathbf{g}_1 r + \mathbf{g}_2 u.$$

where one can think of $r$ as representing a steering velocity input and $u$ being a driving velocity input. All admissible velocities are linear combinations of the vectors $\mathbf{g}_1$ and $\mathbf{g}_2$, which form the columns of $\mathbf{J}(\eta)$.

In order to confirm that the motion of the USV is nonholonomic, we will check that $\dim\Delta_C = 3$. Using the procedure outlined after Definition 9.9, we first take $\Delta_1 = \text{span}\{\mathbf{g}_1, \mathbf{g}_2\}$. Then, $\Delta_2 = \Delta_1 + \text{span}\{[\mathbf{g}_1, \mathbf{g}_2]\}$, where

$$[\mathbf{g}_1, \mathbf{g}_2] = \nabla\mathbf{g}_2 \cdot \mathbf{g}_1 - \underbrace{\nabla\mathbf{g}_1 \cdot \mathbf{g}_2}_{=0}$$
$$= \nabla\mathbf{g}_2 \cdot \mathbf{g}_1$$
$$= \begin{bmatrix} 0 & 0 & -\sin\psi \\ 0 & 0 & \cos\psi \\ 0 & 0 & 0 \end{bmatrix} \begin{bmatrix} 0 \\ 0 \\ 1 \end{bmatrix} \qquad (9.45)$$
$$= \begin{bmatrix} -\sin\psi \\ \cos\psi \\ 0 \end{bmatrix}.$$

Note that here the gradient operator $\nabla$ is given by

$$\nabla = \begin{bmatrix} \dfrac{\partial}{\partial x} & \dfrac{\partial}{\partial y} & \dfrac{\partial}{\partial \psi} \end{bmatrix}^T. \qquad (9.46)$$

Let $\mathbf{g}_3 := [\mathbf{g}_1, \mathbf{g}_2]$ in (9.45). As the second step of computing $\Delta_2 = \Delta_1 +$ span$\{[\mathbf{g}_1, \mathbf{g}_2]\}$ only has to be performed $n - m = 3 - 2$ times, we have that

$$\Delta_C = \Delta_2 = \text{span}\{\mathbf{g}_1, \mathbf{g}_2, \mathbf{g}_3\}. \tag{9.47}$$

The vectors $\mathbf{g}_1$, $\mathbf{g}_2$ and $\mathbf{g}_3$ are linearly independent, so $\dim(\Delta_C) = 3$, which is that same as the dimension of the configuration space. Thus, the system is locally accessible. □

The local accessibility of systems with second order acceleration constraints can also be verified using (9.40).

**Example 9.3** (*Nonholonomic USV with second order constraints*) Consider an unmanned surface vehicle that cannot produce any control forces along its sway axis. Let the pose and velocity vectors be given by $\boldsymbol{\eta} = [x_n \ y_n \ \psi]^T$ and $\mathbf{v} = [u \ v \ r]^T$, respectively. Assume that the inertia tensor is diagonal, $\mathbf{M} = \text{diag}[m_{11} \ m_{22} \ m_{33}]$, the Coriolis and centripetal acceleration tensor is

$$\mathbf{C}(\mathbf{v}) = \begin{bmatrix} 0 & 0 & -m_{22}v \\ 0 & 0 & m_{11}u \\ m_{22}v & -m_{11}u & 0 \end{bmatrix},$$

and that the drag matrix is linear and diagonal, $\mathbf{D} = \text{diag}[d_{11} \ d_{22} \ d_{33}]$. Taking

$$\mathbf{x} = \begin{bmatrix} \boldsymbol{\eta} \\ \mathbf{v} \end{bmatrix} \in \mathbb{R}^6,$$

the kinematic and dynamic equations of motion can be written in a form similar to (9.18),

$$\dot{\mathbf{x}} = \begin{bmatrix} \mathbf{J}(\boldsymbol{\eta})\mathbf{v} \\ -\mathbf{M}^{-1}[\mathbf{C}(\mathbf{v})\mathbf{v} + \mathbf{D}\mathbf{v}] \end{bmatrix} + \begin{bmatrix} \mathbf{0}_{3\times3} \\ \mathbf{M}^{-1} \end{bmatrix} \mathbf{B}\mathbf{u},$$
$$= \mathbf{f}(\mathbf{x}) + \mathbf{G}\boldsymbol{\tau}, \tag{9.48}$$

where $\mathbf{u} = [\tau_x \ \tau_y \ \tau_\psi]^T$,

$$\mathbf{B} = \begin{bmatrix} 1 & 0 & 0 \\ 0 & 0 & 0 \\ 0 & 0 & 1 \end{bmatrix}$$

and $\boldsymbol{\tau} = \mathbf{B}\mathbf{u} = [\tau_x \ \tau_\psi]^T$ is the vector of control forces. The terms involving the control forces on the right hand side of (9.48) can be expanded to obtain

$$\dot{\mathbf{x}} = \mathbf{f}(\mathbf{x}) + \mathbf{g}_1\tau_x + \mathbf{g}_2\tau_\psi,$$

where

$$\mathbf{f} = \begin{bmatrix} u\cos\psi - v\sin\psi \\ u\sin\psi + v\cos\psi \\ r \\ \dfrac{m_{22}}{m_{11}}vr - \dfrac{d_{11}}{m_{11}}u \\ -\dfrac{m_{11}}{m_{22}}ur - \dfrac{d_{22}}{m_{22}}v \\ \dfrac{m_{11} - m_{22}}{m_{33}}uv - \dfrac{d_{33}}{m_{33}}r \end{bmatrix}, \quad \mathbf{g}_1 = \begin{bmatrix} 0 \\ 0 \\ 0 \\ 1 \\ m_{11} \\ 0 \\ 0 \end{bmatrix}, \quad \text{and} \quad \mathbf{g}_2 = \begin{bmatrix} 0 \\ 0 \\ 0 \\ 0 \\ 0 \\ 1 \\ m_{33} \end{bmatrix}.$$

As shown in [22], the vector fields $\mathbf{f}$, $\mathbf{g}_1$ and $\mathbf{g}_2$ satisfy the rank accessibility condition (9.40) of Theorem 9.1. The proof consists of demonstrating that the accessibility distribution $\Delta_C$ is the span of the vector fields $\mathbf{g}_1$, $\mathbf{g}_2$, $[\mathbf{f}, \mathbf{g}_1]$, $[\mathbf{f}, \mathbf{g}_2]$, $[\mathbf{g}_2, [f, \mathbf{g}_1]]$ and $[\mathbf{g}_2, [[\mathbf{g}_1, \mathbf{f}], \mathbf{f}]$ and that $\dim\Delta_C = n = 6$ for all $\eta$ and $\mathbf{v}$. Thus, the system is locally accessible. □

### 9.5.2   Stabilization of Nonholonomic Systems

Consider the general form of a time-invariant nonlinear system (see Sect. 7.2)

$$\dot{\mathbf{x}} = \mathbf{f}(\mathbf{x}, \boldsymbol{\tau}), \quad \mathbf{x} \in \mathbb{R}^n, \boldsymbol{\tau} \in \mathbb{R}^r. \tag{9.49}$$

In the case of an underactuated vehicle, we will assume that system is fully actuated in its workspace $r = m$. We now ask "is there a state-dependent input function $\boldsymbol{\tau} = \boldsymbol{\tau}(\mathbf{x})$ that can be used to asymptotically stabilize the system about some equilibrium point $\mathbf{x}_e$?" This problem is often referred to as the *local stabilizability problem*. A necessary condition for the stabilizability of (9.49) is given by the following theorem.

**Theorem 9.2** (Brockett [6]) *If the system*

$$\dot{\mathbf{x}} = \mathbf{f}(\mathbf{x}, \boldsymbol{\tau}),$$

*where* $\mathbf{f}$ *is continuously differentiable and has an equilibrium point at* $\mathbf{x} = \mathbf{x}_e$ *such that* $\mathbf{f}(\mathbf{x}_e, 0) = 0$, *admits a continuously differentiable state feedback control input* $\boldsymbol{\tau}(\mathbf{x})$ *that makes* $\mathbf{x}_e$ *asymptotically stable, then the image of the function*[4]

$$\mathbf{f} : \mathbb{R}^n \times \mathbb{R}^m \to \mathbb{R}^n$$

*contains an open neighborhood*[5] *of* $\mathbf{x}_e$.

The condition required for the existence of a stabilizing continuously differentiable control input $\boldsymbol{\tau}(\mathbf{x})$ in Theorem 9.2 above, is sometimes simply referred to as

---

[4]The *image of a function* is the set of all of the output values it may produce.

[5]An *open neighborhood* of a point $\mathbf{x}_e$ is an open set containing $\mathbf{x}_e$.

*Brockett's Condition.* It has important implications for the control of underactuated systems. Underactuated systems without drift cannot be stabilized by continuously differentiable, time-invariant feedback control laws. I.e. they can only be stabilized using either discontinuous or time-varying control inputs, $\tau = \tau(\mathbf{x}, t)$ [8].

In other words, there exist configurations of the vehicle (e.g. certain pose values) in the neighborhood of $\mathbf{x} = 0$, which are unreachable using a continuous control input that is a function of the state only, i.e. $\tau = \tau(\mathbf{x})$.

### 9.5.3  Chained Forms

As mentioned above, a common strategy for solving the stabilization problem for a nonholonomic vehicle (i.e. driving the vehicle to any desired final configuration) involves finding a transformation $\mathbf{z} = \mathbf{T}(\mathbf{x})$ that maps the kinematic and dynamic equations of motion of the form (9.19) into a chained form. In the special case of systems with $r = m = 2$ inputs, algorithms for systematically determining such transformations can be found in several works, including [8, 12, 14, 15, 18]. Note that the two-input case is sufficient for covering the stabilization of marine surface vessels. To demonstrate the general approach, the methods proposed in [8, 18] for obtaining transformations that yield two input chained forms for nonholonomic systems with first order (kinematic) constraints will be presented here, with some additional references to systems with second order constraints provided at the end of this section.

Consider the kinematic equations of motion for a marine vehicle with first order kinematic constraints. From (1.58) we have

$$
\begin{aligned}
\dot{\boldsymbol{\eta}} &= \mathbf{J}(\boldsymbol{\eta})\mathbf{v}, \\
&= \sum_{i=1}^{m} \mathbf{j}_i(\boldsymbol{\eta})v_i, \quad m = n - k.
\end{aligned}
\tag{9.50}
$$

The control objective is to drive the system from any initial configuration $\boldsymbol{\eta}_0$ to any desired final configuration $\boldsymbol{\eta}_f$. Note that (9.50) has the form of a driftless control system, see Eq. (9.41). The main control strategies for stabilizing first order nonholonomic systems take advantage of this by treating the components of the body-fixed velocity $\mathbf{v}$ like a set of control inputs. Since, according to Theorem 9.2, an underactuated system without drift cannot be stabilized by a continuously differentiable, time-invariant feedback control law there are generally two main approaches used to control such systems:

(1) *Motion Planning*: Treating the body-fixed velocities like control inputs, determine an appropriate $\mathbf{v}(t)$ to obtain a desired configuration $\boldsymbol{\eta}_f$. Since the kinematic equations of motion are used to generate the motion plan, the resulting trajectory will be *kinematically feasible* and, therefore, the underactuated vehicle should be able to follow it. The resulting controller would be an open loop or feedfor-

ward control input, which cannot depend on the system state or the closed loop error. Unfortunately, this sort of control on its own is not robust to disturbances or modeling uncertainty, but may be combined with a feedback controller in a two degree of freedom configuration (see Sect. 6.7).

(2) Use of either (a) a discontinuous, or (b) a time-varying control law.

To determine the required inputs using either approach, the system will first be transformed into a chained form. A chained form is a system of equations with the form

$$
\begin{aligned}
\dot{z}_1 &= v_1, \\
\dot{z}_2 &= v_2, \\
\dot{z}_3 &= z_2 v_1, \\
\dot{z}_4 &= z_3 v_1, \\
&\vdots \\
\dot{z}_n &= z_{n-1} v_1,
\end{aligned}
\tag{9.51}
$$

which is representative of a broad class of nonholonomic systems (here we focus on two input systems). Systems of this form have two important properties, which make them especially useful for solving the stabilization problem:

(1) The system (9.51) satisfies the controllability conditions of Theorem 9.1 because it is driftless

$$
\dot{\mathbf{z}} =
\begin{pmatrix}
1 \\
0 \\
z_2 \\
z_3 \\
\vdots \\
z_{n-1}
\end{pmatrix}
v_1 +
\begin{pmatrix}
0 \\
1 \\
0 \\
0 \\
\vdots \\
0
\end{pmatrix}
v_2,
$$

$$
= \mathbf{g}_1(\mathbf{z}) v_1 + \mathbf{g}_2 v_2.
$$

and its accessibility distribution is given by

$$
\Delta_C = \operatorname{span} \{ \mathbf{g}_1, \mathbf{g}_2, [\mathbf{g}_1, \mathbf{g}_2], [\mathbf{g}_1, [\mathbf{g}_1, \mathbf{g}_2]], [\mathbf{g}_2, [\mathbf{g}_1, \mathbf{g}_2]] \},
$$

which has dimension $\dim(\Delta_C) = n$.

(2) The state $z_2$ is directly controlled by the input $v_2$, simplying the problem of finding a suitable control input to obtain a desired final value of $z_2$.

Thus, if a system with kinematic constraints can be transformed into a chained form, a motion plan can be determined in terms of the time dependent body-fixed velocities required to drive the system to any desired configuration (pose) from some initial configuration. The process is similar to feedback linearization (Chapter 8) and closely related to the exact feedback linearizability conditions for a general nonlinear system [19]. The technique involves transforming the system into a chained form, finding the required "control inputs" $v_1$ and $v_2$ that drive the transformed system from some transformed initial condition $\mathbf{z}_0 = \mathbf{T}(\mathbf{x}_0)$ to a transformed final desired position $\mathbf{z}_f = \mathbf{T}(\mathbf{x}_f)$ and then transforming the inputs $v_1$ and $v_2$ back into the original coordinates.

A relatively simple three-step algorithm for determining the transformation $\mathbf{z} = \mathbf{T}(\mathbf{x})$ is proposed in [8]:

Step 1:   Verify that the transformation exists. Define the distributions

$$
\begin{aligned}
\Delta_0 &= \mathrm{span}\{\mathbf{g}_1, \mathbf{g}_2, \mathrm{ad}_{\mathbf{g}_1}\mathbf{g}_2, \ldots, \mathrm{ad}_{\mathbf{g}_1}^{n-2}\mathbf{g}_2\}, \\
\Delta_1 &= \mathrm{span}\{\mathbf{g}_2, \mathrm{ad}_{\mathbf{g}_1}\mathbf{g}_2, \ldots, \mathrm{ad}_{\mathbf{g}_1}^{n-2}\mathbf{g}_2\}, \\
\Delta_2 &= \mathrm{span}\{\mathbf{g}_2, \mathrm{ad}_{\mathbf{g}_1}\mathbf{g}_2, \ldots, \mathrm{ad}_{\mathbf{g}_1}^{n-3}\mathbf{g}_2\}.
\end{aligned} \tag{9.52}
$$

If for some open set $U$, $\dim(\Delta_0) = n$ and $\Delta_1$, $\Delta_2$ are involutive, and there exists a smooth function $h_1 : U \to \mathbb{R}^n$ such that

$$
\nabla h_1 \cdot \Delta_1 = 0 \quad \text{and} \quad \nabla h_1 \cdot \mathbf{g}_1 = 1, \tag{9.53}
$$

where $\nabla h_1$ represents the gradient of $h_1$ with respect to the components of $\mathbf{x}$, then a local feedback transformation exists and the change of coordinates that transform the system into chained form can be found.

Step 2:   If a transformation $\mathbf{z} = \mathbf{T}(\mathbf{x})$ exists, the change of coordinates is given by

$$
\begin{aligned}
z_1 &= h_1, \\
z_2 &= L_{\mathbf{g}_1}^{n-2} h_2, \\
&\;\;\vdots \\
z_{n-1} &= L_{\mathbf{g}_1} h_2, \\
z_n &= h_2,
\end{aligned} \tag{9.54}
$$

where $h_2$ is independent of $h_1$ and satisfies the equation

$$
\nabla h_2 \cdot \Delta_2 = 0. \tag{9.55}
$$

When the conditions in Step 1 are satisfied, the existence of independent functions $h_1$ and $h_2$ with the properties listed above is guaranteed by Theorem 8.1, since $\Delta_1$ and $\Delta_2$ are involutive.

Step 3:   Write the equations of the transformed system in chained form. Using the invertible input transformation

$$
\begin{aligned}
v_1 &= \tau_1, \\
v_2 &= \left(L_{\mathbf{g}_1}^{n-1} h_2\right) \tau_1 + \left(L_{\mathbf{g}_2} L_{\mathbf{g}_1}^{n-2} h_2\right) \tau_2,
\end{aligned} \tag{9.56}
$$

the transformed system can be expressed in the chained form

$$\dot{z}_1 = v_1,$$
$$\dot{z}_2 = v_2,$$
$$\dot{z}_3 = z_2 v_1,$$
$$\dot{z}_4 = z_3 v_1,$$

$$\vdots$$

$$\dot{z}_n = z_{n-1} v_1. \tag{9.57}$$

The procedure involves solving the two partial differential equations (9.53) and (9.55). If $\mathbf{g}_1$ and $\mathbf{g}_2$ have the special form

$$\mathbf{g}_1 = \begin{bmatrix} 1 \\ g_{12}(\mathbf{x}) \\ \vdots \\ g_{1n}(\mathbf{x}) \end{bmatrix}, \quad \text{and} \quad \mathbf{g}_2 = \begin{bmatrix} 0 \\ g_{22}(\mathbf{x}) \\ \vdots \\ g_{2n}(\mathbf{x}) \end{bmatrix}, \tag{9.58}$$

with arbitrary $g_{ik}$'s, then it is easy to verify that $\Delta_1$ is always involutive and we can choose $h_1 = x_1$ [8]. In this case, we only have to verify that $\Delta_2$ is involutive, and solve the associated partial differential equation (9.55). Note that it is always possible to cast $\mathbf{g}_1$ and $\mathbf{g}_2$ in the form of (9.58), by reordering variables because the input vector fields are assumed to by independent.

**Example 9.4** (*Motion planning for an underactuated USV*) Returning to the simplified kinematic model of a USV presented in Example 9.2, where we impose the first order (kinematic) constraint $v = 0$, we explore the use of the transformed chained form to generate kinematically feasible trajectories.

Let $\mathbf{x} = \boldsymbol{\eta} = [x \ y \ \psi]^T$. The kinematics of the USV expressed in (9.44) can be converted into a chained form using the vectors $\mathbf{g}_1$, $\mathbf{g}_2$, and $\mathbf{g}_3$ identified in Example 9.2,

$$\mathbf{g}_1 = \begin{bmatrix} 0 \\ 0 \\ 1 \end{bmatrix}, \quad \mathbf{g}_2 = \begin{bmatrix} \cos \psi \\ \sin \psi \\ 0 \end{bmatrix}, \quad \text{and} \quad \mathbf{g}_3 = \begin{bmatrix} -\sin \psi \\ \cos \psi \\ 0 \end{bmatrix}. \tag{9.59}$$

Computing the distributions $\Delta_0$, $\Delta_1$, and $\Delta_2$ using (9.52) gives

$$\Delta_0 = \text{span}\{\mathbf{g}_1, \mathbf{g}_2, \mathbf{g}_3\},$$
$$\Delta_1 = \text{span}\{\mathbf{g}_2, \mathbf{g}_3\}, \tag{9.60}$$
$$\Delta_2 = \text{span}\{\mathbf{g}_2\}.$$

It can be immediately seen from (9.47) that $\Delta_0 = \Delta_C$, so that $\dim(\Delta_0) = \dim(\Delta_C) = n = 3$. In addition, $\Delta_1$ and $\Delta_2$ in (9.60) are involutive.

If we take $h_1 = \psi$ it can be readily verified that the relations in (9.53) are satisfied,

$$\nabla h_1 \cdot \Delta_1 = \begin{bmatrix} \dfrac{\partial h_1}{\partial x} & \dfrac{\partial h_1}{\partial y} & \dfrac{\partial h_1}{\partial \psi} \end{bmatrix} \cdot \Delta_1 = 0 \tag{9.61}$$

and

$$\nabla h_1 \cdot \mathbf{g}_1 = \begin{bmatrix} \dfrac{\partial h_1}{\partial x} & \dfrac{\partial h_1}{\partial y} & \dfrac{\partial h_1}{\partial \psi} \end{bmatrix} \cdot \mathbf{g}_1 = 1. \tag{9.62}$$

Further, $h_2$ can be found from (9.55) by showing that

$$\begin{aligned}
\nabla h_2 \cdot \mathbf{g}_2 &= \begin{bmatrix} \dfrac{\partial h_2}{\partial x} & \dfrac{\partial h_2}{\partial y} & \dfrac{\partial h_2}{\partial \psi} \end{bmatrix} \cdot \mathbf{g}_2 \\
&= \dfrac{\partial h_2}{\partial x} \cos \psi + \dfrac{\partial h_2}{\partial y} \sin \psi \\
&= 0,
\end{aligned} \tag{9.63}$$

is satisfied by taking

$$h_2 = x \sin \psi - y \cos \psi. \tag{9.64}$$

Then, using (9.54), the change of coordinates that transforms (9.44) into chained form is given by

$$\begin{aligned}
z_1 &= h_1 = \psi, \\
z_2 &= L_{\mathbf{g}_1} h_2 = \nabla h_2 \cdot \mathbf{g}_1 = x \cos \psi + y \sin \psi, \\
z_3 &= h_2 = x \sin \psi - y \cos \psi.
\end{aligned} \tag{9.65}$$

This transformation $\mathbf{z} = \mathbf{T}(\mathbf{x})$ can be written in vector form as

$$\mathbf{z} = \begin{bmatrix} 0 & 0 & 1 \\ \cos \psi & \sin \psi & 0 \\ \sin \psi & -\cos \psi & 0 \end{bmatrix} \mathbf{x} \tag{9.66}$$

and the inverse relation is given by taking the transpose of the transformation matrix to get

$$\mathbf{x} = \begin{bmatrix} 0 & \cos \psi & \sin \psi \\ 0 & \sin \psi & -\cos \psi \\ 1 & 0 & 0 \end{bmatrix} \mathbf{z}. \tag{9.67}$$

Using (9.57), the dynamics of the transformed system are given by

$$\begin{aligned}
\dot{z}_1 &= v_1, \\
\dot{z}_2 &= v_2, \\
\dot{z}_3 &= z_2 v_1.
\end{aligned} \tag{9.68}$$

The corresponding transformed control input is

$$\begin{aligned}
v_1 &= \tau_1, \\
v_2 &= \left( L_{\mathbf{g}_1}^2 h_2 \right) \tau_1 + L_{\mathbf{g}_2} \left( L_{\mathbf{g}_1} h_2 \right) \tau_2.
\end{aligned} \tag{9.69}$$

From (9.65), (9.68) and (9.69) we expect $r = \dot\psi = \tau_1$, therefore, take $\tau_2$ to be the remaining control input in the original body-fixed coordinate system, $\tau_2 = u$.

From (9.65) it can be seen that $L_{g_1}h_2 = x\cos\psi + y\sin\psi$, therefore

$$
\begin{aligned}
\left(L_{g_1}^2 h_2\right) &= L_{g_1}\left(L_{g_1}h_2\right), \\
&= \nabla\left(L_{g_1}h_2\right) \cdot g_1, \\
&= \nabla\left(x\cos\psi + y\sin\psi\right) \cdot g_1, \\
&= [\cos\psi \quad \sin\psi \quad (-x\sin\psi + y\cos\psi)] \cdot g_1, \\
&= (-x\sin\psi + y\cos\psi), \\
&= -z_3,
\end{aligned}
\tag{9.70}
$$

and

$$
\begin{aligned}
L_{g_2}\left(L_{g_1}h_2\right) &= \nabla\left(L_{g_1}h_2\right) \cdot g_2, \\
&= [\cos\psi \quad \sin\psi \quad (-x\sin\psi + y\cos\psi)] \cdot g_2, \\
&= \cos^2\psi + \sin^2\psi, \\
&= 1.
\end{aligned}
\tag{9.71}
$$

Then, (9.69) becomes

$$
\begin{aligned}
v_1 &= r, \\
v_2 &= -z_3 r + u.
\end{aligned}
\tag{9.72}
$$

This can be written in vector form as

$$
\begin{bmatrix} v_1 \\ v_2 \end{bmatrix} = \begin{bmatrix} 0 & 1 \\ 1 & -z_3 \end{bmatrix} \begin{bmatrix} u \\ r \end{bmatrix},
\tag{9.73}
$$

which can be solved for the desired velocities as

$$
\begin{bmatrix} u \\ r \end{bmatrix} = \begin{bmatrix} z_3 & 1 \\ 1 & 0 \end{bmatrix} \begin{bmatrix} v_1 \\ v_2 \end{bmatrix}.
\tag{9.74}
$$

A motion plan can be developed using a set of polynomials for the transformed inputs $v_1$ and $v_2$. In order to make the model behave similarly to what one might expect in practice, we take the yaw rate to be $r_f = (\psi_f - \psi_0)/t_f$. This corresponds to a constant yaw rate computed using the difference between the initial heading angle and the final heading angle, divided by an approximation of the anticipated time required for the vehicle to perform the maneuver at a fixed surge speed. Using this approach, we take

$$
\begin{aligned}
v_1 &= r_f \\
v_2 &= c_0 + c_1 t.
\end{aligned}
\tag{9.75}
$$

Given an initial pose $x_0 = \eta_0$ and a desired final pose $x_f = \eta_f$, and their corresponding transformations $z_0 = T(\eta_0)$ and $z_f = T(\eta_f)$, the coefficients $c_0$ and $c_1$ are found by substituting (9.75) into the differential equations for $\dot z_2$ and $\dot z_3$ in (9.68) and integrating them analytically over the time $0 \le t \le t_f$. Thus, we have

$$z_{2f} = c_0 t_f + c_1 \frac{t_f^2}{2} + z_{20},$$

$$z_{3f} = \left( c_0 \frac{t_f^2}{2} + c_1 \frac{t_f^3}{6} \right) r_f + z_{30}. \tag{9.76}$$

By representing these equations in vector form

$$\begin{bmatrix} z_{2f} \\ z_{3f} \end{bmatrix} = \begin{bmatrix} t_f & \dfrac{t_f^2}{2} \\ \dfrac{t_f^2}{2} r_f & \dfrac{t_f^3}{6} r_f \end{bmatrix} \begin{bmatrix} c_0 \\ c_1 \end{bmatrix} + \begin{bmatrix} z_{20} \\ z_{30} \end{bmatrix}, \tag{9.77}$$

they can be solved for $c_0$ and $c_1$ as

$$\begin{bmatrix} c_0 \\ c_1 \end{bmatrix} = \frac{12}{r_f t_f^4} \begin{bmatrix} -\dfrac{t_f^3}{6} r_f & \dfrac{t_f^2}{2} \\ \dfrac{t_f^2}{2} r_f & -t_f \end{bmatrix} \begin{bmatrix} z_{2f} - z_{20} \\ z_{3f} - z_{30} \end{bmatrix}. \tag{9.78}$$

Here we explore the generation of trajectories for two cases:

(1)  $\eta_0 = [0 \ 0 \ 0]^T$, $\eta_f = [15 \ 15 \ \pi/2]^T$ and $t_f = 10$ s; and
(2)  $\eta_0 = [0 \ 0 \ 0]^T$, $\eta_f = [15 \ 15 \ -\pi/2]^T$ and $t_f = 10$ s.

The motion plan for Case 1 is shown in Figs. 9.5, 9.6 and the motion plan for Case 2 is shown in Figs. 9.7, 9.8.

As expected from (9.75), the yaw rate is constant in both cases. In Case 1, the vehicle moves from $\eta_0 = [0 \ 0 \ 0]^T$ to $\eta_f = [15 \ 15 \ \pi/2]^T$ by continuously rotating clockwise while moving forward. An object rotating at a fixed radius about a center point at constant angular velocity would trace a circle with a constant tangential velocity. For the situation at hand, the tangential velocity would correspond to the surge speed of the USV. However, as can be seen in Fig. 9.6, the surge speed varies slightly during the motion plan, so that the trajectory is almost a circular arc, but not quite.

In Case 2, the vehicle moves from $\eta_0 = [0 \ 0 \ 0]^T$ to $\eta_f = [15 \ 15 \ -\pi/2]^T$ by continuously rotating anticlockwise at a constant negative yaw rate. However, in order to reach the final desired configuration, the vehicle initially moves forward turning away from $\eta_f$ as the surge speed decreases, but then starting at around $t = 5$ s the USV starts reversing, while continuing to rotate, to reach $\eta_f$.  □

In Example 9.4 above, the motion is planned using a set of polynomials for the inputs $v_1$ and $v_2$. Alternate motion planning approaches include using sinusoidal or piecewise-constant input functions [8, 18, 19]. As mentioned above, the stabilization of first order nonholonomic systems via feedback control requires the use of either discontinuous or time varying control inputs. In general, for such systems, the feedback stabilization problem is more difficult to solve than the trajectory tracking

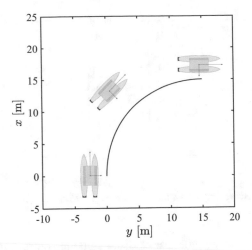

**Fig. 9.5** Case 1 trajectory: $x$-$y$ position and associated heading (indicated by axes affixed to USV)

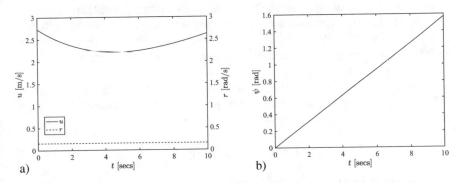

**Fig. 9.6** Case 1 trajectory: Desired **a** surge speed $u$, yaw rate $r$, and **b** heading angle $\psi$

**Fig. 9.7** Case 2 trajectory: $x$-$y$ position and associated heading (indicated by axes affixed to USV)

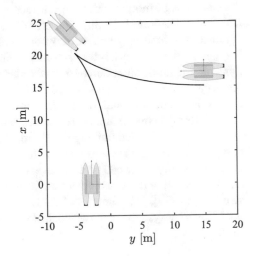

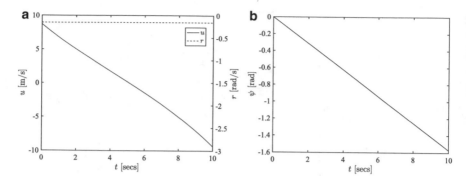

**Fig. 9.8** Case 2 trajectory: Desired **a** surge speed $u$, yaw rate $r$, and **b** heading angle $\psi$

problem [8]. Unfortunately, the use of discontinuous control inputs is often only robust to the initial configuration (and less so to perturbations occurring during the motion) and the use of time-varying feedback is often prone to erratic motion, slow convergence to the desired final configuration and difficult control parameter tuning. Lastly, for information about nonholonomic systems with first order (kinematic) constraints with $r = m \geq 3$ inputs and the corresponding chained forms, the interested reader is referred to [18].

Turning now to an examination of nonholonomic marine vehicles with second order constraints, it can be seen from (9.19) that the kinematic and dynamic equations can be written in the form

$$\dot{\mathbf{x}} = \mathbf{f}(\mathbf{x}) + \mathbf{g}_1 \tau_1 + \mathbf{g}_2 \tau_2,$$

where, unlike the first order nonholonomic systems examined above, the drift term is non zero $\mathbf{f}(\mathbf{x}) \neq 0$. However, even with the presence of a non zero drift term, the equations of motion for marine surface vessels (9.39) do not satisfy Brockett's Condition (Theorem 9.2) [30]. Therefore, stabilization still requires the use of discontinuous or time-varying feedback control. The feedback control design approach is generally the same as it is for first order nonholonomic systems. The kinematic and dynamic equations of motion of second order nonholonomic systems is transformed into a chained form, the transformed control inputs which stabilize the system from a transformed initial condition to a transformed final condition are determined, and the transformed control inputs are mapped back into the original coordinates. However, the process of determining the appropriate chained form is significantly more complex than it is for systems with nonholonomic first order constraints, and so the approach is only described here. In several cases, including underactuated underwater vehicles [9] and surface vehicles that use differential thrust for propulsion and steering [26, 30], it is shown in [12] that the equations describing the motion of nonholonomic systems with two control inputs can be transformed into the chained form

$$\ddot{z}_1 = v_1,$$
$$\ddot{z}_2 = v_2, \qquad\qquad (9.79)$$
$$\ddot{z}_3 = z_2 v_1.$$

**Example 9.5** (*Nonholonomic surface vessel with second order constraints*) Here we consider the stabilization of a surface vehicle that uses differential thrust for steering. The stabilization approach proposed by [14] is implemented on the simplified surface vehicle model developed in [25], which can be written as

$$\ddot{x} = u_1,$$
$$\ddot{y} = u_1 \tan \psi + \frac{c_y}{m} (\dot{x} \tan \psi - \dot{y}),$$
$$\ddot{\psi} = u_2,$$

where $x$ and $y$ are the northward and eastward positions of the vehicle in the NED frame (respectively), $\psi$ is the heading angle,

$$u_1 = \frac{1}{m} \left[ \tau_x \cos \psi - \dot{x} \left( c_y \sin^2 \psi + c_x \cos^2 \psi \right) + (c_y - c_x) \dot{y} \sin \psi \cos \psi \right],$$
$$u_2 = \frac{1}{I} \left( \tau_\psi - c_z \dot{\psi} \right),$$

and $c_x$, $c_y$ and $c_z$ are positive constants related to the hydrodynamic coefficients of the vehicle. Rearranging this latter expression provides explicit expressions for the surge force and the yaw moment

$$\tau_x = \frac{1}{\cos \psi} \left[ m u_1 + \dot{x} (c_y \sin^2 \psi + c_x \cos^2 \psi) - (c_y - c_x) \dot{y} \sin \psi \cos \psi \right],$$
$$\tau_\psi = I u_2 + c_z \dot{\psi}.$$

As shown in [14], these equations can be transformed into the chained form

$$\ddot{z}_1 = v_1,$$
$$\ddot{z}_2 = v_2,$$
$$\ddot{z}_3 = z_2 v_1 + \frac{c_y}{m} (\dot{z}_1 z_2 - \dot{z}_3),$$

where $z_1 = x$, $z_2 = \tan \psi$, $z_3 = y$, $v_1 = u_1$ and

$$v_2 = \frac{u_2}{\cos^2 \psi} + \frac{\dot{\psi}^2 \sin(2\psi)}{\cos^4 \psi}.$$

While this chained form is slightly different from (9.79), the main approach of transforming the system into chained form, designing a controller for the transformed system and transforming the controller back into the original coordinates is generally the same. As shown in [14], the control inputs $u_1$ and $u_2$ can be found to be

$$u_1 = -k\frac{\lambda_1}{\lambda_2}x - \left(k + \frac{\lambda_1}{\lambda_2}\right)\dot{x},$$

$$u_2 = \frac{1}{2}p_1 \sin(2\psi) + p_2\dot{\psi} + p_3\frac{y\cos^2\psi}{\lambda_1 x + \lambda_2\dot{x}} + p_4\frac{\dot{y}\cos^2\psi}{\lambda_1 x + \lambda_2\dot{x}} - 2\dot{\psi}^2\tan\psi,$$

where $0 < k < \lambda_1/\lambda_2$ and $k \neq c_y/m$. The coefficients $p_1, p_2, p_3$ and $p_4$ are selected so that the eigenvalues of the matrix

$$\begin{bmatrix} 0 & 1 & 0 & 0 \\ p_1 & p_2 & p_3 & p_4 \\ 0 & 0 & k & 1 \\ -\dfrac{k(c_y - km)}{m(k\lambda_2 - \lambda_1)} & 0 & 0 & k - \dfrac{c_y}{m} \end{bmatrix}$$

have a set of desired locations in the left half of the complex plane (i.e. so that the real part of each eigenvalue is negative).

The approach taken in [14] is to set all of the desired eigenvalues to the same value. Note that in this case the eigenvalues can not be simply determined from computing the characteristic equation of the matrix. One approach for determining the values of the $p_i$ requires finding the Jordan normal form of the matrix, which will contain the desired eigenvalues along its main diagonal. However, computation of the Jordan normal form tends to be numerically ill conditioned. Small variations in the values of the $p_i$ will change the locations of the eigenvalues fairly substantially and so tuning the controller can be quite challenging. For the purposes of illustrating the method, here most of the same values used in [14] are used: $m = 1$ kg, $I = 1$ kg-m$^2$, $c_x = 1$, $c_y = 1.2$, $c_x = 1.5$, $k = 0.8$, $\lambda_1 = 2.8$, $\lambda_2 = 2$ and the initial conditions

$$[x(0)\ \dot{x}(0)\ y(0)\ \dot{y}(0)\ \psi(0)\ \dot{\psi}(0)] = [-2.5\ 0\ -0.5\ -0.5\ 0.5\ 1].$$

However, here we take the locations of the four closed loop eigenvalues to all be located at $-0.3$ rad/s. Numerically solving for the values of $p_i$ that give these eigenvalues results in

$$p_1 = -1.67999999,$$
$$p_2 = -1.73333333,$$
$$p_3 = -6.18674074,$$
$$p_4 = -5.15555556.$$

Note that with even this degree of precision for the $p_i$, the placement of the eigenvalues is only correct to within approximately $1.0 \times 10^{-2}$ rad/s. For this example, in order to obtain a suitable response it was also necessary to further tune the control input $u_2$. The final value used is

$$u_2 = 0.9955278\left[\frac{p_1}{2}\sin(2\psi) + p_2\dot{\psi} + \frac{p_3 y\cos^2(\psi)}{\lambda_1 x + \lambda_2\dot{x}} + \frac{p_4\dot{y}\cos^2(\psi)}{\lambda_1 x + \lambda_2\dot{x}}\right] - 9.9\dot{\psi}^2\tan\psi.$$

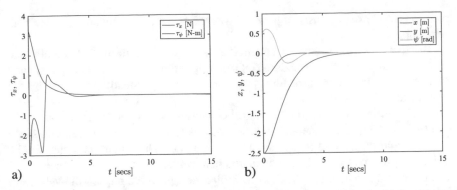

**Fig. 9.9** Stabilization of a surface vehicle **a** Control inputs and **b** $x$-$y$ position and heading $\psi$ in the NED frame

The resulting time response and control inputs are shown in Fig. 9.9. Manual tuning reveals that the system response is very sensitive to even very slight modification of the control parameters and initial conditions. One issue of note is that $u_2 \to 0$ when $\dot{\psi} \to 0$ and $\psi \to \pi/2$, such that when manually tuning the system it was observed that for many values of the coefficients, the system reaches the desired position $x = 0$, $y = 0$, but the vehicle is oriented at a right angle to the desired final pose so that $\psi = \pm \pi/2$, instead of $\psi = 0$.  □

As can be seen from the preceding example, the use of chained forms for stabilizing underactuated marine vehicles has several challenges that severely limit practical implementation of the approach. Despite the significant advances made to our understanding of how to stabilize underactuated systems provided by the development of the chained form approach, this is still an active area of research and further development is required before the existing approaches can be broadly implemented in practice.

Lastly, before proceeding to a discussion of path planning, it should be noted that an alternative to use of the chained form for the stabilization of marine surface vessels has been proposed by [16, 22, 23]. The method relies on a transformation of variables, which gives the transformed system a mathematical property known as homogeneity [17]. Using the approach, time-varying feedback control can be used to make the closed loop system uniformly globally asymptotically stable (see Theorem 7.6). The approach has been implemented both computationally and experimentally.

## 9.6  Path-Following Control for Surface Vessels

Path following requires a vehicle to follow a desired path without having to satisfy explicit time constraints. Here, one of the simplest approaches to underactuated path following control, known as line-of-sight (LOS) guidance [10] is presented. The path to be followed consists of a set of waypoints connected by straight-line path segments.

The motion of the vehicle is assumed to be constrained to the horizontal plane with only three degrees of freedom (DOF): longitudinal (surge), lateral (sway) and yaw. Assume that the control forces can only be independently generated by the vessel's actuators in surge and yaw.

As shown in Sect. 9.4.1, the dynamic equations of motion are given by (9.33) and (9.36)–(9.37). The system is underactuated because it has a configuration space of $n = 3$ and only $r = 2$ independent control inputs. The output variables can be defined so that the system is fully actuated in a workspace of dimension $m = 2$ and can be selected to help simplify the controller design. If we require the vehicle to follow the straight line path segments between waypoints as closely as possible, while maintaining a constant speed, the two output variables can be selected to be:

(1)  the shortest distance from a point on the vessel (e.g. its center of gravity) to the path, which we will call the *cross-track error*; and
(2)  the surge speed of the vessel.

### 9.6.1   Surge Speed Control

Define the surge speed error as

$$\tilde{u} := u - u_d, \tag{9.80}$$

where $u_d$ is the desired speed of the vessel. Taking the derivative of this gives

$$\dot{\tilde{u}} := \dot{u} - \dot{u}_d. \tag{9.81}$$

Then, using (9.33) the equation for the dynamics of the surge speed error is

$$\dot{\tilde{u}} = -\dot{u}_d + f_x + \frac{\tau_x}{m_{11}}. \tag{9.82}$$

Note that this equation is nonlinear, but feedback linearization via inverse dynamics (Sect. 8.2) can be used to design a PID controller by selecting the control input to be

$$\tau_x = -m_{11}\left( -\dot{u}_d + f_x + k_{up}\tilde{u} + k_{ui}\int_0^t \tilde{u}\,\mathrm{d}t + k_{ud}\dot{\tilde{u}} \right), \tag{9.83}$$

which gives the closed loop error dynamics

$$\dot{\tilde{u}} = -k_{up}\tilde{u} - k_{ui}\int_0^t \tilde{u}\,\mathrm{d}t - k_{ud}\dot{\tilde{u}}. \tag{9.84}$$

### 9.6.2  *Control of the Cross-Track Error*

As shown in Fig. 9.10, the position of the vessel in NED coordinates can be written as $\mathbf{p} = [x_n \ y_n]^T \in \mathbb{R}^2$ and the corresponding speed is defined as defined as

$$U := \sqrt{\dot{x}_n^2 + \dot{y}_n^2} := \sqrt{u^2 + v^2} \in \mathbb{R}^+. \tag{9.85}$$

The direction of the velocity vector with respect to the North axis is given by

$$\chi = \tan^{-1}\left(\frac{\dot{y}_n}{\dot{x}_n}\right) \in \mathcal{S} := [-\pi, \pi]. \tag{9.86}$$

Consider a straight-line path defined by two consecutive waypoints $\mathbf{p}_k = [x_k \ y_k]^T \in \mathbb{R}^2$ and $\mathbf{p}_{k+1} = [x_{k+1} \ y_{k+1}]^T \in \mathbb{R}^2$. The path makes an angle of

$$\alpha_k = \tan^{-1}\left(\frac{y_{k+1} - y_k}{x_{k+1} - x_k}\right) \in \mathcal{S} \tag{9.87}$$

with respect to the North axis of the NED frame.

The coordinates of the vessel in the path-fixed reference frame are

$$\boldsymbol{\varepsilon} = \begin{bmatrix} s \\ e \end{bmatrix} = \mathbf{R}_\alpha^T(\alpha_k)(\mathbf{p} - \mathbf{p}_k), \tag{9.88}$$

where $\mathbf{R}_\alpha^T$ is a $2 \times 2$ transformation matrix from the inertial to the path-fixed frame given by

$$\mathbf{R}_\alpha^T(\alpha_k) := \begin{bmatrix} \cos\alpha_k & \sin\alpha_k \\ -\sin\alpha_k & \cos\alpha_k \end{bmatrix}, \tag{9.89}$$

$s$ is the *along track distance*, and $e$ is the cross-track error (see Fig. 9.10). From (9.88), the cross track error can be explicitly written as

$$e = -(x_n - x_k)\sin\alpha_k + (y_n - y_k)\cos\alpha_k. \tag{9.90}$$

Its time derivative, which will be used later, is consequently given by

$$\dot{e} = -\dot{x}_n \sin\alpha_k + \dot{y}_n \cos\alpha_k. \tag{9.91}$$

From (9.21),

$$\begin{aligned} \dot{x}_n &= u\cos\psi - v\sin\psi \\ \dot{y}_n &= u\sin\psi + v\cos\psi \end{aligned} \tag{9.92}$$

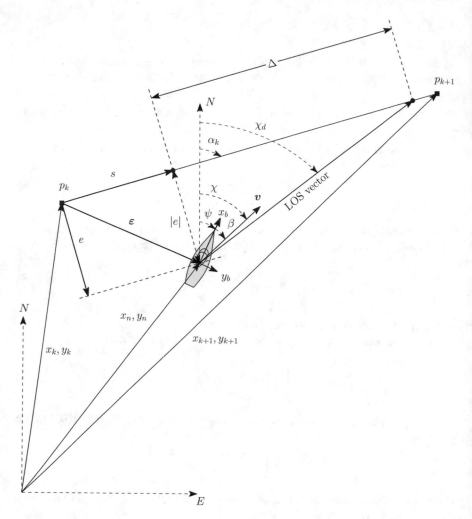

**Fig. 9.10** Line of sight path-following definitions

so that

$$\dot{e} = -(u \cos \psi - v \sin \psi) \sin \alpha_k + (u \sin \psi + v \cos \psi) \cos \alpha_k. \qquad (9.93)$$

The control objective is to drive the cross-track error to zero by steering the vessel. The Lookahead-Based Steering Method will be used [4, 5, 10, 21].

In order to drive the cross track error to zero, a *look-ahead distance* $\Delta$ is defined along the path between $s$ and the next waypoint $p_{k+1}$. The vessel is then steered so that its velocity vector is parallel to the LOS vector (Fig. 9.10). The resulting velocity vector will have a component perpendicular to the path, driving the vessel towards the path until the LOS vector is parallel to the path and $e \rightarrow 0$.

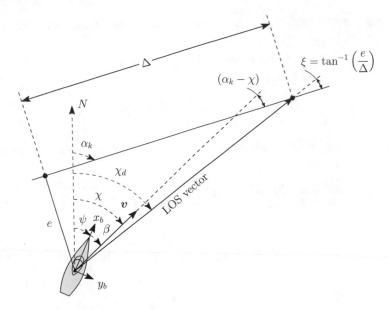

**Fig. 9.11** Relation between $\alpha_k$, $\chi$ and $\chi_d$

From Fig. 9.11 it can be seen that when the velocity vector $\mathbf{v}$ and LOS vector are aligned, so that $\chi = \chi_d$, the angles $(\alpha_k - \chi)$ and $\xi$ will be the same. The change in angle required to achieve this is given by

$$\chi - \chi_d = -(\alpha_k - \chi) + \xi = -(\alpha_k - \chi) + \tan^{-1}\left(\frac{e}{\Delta}\right). \qquad (9.94)$$

Solving for $\chi_d$, gives

$$\chi_d = \alpha_k - \xi = \alpha_k + \tan^{-1}\left(-\frac{e}{\Delta}\right). \qquad (9.95)$$

Define the slide slip angle (see Sect. 1.5.1) to be

$$\beta = \sin^{-1}\left(\frac{v}{U}\right) = \sin^{-1}\left(\frac{v}{\sqrt{u^2 + v^2}}\right), \qquad (9.96)$$

where $U$ is the magnitude of the velocity vector $\mathbf{v}$, which can be expressed in the body-fixed reference frame as $U = \sqrt{u^2 + v^2}$ and in the NED reference frame as $U = \sqrt{\dot{x}_n^2 + \dot{y}_n^2}$

For the purposes of control, it is simpler to define the control input in terms of a desired heading angle $\psi_d$, instead of the desired course angle $\chi_d$. From Fig. 9.11 it can be seen that

$$\chi = \psi + \beta \Rightarrow \chi_d = \psi_d + \beta. \qquad (9.97)$$

The desired course angle and desired heading can then be related using (9.95), as

$$\chi_d = \alpha_k + \tan^{-1}\left(-\frac{e}{\Delta}\right) = \psi_d + \beta, \tag{9.98}$$

so that

$$\psi_d = \alpha_k + \tan^{-1}\left(-\frac{e}{\Delta}\right) - \beta. \tag{9.99}$$

Let the heading error be defined as $\tilde{\psi} := \psi_d - \psi$. Taking the time derivative of the heading error gives

$$\dot{\tilde{\psi}} = \dot{\psi} - \dot{\psi}_d = \dot{\psi} - \frac{d}{dt}\left[\alpha_k + \tan^{-1}\left(-\frac{e}{\Delta}\right) - \beta\right]. \tag{9.100}$$

Let

$$\tan \xi := \left(\frac{e}{\Delta}\right). \tag{9.101}$$

Then

$$\frac{d}{dt}\tan \xi = \frac{1}{\cos^2 \xi}\dot{\xi}$$

and

$$\frac{d}{dt}\left(\frac{e}{\Delta}\right) = \frac{\dot{e}}{\Delta},$$

so that

$$\dot{\xi} = \cos^2 \xi \frac{\dot{e}}{\Delta}. \tag{9.102}$$

From Fig. 9.11 it can be seen that $\cos \xi = \Delta/\sqrt{e^2 + \Delta^2}$. Substituting this expression into (9.100) above yields

$$\frac{d}{dt}\left[\tan^{-1}\left(-\frac{e}{\Delta}\right)\right] = -\dot{\xi},$$

$$= -\frac{\Delta}{(e^2 + \Delta^2)}\dot{e}. \tag{9.103}$$

The the time derivative $\dot{\tilde{\psi}}$ is

$$\dot{\tilde{\psi}} = (r - r_d) + r_d + \frac{\Delta}{(e^2 + \Delta^2)}\dot{e} + \dot{\beta},$$

$$= \tilde{r} + r_d + \frac{\Delta}{(e^2 + \Delta^2)}\dot{e} + \dot{\beta}, \tag{9.104}$$

where $\tilde{r} := r - r_d$ is the yaw rate error and $r_d$ is the the desired yaw rate. Thus, let

$$r_d := -\frac{\Delta}{(e^2 + \Delta^2)}\dot{e} - \dot{\beta} - k_\psi\tilde{\psi}, \tag{9.105}$$

be a *virtual control input* such that the heading error dynamics become

$$\dot{\tilde{\psi}} = -k_\psi\tilde{\psi} + \tilde{r}, \tag{9.106}$$

where $k_\psi > 0$ is a control design parameter.

**Remark 9.2** (*Sideslip and its time derivative*) Note that $\beta$ and $\dot{\beta}$ can often be neglected when computing $\psi_d$ in (9.99) and $r_d$ in (9.105). However, these terms are important when the speed of a vessel is high and significant sideslip is anticipated. In such cases, $\dot{\beta}$ can be computing using an online numerical differentiation technique, such as differentiation by linear filtering (see Remark 10.2), or the use of a higher order sliding mode differentiator (Sect. 12.10).

**Remark 9.3** (*Virtual control inputs*) When using nonlinear control there are many situations, such as this one, where the control input for a controlled output value, e.g. the heading error in (9.104), does not appear explicitly in its equation of motion. A common solution is to find some other quantity, the virtual control input, that can be used to relate the true control input to the desired output using a mapping. Virtual control inputs are often physical quantities like forces, torques, or flows (here it is the desired yaw rate), while the true control input might be the deflection of a steering wheel or the throttle setting on a motor.

Next, the definition of the yaw rate error $\tilde{r} := r - r_d$ can be used in (9.37) to write the equation of motion for the yaw rate error as

$$\dot{\tilde{r}} = -\dot{r}_d + f_\psi + a_\psi\frac{\tau_\psi}{m_{33}}.$$

As done with the surge speed controller, a PID control input is designed using inverse dynamics feedback linearization to get

$$\tau_\psi = -\frac{m_{33}}{a_\psi}\left(-\dot{r}_d + f_\psi + k_{rp}\tilde{r} + k_{ri}\int_0^t \tilde{r}dt + k_{rd}\dot{\tilde{r}}\right). \tag{9.107}$$

Thus, using (9.106), the error dynamics of the coupled heading and yaw rate error system are

$$\dot{\tilde{\psi}} = -k_\psi\tilde{\psi} + \tilde{r},$$
$$\dot{\tilde{r}} = -k_{rp}\tilde{r} - k_{ri}\int_0^t \tilde{r}dt - k_{rd}\dot{\tilde{r}}. \tag{9.108}$$

### 9.6.3  Waypoint Switching

The overall path to be followed consists of a set of $n$ piecewise linear straightline segments, which are connected by $n + 1$ waypoints. When the position of the vessel is within a *circle of acceptance* of radius $R_{k+1}$ around waypoint $p_{k+1}$, such that

$$(x_n - x_{k+1})^2 + (y_n - y_{k+1})^2 \le R_{k+1}^2, \tag{9.109}$$

the system selects the next path segment and its bounding waypoints to compute the heading and cross track errors.

**Example 9.6** (*Path following control of stand up paddle board*) In bathymetric survey applications employing unmanned surface vessels, using an inflatable kayak or stand up paddle board (SUP) as the base USV platform can provide significant operational advantages because these platforms are lightweight and can be easily deflated and folded for transport and storage. Consider the path following control of the stand up paddle board system shown in Fig. 9.12.

The inflatable paddle board has length $L = 3.35$ m, beam $B = 0.897$ m and draft of approximately $T = 1$ cm when fully loaded with instrumentation (total weight of the fully-loaded vehicle is $m = 15.5$ kg). The hydrodynamic coefficients appearing in (9.25)–(9.28) are approximated using the relations given in Tables 9.2, 9.3 and 9.4, where $\rho_w$ is the density of water (approximately 998.1 kg/m$^3$ for freshwater and 1025.9 kg/m$^3$ for saltwater), $L_{CG} = L/2$ is the longitudinal center of gravity of the vessel measured from its aft end and $C_d$ is a drag coefficient (taken as $C_d \approx 1$).

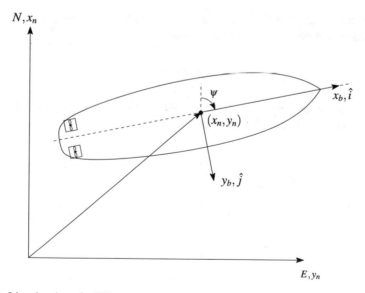

**Fig. 9.12** Line drawing of a SUP unmanned surface vessel

**Table 9.2** Added mass coefficients

| Term | Approximation |
|---|---|
| $N_{\dot{r}}$ | $-1.2\left[\dfrac{4.75}{2}\pi\rho_w\dfrac{B}{2}T^4 + \pi\rho_wT^2\dfrac{(L-L_{\mathrm{CG}})^3+L_{\mathrm{CG}}^3}{3}\right]$ |
| $X_{\dot{u}}$ | $-0.075m$ |
| $Y_{\dot{r}}$ | $-0.4\pi\rho_wT^2\dfrac{(L-L_{\mathrm{CG}})^2+L_{\mathrm{CG}}^2}{2}$ |
| $Y_{\dot{v}}$ | $-1.8\pi\rho_wT^2L$ |

**Table 9.3** Linear drag coefficients

| Term | Approximation |
|---|---|
| $X_u$ | $-1$ |
| $Y_v$ | $-10\rho_w|v|\left[1.1+0.0045\left(\dfrac{L}{T}\right)-0.1\dfrac{B}{T}+0.016\left(\dfrac{B}{T}\right)^2\right]\pi TL$ |
| $N_r$ | $-0.02\pi\rho_w\sqrt{u^2+v^2}\,T^2L^2$ |
| $N_v$ | $-0.06\pi\rho_w\sqrt{u^2+v^2}\,T^2L$ |
| $Y_r$ | $-6\pi\rho_w\sqrt{u^2+v^2}\,T^2L$ |

**Table 9.4** Nonlinear drag coefficients

| Term | Approximation |
|---|---|
| $X_{u|u|}$ | $-5.5$ |
| $Y_{v|v|}$ | $-\rho_wTC_dL$ |
| $Y_{v|r|}$ | $-\rho_wT\dfrac{1.1}{2}\left[(L-L_{\mathrm{CG}})^2+L_{\mathrm{CG}}^2\right]$ |
| $Y_{r|v|}$ | $Y_{v|r|}$ |
| $Y_{r|r|}$ | $-\rho_wT\dfrac{C_d}{3}\left[(L-L_{\mathrm{CG}})^3+L_{\mathrm{CG}}^3\right]$ |
| $N_{v|v|}$ | $Y_{v|r|}$ |
| $N_{v|r|}$ | $Y_{r|r|}$ |
| $N_{r|v|}$ | $Y_{r|r|}$ |
| $N_{r|r|}$ | $-\rho_wT\dfrac{C_d}{4}\left[(L-L_{\mathrm{CG}})^4+L_{\mathrm{CG}}^4\right]$ |

To demonstrate the implementation of a path following controller, a simulation is performed in which the SUP is commanded to follow a *lawn mower pattern* in the NED reference frame at a constant surge speed of 1.0 m/s (dashed line Fig.9.13). For these simulations the path following controller given by (9.83) and (9.108) are implemented using a PI controller (the derivative gains $k_{ud}$ and $k_{rd}$ are set to zero). Here, the slide slip angle $\beta$ in (9.99) and its time derivative $\dot{\beta}$ in (9.105) are assumed to be zero. By manual tuning, the simulation parameters presented in Table 9.5 were found to produce satisfactory performance. The vehicle starts from rest at the origin

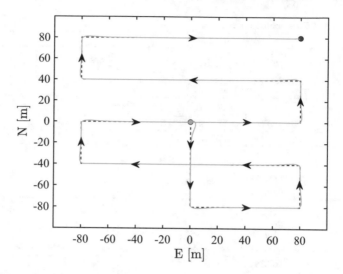

**Fig. 9.13** Desired path (dashed line) and the path followed (green). A green dot indicates the starting waypoint and a red dot shows the location of the final waypoint. Black arrows indicate the direction of desired travel along each segment of the path

**Table 9.5** Path following control parameters

| Parameter | Name | Value |
|---|---|---|
| $\Delta$ | Look-Ahead Distance | $1.5L$ |
| $R_{k+1}$ | Circle of Acceptance | 4 m, $\forall k$ |
| $k_{up}$ | Surge Speed Error Proportional Gain | 10 |
| $k_{ui}$ | Surge Speed Error Integral Gain | 0.5 |
| $k_{\psi}$ | Heading Error Proportional Gain | 0.25 |
| $k_{rp}$ | Yaw Rate Error Proportional Gain | 5 |
| $k_{ri}$ | Yaw Rate Error Integral Gain | 0.5 |

of the NED frame with a heading angle of $\psi = 0°$ and must turn around completely (180°) to reach the first waypoint. The cross track and heading errors are shown in Fig. 9.14. It can be seen that each time the vehicle reaches a waypoint it must turn through a right angle and so the heading error $\tilde{\psi}$ jumps by $\pm 90°$, depending upon whether the vehicle turns left or right. Simultaneously, as the successive path segment is at a right angle to the previous path segment, the cross track error $e$ also discontinuously jumps at each waypoint. The controller is able to drive these rapid changes in $e$ and $\tilde{\psi}$ to zero, so that it follows the path fairly well (green path Fig. 9.13). $\qquad\square$

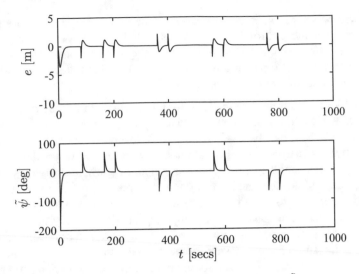

**Fig. 9.14** Time series of the cross track error $e$ and heading angle error $\tilde{\psi}$

### 9.6.4 Other Path Following Approaches

Techniques commonly used for path following include line of sight (as shown above) [2, 11, 32], pure pursuit [7] and vector field methods [13]. Here we have presented an approach for following a path which is fixed in inertial space. Recent work [20, 24, 31] has also explored the *moving path following* control problem involving a path or curve attached to a moving reference frame, which can be useful for applications involving the rendezvous of moving vessels (important for offshore supply problems), or for tracking other vessels or marine life.

## 9.7 Trajectory Tracking for Underactuated Surface Vessels

When planning a trajectory for an underactuated surface vessel, two important considerations must be taken into account.

(1) The trajectory should be *dynamically feasible*. The dynamics of many surface vessels must satisfy second order nonholonomic acceleration constraints. By taking the dynamics of the vehicle into account (in particular its inertial and drag properties, as well as any unactuated directions) when planning a trajectory between a start point $(x(t_0), y(t_0)) = (x_0, y_0)$ and an endpoint $(x(t_f), y(t_f)) = (x_f, y_f)$, where $t_0$ is the time when the vehicle is located at the start point and $t_f$ is the time the vehicle is located at the end point, one can be more certain that the closed loop system will be capable of tracking the planned trajectory.

(2) As will be seen in subsequent chapters, when designing a trajectory tracking
    controller, some control design methods include second order or higher order
    time derivatives of the desired trajectory in the control input. To ensure that the
    associated control inputs are bounded, one must plan trajectories which have
    smooth higher order derivatives in time.

Here, the focus is on trajectory tracking for underactuated surface vessels. We first
review some techniques common in point-to-point motion planning for robotic
manipulators, which can be used to specify a desired time-dependent trajectory
$(x_d(t), y_d(t))$ with a desired level of smoothness between two waypoints. The point-
to-point motion planning approach specifies only the coordinates $(x_d(t), y_d(t))$. The
associated desired heading angle $\psi_d(t)$ at each point can be determined by using
the vehicle dynamics to solve for the heading angle that provides dynamically feasible motion. Here, the approach for computing the dynamically feasible heading
proposed in [1] is presented. An advantage of the method is that it can also be used
to compute the surge force and yaw moment required to achieve the dynamically
feasible heading, which can be used for either open loop control of an underactuated
vessel, or as feedforward control inputs in a two degree of freedom controller (see
Sect. 6.7).

## 9.7.1 Point-to-Point Motion Planning

Point-to-point motion planning involves determining a trajectory from some initial
point $(x_0, y_0)$ to a final point $(x_f, y_f)$. For marine vessels, the points are normally
specified as GPS waypoint coordinates, but over short distances they can be often
be locally represented as $x$, $y$ coordinates in a Cartesian North-East-Down reference
frame. In addition to the starting and ending points, intermediate waypoints (known
as *via points* in robotics) may be used to specify the trajectory. Together with such
a set of points, a set of constraints may be prescribed, such as the initial and final
velocities of the vessel, or its initial and final accelerations. Lastly, as mentioned
above, the use of a given control technique may require higher order derivatives of
the desired trajectory to be bounded. This introduces an additional requirement on
the smoothness of the trajectory.

A common approach to simultaneously satisfying all of the constraints listed
above is to formulate a trajectory using a curve that can be parameterized using a
finite number of coefficients. For example, a polynomial of order $(k - 1)$ can be used
to satisfy a total of $k$ constraints specified at the points $(x_0, y_0)$ and $(x_f, y_f)$ [28].

**Example 9.7** (*Cubic polynomial trajectories*) In some cases it may be necessary
to specify the initial position and velocity and the final position and velocity of
a trajectory. In this cases the number of endpoint constraints is $k = 4$, so that a
$(k - 1) = 3^{\text{rd}}$ order (cubic) polynomial of the form

$$x(t) = a_0 + a_1 t + a_2 t^2 + a_3 t^3 \tag{9.110}$$

can be used to generate a trajectory. The coefficients are computed using the endpoints constraints. The velocity along the trajectory in the $x$ direction can be found by simply taking the time derivative of (9.110) to get

$$\dot{x}(t) = a_1 + 2a_2 t + 3a_3 t^2. \tag{9.111}$$

Let the positions and velocities of the end points be given by

$$x_0 = x(t_0), \quad \dot{x}_0 = \dot{x}(t_0), \quad x_f = x(t_f), \quad \text{and} \quad \dot{x}_f = \dot{x}(t_f), \tag{9.112}$$

where $t_0$ and $t_f$ are the times at the start and end of the trajectory, respectively. Then, using (9.110)–(9.112), a system of equations for the endpoint constraints can be written as

$$
\begin{aligned}
x_0 &= a_0 + a_1 t_0 + a_2 t_0^2 + a_3 t_0^3, \\
\dot{x}_0 &= a_1 + 2a_2 t_0 + 3a_3 t_0^2, \\
x_f &= a_0 + a_1 t_f + a_2 t_f^2 + a_3 t_f^3, \\
\dot{x}_f &= a_1 + 2a_2 t_f + 3a_3 t_f^2.
\end{aligned}
\tag{9.113}
$$

This system of equations can be readily solved by hand one equation at a time, or formulated into the matrix form

$$
\begin{bmatrix} x_0 \\ \dot{x}_0 \\ x_f \\ \dot{x}_f \end{bmatrix}
=
\begin{bmatrix}
1 & t_0 & t_0^2 & t_0^3 \\
0 & 1 & 2t_0 & 3t_0^2 \\
1 & t_f & t_f^2 & t_f^3 \\
0 & 1 & 2t_f & 3t_f^2
\end{bmatrix}
\begin{bmatrix} a_0 \\ a_1 \\ a_2 \\ a_3 \end{bmatrix}
\tag{9.114}
$$

and solved numerically, by using Matlab to invert the $4 \times 4$ matrix above, for example. Note that the determinant of the matrix is $(t_f - t_0)^4$. Thus, it is nonsingular and a unique solution is guaranteed, as long as $t_0 \neq t_f$ [28].

A trajectory requires a pair of points in the NED frame to be specified at each time $t_0 \leq t \leq t_f$. The corresponding $y(t)$ values of the trajectory are determined in the same way using the endpoint constraints $y_0 = y(t_0)$, $\dot{y}_0 = \dot{y}(t_0)$, $y_f = y(t_f)$, and $\dot{y}_f = \dot{y}(t_f)$.  □

The use of cubic polynomials for trajectory planning leads to discontinuous acceleration, known as *jerk* in robotics, at $t_0$ and $t_f$. This can certainly be undesirable for a manned vessel under automatic control, as the vehicle would suddenly lurch forward when starting and feel as if though it was abruptly stopping at the end of the trajectory, which if unexpected, could knock a standing passenger off balance. From a control design standpoint, it can also be unacceptable when the use of some types of nonlinear trajectory tracking controllers is desired. For example, when taking actuator and rate limits into account in the design of a backstepping trajectory tracking controller the first, second and third order time derivatives of the desired trajectory must be smooth and bounded [29]. In such cases, a higher order polynomial can be used to construct the trajectory using the same approach outlined in Example 9.7 above.

When a larger number of waypoints is required to specify a trajectory to ensure that specific locations are traversed, the trajectory can also be constructed using higher order polynomials to satisfy the constraints introduced at multiple waypoints [10, 28]. For example, given three waypoints $x_0$, $x_1$ and $x_f$, constraints can be placed on the velocity and acceleration at $x_0$ and $x_f$ and the position $x_1$. The resulting velocity and acceleration at all points along the trajectory will be smooth and continuous. To satisfy the $k = 7$ constraints a $(k - 1) = 6$th order polynomial of the form

$$x(t) = a_0 + a_1 t + a_2 t^2 + a_3 t^3 + a_4 t^4 + a_5 t^5 + a_6 t^6 \tag{9.115}$$

will be required.

A drawback of this approach is that the order of the polynomial needed to produce a trajectory across a set of points increases as the number of points increases. A way of circumventing this is to split the trajectory into a series of sections divided by the intermediate waypoints and to use a lower order polynomial to model the trajectory along each section. To ensure that the velocity and acceleration are smooth and continuous across the entire trajectory, the initial conditions at the beginning of each section (e.g. the velocity and acceleration) are set equal to the final conditions at the end of the previous section.

Another approach to generating a smooth trajectory is to pass a discontinuous trajectory through a low pass filter (see Sect. 10.3.3). In practice, using a filter to smooth a trajectory can be relatively easy to implement. However, the smoothed trajectory is only approximate. One cannot guarantee that the trajectory will pass exactly through the desired waypoints at specifically desired times.

## 9.7.2  Desired Heading and Feedforward Control Inputs

Once we have a time-dependent set of $x_d(t)$, $y_d(t)$ positions that we would like to follow, we need to determine the corresponding values of $\psi(t)$ that are dynamically feasible. To do this, we start by using the kinematic equations of motion to determine what the corresponding sway acceleration $\dot{v}_d$ at each instant in time would be. While an underactuated vehicle cannot directly apply a force in the sway direction, for most marine vehicles the kinetic equations of motion for sway and yaw are coupled via the added mass terms occurring in their inertia tensors. We take advantage of this coupling to find a virtual control input (see Remark 9.3) relating the time derivative of the yaw rate $\dot{r}$ to $\dot{v}_d$, somewhat akin to the approach used for path following in Sect. 9.6. The method shown here was developed by [1].

From (9.21) and (9.22), the velocity of the vessel in the body-fixed frame can be related to its velocity in the inertial (NED) frame, as

$$\mathbf{v} = \mathbf{R}^{-1}(\psi)\dot{\boldsymbol{\eta}} = \mathbf{R}^T(\psi)\dot{\boldsymbol{\eta}}.$$

From this, one can obtain an expression of the sway speed $v$ in terms of the velocities in the NED frame as

$$v = -\dot{x}_n \sin \psi + \dot{y}_n \cos \psi.$$

Let $(x_d(t), y_d(t))$ be a sufficiently smooth trajectory designed using a point-to-point planning method represented in the NED reference frame. Then, the desired sway speed $v_d(t)$ is given by

$$v_d(t) = -\dot{x}_d \sin \psi + \dot{y}_d \cos \psi.$$

Taking the time derivative of this gives the desired sway acceleration

$$\dot{v}_d(t) = -\ddot{x}_d \sin \psi - \dot{x}_d r \cos \psi + \ddot{y}_d \cos \psi - \dot{y}_d r \sin \psi. \tag{9.116}$$

Next we turn to the kinetic equations of motion to find the relationship between $\dot{r}$ and $\dot{v}_d$. Using (9.37), we can eliminate $\tau_\psi$ in (9.36) to get

$$\dot{v} = f_y + \frac{a_y}{a_\psi} \left[ \dot{r} - f_\psi \right]. \tag{9.117}$$

To find the virtual control input $\dot{r}_d$ that gives the heading angle required for a dynamically feasible trajectory, define the sway acceleration error as $\overset{\cdot}{\tilde{v}} := \dot{v} - \dot{v}_d$, where $\dot{v}_d$ is given by (9.116). Then, using (9.117) we have

$$\overset{\cdot}{\tilde{v}} = f_y + \frac{a_y}{a_\psi} \left[ \dot{r} - f_\psi \right] - \dot{v}_d,$$

so that the virtual control input $\dot{r}_d$ that makes $\overset{\cdot}{\tilde{v}} = 0$ is given by

$$\dot{r}_d = \frac{a_\psi}{a_y} \left[ \dot{v}_d - f_y \right] + f_\psi. \tag{9.118}$$

Thus, given a desired trajectory $(x_d(t), y_d(t))$ one can completely specify the dynamically feasible time-dependent pose $\eta_d(t) = [x_d \ y_d \ \psi_d]^T$ of an underactuated surface vessel using the associated sway acceleration $\dot{v}_d(t)$ from (9.116) and integrating (9.118) twice to obtain the desired heading angle $\psi_d(t)$, subject to the initial conditions $\psi_d(t = t_0)$ and $\dot{\psi}_d(t = t_0)$.

Further, (9.33) and (9.37) can be used to find a set of feedforward control inputs that can be used in a two degree of freedom control system architecture for trajectory tracking (see Sect. 6.7). To do this, use (9.22) to obtain an expression for the desired surge speed $u_d$ in terms of the desired velocities in the NED frame, as

$$u_d = \dot{x}_d \cos \psi + \dot{y}_d \sin \psi.$$

The desired surge acceleration is then

$$\dot{u}_d = \ddot{x}_d \cos \psi - \dot{x}_d r \sin \psi + \ddot{y}_d \sin \psi + \dot{y}_d r \cos \psi.$$

Replacing $\dot{u}$ with $\dot{u}_d$ in (9.33) gives the feedforward surge control input

$$\tau_x = m_{11} (\dot{u}_d - f_x), \tag{9.119}$$

and similarly replacing $\dot{r}$ with $\dot{r}_d$ in (9.37) gives the feedforward yaw moment control input

$$\tau_\psi = \frac{m_{33}}{a_\psi} (\dot{r}_d - f_\psi), \tag{9.120}$$

that will produce the desired dynamically feasible trajectory in the absence of disturbances. A separate feedback controller can be used to ensure that the system is robust to disturbances and modeling uncertainties.

**Example 9.8** (*USV trajectory planning*) Consider the problem of constructing dynamically feasible trajectories for a USV with the following inertial properties

$$M = \begin{bmatrix} 189.0 & 0.0 & 0.0 \\ 0.0 & 1036.4 & -543.5 \\ 0.0 & -543.5 & 2411.1 \end{bmatrix}.$$

The drag is assumed to be linear and is modeled with the following coefficients

$$D = \begin{bmatrix} 50.0 & 0.0 & 0.0 \\ 0.0 & 948.2 & 385.4 \\ 0.0 & 385.4 & 1926.9 \end{bmatrix}.$$

Using these values in (9.38) $a_y$ and $a_\psi$ are found to be

$$a_y = 0.595 \quad \text{and} \quad a_\psi = 1.134.$$

Here we will examine three cases:

(a) The vehicle moves in a straight line from an initial point $(x_0, y_0) = (0, 0)$ at $t_0 = 0$ to a final point $(x_f, y_f) = (-30, -30)$ at $t_f = 50$ s, where the $(x, y)$ coordinates are specified in meters. The initial velocity of the USV is 1 m/s in the northward $(x)$ direction. Here, the trajectory is designed using a cubic polynomial.

(b) The vehicle moves in a straight line from an initial point $(x_0, y_0) = (0, 0)$ at $t_0 = 0$ to a final point $(x_f, y_f) = (-30, 0)$ at $t_f = 50$ s, where the $(x, y)$ coordinates are specified in meters. The initial velocity of the USV is 1 m/s in the northward $(x)$ direction. Here, the trajectory will designed using a quintic (5th order) polynomial.

(c) The vehicle follows a circular trajectory of radius $R = 10$ m in the clockwise direction, starting from $(x_0, y_0) = (0, 0)$ at $t_0 = 0$ and finishing at the same position $(x_f, y_f) = (0, 0)$ at $t_f = 50$ s, where the $(x, y)$ coordinates are specified in meters. The center of the circle is located at the point $(x_c, y_c) = (0, 10)$.

**Case a:** In the first case, a cubic polynomial is selected to represent the trajectory. Using the approach outlined in Example 9.7 above, the $x_d(t)$ and $y_d(t)$ positions along the trajectory are given by

$$x_d(t) = t - 0.076t^2 + 8.8 \times 10^{-4}t^3,$$
$$y_d(t) = -0.036t^2 + 4.8 \times 10^{-4}t^3.$$

These $x_d(t)$ and $y_d(t)$ positions and the corresponding desired heading angle $\psi_d(t)$ are shown in Fig. 9.15. Since the USV has an initial heading and speed northward, it must rotate counterclockwise to reach $(x_f, y_f) = (-30, -30)$. As can be seen in Fig. 9.16 the vehicle dynamics predict that the vehicle will have a sway speed as it executes the turn to the southwest and that there will be some oscillation in both the surge speed and yaw rate. The corresponding control inputs oscillate during the turn. As a cubic polynomial is used for trajectory planning, even though the desired surge speed is zero at $(x_f, y_f)$, the control input is nonzero at the end of the trajectory. This will result in a jerk at the end of the trajectory.

**Case b:** In the second case, a 5th order polynomial of the form

$$x(t) = a_0 + a_1 t + a_2 t^2 + a_3 t^3 + a_4 t^4 + a_5 t^5 \tag{9.121}$$

is used to represent the trajectory. The coefficients can be determined following the approach shown in Example 9.7 in which the system of equations describing the endpoint constraints is written in matrix form and the coefficients are solved as a vector. The resulting desired trajectory is

$$x_d(t) = t - 4.80 \times 10^{-3}t^3 + 1.36 \times 10^{-4}t^4 - 1.06 \times 10^{-6}t^5,$$
$$y_d(t) = 0.$$

As can be seen in Figs. 9.17, 9.18, the vehicle does not need to turn to perform this maneuver and $\psi_d(t) = 0$ throughout the trajectory. Since the $x_d(t)$ and $y_d(t)$ positions determined using the polynomial require the vehicle to move forward for several seconds, the $\tau_x$ control input remains positive for about 10 s, before reversing to drive the vehicle backwards towards the final point at $(x_f, y_f) = (-30, 0)$. Note that the desired surge speed and the surge control input both smoothly approach zero as the vehicle approaches the end of the trajectory.

**Case c:** In the third case, the trajectory is given by

$$x_d(t) = R \sin(\omega_d t),$$
$$y_d(t) = R[1 - \cos(\omega_d t)],$$

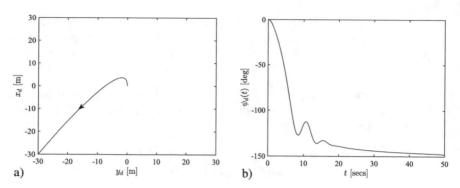

**Fig. 9.15**  Case a: **a** trajectory and **b** desired heading angle

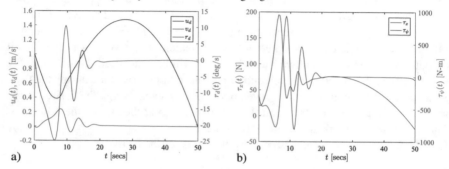

**Fig. 9.16**  Case a: **a** desired body-fixed velocities and **b** control inputs

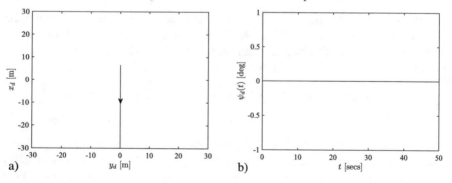

**Fig. 9.17**  Case b: **a** trajectory and **b** desired heading angle

where $\omega_d = 2\pi/t_f$. After a small initial transient, the desired yaw rate becomes constant so that $r_d = \omega_d$ (see Figs. 9.19, 9.20). In order to maintain the turn, vehicle has a small desired sway speed $v_d$ yaw moment control input $\tau_\psi$ throughout the trajectory.                                                                                        □

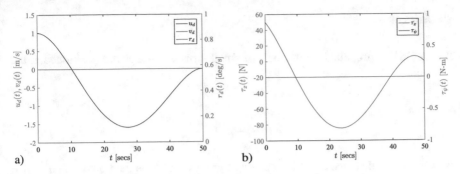

**Fig. 9.18** Case b: **a** desired body-fixed velocities and **b** control inputs

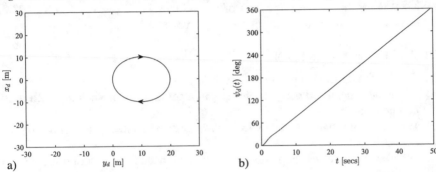

**Fig. 9.19** Case c: **a** trajectory and **b** desired heading angle

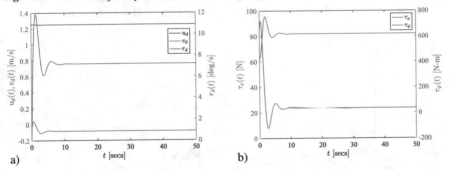

**Fig. 9.20** Case c: **a** desired body-fixed velocities and **b** control inputs

## Problems

**9.1** A USV is propelled using a pair of small thrusters, which are positioned at the locations

$$\mathbf{r}_p := -l_t \hat{i} - \frac{b_t}{2}\hat{j},$$

$$\mathbf{r}_s := -l_t \hat{i} + \frac{b_t}{2}\hat{j},$$

where $l_t$ is the distance aft of the center of gravity and $b_t$ is the transverse distance between the thrusters (see Fig. 9.21). Each thruster is oriented so that the thrust produced is parallel to the surge axis of the vehicle, i.e. $\mathbf{T}_p = T_p\hat{i}$ and $\mathbf{T}_s = T_s\hat{i}$. Write the three component vector of control forces $\boldsymbol{\tau} = [\tau_x \ \tau_y \ \tau_\psi]^T$.

**9.2** An AUV operates in 6 DOF.

(a) What is the dimension of its configuration space $n$?
(b) Write the generalized position (pose) $\boldsymbol{\eta}$ as a column vector, showing each of its components.
(c) Write the body-fixed velocity $\mathbf{v}$ as a column vector, showing each of its components.
(d) What is the order of the system of equations needed to describe the AUV's motion?
(e) Suppose the AUV employs a set of 4 independently controlled fore-aft tunnel thrusters that produce forces in the heave and sway directions, a moving mass

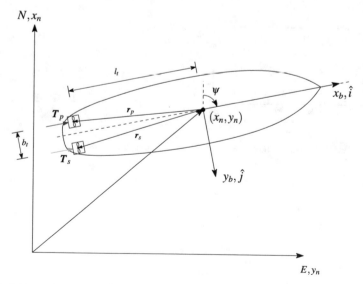

**Fig. 9.21** USV thruster configuration (view from top)

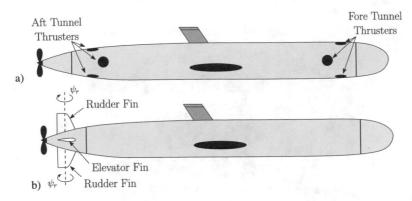

**Fig. 9.22** An AUV steered using **a** fore-aft tunnel thrusters and **b** rudder and elevator planes (fins) at the stern

system to produce a torque about its roll axis and a propeller at its stern to produce thrust along its surge axis (Fig. 9.22a). What is the number of independent control inputs $r$? Is the vehicle fully-actuated or underactuated?

(f) Suppose the stern of the AUV has a propeller, a set of elevator planes that move in unison (i.e. they are connected by a rigid shaft and are constrained to move by the same amount in the same direction), and a set of rudder planes that move in unison (Fig. 9.22b). What is the number of independent control inputs $r$? Is the vehicle fully-actuated or underactuated?

**9.3** A surface vessel restricted to operate in the horizontal plane (surge, sway and yaw).

(a) What is the dimension of its configuration space $n$?
(b) Write the generalized position (pose) $\eta$ as a column vector, showing each of its components.
(c) Write the body-fixed velocity $\mathbf{v}$ as a column vector, showing each of its components.
(d) What is the order of the system of equations needed to describe the surface vessel's motion?
(e) Suppose that the surface vessel uses two fixed thrusters at its stern, like that shown in Fig. 9.21. What is the number of independent control inputs $r$? Is the vehicle fully-actuated or underactuated?
(f) Suppose the surface vessel uses two azimuthing thrusters at its stern, like that shown in Fig. 9.23. What is the number of independent control inputs $r$? Is the vehicle fully-actuated or underactuated?

**9.4** You are designing a trajectory tracking controller for an underactuated surface vessel, as shown in Fig. 9.24. The desired trajectory is a circle of constant radius. Is

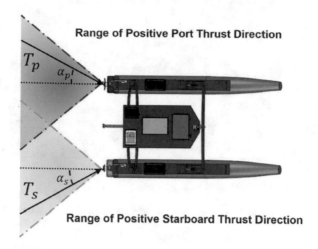

**Fig. 9.23** Top view of a surface vessel with azimuthing thrusters. ©[2009] IEEE. Reprinted, with permission, from [3]

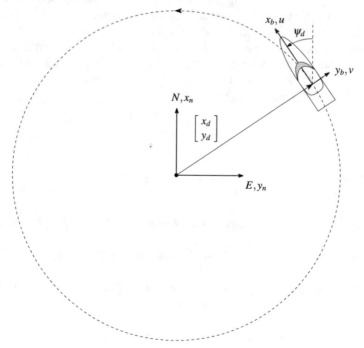

**Fig. 9.24** A surface vessel tracking a circular trajectory

it possible to control the vehicle so that the desired heading $\psi_d(t)$ is always tangent to the circle? Explain your reasoning.

[Hint: you may want to consider the results in Case c of Example 9.8 and Figs. 9.19–9.20 while formulating your response.]

**9.5** (Adopted from [8]) Let $\mathbf{x} := [x_1 \ x_2]^T$. Consider the system

$$\dot{\mathbf{x}} = \begin{bmatrix} x_2^2 \\ 0 \end{bmatrix} + \begin{bmatrix} 0 \\ 1 \end{bmatrix} u.$$

(a) Show that the system is locally accessible.
(b) Is the system small-time locally controllable? Explain why or why not.

**9.6** Consider a driftless two-input control system with the chained form

$$\begin{aligned}
\dot{\eta}_1 &= u_1, \\
\dot{\eta}_2 &= u_2, \\
\dot{\eta}_3 &= \eta_2 u_1, \\
\dot{\eta}_4 &= \eta_3 u_1, \\
&\ \vdots \\
\dot{\eta}_n &= \eta_{n-1} u_1.
\end{aligned}$$

(a) Show that the system is driftless and can be written in vector form

$$\dot{\eta} = \mathbf{j}_1(\eta) u_1 + \mathbf{j}_2(\eta) u_2.$$

Explicitly write out the components of $\mathbf{j}_1$ and $\mathbf{j}_2$.

(b) Show that if $u_1 = 1$, the system becomes linear and behaves like a chain of integrators from $\eta_2$ to $\eta_n$, driven by the input $u_2$.

(c) Show that the (repeated) Lie brackets $\mathrm{ad}_{\mathbf{j}_1}^k \mathbf{j}_2$ are given by

$$\mathrm{ad}_{\mathbf{j}_1}^k \mathbf{j}_2 = \begin{bmatrix} 0 \\ 0 \\ \vdots \\ (-1)^k \\ 0 \end{bmatrix}, \quad 1 \le k \le n-2,$$

where the only nonzero element is the $(k+2)^{\text{nd}}$ entry.

(d) Using the result from part c), construct the accessibility distribution

$$\Delta_C = \mathrm{span}\left\{ \mathbf{j}_1, \mathbf{j}_2, \mathrm{ad}_{\mathbf{j}_1} \mathbf{j}_2, \dots, \mathrm{ad}_{\mathbf{j}_1}^{n-2} \mathbf{j}_2 \right\}$$

and use Theorem 9.1 to prove that the system is controllable.

**9.7** Show that the USV with acceleration constraints of Example 9.3 is locally accessible by computing the dimension of the span of

$$\Delta_C = \text{span}\,\{\mathbf{g}_1, \mathbf{g}_2, [\mathbf{f}, \mathbf{g}_1], [\mathbf{f}, \mathbf{g}_2], [\mathbf{g}_2, [f, \mathbf{g}_1]], [\mathbf{g}_2, [[\mathbf{g}_1, \mathbf{f}], \mathbf{f}]]\}\,.$$

**Hint:** One approach is to use Matlab to compute the Lie brackets appearing in $\Delta_C$, as shown in Example 8.10. The $\dim\Delta_C$ can then be computed by constructing a matrix with the vectors $\mathbf{g}_1, \mathbf{g}_2, [\mathbf{f}, \mathbf{g}_1], [\mathbf{f}, \mathbf{g}_2], [\mathbf{g}_2, [f, \mathbf{g}_1]]$ and $[\mathbf{g}_2, [[\mathbf{g}_1, \mathbf{f}], \mathbf{f}]]$ as its columns and using Matlab to find its rank.

**9.8** A desired trajectory has the endpoint constraints $x_0 = 10$ m and $\dot{x}_0 = 0$ m/s at $t_0 = 0$ s and $x_f = 30$ m and $\dot{x}_f = 0$ at $t_f = 10$ s.

(a) What is the order of the polynomial needed to construct this trajectory?
(b) Determine the coefficients of the polynomial.
(c) Plot the position $x(t)$ and the speed $\dot{x}(t)$ versus time $t$ to confirm that the trajectory works as planned.

**9.9** A desired trajectory has the endpoint constraints $x_0 = 0$ m, $\dot{x}_0 = 1$ m/s and $\ddot{x}_0 = 0$ m/s$^2$ at $t_0 = 5$ s and $x_f = -30$ m, $\dot{x}_f = 0$ m/s and $\ddot{x}_f = 0$ m/s$^2$ at $t_f = 20$ s.

(a) What is the order of the polynomial needed to construct this trajectory?
(b) Determine the coefficients of the polynomial.
(c) Plot the position $x(t)$, speed $\dot{x}(t)$ and acceleration $\ddot{x}(t)$ versus time $t$ to confirm that the trajectory works as planned.

**9.10** A trajectory is specified by three waypoints and times, $x_0(t_0) = 0$ m at $t_0 = 0$ s, $x_1(t_1) = 5$ m at $t_1 = 10$ s and $x_2(t_2) = 15$ m at $t_2 = 20$ s. The initial and final velocities are required to be $\dot{x}_0(t_0) = 0$ m/s and $\dot{x}_2(t_2) = 0$ m/s, respectively.

(a) If a single polynomial is used to construct the trajectory, what is the order of the polynomial needed?
(b) Determine the coefficients of the polynomial.
(c) Plot the position $x(t)$ and speed $\dot{x}(t)$ versus time $t$ to confirm that the trajectory works as planned.

**9.11** A trajectory is specified by three waypoints and times, $x_0(t_0) = 0$ m at $t_0 = 0$ s, $x_1(t_1) = 5$ m at $t_1 = 10$ s and $x_2(t_2) = 15$ m at $t_2 = 20$ s. The initial and final velocities are required to be $\dot{x}_0(t_0) = 0$ m/s and $\dot{x}_2(t_2) = 0$ m/s, respectively. Divide the trajectory into two sections divided by the waypoint $x_1$ and use a cubic polynomial to model the trajectory along each section.

(a) What endpoint conditions must be matched between the two sections at $x_1$?
(b) Determine the coefficients of the two cubic polynomials.
(c) Plot the position $x(t)$ and speed $\dot{x}(t)$ versus time $t$ to confirm that the trajectory works as planned.

**9.12** Create a Matlab/Simulink simulation of the SUP using the parameters given in Example 9.6.

(a) Can you reproduce the results shown in Example 9.6?
(b) The controllers in Example 9.6 were manually tuned. Try additional tuning of the PI controller gains to improve the performance as measured using the ITAE criterion for $e$ and $\tilde{\psi}$ (Sect. 3.5).
(c) Add a derivative term to the controller. Are you able to further improve the performance?

**9.13** Create a simulation to reproduce the results shown in case a of Example 9.8. What happens to the results if the initial velocity is given by $\dot{x} = 0$ and $\dot{y} = 0$ at time $t_0 = 0$? Explain why.

**9.14** Using the values provided in case c of Example 9.8, create a simulation to determine the desired heading, body-fixed velocity and control inputs for the trajectory given by the points

$$x(t) = R\sin(\omega t),$$
$$y(t) = -R[1 - \cos(\omega t)].$$

# References

1. Ashrafiuon, H., Nersesov, S., Clayton, G.: Trajectory tracking control of planar underactuated vehicles. IEEE Trans. Autom. Control **62**(4), 1959–1965 (2017)
2. Belleter, D., Maghenem, M.A., Paliotta, C., Pettersen, K.Y.: Observer based path following for underactuated marine vessels in the presence of ocean currents: a global approach. Automatica **100**, 123–134 (2019)
3. Bertaska, I., von Ellenrieder, K.: Experimental evaluation of supervisory switching control for unmanned surface vehicles. IEEE J. Oceanic Eng. **44**(1), 7–28 (2019)
4. Borhaug, E., Pavlov, A., Pettersen, K.Y.: Cross-track formation control of underactuated surface vessels. In: Proceedings of the 45th IEEE Conference on Decision and Control, pp. 5955–5961. IEEE (2006)
5. Breivik, M., Fossen, T.I.: Guidance laws for autonomous underwater vehicles. Underwater Veh. 51–76 (2009)
6. Brockett, R.W.: Asymptotic stability and feedback stabilization. In: Brockett, R.W., Millman, R.S., Sussmann, H.J., (eds.) Differential Geometric Control Theory: Proceedings of the conference held at Michigan Technological University, June 28–July 2, 1982, vol. 27, pp. 181–191. Birkhauser, Boston (1983)
7. Cho, N., Kim, Y., Park, S.: Three-dimensional nonlinear differential geometric path-following guidance law. J. Guid. Control. Dyn. **38**(12), 2366–2385 (2015)
8. De Luca, A., Oriolo, G.: Modelling and control of nonholonomic mechanical systems. In: Kinematics and Dynamics of Multi-body Systems, pp. 277–342. Springer, Berlin (1995)
9. Egeland, O., Berglund, E.: Control of an underwater vehicle with nonholonomic acceleration constraints. IFAC Proc. **27**(14), 959–964 (1994)
10. Fossen, T.I.: Handbook of Marine Craft Hydrodynamics and Motion Control. Wiley, New York (2011)
11. Fossen, T.I., Pettersen, K.Y.: On uniform semiglobal exponential stability (usges) of proportional line-of-sight guidance laws. Automatica **50**(11), 2912–2917 (2014)
12. Ge, S.S., Sun, Z., Lee, T.H., Spong, M.W.: Feedback linearization and stabilization of second-order non-holonomic chained systems. Int. J. Control **74**(14), 1383–1392 (2001)

13. Yuri A Kapitanyuk, Anton V Proskurnikov, and Ming Cao. A guiding vector-field algorithm for path-following control of nonholonomic mobile robots. IEEE Trans. Control Syst. Technol. **26**(4), 1372–1385 (2017)
14. Laiou, M.-C., Astolfi, A.: Discontinuous control of high-order generalized chained systems. Syst. Control Lett. **37**(5), 309–322 (1999)
15. Laiou, M.-C., Astolfi, A.: Local transformations to generalized chained forms. Process Syst. Eng. RWTH Aachen Univ. Templergraben **55**, 1–14 (2004)
16. Mazenc, F., Pettersen, K., Nijmeijer, H.: Global uniform asymptotic stabilization of an under-actuated surface vessel. IEEE Trans. Autom. Control **47**(10), 1759–1762 (2002)
17. M'Closkey, R.T., Murray, R.M.: Nonholonomic systems and exponential convergence: some analysis tools. In Proceedings of 32nd IEEE Conference on Decision and Control, pp. 943–948. IEEE (1993)
18. Murray, R.M., Sastry, S.S.: Nonholonomic motion planning: steering using sinusoids. IEEE Trans. Autom. Control **38**(5), 700–716 (1993)
19. Murray, R.M., Sastry, S.S.: Steering nonholonomic systems in chained form. In Proceedings of the 30th IEEE Conference on Decision and Control, pp. 1121–1126. IEEE (1991)
20. Oliveira, T., Aguiar, A.P., Encarnacao, P.: Moving path following for unmanned aerial vehicles with applications to single and multiple target tracking problems. IEEE Trans. Robot. **32**(5), 1062–1078 (2016)
21. Papoulias, F.A.: Bifurcation analysis of line of sight vehicle guidance using sliding modes. Int. J. Bifurc. Chaos **1**(04), 849–865 (1991)
22. Pettersen, K.Y., Egeland, O.: Exponential stabilization of an underactuated surface vessel. In Proceedings of 35th IEEE Conference on Decision and Control, vol. 1, pp. 967–972. IEEE (1996)
23. Pettersen, K.Y., Mazenc, F., Nijmeijer, H.: Global uniform asymptotic stabilization of an under-actuated surface vessel: Experimental results. IEEE Trans. Control Syst. Technol. **12**(6), 891–903 (2004)
24. Reis, M.F., Jain, R.P., Aguiar, A.P., de Sousa, J.B.: Robust moving path following control for robotic vehicles: Theory and experiments. IEEE Robot. Autom. Lett. **4**(4), 3192–3199 (2019)
25. Mahmut Reyhanoglu, Arjan van der Schaft, N Harris McClamroch, and Ilya Kolmanovsky. Nonlinear control of a class of underactuated systems. In: Proceedings of the 35th IEEE Conference on Decision and Control, vol. 2, pp. 1682–1687. IEEE (1996)
26. Reyhanoglu, M., van der Schaft, A., McClamroch, N.H., Kolmanovsky, I.: Dynamics and control of a class of underactuated mechanical systems. IEEE Trans. Autom. Control **44**(9), 1663–1671 (1999)
27. SNAME. Nomenclature for treating the motion of a submerged body through a fluid: Report of the American Towing Tank Conference. Technical and Research Bulletin 1–5, Society of Naval Architects and Marine Engineers (1950)
28. Spong, M.W., Hutchinson, S., Vidyasagar, M.: Robot Modeling and Control. Wiley, New York (2006)
29. von Ellenrieder, K.D.: Stable backstepping control of marine vehicles with actuator rate limits and saturation. IFAC-PapersOnLine **51**(29), 262–267 (2018)
30. Wichlund, K.Y., Sørdalen, O.J., Egeland, O.: Control of vehicles with second-order nonholonomic constraints: underactuated vehicles. In: Proceedings of the 3rd European Control Conference. Citeseer (1995)
31. Zheng, Z.: Moving path following control for a surface vessel with error constraint. Automatica **118**, 109040 (2020)
32. Zheng, Z., Sun, L., Xie, L.: Error-constrained LOS path following of a surface vessel with actuator saturation and faults. IEEE Trans. Syst., Man, Cybernet.: Syst. **48**(10), 1794–1805 (2017)

# Chapter 10
# Integrator Backstepping and Related Techniques

## 10.1 Introduction

Nonlinear backstepping is a recursive design procedure for constructing feedback control laws and their associated Lyapunov functions. The approach is systematic, flexible and can be applied in vectorial form to MIMO systems. Backstepping methods permit a control designer to exploit "good" nonlinearities, while "bad" nonlinearities can be dominated by adding nonlinear damping, for instance. Hence, it is often possible to obtain additional robustness to model uncertainty and exogenous (externally-generated) disturbances, which is important for the control of marine vehicles because it is typically difficult to obtain precise models in practice. As with pole placement techniques, the most basic form of backstepping generally leads to proportional-derivative-like controllers. However, similar to state space techniques, backstepping permits one to use integrator augmentation to design controllers that also include an integral-like term. Because of its versatility, backstepping is often combined with other control methods, such as optimal control and adaptive control. The backstepping methodology also makes it fairly straightforward to handle actuator constraints. Lastly, a significant advantage of backstepping controllers is that they have globally bounded tracking errors. Because of these advantages, backstepping is one of the most widely used type of control for marine vehicles.

However, a serious drawback is that their implementation often involves an *explosion of complexity*, whereby differentiation of the plant model requires that a very large number of terms be included in the computation of the control law. Dynamic surface control (DSC) is an approach similar to backstepping, which allows one to avoid the "explosion of complexity" by using first order filters to compute the derivatives of the plant model [17].

© Springer Nature Switzerland AG 2021
K. D. von Ellenrieder, *Control of Marine Vehicles*, Springer Series on Naval Architecture, Marine Engineering, Shipbuilding and Shipping 9,
https://doi.org/10.1007/978-3-030-75021-3_10

## 10.2  Integrator Backstepping

Consider the dynamic equations of a marine vehicle in body-fixed coordinates (1.58)–(1.59), which are reproduced here for convenience

$$\dot{\eta} = \mathbf{J}(\eta)\mathbf{v},$$
$$\mathbf{M}\dot{\mathbf{v}} = -\mathbf{C}(\mathbf{v})\mathbf{v} - \mathbf{D}(\mathbf{v})\mathbf{v} - \mathbf{g}(\eta) + \tau + \mathbf{w}_d.$$

Note the structure of the equations. The terms $\eta$ and $\mathbf{v}$ are state variables. The equations are coupled and there are functions of the state variables (i.e. $\mathbf{J}(\eta)$, $\mathbf{C}(\mathbf{v})$, and $\mathbf{D}(\mathbf{v})$) that are multiplied by the state variable $\mathbf{v}$. The coupling created by the products of nonlinear functions of the state variables multiplying the state variables makes it difficult to identify a control input that can be used to stabilize the system around a desired equilibrium point. Backstepping techniques provide means of decoupling the dynamic equations so that they can be stabilized.

For many applications in marine vehicles it will be necessary to use a vector form of backstepping. The vector form contains many terms, such as stabilizing nonlinearities and Lie derivatives of functions that are unlikely to be familiar to engineers who have not already used these methods. Therefore, the backstepping concepts are introduced gradually here. Starting with a simple two-state SISO system, terms are gradually added to create a three-state SISO system, then an $n$-state SISO system is examined, until finally arriving at the full vectorial form of backstepping.

### 10.2.1  A Simple 2-State SISO System

Consider the single-input single-output (SISO) system

$$\dot{x}_1 = f_1(x_1) + g_1(x_1)x_2, \tag{10.1}$$
$$\dot{x}_2 = u, \tag{10.2}$$
$$y = x_1,$$

where $f_1(x_1)$ and $g_1(x_1)$ are known functions. It is assumed that $g_1(x_1) \neq 0$ for all $x_1$ in the domain of interest. The control objective is to design a state feedback control law that tracks a desired output $y_d(t)$.

The system can be viewed as a cascade connection of two components. The first component (10.1) can be thought of as having the state $x_2$ as a *virtual control input* and the second component (10.2) is an integrator with $u$ as input (see Fig. 10.1). Since there are two states $x_1$ and $x_2$, the design will be conducted recursively in 2 steps.

Suppose that we can find a smooth *stabilizing function* $\alpha_1(x_1)$ that permits us to decouple the system by driving $x_2 \rightarrow \alpha_1$. To see how this can work, let us add and subtract $g_1(x_1)\alpha_1(x_1)$ to the right hand side of (10.1) to obtain a new system

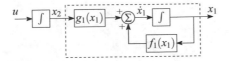

**Fig. 10.1** Block diagram of system (10.1)–(10.2)

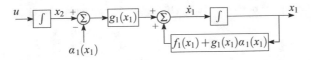

**Fig. 10.2** Block diagram of system (10.3)–(10.4). The stabilizing function $\alpha_1(x_1)$ is introduced to stabilize the $x_1$ system (10.3) at $x_1 = 0$

$$\dot{x}_1 = [f_1(x_1) + g_1(x_1)\alpha_1(x_1)] + g_1(x_1)[x_2 - \alpha_1(x_1)] \tag{10.3}$$

$$\dot{x}_2 = u. \tag{10.4}$$

Define the stabilization errors as

$$z_1 := y - y_d = x_1 - y_d \tag{10.5}$$

and

$$z_2 := x_2 - \alpha_1. \tag{10.6}$$

Then (10.1) and (10.2) can be rewritten in terms of the stabilizations errors as

$$\dot{z}_1 = \dot{x}_1 - \dot{y}_d = [f_1(x_1) + g_1(x_1)\alpha_1] + g_1(x_1)z_2 - \dot{y}_d, \tag{10.7}$$

$$\dot{z}_2 = u - \dot{\alpha}_1 = v. \tag{10.8}$$

From (10.7), it can be seen that when $x_2 \to \alpha_1$, so that $z2 \to 0$, the resulting subsystem $\dot{z}_1 = f_1(x_1) + g_1(x_1)\alpha_1(x_1) - \dot{y}_d$ can be stabilized independently of $z_2$. The stabilizing function needed to achieve this will be determined as the first step of the backstepping control design process. As can be seen by comparing Fig. 10.3 with Fig. 10.1, the new system (10.7)–(10.8) is similar to the one we started with (10.1)–(10.2). However, when the input of this new system is $v = 0$, it can be asymptotically stabilized to its origin ($z_1 = 0, z_2 = 0$).

Also note that by comparing Fig. 10.3 with Fig. 10.2, we can see that in the process of stabilizing the $x_2$ function we are stepping $\alpha_1(x_1)$ back through an integrator block to compute $v = u - \dot{\alpha}_1$. This is why this control design technique is known as *integrator backstepping*.

The stabilizing function $\alpha_1$ and the control input $u$ will be determined by selecting a suitable *control Lyapunov function* (CLF) $V(z_1, z_2, u)$. Recall from Sect. 7.2.2 that a *Lyapunov function* $V$, is an energy-like function that can be used to determine the stability of a system. If we can find a positive definite $V > 0$ that always decreases

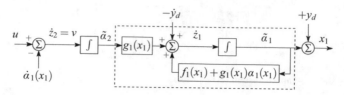

**Fig. 10.3** Block diagram of system (10.7)–(10.8). The control input $u$ is selected to stabilize the $x_2$ system

along trajectories of the system, e.g.

$$\dot{V} = \frac{\partial V}{\partial z_1}\dot{z}_1 + \frac{\partial V}{\partial z_2}\dot{z}_2 < 0 \quad \forall z_1, z_2 \neq 0,$$

we can conclude that the minimum of the function is a stable equilibrium point. According to Theorem 7.1, Lyapunov functions can be used to test whether a dynamical system is *stable*, e.g. whether the system will remain within some domain $D$ when it starts at some initial state $[z_1(0) \quad z_2(0)]^T \in D$, or whether a system is *asymptotically stable*, meaning that when it starts at some initial state $[z_1(0) \quad z_2(0)]^T$, $\lim_{t\to\infty}[z_1(t) \quad z_2(t)]^T \to \mathbf{0}$. Similarly, CLFs are used to test whether a closed loop system can be feedback stabilized — e.g. whether or not a control input $u(z_1, z_2, t)$ exists that can drive the system from any initial state $[z_1(0) \quad z_2(0)]^T$ to $[z_1(t) \quad z_2(t)]^T = 0$.

Here, we explore use of the candidate CLF

$$V = \frac{1}{2}z_1^2 + \frac{1}{2}z_2^2. \tag{10.9}$$

Taking its time derivative gives

$$\begin{aligned}
\dot{V} &= z_1\dot{z}_1 + z_2\dot{z}_2, \\
&= z_1\left[f_1(x_1) + g_1(x_1)\alpha_1 + g_1(x_1)z_2 - \dot{y}_d\right] + z_2(u - \dot{\alpha}_1). \tag{10.10}
\end{aligned}$$

Noting that $z_1 g_1(x_1)z_2 = z_2 g_1(x_1)z_1$, this term can be moved from the first expression in square brackets to the second one so that the first one is a function of $x_1$ only, so that (10.10) can be rewritten as

$$\dot{V} = z_1\left[f_1(x_1) + g_1(x_1)\alpha_1 - \dot{y}_d\right] + z_2[u - \dot{\alpha}_1 + g_1(x_1)z_1]. \tag{10.11}$$

**Step 1:** With our assumption that $g_1(x_1) \neq 0$ in our domain of interest, select $\alpha_1$ as

$$\alpha_1 = \frac{1}{g_1(x_1)}\left[-k_1 z_1 + \dot{y}_d - f_1(x_1)\right], \tag{10.12}$$

where $k_1 > 0$. Then, using (10.11), (10.7) and (10.8) gives

$$\dot{V} = -k_1 z_1^2 + z_2[u - \dot{\alpha}_1 + g_1(x_1)z_1]. \tag{10.13}$$

As the first term on the right hand side of this equation is negative definite, the $z_1$ subsystem is stabilized with our choice of $\alpha_1$.

**Step 2:** Next, the control input $u$ is selected to stabilize the $z_2$ system by making the remaining terms in (10.13) negative definite. Let

$$u = -k_2 z_2 + \dot{\alpha}_1 - g_1(x_1)z_1 \tag{10.14}$$

with $k_2 > 0$, then

$$\dot{V} = -k_1 z_1^2 - k_2 z_2^2 < 0 \quad \forall z_1, z_2 \neq 0. \tag{10.15}$$

As $\dot{V}$ is negative definite $\forall z_1, z_2 \neq 0$, the full system is now stabilized. Further, using (10.9) and (10.15) it can be seen that

$$\dot{V} \leq -2\mu V \tag{10.16}$$

where $\mu := \min\{k_1, k_2\}$. Let the value of $V$ at time $t = 0$ be $V_0$, then integrating (10.15) gives

$$V \leq V_0 e^{-2\mu t}, \tag{10.17}$$

where

$$V_0 := \frac{1}{2}\left[z_1(0)^2 + z_2(0)^2\right].$$

Thus, using (10.17) and (10.9), we find that both $z_1(t)$ and $z_2(t)$ decrease exponentially in time and that the controller is globally exponentially stable (see Sect. 7.2.1).

In terms of the states $x_1$, $x_2$ and the desired time-dependent trajectory $y_d(t)$, the final control law can be rewritten as

$$u = -k_2(x_2 - \alpha_1) + \dot{\alpha}_1 - g_1(x_1)[x_1 - y_d], \tag{10.18}$$

where

$$\alpha_1 = -\frac{1}{g_1(x_1)}[f_1(x_1) + k_1(x_1 - y_d) - \dot{y}_d]. \tag{10.19}$$

The computation of the control input $u$ requires one to take the time derivative of the stabilizing function, $\dot{\alpha}_1$. There are two important implementation issues associated with this:

(1) **The Explosion of Complexity:** The computation of $\dot{\alpha}_1$ involves taking time derivatives of the states and plant model, which in turn can lead to a situation sometimes called an *explosion of complexity* [17, 18], where the number of terms required to compute the time derivative of the stabilizing function becomes very large. In general, one should avoid taking the time derivatives of the states directly, instead using the original state equation whenever possible. For example,

$\dot{\alpha}_1$ can be computed as

$$\dot{\alpha}_1 = \frac{\partial \alpha_1}{\partial x_1}\dot{x}_1 = \frac{\partial \alpha_1}{\partial x_1}[f_1(x_1) + g_1(x_1)x_2]. \tag{10.20}$$

The Dynamic Surface Control Technique, which is presented in Sect. 10.5, is an approach developed to avoid this problem.

(2) As can be seen from (10.19), the computation of $\dot{\alpha}_1$ also involves the second derivative in time of the desired trajectory. In order for the control input to be bounded, the desired trajectory must be smooth and continuous to second order. In practice, it is common to pass the desired trajectory through a linear filter to ensure the smoothness of $\ddot{y}_d(t)$. The use of linear filters for smoothing a desired trajectory is discussed in Sect. 10.3.3.

**Example 10.1**  A buoyancy-driven automatic profiler is used to collect data by vertically traversing the water column in the open ocean (Fig. 10.4). Careful regulation of the profiler's vertical speed is desired to ensure the effectiveness of the sampling instrumentation, which could be a conductivity-temperature-depth (CTD) sensor, for example. The vertical ascent/descent rate is controlled using a piston to regulate the system's buoyancy by changing the volume of an oil-filled bladder. The equation of motion of the system is

$$\dot{x}_1 = -\frac{\rho A C_d}{2m}x_1|x_1| + \frac{g(\rho - \rho_f)}{m}x_2 \tag{10.21}$$

where $x_1$ is the ascent/descent rate of the profiler, $m$ is the mass and added mass, $A$ is the frontal area, $C_d$ is the drag coefficient, $x_2$ is the change in piston volume, $g$ is gravity, $\rho$ is the density of seawater and $\rho_f$ is the density of the oil. The control input is the piston's rate of volume change (the product of the piston velocity and the circular cross sectional area of the piston head),

$$\dot{x}_2 = u. \tag{10.22}$$

The system output is the ascent/descent rate $y = x_1$. Comparing (10.21) and (10.22) to (10.1) and (10.2), we see that

$$f_1(x_1) = -\frac{\rho A C_d}{2m}x_1|x_1|, \quad \text{and} \quad g_1(x_1) = \frac{g(\rho - \rho_f)}{m}.$$

From (10.19) we have

$$\begin{aligned}
\alpha_1 &= -\frac{m}{g(\rho - \rho_f)}\left[-\frac{\rho A C_d}{2m}x_1|x_1| + k_1(x_1 - y_d) - \dot{y}_d\right] \\
&= -\frac{m}{g(\rho - \rho_f)}\left[-\frac{\rho A C_d}{2m}x_1^2\mathrm{sgn}(x_1) + k_1(x_1 - y_d) - \dot{y}_d\right]
\end{aligned}$$

**Fig. 10.4** The *cyclesonde*, an example of a buoyancy-controlled automatic profiler [10]

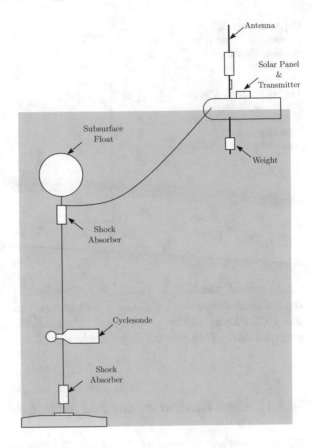

and from (10.20)

$$\dot{\alpha}_1 = \frac{\partial \alpha_1}{\partial x_1}\dot{x}_1,$$

$$= -\frac{m}{g(\rho - \rho_f)}\left\{-\frac{\rho A C_d}{2m}\frac{\partial}{\partial x_1}\left[x_1^2 \mathrm{sgn}(x_1)\right] + k_1\right\}\dot{x}_1,$$

$$= -\frac{m}{g(\rho - \rho_f)}\left\{-\frac{\rho A C_d}{2m}\left[2x_1 \mathrm{sgn}(x_1) + 2x_1^2 \delta(x_1)\right] + k_1\right\}\dot{x}_1,$$

$$= -\frac{m}{g(\rho - \rho_f)}\left[-\frac{\rho A C_d}{m}|x_1| + k_1\right]\dot{x}_1,$$

Then from (10.18) the control input is

$$u = -k_2(x_2 - \alpha_1) + \dot{\alpha}_1 - \frac{g(\rho - \rho_f)}{m}[x_1 - y_d].$$

**Fig. 10.5** Vertical ascent/descent rate $x_1$ of the buoyancy-driven profiler when tracking a desired time-dependent input $y_d(t)$. The vertical axis of the plot is reversed, as the downward direction (increasing depth) is usually taken to be positive for marine vehicles

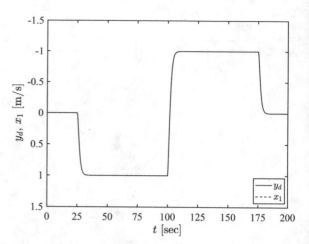

Let the control gains be $k_1 = 1$ and $k_2 = 1$. The physical characteristics of the profiler and the other constants used are $m = 50$ kg, $A = 1$ m$^2$, $C_d = 1.0$, $g = 9.81$ m/s$^2$, $\rho = 1025.9$ kg/m$^3$ and $\rho_f = 850$ kg/m$^3$. The ascent/descent rate of the closed loop system is shown for a time varying $y_d(t)$ in Fig. 10.5.                                        □

### 10.2.2  More General 2-State and 3-State SISO Systems

The term $\dot{x}_2 = u$ in the system (10.1)–(10.2) is a pure integrator. Next, let's consider the more general SISO system

$$\dot{x}_1 = f_1(x_1) + g_1(x_1)x_2,  \tag{10.23}$$
$$\dot{x}_2 = f_2(x_1, x_2) + g_2(x_1, x_2)u,  \tag{10.24}$$
$$y = x_1,$$

where $f_2(x_1, x_2)$ and $g_2(x_1, x_2)$ are smooth, known functions. If $g_2(x_1, x_2) \neq 0$ over the domain of interest, the input transformation

$$u = \frac{1}{g_2(x_1, x_2)} [u_2 - f_2(x_1, x_2)]  \tag{10.25}$$

will reduce (10.24) to the integrator $\dot{x}_2 = u_2$. Then, if a stabilizing function $\alpha_1(x_1)$ (10.19) for (10.23) and a Lyapunov function $V(z_1, z_2)$ (10.9) exist, then from (10.24) the control law can be written as

$$u = \frac{1}{g_2(x_1, x_2)} [-k_2 z_2 + \dot{\alpha}_1 - f_2(x_1, x_2) - g_1(x_1)z_1],  \tag{10.26}$$

The second order backstepping approach presented here can be extended to a third order system. Consider the third order SISO system

$$\dot{x}_1 = f_1(x_1) + g_1(x_1)x_2, \tag{10.27}$$
$$\dot{x}_2 = f_2(x_1, x_2) + g_2(x_1, x_2)x_3, \tag{10.28}$$
$$\dot{x}_3 = f_3(x_1, x_2, x_3) + g_3(x_1, x_2, x_3)u, \tag{10.29}$$
$$y = x_1,$$

where the functions $f_n$ and $g_n$ are smooth and known, and $g_n \neq 0$ over the domain of interest for $n = \{1, 2, 3\}$. As above, we define the stabilizing functions $\alpha_1(x_1)$ and $\alpha_2(x_1, x_2)$, which can be used to decouple the system by adding and subtracting $\alpha_1$ to (10.27) and by adding and subtracting $\alpha_2$ to (10.28) to obtain

$$\dot{x}_1 = f_1(x_1) + g_1(x_1)\alpha_1 + g_1(x_1)[x_2 - \alpha_1], \tag{10.30}$$
$$\dot{x}_2 = f_2(x_1, x_2) + g_2(x_1, x_2)\alpha_2 + g_2(x_1, x_2)[x_3 - \alpha_2], \tag{10.31}$$
$$\dot{x}_3 = f_3(x_1, x_2, x_3) + g_3(x_1, x_2, x_3)u, \tag{10.32}$$
$$y = x_1.$$

Defining the stabilization errors as

$$z_1 := y - y_d = x_1 - y_d \tag{10.33}$$
$$z_2 := x_2 - \alpha_1 \tag{10.34}$$
$$z_3 := x_3 - \alpha_2 \tag{10.35}$$

the error system can be written as

$$\dot{z}_1 = f_1(x_1) + g_1(x_1)\alpha_1 + g_1(x_1)z_2 - \dot{y}_d, \tag{10.36}$$
$$\dot{z}_2 = f_2(x_1, x_2) + g_2(x_1, x_2)\alpha_2 + g_2(x_1, x_2)z_3 - \dot{\alpha}_1, \tag{10.37}$$
$$\dot{z}_3 = f_3(x_1, x_2, x_3) + g_3(x_1, x_2, x_3)u - \dot{\alpha}_2. \tag{10.38}$$

Consider the CLF

$$V = \frac{1}{2}z_1^2 + \frac{1}{2}z_2^2 + \frac{1}{2}z_3^2. \tag{10.39}$$

Taking the time derivative gives

$$
\begin{aligned}
\dot{V} =& z_1\dot{z}_1 + z_2\dot{z}_2 + z_3\dot{z}_3, && (10.40) \\
=& z_1[f_1(x_1) + g_1(x_1)\alpha_1 + g_1(x_1)z_2 - \dot{y}_d] \\
& + z_2[f_2(x_1, x_2) + g_2(x_1, x_2)\alpha_2 + g_2(x_1, x_2)z_3 - \dot{\alpha}_1] \\
& + z_3[f_3(x_1, x_2, x_3) + g_3(x_1, x_2, x_3)u - \dot{\alpha}_2], \\
=& z_1[f_1(x_1) + g_1(x_1)\alpha_1 - \dot{y}_d] \\
& + z_2[f_2(x_1, x_2) + g_2(x_1, x_2)\alpha_2 + g_1(x_1)z_1 - \dot{\alpha}_1] \\
& + z_3[f_3(x_1, x_2, x_3) + g_3(x_1, x_2, x_3)u + g_2(x_1, x_2)z_2 - \dot{\alpha}_2].
\end{aligned}
$$

We stabilize the system in steps, one subsystem at a time.
**Step 1:** The $z_1$ system can be stabilized by setting

$$
z_1[f_1(x_1) + g_1(x_1)\alpha_1 - \dot{y}_d] = -k_1 z_1^2, \tag{10.41}
$$

where $k_1 > 0$. Then

$$
\alpha_1 = \frac{1}{g_1(x_1)}[-k_1 z_1 + \dot{y}_d - f_1(x_1)]. \tag{10.42}
$$

**Step 2:** Next, the $z_2$ subsystem is stabilized by setting

$$
z_2[f_2(x_1, x_2) + g_2(x_1, x_2)\alpha_2 + g_1(x_1)z_1 - \dot{\alpha}_1] = -k_2 z_2^2, \tag{10.43}
$$

where $k_2 > 0$. Then

$$
\alpha_2 = \frac{1}{g_2(x_1, x_2)}[-k_2 z_2 + \dot{\alpha}_1 - g_1(x_1)z_1 - f_2(x_1, x_2)]. \tag{10.44}
$$

**Step 3:** Lastly, the $z_3$ subsystem is stabilized by selecting the control input so that

$$
z_3[f_3(x_1, x_2, x_3) + g_3(x_1, x_2, x_3)u + g_2(x_1, x_2)z_2 - \dot{\alpha}_2] = -k_3 z_3^2, \tag{10.45}
$$

where $k_3 > 0$. Then

$$
u = \frac{1}{g_3(x_1, x_2, x_3)}[-k_3 z_3 + \dot{\alpha}_2 - g_2(x_1, x_2)z_2 - f_3(x_1, x_2, x_3)]. \tag{10.46}
$$

Solving for $\dot{y}_d$, $\dot{\alpha}_1$ and $\dot{\alpha}_2$ from the expressions obtained for $\alpha_1$, $\alpha_2$ and $u$ in (10.42), (10.44) and (10.46), respectively, and using (10.36)–(10.38), the closed loop error system can be written as

$$
\begin{aligned}
\dot{z}_1 &= -k_1 z_1 + g_1(x_1)z_2, && (10.47) \\
\dot{z}_2 &= -k_2 z_2 + g_2(x_1, x_2)z_3 - g_1(x_1)z_1, && (10.48) \\
\dot{z}_3 &= -k_3 z_3 - g_2(x_1, x_2)z_2. && (10.49)
\end{aligned}
$$

The closed loop error system is globally exponentially stable with a stable equilibrium point $(\dot{z}_1, \dot{z}_2, \dot{z}_3) = (0, 0, 0)$ at its origin $(z_1, z_2, z_3) = (0, 0, 0)$.

**Example 10.2** Here, we reconsider control of the automatic profiler investigated in Example 10.1. Suppose that a different sampling mission requires precise control of the profiler depth. In this case the equations of motion are

$$\dot{x}_1 = x_2, \tag{10.50}$$

$$\dot{x}_2 = -\frac{\rho A C_d}{2m} x_2 |x_2| + \frac{g(\rho - \rho_f)}{m} x_3, \tag{10.51}$$

$$\dot{x}_3 = u, \tag{10.52}$$

$$y = x_1, \tag{10.53}$$

where $u$ is again the piston's rate of volume change, but now $x_1$ is the profiler depth and $x_2$ is its ascent/descent rate. The system output is the depth of the profiler $y = x_1$.

Comparing (10.50)–(10.52) to (10.27)–(10.29), we see that

$$f_1(x_1) = 0, \quad f_2(x_1, x_2) = -\frac{\rho A C_d}{2m} x_2 |x_2|, \quad f_3(x_1, x_2, x_3) = 0,$$

$$g_1(x_1) = 1, \quad g_2(x_1, x_2) = \frac{g(\rho - \rho_f)}{m}, \quad g_3(x_1, x_2, x_3) = 1.$$

From (10.42), (10.44) and (10.46), respectively, it can be seen that the stabilizing functions are

$$\alpha_1 = -k_1 z_1 + \dot{y}_d = -k_1(x_1 - y_d) + \dot{y}_d$$

$$\alpha_2 = \frac{m}{g(\rho - \rho_f)} \left[ -k_2 z_2 + \dot{\alpha}_1 - z_1 + \frac{\rho A C_d}{2m} x_2 |x_2| \right]$$

$$= \frac{m}{g(\rho - \rho_f)} \left[ -k_2(x_2 - \alpha_1) + \dot{\alpha}_1 - (x_1 - y_d) + \frac{\rho A C_d}{2m} x_2^2 \mathrm{sgn}(x_2) \right]$$

and the control input is

$$u = -k_3 z_3 + \dot{\alpha}_2 - \frac{g(\rho - \rho_f)}{m} z_2,$$

$$= -k_3(x_3 - \alpha_2) + \dot{\alpha}_2 - \frac{g(\rho - \rho_f)}{m} (x_2 - \alpha_1),$$

where

$$\dot{\alpha}_1 = \frac{\partial \alpha_1}{\partial x_1} \dot{x}_1 = -k_1 \dot{x}_1$$

and

**Fig. 10.6** Depth $x_1$ of the buoyancy-driven profiler when tracking a sinusoidal input $y_d(t)$. The vertical axis of the plot is reversed, as the downward direction (increasing depth) is normally taken to be positive for marine vehicles

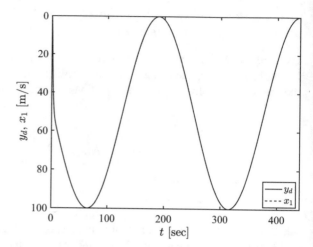

$$\dot{\alpha}_2 = \frac{\partial \alpha_2}{\partial x_1} \dot{x}_1 + \frac{\partial \alpha_2}{\partial x_2} \dot{x}_2,$$

$$= -\frac{m}{g(\rho - \rho_f)} \left[ (k_1 k_2 + 1) \dot{x}_1 + \left( k_2 - \frac{\rho A C_d}{m} |x_2| \right) \dot{x}_2 \right].$$

Let the control gains be $k_1 = 1$, $k_2 = 1$ and $k_3 = 1$. The physical constants are taken to be the same as in Example 10.1. The depth of the profiler when the closed loop system is used to track a time varying sinusoidal input $y_d(t)$ is shown in Fig. 10.6. $\square$

By comparing (10.12) and (10.26) with (10.42), (10.44) and (10.46), one can see that there is a pattern in how the stabilizing functions and control input are defined. In fact, this pattern hints at the fact that the procedure can be extended to a generalized SISO system of order $n$.

### 10.2.3   Generalized n-State SISO Systems: Recursive Backstepping

As above, let the trajectory tracking error be defined as $z_1 = y - y_d$, where the output is taken to be a function of the $x_1$ state variable, $y = h(x_1)$. Then its time derivative is given by $\dot{z}_1 = \dot{y} - \dot{y}_d$, where $\dot{y}$ can be computed using the Lie derivative as

$$\dot{y} = \frac{\partial y}{\partial x_1} \dot{x}_1 = \frac{\partial h}{\partial x_1} [f_1(x_1) + g_1(x_1)x_2] = L_{f_1}h(x_1) + L_{g_1}h(x_1)x_2, \qquad (10.54)$$

with

$$L_{f_1}h(x_1) := \frac{\partial h}{\partial x_1} f_1(x_1) \quad \text{and} \quad L_{g_1}h(x_1) := \frac{\partial h}{\partial x_1} g_1(x_1).$$

The systems we have analyzed thus far, e.g. (10.1)–(10.2), (10.23)–(10.24) and (10.27)–(10.29), are known as *strict-feedback nonlinear systems* (SFNSs) because the nonlinearity $f_i(x_1, \ldots, x_i)$ is fed back in the $\dot{x}_i$ equation. Through the recursive application of backstepping, we can develop trajectory tracking controllers for SISO SFNSs of the form

$$
\begin{aligned}
\dot{x}_1 &= f_1(x_1) + g_1(x_1)x_2, \\
\dot{x}_2 &= f_2(x_1, x_2) + g_2(x_1, x_2)x_3, \\
\dot{x}_3 &= f_3(x_1, x_2, x_3) + g_3(x_1, x_2, x_3)x_4, \\
&\;\;\vdots \\
\dot{x}_{n-1} &= f_{n-1}(x_1, x_2, x_3, \ldots, x_{n-1}) + g_{n-1}(x_1, x_2, x_3, \ldots, x_{n-1})x_n, \\
\dot{x}_n &= f_n(x_1, x_2, x_3, \ldots, x_n) + g_n(x_1, x_2, x_3, \ldots, x_n)u, \\
y &= h(x_1),
\end{aligned}
\tag{10.55}
$$

where $u \in \mathbb{R}$ is the single input, $x_i \in \mathbb{R} \; \forall i$, and $y \in \mathbb{R}$ is the single controlled output. The terms $f_i(x_1, \ldots, x_i)$ and $g_i(x_1, \ldots, x_i)$, where $g_i(x_1, \ldots, x_i) \neq 0$, are known functions for $i \in \{1, \ldots, n\}$. The reference trajectory is assumed to be bounded and $r$ times differentiable (smooth) $y_d \in \mathbb{C}^r$; $f_i(x_1, \ldots, x_i)$ and $g_i(x_1, \ldots, x_i)$ are assumed to be $n - 1$ times differentiable (smooth) $f_i, g_i \in \mathbb{C}^{n-1}$ for $i \in \{2, \ldots, n\}$.

Define the tracking errors and their time derivatives to be

$$
\begin{aligned}
z_1 := y - y_d \quad &\rightarrow \quad \dot{z}_1 = \dot{y} - \dot{y}_d = L_{f1}h + L_{g1}hx_2 - \dot{y}_d \\
&\qquad\qquad = L_{f1}h + L_{g1}hz_2 + L_{g1}h\alpha_1 - \dot{y}_d \\
z_i := x_i - \alpha_{i-1} \quad &\rightarrow \quad \dot{z}_i = \dot{x}_i - \dot{\alpha}_{i-1} = f_i + g_i x_{i+1} - \dot{\alpha}_{i-1} \\
&\qquad\qquad = f_i + g_i z_{i+1} + g_i \alpha_i - \dot{\alpha}_{i-1} \\
&\qquad\quad \text{for } i \in \{2, \ldots, n-1\} \\
z_n := x_n - \alpha_{n-1} \quad &\rightarrow \quad \dot{z}_n = \dot{x}_n - \dot{\alpha}_{n-1} = f_n + g_n u - \dot{\alpha}_{n-1}
\end{aligned}
\tag{10.56}
$$

where (10.54) is used to compute the time derivative of $y$ and the $\alpha_i(t)$ are stabilizing functions.

The stabilizing functions are determined using a CLF of the form

$$
V = \frac{1}{2} \sum_{i=1}^{n} z_i^2 = \frac{1}{2} z_1^2 + \cdots + \frac{1}{2} z_n^2, \quad \text{for } i = 1, \ldots, n,
\tag{10.57}
$$

The time derivative of (10.57) is

$$
\begin{aligned}
\dot{V} &= \sum_{i=1}^{n} z_i \dot{z}_i \\
&= z_1 \left\{ L_{f1}h + L_{g1}hz_2 + L_{g1}h\alpha_1 - \dot{y}_d \right\} \\
&\quad + z_2 \left\{ f_2 + g_2 z_3 + g_2 \alpha_2 - \dot{\alpha}_1 \right\} \\
&\quad + \cdots \\
&\quad + z_n \left\{ f_n + g_n u - \dot{\alpha}_{n-1} \right\}.
\end{aligned}
\tag{10.58}
$$

Rearranging terms gives

$$
\begin{aligned}
\dot{V} = z_1 &\left\{ L_{f1}h + L_{g1}h\alpha_1 - \dot{y}_d \right\} \\
+z_2 &\left\{ f_2 + L_{g1}hz_1 + g_2\alpha_2 - \dot{\alpha}_1 \right\} \\
+z_3 &\left\{ f_3 + g_2z_2 + g_3\alpha_3 - \dot{\alpha}_2 \right\} \\
+ &\cdots \\
+z_n &\left\{ f_n + g_{n-1}z_{n-1} + g_n u - \dot{\alpha}_{n-1} \right\}.
\end{aligned}
\tag{10.59}
$$

The time derivative of the Lyapunov function (10.57) can be made to be

$$
\dot{V} = -\sum_{i=1}^{n} k_i z_i^2 = -k_1 z_1^2 - \cdots - k_n z_n^2, \quad \text{for } i = 1, \ldots, n,
\tag{10.60}
$$

where the $k_i > 0$ are control gains, by selecting the stabilizing functions and control input to be

$$
\begin{aligned}
\alpha_1 &= \frac{1}{L_{g1}h(x_1)} \left[ -k_1 z_1 + \dot{y}_d - L_{f1}h(x_1) \right], \\
\alpha_2 &= \frac{1}{g_2(x_1, x_2)} \left[ -k_2 z_2 + \dot{\alpha}_1 - f_2(x_1, x_2) - L_{g1}h(x_1)z_1 \right], \\
\alpha_i &= \frac{1}{g_i(x_1, \ldots, x_i)} [-k_i z_i + \dot{\alpha}_{i-1} \\
&\quad - f_i(x_1, \ldots, x_i) - g_{i-1}(x_1, \ldots, x_{i-1})z_{i-1}], \quad i \in \{3, \ldots, n-1\}, \\
u &= \frac{1}{g_n(x_1, \ldots, x_n)} [-k_n z_n + \dot{\alpha}_{n-1} - f_n(x_1, \ldots, x_n) - g_{n-1}(x_1, \ldots, x_{n-1})z_{n-1}].
\end{aligned}
\tag{10.61}
$$

Applying (10.61) to (10.55) yields the closed loop tracking error dynamics

$$
\begin{aligned}
\dot{z}_1 &= -k_1 z_1 + L_{g1}h(x_1)z_2, \\
\dot{z}_2 &= -k_2 z_2 + g_2(x_1, x_2)z_3 - L_{g1}h(x_1)z_1, \\
\dot{z}_i &= -k_i z_i + g_i(x_1, \ldots, x_i)z_{i+1} - g_{i-1}(x_1, \ldots, x_{i-1})z_{i-1}, \quad \text{for } i \in \{3, \ldots, n-1\}, \\
\dot{z}_n &= -k_n z_n - g_{n-1}(x_1, \ldots, x_{n-1})z_{n-1}.
\end{aligned}
\tag{10.62}
$$

For the system (10.55), if the control input $u$ is determined by (10.61), then a CLF of the form

$$
V = \frac{1}{2} \sum_{i=1}^{n} z_i^2
$$

can be used to show that the closed loop tracking error dynamics (10.62) are globally exponentially stable.

Since $\dot{V}$ in (10.60) is negative definite $\forall z_i \neq 0$, the full system is stabilized. Further, using (10.57) and (10.60) it can be seen that

$$
\dot{V} \leq -2\mu V
\tag{10.63}
$$

where $\mu := \min\{k_i\}$ for $i \in \{1, \ldots, n\}$ is the minimum control gain. Let the value of $V$ at time $t = 0$ be $V_0$, then the solution to (10.63) is

$$
V \leq V_0 e^{-2\mu t}.
\tag{10.64}
$$

Thus, using (10.57) and (10.64), it can be seen that the $z_i(t)$ decrease exponentially in time so that the controller is globally exponentially stable (see Sect. 7.2.1). From (10.62) it can be seen that the origin of the error system $z_i = 0 \; \forall i \in \{1, \ldots, n\}$ is the equilibrium point of the system. As a result, the control objective is satisfied so that the vehicle can track a desired trajectory $y_d$ with (theoretically) zero error.

Note that the tracking problem, i.e. finding a control input $u$ that drives $y$ to $y_d$, can be reduced to a stabilization problem (regulation) problem, i.e. driving $x_1$ to zero, by taking $y_d = \dot{y}_d = 0$.

Most marine vehicles are MIMO systems. For these systems each state variable is naturally represented as a vector. Thus, a vectorial form of the backstepping technique is convenient for applications in marine vehicleics.

### 10.2.4 Vectorial Backstepping for MIMO Systems

The backstepping approach shown for SISO SFNSs can be extended to multiple input multiple output (MIMO) systems in strict feedback nonlinear form using an approach known as *vectorial backstepping* [7–9] or *block backstepping* [11], provided that certain nonsingularity conditions are satisfied. Consider the vector equivalent of (10.55)

$$
\begin{aligned}
\dot{\mathbf{x}}_1 &= \mathbf{f}_1(\mathbf{x}_1) + \mathbf{G}_1(\mathbf{x}_1)\mathbf{x}_2, \\
\dot{\mathbf{x}}_2 &= \mathbf{f}_2(\mathbf{x}_1, \mathbf{x}_2) + \mathbf{G}_2(\mathbf{x}_1, \mathbf{x}_2)\mathbf{x}_3, \\
\dot{\mathbf{x}}_i &= \mathbf{f}_i(\mathbf{x}_1, \ldots, \mathbf{x}_i) + \mathbf{G}_i(\mathbf{x}_1, \ldots, \mathbf{x}_i)\mathbf{x}_{i+1}, \quad \text{for } i \in \{3, \ldots, n-2\}, \\
\dot{\mathbf{x}}_{n-1} &= \mathbf{f}_{n-1}(\mathbf{x}_1, \ldots, \mathbf{x}_{n-1}) + \mathbf{G}_{n-1}(\mathbf{x}_1, \ldots, \mathbf{x}_{n-1})\mathbf{x}_n, \\
\dot{\mathbf{x}}_n &= \mathbf{f}_n(\mathbf{x}_1, \ldots, \mathbf{x}_n) + \mathbf{G}_n(\mathbf{x}_1, \ldots, \mathbf{x}_n)\mathbf{u}, \\
\mathbf{y} &= \mathbf{h}(\mathbf{x}_1),
\end{aligned}
\tag{10.65}
$$

where the $\mathbf{f}_i$ are smooth vector functions, the $\mathbf{G}_i$ are smooth, nonsingular matrix functions,

$$
\dim \mathbf{y}(\mathbf{x}_1) \leq \dim \mathbf{x}_1(t) \leq \dim \mathbf{x}_2(t) \leq \cdots \leq \dim \mathbf{x}_n(t) \leq \dim \mathbf{u}(\mathbf{x}, t),
$$

and $\mathbf{x} = [\mathbf{x}_1^T, \cdots, \mathbf{x}_n^T]$.

Let $\mathbf{h}(\mathbf{x}_1)$ be a smooth vector field. The Jacobian of $\mathbf{h}(\mathbf{x}_1)$ is a matrix of partial derivatives

$$
\nabla \mathbf{h}(\mathbf{x}_1) = \frac{\partial \mathbf{h}}{\partial \mathbf{x}_1},
\tag{10.66}
$$

which can be written using indicial notation as $\nabla \mathbf{h}(\mathbf{x})_{ij} = \partial h_i / \partial x_j$. Note that the time derivative of the output $\mathbf{y}$ can be written as

$$
\dot{\mathbf{y}} = \nabla \mathbf{h}(\mathbf{x}_1)\dot{\mathbf{x}}_1 = \nabla \mathbf{h}(\mathbf{x}_1)[\mathbf{f}_1(\mathbf{x}_1) + \mathbf{G}_1(\mathbf{x}_1)\mathbf{x}_2].
\tag{10.67}
$$

It is assumed that $\nabla \mathbf{h} \cdot \mathbf{G}_1 (\nabla \mathbf{h} \cdot \mathbf{G}_1)^T$ and $\mathbf{G}_i \mathbf{G}_i^T$ for $(i = 2 \ldots n)$ are invertible for all $\mathbf{x}_i$ such that

$$(\nabla \mathbf{h} \cdot \mathbf{G}_1)^\dagger = (\nabla \mathbf{h} \cdot \mathbf{G}_1)^T \left[ \nabla \mathbf{h} \cdot \mathbf{G}_1 (\nabla \mathbf{h} \cdot \mathbf{G}_1)^T \right]^{-1} \tag{10.68}$$

$$\mathbf{G}_1^\dagger = \mathbf{G}_1^T (\mathbf{G}_1 \mathbf{G}_1^T)^{-1} \tag{10.69}$$

exist, where the superscript † indicates the *pseudo-inverse*, also known as the *Moore–Penrose inverse*, see (9.3).

Define the tracking errors and their time derivatives to be

$$\begin{aligned}
\mathbf{z}_1 := \mathbf{y} - \mathbf{y}_d \quad &\to \dot{\mathbf{z}}_1 = \nabla \mathbf{h}(\mathbf{x}_1)[\mathbf{f}_1(\mathbf{x}_1) + \mathbf{G}_1(\mathbf{x}_1)\mathbf{x}_2] - \dot{\mathbf{y}}_d \\
&= \nabla \mathbf{h} \mathbf{f}_1 + \nabla \mathbf{h} \mathbf{G}_1 \mathbf{z}_2 + \nabla \mathbf{h} \mathbf{G}_1 \alpha_1 - \dot{\mathbf{y}}_d \\
\mathbf{z}_i := \mathbf{x}_i - \alpha_{i-1} \quad &\to \dot{\mathbf{z}}_i = \mathbf{f}_i(\mathbf{x}_1, \ldots, \mathbf{x}_i) + \mathbf{G}_i(\mathbf{x}_1, \ldots, \mathbf{x}_i)\mathbf{x}_{i+1} - \dot{\alpha}_{i-1} \\
&= \mathbf{f}_i + \mathbf{G}_i \mathbf{z}_{i+1} + \mathbf{G}_i \alpha_i - \dot{\alpha}_{i-1} \\
&\text{for } i \in \{2, \ldots, n-1\} \\
\mathbf{z}_n := \mathbf{x}_n - \alpha_{n-1} \quad &\to \dot{\mathbf{z}}_n = \mathbf{f}_n(\mathbf{x}_1, \ldots, \mathbf{x}_n) + \mathbf{G}_n(\mathbf{x}_1, \ldots, \mathbf{x}_n)\mathbf{u} - \dot{\alpha}_{n-1}
\end{aligned} \tag{10.70}$$

where (10.67) is used to compute the time derivative of $\mathbf{y}$ and the $\alpha_{i-1}(t)$ are stabilizing vector functions.

As with the SISO systems studied above, the stabilizing functions are selected using a CLF of the form

$$V = \frac{1}{2} \sum_{i=1}^{n} \mathbf{z}_i^T \mathbf{z}_i = \frac{1}{2} \mathbf{z}_1^T \mathbf{z}_1 + \cdots + \frac{1}{2} \mathbf{z}_n^T \mathbf{z}_n, \quad \text{for } i = 1, \ldots, n. \tag{10.71}$$

Thus, $V$ is positive definite and radially-unbounded (Sect. 7.2.2). The time derivative of (10.71) is

$$\begin{aligned}
\dot{V} &= \sum_{i=1}^{n} \mathbf{z}_i^T \dot{\mathbf{z}}_i \\
&= \mathbf{z}_1^T \{\nabla \mathbf{h} \mathbf{f}_1 + \nabla \mathbf{h} \mathbf{G}_1 \mathbf{z}_2 + \nabla \mathbf{h} \mathbf{G}_1 \alpha_1 - \dot{\mathbf{y}}_d\} \\
&\quad + \mathbf{z}_2^T \{\mathbf{f}_2 + \mathbf{G}_2 \mathbf{z}_3 + \mathbf{G}_2 \alpha_2 - \dot{\alpha}_1\} \\
&\quad + \cdots \\
&\quad + \mathbf{z}_n^T \{\mathbf{f}_n + \mathbf{G}_n \mathbf{u} - \dot{\alpha}_{n-1}\}.
\end{aligned} \tag{10.72}$$

The terms $\mathbf{z}_1^T (\nabla \mathbf{h} \mathbf{G}_1)\mathbf{z}_2$ and $\mathbf{z}_i^T \mathbf{G}_i \mathbf{z}_{i+1}$ are scalar quantities so that $\mathbf{z}_1^T (\nabla \mathbf{h} \mathbf{G}_1)\mathbf{z}_2 = \mathbf{z}_2^T (\nabla \mathbf{h} \mathbf{G}_1)^T \mathbf{z}_1$ and $\mathbf{z}_i^T \mathbf{G}_i \mathbf{z}_{i+1} = \mathbf{z}_{i+1}^T \mathbf{G}_i^T \mathbf{z}_i$. Rearranging terms gives

$$\begin{aligned}
\dot{V} &= \mathbf{z}_1^T \{\nabla \mathbf{h} \mathbf{f}_1 + \nabla \mathbf{h} \mathbf{G}_1 \alpha_1 - \dot{\mathbf{y}}_d\} \\
&\quad + \mathbf{z}_2^T \{\mathbf{f}_2 + (\nabla \mathbf{h} \mathbf{G}_1)^T \mathbf{z}_1 + \mathbf{G}_2 \alpha_2 - \dot{\alpha}_1\} \\
&\quad + \cdots \\
&\quad + \mathbf{z}_n^T \{\mathbf{f}_n + \mathbf{G}_{n-1}^T \mathbf{z}_{n-1} + \mathbf{G}_n \mathbf{u} - \dot{\alpha}_{n-1}\}.
\end{aligned} \tag{10.73}$$

Let $\mathbf{K}_i = \mathbf{K}_i^T > 0$ be a positive definite matrix of control gains, so that $\mathbf{z}_i^T \mathbf{K}_i \mathbf{z}_i > 0$ for any $\mathbf{z}_i \neq 0$ and for all $i$. The time derivative of the Lyapunov function (10.71) can be made to be

$$\dot{V} = -\sum_{i=1}^{n} \mathbf{z}_i^T \mathbf{K}_i \mathbf{z}_i = -\mathbf{z}_1^T \mathbf{K}_1 \mathbf{z}_1 - \cdots - \mathbf{z}_n^T \mathbf{K}_n \mathbf{z}_n, \quad \text{for } i = 1, \ldots, n, \quad (10.74)$$

by selecting the stabilizing vector functions and control input to be

$$\begin{aligned}
\alpha_1 &= [\nabla \mathbf{h}(\mathbf{x}_1) \cdot \mathbf{G}_1(\mathbf{x}_1)]^\dagger \{-\mathbf{K}_1 \mathbf{z}_1 - \nabla \mathbf{h}(\mathbf{x}_1) \mathbf{f}_1(\mathbf{x}_1) + \dot{\mathbf{y}}_d\}, \\
\alpha_2 &= \mathbf{G}_2^\dagger(\mathbf{x}_1, \mathbf{x}_2) \{-\mathbf{K}_2 \mathbf{z}_2 - \mathbf{f}_2(\mathbf{x}_1, \mathbf{x}_2) - [\nabla \mathbf{h}(\mathbf{x}_1) \mathbf{G}_1(\mathbf{x}_1)]^T \mathbf{z}_1 + \dot{\alpha}_1\}, \\
\alpha_i &= \mathbf{G}_i^\dagger(\mathbf{x}_1, \ldots, \mathbf{x}_i) \{-\mathbf{K}_i \mathbf{z}_i - \mathbf{f}_i(\mathbf{x}_1, \ldots, \mathbf{x}_i) \\
&\qquad\qquad - \mathbf{G}_{i-1}^T(\mathbf{x}_1, \ldots, \mathbf{x}_{i-1}) \mathbf{z}_{i-1} + \dot{\alpha}_{i-1}\}, \quad (10.75) \\
&\quad i \in \{3, \ldots, n-1\}, \\
\mathbf{u} &= \mathbf{G}_n^\dagger(\mathbf{x}_1, \ldots, \mathbf{x}_n) \{-\mathbf{K}_n \mathbf{z}_n - \mathbf{f}_n(\mathbf{x}_1, \ldots, \mathbf{x}_n) \\
&\qquad\qquad - \mathbf{G}_{n-1}^T(\mathbf{x}_1, \ldots, \mathbf{x}_{n-1}) \mathbf{z}_{n-1} + \dot{\alpha}_{n-1}\}.
\end{aligned}$$

Applying (10.75) to (10.70) yields the closed loop tracking error dynamics

$$\begin{aligned}
\dot{\mathbf{z}}_1 &= -\mathbf{K}_1 \mathbf{z}_1 + \nabla \mathbf{h}(\mathbf{x}_1) \mathbf{G}_1(\mathbf{x}_1) \mathbf{z}_2, \\
\dot{\mathbf{z}}_2 &= -\mathbf{K}_2 \mathbf{z}_2 + \mathbf{G}_2(x_1, x_2) \mathbf{z}_3 - [\nabla \mathbf{h}(\mathbf{x}_1) \mathbf{G}_1(\mathbf{x}_1)]^T \mathbf{z}_1, \\
\dot{\mathbf{z}}_i &= -\mathbf{K}_i \mathbf{z}_i + \mathbf{G}_i(x_1, \ldots, x_i) \mathbf{z}_{i+1} - \mathbf{G}_{i-1}^T(x_1, \ldots, x_{i-1}) \mathbf{z}_{i-1}, \quad (10.76) \\
&\quad i \in \{3, \ldots, n-1\}, \\
\dot{\mathbf{z}}_n &= -\mathbf{K}_n \mathbf{z}_n - \mathbf{G}_{n-1}^T(x_1, \ldots, x_{n-1}) \mathbf{z}_{n-1}.
\end{aligned}$$

Since $\dot{V}$ in (10.74) is negative definite $\forall \mathbf{z}_i \neq 0$, the full system is stabilized. Further, using (10.71) and (10.74) it can be seen that

$$\dot{V} \leq -2\mu V \quad (10.77)$$

where $\mu := \min\{\lambda_{\min}(\mathbf{K}_i)\}$ for $i \in \{1, \ldots, n\}$ is the minimum eigenvalue of all of the control gain matrices. Let the value of $V$ at time $t = 0$ be $V_0$, then the solution to (10.15) is

$$V \leq V_0 e^{-2\mu t}. \quad (10.78)$$

Thus, using (10.71) and (10.78), it can be seen that the $\mathbf{z}_i(t)$ decrease exponentially in time so that the controller is globally exponentially stable (Sect. 7.2.1). From (10.76) it can be seen that the origin of the error system $\mathbf{z}_i = 0 \, \forall i \in \{1, \ldots, n\}$ is the equilibrium point of the system. As a result, the control objective is satisfied so that the vehicle can track a desired trajectory $\mathbf{y}_d$ with (theoretically) zero error.

As for SISO systems, the MIMO tracking problem can be reduced to a stabilization problem (regulation) problem by taking $\mathbf{y}_d = \dot{\mathbf{y}}_d = 0$. For marine vehicles, this is also sometimes referred to as *station–keeping*.

## 10.3   Backstepping for Trajectory Tracking Marine Vehicles

Recall the equations of motion for a marine vehicle (1.58)–(1.59), which are represented as

$$\dot{\eta} = \mathbf{J}(\eta)\mathbf{v}$$

and

$$\mathbf{M}\dot{\mathbf{v}} + \mathbf{C}(\mathbf{v})\mathbf{v} + \mathbf{D}(\mathbf{v})\mathbf{v} + \mathbf{g}(\eta) = \tau$$

in body-fixed coordinates [6]. For the time being, any actuator constraints and the effects of external disturbances are neglected.

In the following formulation, the coordinate transformation matrix $\mathbf{J}(\eta)$, which is used to convert the representation of vectors between a body-fixed coordinate system and an Earth-fixed North-East-Down coordinate system, is based on the use of Euler angles. The transformation matrix $\mathbf{J}(\eta)$ has singularities at pitch angles of $\theta = \pm\pi/2$. Here, it is assumed that $|\theta| < \pi/2$.

Note that the singularities at $\theta = \pm\pi/2$ are not generally a problem for unmanned ground vehicles or unmanned surface vessels. However, unmanned aerial vehicles and underwater vehicles may occasionally approach these singularities if performing extreme maneuvers. In such cases, as suggested in [6], the kinematic equations could be described by two Euler angle representations with different singularities and the singular points can be avoided by switching between the representations.

The control objective is to make the system track a desired trajectory $\eta_d$. The trajectory $\eta_d$ and its derivatives $\eta_d^{(3)}$, $\ddot{\eta}_d$, and $\dot{\eta}_d$ are assumed to be smooth and bounded.

Two approaches, often used for backstepping control design in marine vehicles, are presented here. The first approach uses a straight-forward implementation of vectorial backstepping, as described above in Sect. 10.2.4. The second approach presented relies upon the definition of a virtual reference velocity and requires that the equations of motion be reformulated in a NED, Earth-fixed inertial reference frame where the inertia tensor becomes a function of vehicle orientation $\mathbf{M}(\eta)$. This second approach is similar to that often used for designing controllers for robotic manipulators and is based on the concept of passivity [15, 16]. The passivity approach is slightly more complicated, but has advantages when using backstepping for adaptive or robust control. Further, the passivity approach permits one to more heavily weight either position tracking or velocity tracking, as desired. On the other hand, with the straight-forward vectorial backstepping approach the identification of a control law is simpler. Both approaches ensure the global exponential stability of the closed loop system in the absence of model uncertainty and disturbance.

### 10.3.1 Straight-Forward Backstepping

Let

$$\tilde{\eta} := \eta - \eta_d, \quad \tilde{\eta} \in \mathbb{R}^n \tag{10.79}$$

be the Earth-fixed tracking surface error and define the body-fixed velocity surface error vector as

$$\tilde{\mathbf{v}} := \mathbf{v} - \alpha_1, \quad \tilde{\mathbf{v}} \in \mathbb{R}^n \tag{10.80}$$

Using (10.79) and (10.80), (1.58)–(1.59) can be rewritten as

$$\begin{aligned}
\dot{\tilde{\eta}} &= \mathbf{J}(\eta)\mathbf{v} - \dot{\eta}_d = \mathbf{J}(\eta)\tilde{\mathbf{v}} + \mathbf{J}(\eta)\alpha_1 - \dot{\eta}_d, \\
\mathbf{M}\dot{\tilde{\mathbf{v}}} &= -\mathbf{C}(\mathbf{v})\mathbf{v} - \mathbf{D}(\mathbf{v})\mathbf{v} - \mathbf{g}(\eta) + \tau - \mathbf{M}\dot{\alpha}_1.
\end{aligned} \tag{10.81}$$

Consider the candidate Lyapunov function

$$V = \frac{1}{2}\tilde{\eta}^T\tilde{\eta} + \frac{1}{2}\tilde{\mathbf{v}}^T\mathbf{M}\tilde{\mathbf{v}}. \tag{10.82}$$

Taking the time derivative gives

$$\dot{V} = \tilde{\eta}^T\dot{\tilde{\eta}} + \tilde{\mathbf{v}}^T\mathbf{M}\dot{\tilde{\mathbf{v}}}, \tag{10.83}$$

where it is assumed that $\mathbf{M} = \mathbf{M}^T > 0$. Using (10.81), $\dot{V}$ can be written as

$$\begin{aligned}
\dot{V} &= \tilde{\eta}^T\left[\mathbf{J}\tilde{\mathbf{v}} + \mathbf{J}\alpha_1 - \dot{\eta}_d\right] + \tilde{\mathbf{v}}^T\left[-\mathbf{C}\mathbf{v} - \mathbf{D}\mathbf{v} - \mathbf{g} + \tau - \mathbf{M}\dot{\alpha}_1\right], \\
&= \tilde{\eta}^T\left[\mathbf{J}\alpha_1 - \dot{\eta}_d\right] + \tilde{\mathbf{v}}^T\left[\mathbf{J}^T\tilde{\eta} - \mathbf{C}\mathbf{v} - \mathbf{D}\mathbf{v} - \mathbf{g} + \tau - \mathbf{M}\dot{\alpha}_1\right].
\end{aligned} \tag{10.84}$$

The stabilizing function $\alpha_1$ and control input $\tau$ can be selected so that

$$\dot{V} = -\tilde{\eta}^T\mathbf{K}_p\tilde{\eta} - \tilde{\mathbf{v}}^T\mathbf{K}_d\tilde{\mathbf{v}} < 0, \tag{10.85}$$

where $\mathbf{K}_p = \mathbf{K}_p^T > 0 \in \mathbb{R}^{n \times n}$ and $\mathbf{K}_d = \mathbf{K}_d^T > 0 \in \mathbb{R}^{n \times n}$ are positive definite matrices of control gains. To accomplish this take

$$\mathbf{J}\alpha_1 - \dot{\eta}_d = -\mathbf{K}_p\tilde{\eta}$$

so that

$$\alpha_1 := \mathbf{J}^{-1}\left[-\mathbf{K}_p\tilde{\eta} + \dot{\eta}_d\right] \tag{10.86}$$

and

$$\mathbf{J}^T\tilde{\eta} - \mathbf{C}\mathbf{v} - \mathbf{D}\mathbf{v} - \mathbf{g} + \tau - \mathbf{M}\dot{\alpha}_1 = -\mathbf{K}_d\tilde{\mathbf{v}},$$

which gives

$$\tau := -\mathbf{K}_d\tilde{\mathbf{v}} + \mathbf{M}\dot{\alpha}_1 + \mathbf{C}\mathbf{v} + \mathbf{D}\mathbf{v} + \mathbf{g} - \mathbf{J}^T\tilde{\eta}. \tag{10.87}$$

The closed loop error dynamics can be obtained using (10.81), (10.86) and (10.87) to get

$$
\begin{aligned}
\dot{\tilde{\eta}} &= -\mathbf{K}_p \tilde{\eta} + \mathbf{J}\tilde{\mathbf{v}}, \\
\mathbf{M}\dot{\tilde{\mathbf{v}}} &= -\mathbf{K}_d \tilde{\mathbf{v}} - \mathbf{J}^T \tilde{\eta},
\end{aligned}
\tag{10.88}
$$

which has a stable equilibrium point at $(\tilde{\eta}, \tilde{\mathbf{v}}) = (0, 0)$.

From (10.85), it can be seen that

$$
\dot{V} \le -\Lambda_{\min}(\mathbf{K}_p)\tilde{\eta}^T \tilde{\eta} - \Lambda_{\min}(\mathbf{K}_d)\tilde{\mathbf{v}}^T \tilde{\mathbf{v}},
$$

where $\Lambda_{\min}(\mathbf{K})$ denotes the minimum eigenvalue of matrix $\mathbf{K}$. Noting that $\mathbf{M} = \mathbf{M}^T > 0$ implies that $\tilde{\mathbf{v}}^T \tilde{\mathbf{v}} \le \tilde{\mathbf{v}}^T \mathbf{M} \tilde{\mathbf{v}}$ so that

$$
\dot{V} \le -\Lambda_{\min}(\mathbf{K}_p)\tilde{\eta}^T \tilde{\eta} - \Lambda_{\min}(\mathbf{K}_d)\tilde{\mathbf{v}}^T \mathbf{M}\tilde{\mathbf{v}} \le -2\mu V,
$$

where $\mu = \min\{\Lambda_{\min}(\mathbf{K}_p), \Lambda_{\min}(\mathbf{K}_d)\}$. Taking $V_0$ to be the value of $V$ at time $t = 0$, $V$ can therefore be expressed as

$$
V \le V_0 e^{-2\mu t}.
$$

Since $V$ is radially unbounded (i.e. $V \to \infty$ if $\tilde{\eta} \to \infty$ or $\tilde{\mathbf{v}} \to \infty$) and exponentially decaying with a stable equilibrium point at the origin, the closed loop system is globally exponentially stable.

### 10.3.2  Passivity-Based Backstepping

Let

$$
\tilde{\eta} := \eta - \eta_d
\tag{10.89}
$$

be the Earth-fixed tracking error. Then define an Earth-fixed *reference velocity* [15], also called a *virtual reference trajectory* [6], as

$$
\dot{\eta}_r := \dot{\eta}_d - \Lambda\tilde{\eta},
\tag{10.90}
$$

where $\Lambda > 0$ is a diagonal control design matrix. In body-fixed (NED-fixed) coordinates the reference velocity is

$$
\mathbf{v}_r := \mathbf{J}^{-1}(\eta)\dot{\eta}_r.
\tag{10.91}
$$

Let s be a *measure of tracking* [9, 14] (also referred to as a *velocity error* [15]) defined as

$$
\mathbf{s} := \dot{\eta} - \dot{\eta}_r = \dot{\tilde{\eta}} + \Lambda\tilde{\eta}.
\tag{10.92}
$$

**Table 10.1** Relations between the inertia tensor, Coriolis/centripetal acceleration tensor, drag tensor and hydrostatic force vector in Earth-fixed and body-fixed coordinate systems

$$\mathbf{M}_\eta(\eta) := \mathbf{J}^{-T}(\eta)\mathbf{M}\mathbf{J}^{-1}(\eta)$$

$$\mathbf{C}_\eta(\mathbf{v},\eta) := \mathbf{J}^{-T}(\eta)[\mathbf{C}(\mathbf{v}) - \mathbf{M}\mathbf{J}^{-1}(\eta)\dot{\mathbf{J}}(\eta)]\mathbf{J}^{-1}(\eta)$$

$$\mathbf{D}_\eta(\mathbf{v},\eta) := \mathbf{J}^{-T}(\eta)\mathbf{D}(\mathbf{v})\mathbf{J}^{-1}(\eta)$$

$$\mathbf{g}_\eta(\eta) := \mathbf{J}^{-T}(\eta)\mathbf{g}(\eta)$$

As shown in [6], the equations of motion (both kinematic and kinetic) can be rewritten in the combined form

$$\mathbf{M}_\eta(\eta)\ddot{\eta} + \mathbf{C}_\eta(\mathbf{v},\eta)\dot{\eta} + \mathbf{D}_\eta(\mathbf{v},\eta)\dot{\eta} + \mathbf{g}_\eta(\eta) = \mathbf{J}^{-T}(\eta)\tau, \qquad (10.93)$$

where

$$\begin{aligned}
\mathbf{M}_\eta(\eta) &= \mathbf{M}_\eta^T(\eta) > 0, \\
\mathbf{s}^T&\left[\frac{1}{2}\dot{\mathbf{M}}_\eta(\eta) - \mathbf{C}_\eta(\mathbf{v},\eta)\right]\mathbf{s} = 0, \quad \forall \mathbf{v},\eta,\mathbf{s} \\
\mathbf{D}_\eta(\mathbf{v},\eta) &> 0,
\end{aligned} \qquad (10.94)$$

and the relationships between the terms $\mathbf{M}$, $\mathbf{C}(\mathbf{v})$, $\mathbf{D}(\mathbf{v})$ and $\mathbf{g}(\eta)$ in the body-fixed coordinate system and $\mathbf{M}_\eta(\eta)$, $\mathbf{C}_\eta(\mathbf{v},\eta)$, $\mathbf{D}_\eta(\mathbf{v},\eta)$ and $\mathbf{g}_\eta(\eta)$ in the Earth-fixed coordinate system shown Table 9.1 are provided again for convenience in Table 10.1. Note that the second property in (10.94) results from the fact that $\left[\frac{1}{2}\dot{\mathbf{M}}_\eta(\eta) - \mathbf{C}_\eta(\mathbf{v},\eta)\right]$ is skew symmetric. The skew-symmetry of this term can be viewed as an expression of the conservation of energy [15].

**Remark 10.1** As shown in (10.94), the drag tensor $\mathbf{D}(\mathbf{v})$ is strictly positive, such that

$$\mathbf{D}(\mathbf{v}) > 0 \Rightarrow \frac{1}{2}\mathbf{x}^T[\mathbf{D}(\mathbf{v}) + \mathbf{D}^T(\mathbf{v})]\mathbf{x} > 0, \forall \mathbf{x} \neq 0.$$

This assumption implies that $\mathbf{D}_\eta$ is also strictly positive. We can see this by examining the symmetric part of $\mathbf{D}_\eta$,

$$\begin{aligned}
\frac{1}{2}\mathbf{x}^T\left[\mathbf{D}_\eta + \mathbf{D}_\eta^T\right]\mathbf{x} &= \frac{1}{2}\mathbf{x}^T\left\{\mathbf{J}^{-T}(\eta)\mathbf{D}(\mathbf{v})\mathbf{J}^{-1}(\eta) + \left[\mathbf{J}^{-T}(\eta)\mathbf{D}(\mathbf{v})\mathbf{J}^{-1}(\eta)\right]^T\right\}\mathbf{x}, \\
&= \frac{1}{2}\mathbf{x}^T\left[\mathbf{J}^{-T}(\eta)\mathbf{D}(\mathbf{v})\mathbf{J}^{-1}(\eta) + \mathbf{J}^{-T}(\eta)\mathbf{D}^T(\mathbf{v})\mathbf{J}^{-1}(\eta)\right]\mathbf{x}, \\
&= \frac{1}{2}\mathbf{x}^T\mathbf{J}^{-T}(\eta)\left[\mathbf{D}(\mathbf{v}) + \mathbf{D}^T(\mathbf{v})\right]\mathbf{J}^{-1}(\eta)\mathbf{x},
\end{aligned}$$

Let $\bar{\mathbf{x}}^T := \mathbf{x}^T\mathbf{J}^{-T}(\eta)$, then $\bar{\mathbf{x}} = \mathbf{J}^{-1}(\eta)\mathbf{x}$, which implies

$$\frac{1}{2}\mathbf{x}^T\left[\mathbf{D}_\eta + \mathbf{D}_\eta^T\right]\mathbf{x} = \frac{1}{2}\bar{\mathbf{x}}^T\left[\mathbf{D}(\mathbf{v}) + \mathbf{D}^T(\mathbf{v})\right]\bar{\mathbf{x}} > 0, \forall\bar{\mathbf{x}} \neq 0,$$

so that $\mathbf{D}_\eta(\mathbf{v}, \eta)$ is also strictly positive. This result will be used in the stability analysis below.

The product $\mathbf{M}_\eta(\eta)\dot{\mathbf{s}}$ is an important quantity that will be used in the controller derivation. It can be written as

$$
\begin{aligned}
\mathbf{M}_\eta(\eta)\dot{\mathbf{s}} = &-\mathbf{C}_\eta(\mathbf{v}, \eta)\mathbf{s} - \mathbf{D}_\eta(\mathbf{v}, \eta)\mathbf{s} \\
&+ \mathbf{J}^{-T}(\eta)\left[\tau - \mathbf{M}\dot{\mathbf{v}}_r - \mathbf{C}(\mathbf{v})\mathbf{v}_r - \mathbf{D}(\mathbf{v})\mathbf{v}_r - \mathbf{g}(\eta)\right],
\end{aligned}
\tag{10.95}
$$

where $\dot{\mathbf{v}}_r = \dot{\mathbf{J}}^{-1}(\eta)\dot{\eta}_r + \mathbf{J}^{-1}(\eta)\ddot{\eta}_r = \dot{\mathbf{J}}^{-1}(\eta)\dot{\eta}_r + \mathbf{J}^{-1}(\eta)\left[\ddot{\eta}_d - \Lambda(\dot{\eta} - \dot{\eta}_d)\right]$.

**Step 1:** Define the virtual control signal

$$
\dot{\eta} = \mathbf{J}(\eta)\mathbf{v} := \mathbf{s} + \alpha_1,
\tag{10.96}
$$

where $\alpha_1$ is a smoothly continuous function that stabilizes the system at the origin, which can be chosen as

$$
\alpha_1 = \dot{\eta}_r = \dot{\eta}_d - \Lambda\tilde{\eta}.
\tag{10.97}
$$

Then, (10.96) can be written as

$$
\dot{\tilde{\eta}} = -\Lambda\tilde{\eta} + \mathbf{s}.
\tag{10.98}
$$

Next, consider the Lyapunov function candidate

$$
V_1 = \frac{1}{2}\tilde{\eta}^T\mathbf{K}_p\tilde{\eta}.
\tag{10.99}
$$

Its time derivative is

$$
\dot{V}_1 = \tilde{\eta}^T\mathbf{K}_p\dot{\tilde{\eta}} = -\tilde{\eta}^T\mathbf{K}_p\Lambda\tilde{\eta} + \tilde{\eta}^T\mathbf{K}_p\mathbf{s}
\tag{10.100}
$$

where $\mathbf{K}_p = \mathbf{K}_p^T > 0$ is a design matrix.

**Step 2:** Consider the Lyapunov function candidate

$$
V_2 = V_1 + \frac{1}{2}\mathbf{s}^T\mathbf{M}_\eta(\eta)\mathbf{s}.
\tag{10.101}
$$

Its time derivative is

$$
\dot{V}_2 = -\tilde{\eta}^T\mathbf{K}_p\Lambda\tilde{\eta} + \mathbf{s}^T\left[\mathbf{K}_p\tilde{\eta} + \mathbf{M}_\eta(\eta)\dot{\mathbf{s}} + \frac{1}{2}\dot{\mathbf{M}}_\eta(\eta)\mathbf{s}\right].
\tag{10.102}
$$

Using (10.95) and then applying the skew symmetry property of $\left[\frac{1}{2}\dot{\mathbf{M}}_\eta(\eta) - \mathbf{C}_\eta(\mathbf{v}, \eta)\right]$ in (10.94), $\dot{V}_2$ can be rewritten as

$$\dot{V}_2 = -\tilde{\eta}^T \mathbf{K}_p \Lambda \tilde{\eta} - \mathbf{s}^T \mathbf{D}_\eta(\mathbf{v}, \eta)\mathbf{s}$$
$$+\mathbf{s}^T \mathbf{J}^{-T}(\eta) \left[ \mathbf{J}^T(\eta)\mathbf{K}_p \tilde{\eta} + \tau - \mathbf{M}\dot{\mathbf{v}}_r - \mathbf{C}(\mathbf{v})\mathbf{v}_r - \mathbf{D}(\mathbf{v})\mathbf{v}_r - \mathbf{g}(\eta) \right].$$
$$(10.103)$$

Select the control law to be

$$\tau = \mathbf{M}\dot{\mathbf{v}}_r + \mathbf{C}(\mathbf{v})\mathbf{v}_r + \mathbf{D}(\mathbf{v})\mathbf{v}_r + \mathbf{g}(\eta) - \mathbf{J}^T(\eta)\mathbf{K}_d \mathbf{s} - \mathbf{J}^T(\eta)\mathbf{K}_p \tilde{\eta}. \qquad (10.104)$$

Then $\dot{V}_2$ becomes

$$\dot{V}_2 = -\tilde{\eta}^T \mathbf{K}_p \Lambda \tilde{\eta} - \mathbf{s}^T \left[ \mathbf{D}_\eta(\mathbf{v}, \eta) + \mathbf{K}_d \right] \mathbf{s}. \qquad (10.105)$$

Since $\Lambda > 0$, $\mathbf{K}_p > 0$ and $\mathbf{D}_\eta(\mathbf{v}, \eta) > 0$, $\dot{V}_2 < 0$, $\forall \mathbf{v}, \eta, \mathbf{s} \neq 0$. Note that the closed loop error system is given by

$$\dot{\tilde{\eta}} = -\Lambda \tilde{\eta} + \mathbf{s},$$
$$\mathbf{M}_\eta(\eta)\dot{\mathbf{s}} = -\mathbf{C}_\eta(\mathbf{v}, \eta)\mathbf{s} - \mathbf{D}_\eta(\mathbf{v}, \eta)\mathbf{s} - \mathbf{K}_d \mathbf{s} - \mathbf{K}_p \Lambda \tilde{\eta}.$$

Thus, it can be seen that the use of passivity does not lead to a linear closed loop error system, as straight-forward backstepping does in (10.88). However, the trajectory tracking control law developed with the use of the passivity (10.104) can be shown to render the closed loop system globally exponentially stable (see [6]).

## 10.3.3 Backstepping Implementation Issues

The first two implementation issues are associated with the computation of the time derivatives of the stabilizing functions $\alpha_i$. Both of these issues become more complicated as the order of the system $n$ increases.

(1) Computing the time derivatives of the stabilizing functions $\alpha_i$ involves taking time derivatives of the state variables, which appear in the dynamic model of the system. These computations can involve a very large number of terms, leading to the *explosion of complexity* problem discussed in Sect. 10.2.1.

(2) Computing the time derivatives of the stabilizing functions $\alpha_i$ also requires taking higher order time derivatives of the desired trajectory. Generally, these time derivatives must be smooth and bounded to at an order corresponding to at least the number of integrators in the dynamic system in order to ensure that the stabilizing functions and control input are bounded.

If a desired trajectory is not smooth to the required order, a simple approach for using backstepping is to generate an approximate trajectory by passing the desired trajectory through a low pass linear filter [6]. For example, the square wave like trajectory used in Example 10.1 above was generated by passing a square wave trajectory through a second order linear filter.

To ensure that the smoothed trajectory is feasible, the choice of filter parameters should be based on the dynamics of the vehicle, taking into account speed and acceleration limits, as well as actuator constraints. The bandwidth of the filter used to generate the reference trajectory should be lower than the bandwidth of the motion control system.

Let $r(t)$ be the unfiltered trajectory. The approximate desired trajectory $y_d(t)$ to be used for backstepping control can be determined as

$$y_d(t) = G_f(s)R(s) \tag{10.106}$$

where $s$ is the Laplace transform variable and $G_f(s)$ is the low pass filter transfer function. A convenient filter function that can be used for marine vessels is the canonical form of a second order system

$$G_f(s) = \frac{\omega_n^2}{s^2 + 2\zeta\omega_n s + \omega_n^2}, \tag{10.107}$$

where the natural frequency $\omega_n$ and damping ratio $\zeta$ can be tuned to specify the phase lag and overshoot between the desired and approximate trajectories. They should also be tuned bearing in mind the physical capabilities of the vehicle to be controlled, as discussed above. If needed, higher order filters can be obtained by taking powers of $G_f(s)$. For example a fourth order filter can be constructed with

$$G_4(s) = G_f(s) \cdot G_f(s) = \frac{\omega_n^4}{\left[s^2 + 2\zeta\omega_n s + \omega_n^2\right]^2}. \tag{10.108}$$

During computational implementation, it is often convenient to express these filters in state space form. For example, the second order differential equation corresponding to the application of (10.107) to a nonsmooth input trajectory $r(t)$ to produce an approximate trajectory $y_d(t)$ is given by

$$\ddot{y}_d + 2\zeta\omega_n \dot{y}_d + \omega_n^2 y_d = \omega_n^2 r, \tag{10.109}$$

so that

$$\ddot{y}_d = -2\zeta\omega_n \dot{y}_d - \omega_n^2 y_d + \omega_n^2 r.$$

Let $y_{d1} = y_d$, $y_{d2} = \dot{y}_d$ and $\mathbf{y}_d = [y_{d1} \quad y_{d2}]^T$, then

$$\dot{\mathbf{y}}_d = \begin{bmatrix} 0 & 1 \\ -\omega_n^2 & -2\zeta\omega_n \end{bmatrix} \mathbf{y}_d + \begin{bmatrix} 0 \\ \omega_n^2 \end{bmatrix} r. \tag{10.110}$$

With this system, given a nonsmooth $r(t)$ one can solve for a smooth $y_d(t)$ and $\dot{y}_d(t)$ (note that $\ddot{y}_d(t)$ will not be smooth because it has $r(t)$ as an input).

**Remark 10.2** (*Numerical differentiation of a signal*) Note that linear state space filtering of a signal of the form shown in (10.110) can be used to compute the time derivative of any input signal and can be extended to compute higher order derivatives by using the state space representation of higher order filters, such as the 4$^{\text{th}}$ order filter given by (10.108). This can be useful in many practical circumstances, for example when it is necessary to compute the time derivative of a signal measured by a feedback sensor, or the time derivative of a stabilizing function. By formulating the state space equations as a set of finite difference equations, it is fairly easy to implement such a differentiator on a control computer (e.g. a microcontroller or single board computer). Normally, the highest order time derivatives will contain the most noise, so the order of the filter should be higher than the order of the desired derivative. Note that some phase lag and attenuation of the filtered signal and its time derivatives will occur, depending on the values of $\zeta$ and $\omega_n$. These values should be selected based on knowledge of how fast the input signal tends to fluctuate.

(3) In the computation of the stabilizing functions and control input it is implicitly assumed that the terms $\mathbf{f}_i(\mathbf{x}_i)$ and $\mathbf{G}_i(\mathbf{x}_i)$ are precisely known. However, there will generally be some uncertainty associated with the dynamical model of any physical system. By including the linear terms $-k_i z_i$ in the stabilizing functions and control input, it can be ensured that the control Lyapunov function of the closed loop system is positive definite and that its derivative is negative definite when model uncertainties are small. The terms $-k_i z_i$ can be thought of as linear damping terms. However, owing to the nonlinearity of most systems, model uncertainty can adversely affect the stability of the system when it is operated away from its closed loop equilibrium point $\tilde{\alpha}_i = 0$. The nonlinearities associated with model uncertainty are often referred to as *bad nonlinearities*.

With backstepping it is relatively straightforward to make the closed loop system more robust to model uncertainty by including additional *nonlinear damping* terms in the definitions of the stabilizing functions and control input. As mentioned in Sect. 10.1, this nonlinear damping is intended to dominate any "bad nonlinearities". By suitable selection of the control gains multiplying these nonlinear damping terms, it is (in principle) possible to ensure that the system is stable.

**Definition 10.1** (*Scalar nonlinear damping function*) A function $n_i(z_i) : \mathbb{R} \to \mathbb{R}$ is a scalar nonlinear damping function, if it is continuous, odd $n_i(-z_i) = -n_i(z_i)$, zero at zero $n_i(0) = 0$, and strictly increasing.

Nonlinear damping can be implemented in SISO backstepping by modifying the equations for the stabilizing functions and control law given above to include nonlinear damping functions $n_i(z_i)$, which use the the error surfaces $z_i$ as their arguments. Specifically, for $n$-state SISO systems, (10.61) can be modified to give the stabilizing functions

$$
\begin{aligned}
\alpha_1 &= \frac{1}{L_{g1}h(x_1)} \left\{ -k_1 z_1 - n_1(z_1) + \dot{y}_d - L_{f1}h(x_1) \right\}, \\
\alpha_2 &= \frac{1}{g_2(x_1, x_2)} \left\{ -k_2 z_2 - n_2(z_2) - f_2(x_1, x_2) - L_{g1}h(x_1)z_1 + \dot{\alpha}_1 \right\}, \\
\alpha_i &= \frac{1}{g_i(x_1, \ldots, x_i)} \left\{ -k_i z_i - n_i(z_i) - f_i(x_1, \ldots, x_i) \right. \\
&\qquad\qquad\qquad \left. - g_{i-1}(x_1, \ldots, x_{i-1})z_{i-1} + \dot{\alpha}_{i-1} \right\}, \\
&\quad i \in \{3, \ldots, n-1\}, \\
u &= \frac{1}{g_n(x_1, \ldots, x_n)} \left\{ -k_n z_n - n_n(z_n) - f_n(x_1, \ldots, x_n) \right. \\
&\qquad\qquad\qquad \left. - g_{n-1}(x_1, \ldots, x_{n-1})z_{n-1} + \dot{\alpha}_{n-1} \right\}.
\end{aligned} \tag{10.111}
$$

The closed loop tracking error dynamics then become

$$
\begin{aligned}
\dot{z}_1 &= -k_1 z_1 - n_1(z_1) + L_{g1}h(x_1)z_2, \\
\dot{z}_2 &= -k_2 z_2 - n_2(z_2) + g_2(x_1, x_2)z_3 - L_{g1}h(x_1)z_1, \\
\dot{z}_i &= -k_i z_i - n_i(z_i) + g_i(x_1, \ldots, x_i)z_{i+1} - g_{i-1}(x_1, \ldots, x_{i-1})z_{i-1}, \\
&\qquad\qquad \text{for } i \in \{3, \ldots, n-1\}, \\
\dot{z}_n &= -k_n z_n - n_n(z_n) - g_{n-1}(x_1, \ldots, x_{n-1})z_{n-1}.
\end{aligned} \tag{10.112}
$$

**Definition 10.2** (*Vectorized nonlinear damping function*) A vector function $\mathbf{n}_i(\mathbf{z}_i) : \mathbb{R}^n \to \mathbb{R}^n$ is a vectorized nonlinear damping function, if it operates component-wise on its vector argument and is continuous, odd $\mathbf{n}_i(-\mathbf{z}_i) = -\mathbf{n}_i(\mathbf{z}_i)$, zero at zero $\mathbf{n}_i(\mathbf{0}) = \mathbf{0}$, and strictly increasing.

Similarly, for $n$-state MIMO systems, vectorized nonlinear damping can be added to (10.75) so that it becomes

$$
\begin{aligned}
\alpha_1 &= \left[ \nabla\mathbf{h}(\mathbf{x}_1) \cdot \mathbf{G}_1(\mathbf{x}_1) \right]^{\dagger} \left\{ -\mathbf{K}_1\mathbf{z}_1 - \mathbf{N}_1(\mathbf{z}_1) - \nabla\mathbf{h}(\mathbf{x}_1)\mathbf{f}_1(\mathbf{x}_1) + \dot{\mathbf{y}}_d \right\}, \\
\alpha_2 &= \mathbf{G}_2^{\dagger}(\mathbf{x}_1, \mathbf{x}_2) \left\{ -\mathbf{K}_2\mathbf{z}_2 - \mathbf{N}_2(\mathbf{z}_2) - \mathbf{f}_2(\mathbf{x}_1, \mathbf{x}_2) \right. \\
&\qquad\qquad\qquad \left. - \left[ \nabla\mathbf{h}(\mathbf{x}_1)\mathbf{G}_1(\mathbf{x}_1) \right]^T \mathbf{z}_1 + \dot{\alpha}_1 \right\}, \\
\alpha_i &= \mathbf{G}_i^{\dagger}(\mathbf{x}_1, \ldots, \mathbf{x}_i) \left\{ -\mathbf{K}_i\mathbf{z}_i - \mathbf{N}_i(\mathbf{z}_i) - \mathbf{f}_i(\mathbf{x}_1, \ldots, \mathbf{x}_i) \right. \\
&\qquad\qquad\qquad \left. - \mathbf{G}_{i-1}^T(\mathbf{x}_1, \ldots, \mathbf{x}_{i-1})\mathbf{z}_{i-1} + \dot{\alpha}_{i-1} \right\}, \\
&\quad i \in \{3, \ldots, n-1\}, \\
\mathbf{u} &= \mathbf{G}_n^{\dagger}(\mathbf{x}_1, \ldots, \mathbf{x}_n) \left\{ -\mathbf{K}_n\mathbf{z}_n - \mathbf{N}_n(\mathbf{z}_n) - \mathbf{f}_n(\mathbf{x}_1, \ldots, \mathbf{x}_n) \right. \\
&\qquad\qquad\qquad \left. - \mathbf{G}_{n-1}^T(\mathbf{x}_1, \ldots, \mathbf{x}_{n-1})\mathbf{z}_{n-1} + \dot{\alpha}_{n-1} \right\},
\end{aligned} \tag{10.113}
$$

with $\mathbf{N}_i(\mathbf{z}_i) \geq 0$ for $i = 1 \ldots n$. The resulting closed loop tracking error dynamics are

$$\dot{\mathbf{z}}_1 = -\mathbf{K}_1\mathbf{z}_1 - \mathbf{N}_1(\mathbf{z}_1) + \nabla\mathbf{h}(\mathbf{x}_1)\mathbf{G}_1(\mathbf{x}_1)\mathbf{z}_2,$$
$$\dot{\mathbf{z}}_2 = -\mathbf{K}_2\mathbf{z}_2 - \mathbf{N}_2(\mathbf{z}_2) + \mathbf{G}_2(x_1, x_2)\mathbf{z}_3 - [\nabla\mathbf{h}(\mathbf{x}_1)\mathbf{G}_1(\mathbf{x}_1)]^T\mathbf{z}_1,$$
$$\dot{\mathbf{z}}_i = -\mathbf{K}_i\mathbf{z}_i - \mathbf{N}_i(\mathbf{z}_i) + \mathbf{G}_i(x_1, \ldots, x_i)\mathbf{z}_{i+1} - \mathbf{G}_{i-1}^T(x_1, \ldots, x_{i-1})\mathbf{z}_{i-1}, \quad (10.114)$$
$$i \in \{3, \ldots, n-1\},$$
$$\dot{\mathbf{z}}_n = -\mathbf{K}_n\mathbf{z}_n - \mathbf{N}_n(\mathbf{z}_n) - \mathbf{G}_{n-1}^T(x_1, \ldots, x_{n-1})\mathbf{z}_{n-1}.$$

Since it can be difficult to characterize the bad nonlinearities, an obvious drawback to the use of nonlinear damping is that the control effort used may be more than is actually required. An approach to mitigating this problem is the use of nonlinear disturbance observer based control, which is discussed in Sect. 10.7.

## 10.4 Augmented Integrator Backstepping

When there is no model uncertainty, the use of backstepping for controller design results in a nonlinear proportional derivative (PD) control law. If one is interested in ensuring that the steady state error of the closed loop system can be driven to zero when constant disturbances are present, it would be preferable to use a nonlinear proportional integral derivative (PID) controller. A nonlinear PID control law can be designed by augmenting the system with an additional integrator. Here, the use of augmented integrator backstepping is illustrated with an example.

**Example 10.3** Consider the straight-line control of a surface vessel in a headwind (Fig. 10.7), similar to that explored in Example 2.3. The equations of motion can be expressed in the form

$$\dot{x} = u,$$
$$\dot{u} = -\frac{\tilde{d}(u)}{m}u + \frac{\tilde{\tau}}{m} + \frac{\tilde{w}_d}{m} = -d(u)u + \tau + w_d, \quad (10.115)$$
$$y = x,$$

where $y = x$ is the distance traveled by the vessel along its surge direction, $m$ is its mass, $u$ is its speed, $\tilde{\tau}$ is the thrust generated by the propeller, $\tilde{d}(u)$ is a nonlinear, speed-dependent drag and $\tilde{w}_d$ is the exogenous disturbance force of the wind.

First, it is shown that the backstepping technique without integrator augmentation leads to nonlinear PD control. Let us take $w_d = 0$. Using the backstepping technique, we define the tracking errors and their time derivatives as

$$z_1 := y - y_d \rightarrow \dot{z}_1 = \dot{y} - \dot{y}_d = \dot{x} - \dot{y}_d = u - \dot{y}_d$$
$$= z_2 + \alpha_1 - \dot{y}_d \quad (10.116)$$
$$z_2 := u - \alpha_1 \rightarrow \dot{z}_2 = \dot{u} - \dot{\alpha}_1 = -d(u)u + \tau - \dot{\alpha}_1.$$

The stabilizing functions are determined using a CLF of the form

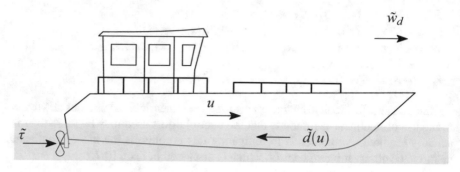

**Fig. 10.7**  Straight-line trajectory-tracking control of a surface vessel in a headwind

$$V = \frac{1}{2}z_1^2 + \frac{1}{2}z_2^2. \tag{10.117}$$

Its time derivative is

$$\begin{aligned}
\dot{V} &= z_1 \dot{z}_1 + z_2 \dot{z}_2 \\
&= z_1 \{z_2 + \alpha_1 - \dot{y}_d\} + z_2 \{-d(u)u + \tau - \dot{\alpha}_1\}, \\
&= z_1 \{\alpha_1 - \dot{y}_d\} + z_2 \{-d(u)u + z_1 + \tau - \dot{\alpha}_1\}.
\end{aligned} \tag{10.118}$$

The system can be stabilized such that $\dot{V} = -k_1 z_1^2 - k_2 z_2^2$, for $k_1, k_2 < 0$ by selecting the stabilizing function and control input to be

$$\begin{aligned}
\alpha_1 &= -k_1 z_1 + \dot{y}_d, \\
\tau &= -k_2 z_2 - z_1 + d(u)u + \dot{\alpha}_1.
\end{aligned}$$

Note that the terms in the control input can be expanded to give

$$\begin{aligned}
\tau &= -k_2(u - \alpha_1) - (y - y_d) + d(u)u - k_1 \dot{z}_1 + \ddot{y}_d, \\
&= -k_2(u + k_1 z_1 - \dot{y}_d) - (y - y_d) + d(u)u - k_1(u - \dot{y}_d) + \ddot{y}_d, \\
&= -\underbrace{(1 + k_1 k_2)(y - y_d)}_{\text{proportional term}} - \underbrace{(k_1 + k_2)(u - \dot{y}_d)}_{\text{derivative term}} + \underbrace{d(u)u + \ddot{y}_d}_{\text{linearization terms}},
\end{aligned}$$

where, as indicated, the first term is proportional to the error, the second term is proportional to the derivative of the error and the final term provides feedback that linearizes the system. Thus, the closed loop controller acts like a PD controller, plus feedback linearizing terms.

The corresponding error system is given by

$$\begin{aligned}
\dot{z}_1 &= -k_1 z_1 + z_2 \\
\dot{z}_2 &= -k_2 z_2 - z_1,
\end{aligned} \tag{10.119}$$

which has a stable equilibrium point when the tracking errors are zero, e.g. at its origin $(z_1, z_2) = (0, 0)$.

Next, consider the case when the controller derived is used to control the plant with $w_d \neq 0$. The closed loop error system becomes

$$\begin{aligned} \dot{z}_1 &= -k_1 z_1 + z_2 \\ \dot{z}_2 &= -k_2 z_2 - z_1 + w_d. \end{aligned} \tag{10.120}$$

At steady state, when $\dot{z}_1 = \dot{z}_2 = 0$, the error system (10.120) can be solved to show that the stable equilibrium point $(z_{1e}, z_{2e})$ is located at

$$\begin{aligned} z_{1e} &= \frac{w_d}{(1 + k_1 k_2)}, \\ z_{2e} &= \frac{k_1}{(1 + k_1 k_2)} w_d. \end{aligned} \tag{10.121}$$

Thus, although the tracking errors can be mitigated by selecting $k_1$ and $k_2$ to be large, the steady state tracking errors are nonzero. Now, *augment* the system in (10.115) with an additional differential equation to show how the integrator-augmented system leads to a nonlinear PID controller. Let

$$\begin{aligned} \dot{e}_1 &= e = y - y_d \\ \dot{e} &= u - \dot{y}_d \\ \dot{u} &= -d(u)u + \tau. \end{aligned} \tag{10.122}$$

Then, define the tracking errors and their time derivatives as

$$\begin{aligned} z_1 &:= e_1 &&\rightarrow \dot{z}_1 = e = z_2 + \alpha_1 \\ z_2 &:= e - \alpha_1 &&\rightarrow \dot{z}_2 = \dot{e} - \dot{\alpha}_1 = u - \dot{y}_d - \dot{\alpha}_1 \\ & && \qquad = z_3 + \alpha_2 - \dot{y}_d - \dot{\alpha}_1 \\ z_3 &:= u - \alpha_2 &&\rightarrow \dot{z}_3 = \dot{u} - \dot{\alpha}_2 = -d(u)u + \tau - \dot{\alpha}_2. \end{aligned} \tag{10.123}$$

Use a CLF of the form

$$V = \frac{1}{2} z_1^2 + \frac{1}{2} z_2^2 + \frac{1}{2} z_3^2 \tag{10.124}$$

to determine the stabilizing functions. Its time derivative is

$$\begin{aligned} \dot{V} &= z_1 \dot{z}_1 + z_2 \dot{z}_2 + z_3 \dot{z}_3 \\ &= z_1 \{z_2 + \alpha_1\} + z_2 \{z_3 + \alpha_2 - \dot{y}_d - \dot{\alpha}_1\} + z_3 \{-d(u)u + \tau - \dot{\alpha}_2\}, \quad (10.125) \\ &= z_1 \alpha_1 + z_2 \{z_1 + \alpha_2 - \dot{y}_d - \dot{\alpha}_1\} + z_3 \{z_2 - d(u)u + \tau - \dot{\alpha}_2\}. \end{aligned}$$

The system can be stabilized such that $\dot{V} = -k_1 z_1^2 - k_2 z_2^2 - k_3 z_3^2$, where $k_1, k_2, k_3 > 0$ by selecting the stabilizing function and control input to be

$$\alpha_1 = -k_1 z_1,$$
$$\alpha_2 = -k_2 z_2 - z_1 + \dot{y}_d + \dot{\alpha}_1,$$
$$\tau = -k_3 z_3 - z_2 + d(u)u + \dot{\alpha}_2.$$

Expanding the latter equation, the control input can be shown to be composed of three parts

$$\tau = -\underbrace{(2 + k_1 k_2 + k_2 k_3 + k_1 k_3)e}_{\text{proportional term}} - \underbrace{(k_1 + k_3 + k_1 k_2 k_3)e_1}_{\text{integral term}}$$
$$-\underbrace{(k_1 + k_2 + k_3)\dot{e}}_{\text{derivative term}} + \underbrace{d(u)u + \ddot{y}_d}_{\text{linearization terms}} \; .$$

Thus, the closed loop controller acts like a proportional-integral-derivative controller, plus feedback linearizing terms.

Now, if $w_d \neq 0$, the closed loop error system is given by

$$\begin{aligned}
\dot{z}_1 &= -k_1 z_1 + z_2, \\
\dot{z}_2 &= -k_2 z_2 - z_1 + z_3, \\
\dot{z}_3 &= -k_3 z_3 - z_2 + w_d.
\end{aligned} \tag{10.126}$$

Now, the steady state tracking error becomes zero $e = \dot{z}_1 = 0$, even though the equilibrium point of the augmented system is nonzero

$$\begin{aligned}
z_{1e} &= \frac{w_d}{(k_1 + k_1 k_2 k_3 + k_3)}, \\
z_{2e} &= k_1 z_{1e} = \frac{k_1 w_d}{(k_1 + k_1 k_2 k_3 + k_3)}, \\
z_{3e} &= (1 + k_1 k_2) z_{1e} = \frac{(1 + k_1 k_2) w_d}{(k_1 + k_1 k_2 k_3 + k_3)}.
\end{aligned} \tag{10.127}$$

□

Here, it has been shown by example that the dynamic equations of a system can be augmented with a first order differential equation describing the evolution of an additional state of interest. Augmenting a system in this way permits the effect of the augmented state to be included in the control design. In the preceding example, the additional state of interest is the integral of the error $e_1$ and it permits us to include the effect of an integral term in the controller. In general, the dynamic equations of a system can be augmented with multiple differential equations describing the evolution of additional states of interest. In the following sections, integrator augmentation is used to: (1) compute approximations of the backstepping stabilizing functions $\alpha_i$ in the Dynamic Surface Control Technique, (2) incorporate the effects of actuator rate limits in control design, and (3) to design nonlinear disturbance observer based controllers.

**Fig. 10.8** Frequency response of the transfer function given by (10.128). Note that both axes are represented on a logarithmic scale

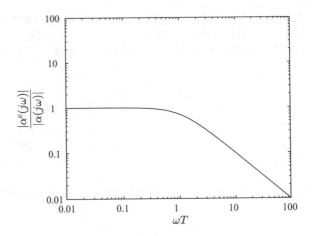

## 10.5  Dynamic Surface Control

Dynamic surface control (DSC) is an approach similar to backstepping, which allows one to avoid the "explosion of complexity" issue associated with traditional backstepping by using first order filters to compute the time derivatives of the plant model [17] (see Remark 10.2). As shown in [12], the DSC technique can be extended to strict-feedback nonlinear systems of the form presented in Sect. 10.2.3.

Consider the transfer function

$$\frac{\alpha^c(s)}{\alpha(s)} = \frac{1}{Ts+1},\qquad(10.128)$$

where, as usual, $\alpha^c(s)$ and $\alpha(s)$ are the Laplace transforms of $\alpha^c(t)$ and $\alpha(t)$, respectively, and $s$ is the Laplace transform variable. The frequency response of this transfer function is the frequency-dependent ratio of the magnitude of the output signal to the magnitude of the input signal. In Fig. 10.8, it can be seen that below the corner frequency where $\omega T = 1$ the ratio of the output to the input magnitudes is approximately 1, but at higher frequencies the ratio of magnitudes decreases as frequency increases. Thus, the transfer function (10.128) acts like a first-order low pass filter.

The first order differential equation corresponding to (10.128) is

$$T\dot{\alpha}^c = -\alpha^c + \alpha,\qquad(10.129)$$

where $T$ can be thought of as the filter time constant. Thus, $\alpha$ acts as a forcing term for the time evolution of $\alpha^c$, so that at steady state $\alpha^c = \alpha$. Therefore, by augmenting the dynamic equations describing a marine vehicle with (10.129), $\alpha^c$ and $\dot{\alpha}^c$ can be computed. The basic idea behind the DSC Technique is to avoid the explosion of complexity associated with traditional backstepping by approximating $\dot{\alpha}$ with $\dot{\alpha}^c$, which substantially simplifies the computation of the stabilizing functions and control law.

To see how the DSC approach can be applied to an $n$th order SISO system in SFNS form (10.55), let $\alpha_i^c \in \mathbb{R}$ be the first order filtered approximations of the stabilizing functions $\alpha_i(t)$. Redefine the tracking errors and their time derivatives to be

$$
\begin{aligned}
z_1 := y - y_d \quad &\rightarrow \dot{z}_1 = \dot{y} - \dot{y}_d = L_{f1}h + L_{g1}hx_2 - \dot{y}_d \\
&= L_{f1}h + L_{g1}hz_2 + L_{g1}h\alpha_1^c - \dot{y}_d \\
&= L_{f1}h + L_{g1}hz_2 + L_{g1}h\tilde{\alpha}_1 + L_{g1}h\alpha_1 - \dot{y}_d \\
z_i := x_i - \alpha_{i-1}^c \quad &\rightarrow \dot{z}_i = \dot{x}_i - \dot{\alpha}_{i-1}^c = f_i + g_i x_{i+1} - \dot{\alpha}_{i-1}^c \\
&= f_i + g_i z_{i+1} + g_i \alpha_i^c - \dot{\alpha}_{i-1}^c \\
&= f_i + g_i z_{i+1} + g_i \tilde{\alpha}_i + g_i \alpha_i - \dot{\alpha}_{i-1}^c \\
&\quad \text{for} \quad i \in \{2, \ldots, n-1\} \\
z_n := x_n - \alpha_{n-1}^c \quad &\rightarrow \dot{z}_n = \dot{x}_n - \dot{\alpha}_{n-1}^c = f_n + g_n u - \dot{\alpha}_{n-1}^c
\end{aligned}
\tag{10.130}
$$

where (10.54) is used to compute the time derivative of $y$, the $\alpha_i(t)$ are stabilizing functions and the $\tilde{\alpha}_i := \alpha_i^c - \alpha_i$ for $i = 1, 2, \ldots, n-1$ are stabilizing function error surfaces. As explained above, the $\dot{\alpha}_i^c$ are approximated using a first order filter equation of the form

$$
T_i \dot{\alpha}_i^c = -\alpha_i^c + \alpha_i, \quad \alpha_i^c(0) = \alpha_i(0),
\tag{10.131}
$$

where the filter time constants are taken to lie in the range $0 < T_i < 1$.

To compute the stabilizing functions and control input, the $\dot{\alpha}_i$ terms in (10.111) are now approximated by $\dot{\alpha}_i^c$ to obtain

$$
\begin{aligned}
\alpha_1 &= \frac{1}{L_{g1}h(x_1)} \left\{ -[k_1 + n_1(z_1)]z_1 - L_{f1}h(x_1) + \dot{y}_d \right\}, \\
\alpha_2 &= \frac{1}{g_2(x_1, x_2)} \left\{ -[k_2 + n_2(z_2)]z_2 - f_2(x_1, x_2) - L_{g1}h(x_1)z_1 + \dot{\alpha}_1^c \right\}, \\
\alpha_i &= \frac{1}{g_i(x_1, \ldots, x_i)} \{ -[k_i + n_i(z_i)]z_i - f_i(x_1, \ldots, x_i) \\
&\qquad\qquad\qquad - g_{i-1}(x_1, \ldots, x_{i-1})z_{i-1} + \dot{\alpha}_{i-1}^c \}, \\
&\quad i \in \{3, \ldots, n-1\}, \\
u &= \frac{1}{g_n(x_1, \ldots, x_n)} \{ -[k_n + n_n(z_n)]z_n - f_n(x_1, \ldots, x_n) \\
&\qquad\qquad\qquad - g_{n-1}(x_1, \ldots, x_{n-1})z_{n-1} + \dot{\alpha}_{n-1}^c \},
\end{aligned}
\tag{10.132}
$$

where the $k_i > 0$ are control gains and the $n_i(z_i) > 0$ are good nonlinearities.

Combining (10.130) with (10.132), one gets the closed loop tracking error dynamics

$$\dot{z}_1 = -[k_1 + n_1(z_1)]z_1 + L_{g1}h(x_1)[z_2 + \tilde{\alpha}_1],$$
$$\dot{z}_2 = -[k_2 + n_2(z_2)]z_2 + g_2(x_1, x_2)[z_3 + \tilde{\alpha}_2] - L_{g1}h(x_1)z_1,$$
$$\dot{z}_i = -[k_i + n_i(z_i)]z_i + g_i(x_1, \ldots, x_i)[z_{i+1} + \tilde{\alpha}_i] - g_{i-1}(x_1, \ldots, x_{i-1})z_{i-1},$$
$$\text{for } i \in \{3, \ldots, n-1\},$$
$$\dot{z}_n = -[k_n + n_n(z_n)]z_n - g_{n-1}(x_1, \ldots, x_{n-1})z_{n-1}.$$

$$(10.133)$$

It can be proven that, when the control gains and filter time constants are suitably chosen, the closed loop errors are bounded so that $\dot{V} \le -\mu V + \delta$, where $\delta > 0$ is a constant [17]. Some guidance for selecting suitable values of the filter time constants can be found in [12].

The DSC Method can also be extended for use in vectorial backstepping on MIMO systems in SFNS form (10.65). To see how, define the tracking errors and their time derivatives to be

$$
\begin{aligned}
\mathbf{z}_1 := \mathbf{y} - \mathbf{y}_d \quad &\rightarrow \dot{\mathbf{z}}_1 = \nabla \mathbf{h}(\mathbf{x}_1)[\mathbf{f}_1(\mathbf{x}_1) + \mathbf{G}_1(\mathbf{x}_1)\mathbf{x}_2] - \dot{\mathbf{y}}_d \\
&= \nabla \mathbf{h}\mathbf{f}_1 + \nabla \mathbf{h}\mathbf{G}_1\mathbf{z}_2 + \nabla \mathbf{h}\mathbf{G}_1\alpha_1^c - \dot{\mathbf{y}}_d \\
&= \nabla \mathbf{h}\mathbf{f}_1 + \nabla \mathbf{h}\mathbf{G}_1\mathbf{z}_2 + \nabla \mathbf{h}\mathbf{G}_1\tilde{\alpha}_1 + \nabla \mathbf{h}\mathbf{G}_1\alpha_1 - \dot{\mathbf{y}}_d \\
\mathbf{z}_i := \mathbf{x}_i - \alpha_{i-1}^c \quad &\rightarrow \dot{\mathbf{z}}_i = \mathbf{f}_i(\mathbf{x}_1, \ldots, \mathbf{x}_i) + \mathbf{G}_i(\mathbf{x}_1, \ldots, \mathbf{x}_i)\mathbf{x}_{i+1} - \dot{\alpha}_{i-1}^c \\
&= \mathbf{f}_i + \mathbf{G}_i\mathbf{z}_{i+1} + \mathbf{G}_i\alpha_i^c - \dot{\alpha}_{i-1}^c \\
&= \mathbf{f}_i + \mathbf{G}_i\mathbf{z}_{i+1} + \mathbf{G}_i\tilde{\alpha}_i + \mathbf{G}_i\alpha_i - \dot{\alpha}_{i-1}^c \\
&\quad \text{for } i \in \{2, \ldots, n-1\} \\
\mathbf{z}_n := \mathbf{x}_n - \alpha_{n-1}^c \quad &\rightarrow \dot{\mathbf{z}}_n = \mathbf{f}_n(\mathbf{x}_1, \ldots, \mathbf{x}_n) + \mathbf{G}_n(\mathbf{x}_1, \ldots, \mathbf{x}_n)\mathbf{u} - \dot{\alpha}_{n-1}^c
\end{aligned}
$$

$$(10.134)$$

where (10.67) is used to compute the time derivative of $\mathbf{y}$, the $\alpha_i(t)$ are stabilizing vector functions and $\tilde{\alpha}_i := \alpha_i^c - \alpha_i$ for $i = 1, 2, \ldots, n-1$. The $\dot{\alpha}_i^c$ are now vectors, which are approximated using a set of first order filter equations of the form

$$\mathbf{T}_i\dot{\alpha}_i^c = -\alpha_i^c + \alpha_i, \quad \alpha_i^c(0) = \alpha_i(0),$$

$$(10.135)$$

where the $\mathbf{0}_{n \times n} < \mathbf{T}_i < \mathbf{1}_{n \times n}$ are now diagonal matrices of filter time constants.

The stabilizing functions and control input are selected to be

$$
\begin{aligned}
\alpha_1 &= [\nabla \mathbf{h}(\mathbf{x}_1) \cdot \mathbf{G}_1(\mathbf{x}_1)]^\dagger \{-[\mathbf{K}_1 + \mathbf{N}_1(\mathbf{z}_1)]\mathbf{z}_1 - \nabla \mathbf{h}(\mathbf{x}_1)\mathbf{f}_1(\mathbf{x}_1) + \dot{\mathbf{y}}_d\}, \\
\alpha_2 &= \mathbf{G}_2^\dagger(\mathbf{x}_1, \mathbf{x}_2)\{-[\mathbf{K}_2 + \mathbf{N}_2(\mathbf{z}_2)]\mathbf{z}_2 - \mathbf{f}_2(\mathbf{x}_1, \mathbf{x}_2) \\
&\qquad - [\nabla \mathbf{h}(\mathbf{x}_1)\mathbf{G}_1(\mathbf{x}_1)]^T \mathbf{z}_1 + \dot{\alpha}_1^c\}, \\
\alpha_i &= \mathbf{G}_i^\dagger(\mathbf{x}_1, \ldots, \mathbf{x}_i)\{-[\mathbf{K}_i + \mathbf{N}_i(\mathbf{z}_i)]\mathbf{z}_i - \mathbf{f}_i(\mathbf{x}_1, \ldots, \mathbf{x}_i) \\
&\qquad - \mathbf{G}_{i-1}^T(\mathbf{x}_1, \ldots, \mathbf{x}_{i-1})\mathbf{z}_{i-1} + \dot{\alpha}_{i-1}^c\}, \\
&\quad i \in \{3, \ldots, n-1\}, \\
\mathbf{u} &= \mathbf{G}_n^\dagger(\mathbf{x}_1, \ldots, \mathbf{x}_n)\{-[\mathbf{K}_n + \mathbf{N}_n(\mathbf{z}_n)]\mathbf{z}_n - \mathbf{f}_n(\mathbf{x}_1, \ldots, \mathbf{x}_n) \\
&\qquad - \mathbf{G}_{n-1}^T(\mathbf{x}_1, \ldots, \mathbf{x}_{n-1})\mathbf{z}_{n-1} + \dot{\alpha}_{n-1}^c\},
\end{aligned}
$$

$$(10.136)$$

Lastly, the vector form of the closed loop tracking error dynamics is given by the relations

$$\dot{\mathbf{z}}_1 = -\left[\mathbf{K}_1 + \mathbf{N}_1(\mathbf{z}_1)\right]\mathbf{z}_1 + \nabla\mathbf{h}(\mathbf{x}_1)\mathbf{G}_1(\mathbf{x}_1)\left[\mathbf{z}_2 + \tilde{\alpha}_1\right],$$
$$\dot{\mathbf{z}}_2 = -\left[\mathbf{K}_2 + \mathbf{N}_2(\mathbf{z}_2)\right]\mathbf{z}_2 + \mathbf{G}_2(x_1, x_2)\left[\mathbf{z}_3 + \tilde{\alpha}_2\right] - \left[\nabla\mathbf{h}(\mathbf{x}_1)\mathbf{G}_1(\mathbf{x}_1)\right]^T\mathbf{z}_1,$$
$$\dot{\mathbf{z}}_i = -\left[\mathbf{K}_i + \mathbf{N}_i(\mathbf{z}_i)\right]\mathbf{z}_i + \mathbf{G}_i(x_1, \dots, x_i)\left[\mathbf{z}_{i+1} + \tilde{\alpha}_i\right] - \mathbf{G}_{i-1}^T(x_1, \dots, x_{i-1})\mathbf{z}_{i-1},$$
$$\qquad i \in \{3, \dots, n-1\},$$
$$\dot{\mathbf{z}}_n = -\left[\mathbf{K}_n + \mathbf{N}_n(\mathbf{z}_n)\right]\mathbf{z}_n - \mathbf{G}_{n-1}^T(x_1, \dots, x_{n-1})\mathbf{z}_{n-1}.$$

$$(10.137)$$

### 10.5.1   DSC for Trajectory Tracking Marine Vehicles

Recall the equations of motion for a marine vehicle

$$\dot{\eta} = \mathbf{J}(\eta)\mathbf{v},$$

$$\mathbf{M}\dot{\mathbf{v}} + \mathbf{C}(\mathbf{v})\mathbf{v} + \mathbf{D}(\mathbf{v})\mathbf{v} + \mathbf{g}(\eta) = \tau.$$

Let

$$\tilde{\eta} := \eta - \eta_d, \quad \tilde{\eta} \in \mathbb{R}^n \qquad (10.138)$$

be the Earth-fixed tracking surface error and

$$\tilde{\mathbf{v}} := \mathbf{v} - \alpha_1^c, \quad \tilde{\mathbf{v}} \in \mathbb{R}^n, \qquad (10.139)$$

where $\alpha_1^c$ is the first order filtered approximation of the stabilizing function $\alpha_1$, be the body-fixed velocity surface error. Augment the marine vehicle equations of motion above with the first order differential equation

$$\mathbf{T}_1\dot{\alpha}_1^c = -\alpha_1^c + \alpha_1, \qquad (10.140)$$

where $\mathbf{T}_1 > 0 \in \mathbb{R}^{n\times n}$ is a diagonal matrix of filter time constants and the initial conditions for integrating $\dot{\alpha}_1^c$ are $\alpha_1^c(0) = \alpha_1(0)$. Define an additional error surface for the estimate of the stabilizing function as

$$\tilde{\alpha}_1 = \alpha_1^c - \alpha_1. \qquad (10.141)$$

With the DSC approach, the undifferentiated value of the stabilizing function $\alpha_1$ is computed, as per normal, using Lyapunov stability (essentially assuming that $\alpha_1 = \alpha_1^c$ for the stability analysis), but the approximation $\dot{\alpha}_1^c$ is still used for the time derivative of the stabilizing function. With this in mind take the Lyapunov function to be

$$V = \frac{1}{2}\tilde{\eta}^T\tilde{\eta} + \frac{1}{2}\tilde{\mathbf{v}}^T\mathbf{M}\tilde{\mathbf{v}} + \frac{1}{2}\tilde{\alpha}_1^T\tilde{\alpha}_1. \qquad (10.142)$$

Its time derivative is

$$\dot{V} = \tilde{\eta}^T \dot{\tilde{\eta}} + \tilde{\mathbf{v}}^T \mathbf{M} \dot{\tilde{\mathbf{v}}} + \tilde{\alpha}_1^T \dot{\tilde{\alpha}}_1. \tag{10.143}$$

Using the equations of motion shown above, it can seen that

$$\dot{\tilde{\eta}} = \mathbf{J}(\eta)\mathbf{v} - \dot{\eta}_d = \mathbf{J}\tilde{\mathbf{v}} + \mathbf{J}\alpha_1 - \dot{\eta}_d,$$

where we take $\alpha_1 = \alpha_1^c$ for the purposes of the stability analysis, as mentioned above. Similarly, we compute $\mathbf{M}\dot{\tilde{\mathbf{v}}}$ from the equations of motion as

$$\mathbf{M}\dot{\tilde{\mathbf{v}}} = -\mathbf{C}(\mathbf{v})\mathbf{v} - \mathbf{D}(\mathbf{v})\mathbf{v} - \mathbf{g}(\eta) + \tau - \mathbf{M}\dot{\alpha}_1^c,$$

where the approximation for the time derivative of the stabilizing function $\dot{\alpha}_1^c$ is used. From (10.141) and (10.140)

$$\dot{\tilde{\alpha}}_1 = \dot{\alpha}_1^c - \dot{\alpha}_1 = -\mathbf{T}_1^{-1}\tilde{\alpha}_1 - \dot{\alpha}_1.$$

Therefore, the time derivative of the Lyapunov function (10.143) becomes

$$\begin{aligned}
\dot{V} &= \tilde{\eta}^T \left[ \mathbf{J}\tilde{\mathbf{v}} + \mathbf{J}\alpha_1 - \dot{\eta}_d \right] + \tilde{\mathbf{v}}^T \left[ -\mathbf{C}(\mathbf{v})\mathbf{v} - \mathbf{D}(\mathbf{v})\mathbf{v} - \mathbf{g}(\eta) + \tau - \mathbf{M}\dot{\alpha}_1^c \right] \\
&\quad + \tilde{\alpha}_1^T \left[ -\mathbf{T}_1^{-1}\tilde{\alpha}_1 - \dot{\alpha}_1 \right] \\
&= \tilde{\eta}^T \left[ \mathbf{J}\alpha_1 - \dot{\eta}_d \right] + \tilde{\mathbf{v}}^T \left[ \mathbf{J}^T \tilde{\eta} - \mathbf{C}(\mathbf{v})\mathbf{v} - \mathbf{D}(\mathbf{v})\mathbf{v} - \mathbf{g}(\eta) + \tau - \mathbf{M}\dot{\alpha}_1^c \right] \\
&\quad - \tilde{\alpha}_1^T \mathbf{T}_1^{-1}\tilde{\alpha}_1 - \tilde{\alpha}_1^T \dot{\alpha}_1.
\end{aligned} \tag{10.144}$$

To ensure stability, take the stabilizing function $\alpha_1$ to be

$$\alpha_1 = \mathbf{J}^{-1} \left[ -\mathbf{K}_1 \tilde{\eta} + \dot{\eta}_d \right], \tag{10.145}$$

and the control input $\tau$ to be

$$\tau = -\mathbf{K}_2 \tilde{\mathbf{v}} - \mathbf{J}^T \tilde{\eta} + \mathbf{C}(\mathbf{v})\mathbf{v} + \mathbf{D}(\mathbf{v})\mathbf{v} + \mathbf{g}(\eta) + \mathbf{M}\dot{\alpha}_1^c, \tag{10.146}$$

where $\mathbf{K}_1 = \mathbf{K}_1^T > 0 \in \mathbb{R}^{n \times n}$ and $\mathbf{K}_2 = \mathbf{K}_2^T > 0 \in \mathbb{R}^{n \times n}$ are positive definite design matrices. With (10.145) and (10.146), (10.144) becomes

$$\dot{V} = -\tilde{\eta}^T \mathbf{K}_1 \tilde{\eta} - \tilde{\mathbf{v}}^T \mathbf{K}_2 \tilde{\mathbf{v}} - \tilde{\alpha}_1^T \mathbf{T}_1^{-1}\tilde{\alpha}_1 - \tilde{\alpha}_1^T \dot{\alpha}_1. \tag{10.147}$$

Here, it can be seen that the first three terms of $\dot{V}$ are negative definite. We can expand the remaining term $-\tilde{\alpha}_1^T \dot{\alpha}_1$ by applying Young's Inequality as

$$-\tilde{\alpha}_1^T \dot{\alpha}_1 \leq \frac{1}{2}\tilde{\alpha}_1^T \tilde{\alpha}_1 + \frac{1}{2}\dot{\alpha}_1^T \dot{\alpha}_1 \tag{10.148}$$

so that

$$
\begin{aligned}
\dot{V} &\leq -\tilde{\eta}^T \mathbf{K}_1 \tilde{\eta} - \tilde{\mathbf{v}}^T \mathbf{K}_2 \tilde{\mathbf{v}} - \tilde{\alpha}_1^T \mathbf{T}_1^{-1} \tilde{\alpha}_1 + \frac{1}{2} \tilde{\alpha}_1^T \tilde{\alpha}_1 + \frac{1}{2} \dot{\alpha}_1^T \dot{\alpha}_1, \\
&\leq -\tilde{\eta}^T \mathbf{K}_1 \tilde{\eta} - \tilde{\mathbf{v}}^T \mathbf{K}_2 \tilde{\mathbf{v}} - \left[ \lambda_{\min}(\mathbf{T}_1^{-1}) - \frac{1}{2} \right] \tilde{\alpha}_1^T \tilde{\alpha}_1 + \frac{1}{2} \dot{\alpha}_1^T \dot{\alpha}_1,
\end{aligned}
\tag{10.149}
$$

where $\lambda_{\min}(\mathbf{T}_1^{-1})$ is the minimum eigenvalue of $\mathbf{T}_1^{-1}$. As long as the filter time constants are chosen such that $\lambda_{\min}(\mathbf{T}_1^{-1}) > 1/2$ and the time derivative of the stabilizing function is bounded so that $\dot{\alpha}_1^T \dot{\alpha}_1 / 2 \leq \delta$ for some $\delta > 0$, the time derivative of the Lyapunov function has the form $\dot{V} \leq -\mu V + \delta$ and the closed loop system will converge to within a bounded error, as mentioned above for SISO systems in strict feedback nonlinear system form (see [20] for additional details).

## 10.6   Actuator Constraints

Control design for marine vehicles is often done under the assumption that the dynamic response of actuators can be neglected. However, in a real-world implementation the actuators may not be capable of providing the control forces/moments commanded by the control system. Generally, the actuator outputs are constrained by both the magnitude of the forces/moments that the actuators can produce (referred to as *magnitude limits* or *actuator saturation*), as well as by how quickly the actuators can generate the required forces/moments (*rate limits*). As shown in Sect. 3.3.4, if actuator constraints are not included in the control design, the closed loop system may exhibit poor performance or instability, especially in situations where the forces generated by environmental disturbances approach the output capabilities of its actuators [13] or when a vehicle's mission requires it to operate in a wide range of different orientations for which its actuator configuration may not be optimal [1].

The saturation function

$$
u_{iM}\mathrm{sat}\left(\frac{u_i}{u_{iM}}\right) := \begin{cases} u_i, & \left|\dfrac{u_i}{u_{iM}}\right| < 1, \\ u_{iM}\mathrm{sgn}\left(\dfrac{u_i}{u_{iM}}\right), & \left|\dfrac{u_i}{u_{iM}}\right| \geq 1, \end{cases}
\tag{10.150}
$$

is often used to model actuator saturation in linear systems. However, it is discontinuous and integrator backstepping, requires smoothly differentiable functions.

Here, an approach for modeling actuator saturation with a hyperbolic tangent function is presented. Let $\mathbf{u}$ be a vector representing the state of the actuators and $\tau(\mathbf{u})$ be a smooth, upper- and lower-bounded, vector function. Assume that the Jacobian of $\tau(\mathbf{u})$ is smooth and nonsingular, such that its inverse exists. Each component of $\tau(\mathbf{u})$ is taken to only depend on the corresponding component of the actuator states, i.e. $\partial \tau_i / \partial u_j = 0, \forall i \neq j$ such that $\partial \tau / \partial \mathbf{u}$ is diagonal. With these assumptions, the

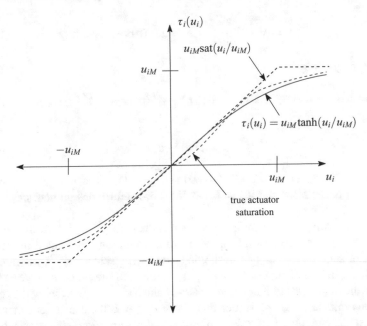

**Fig. 10.9** Use of the hyperbolic tangent function to model actuator saturation. Compared with the saturation function, $\tanh(\cdot)$ is continuous. Unlike true actuator saturation, it does not include a deadband region near the origin $u_i = 0$

saturation limits and the rate limit of each component of $\tau(\mathbf{u})$ are independent.

To model rate limits, the dynamic equations governing the time evolution of the system are augmented with the first order differential equation

$$\mathbf{T}\dot{\mathbf{u}} = -\mathbf{u} + \mathbf{u}_c \tag{10.151}$$

where $\mathbf{T} > 0 \in \mathbb{R}^n \times \mathbb{R}^n$ is a diagonal matrix of time constants (see [7], for example).

A convenient function for modeling actuator saturation is the hyperbolic tangent function (Fig. 10.9)

$$\tau_i(u_i) \approx u_{iM} \tanh\left(\frac{u_i}{u_{iM}}\right), \tag{10.152}$$

where the function $\tanh(\cdot)$ is extended element-wise to vector arguments.

A backstepping controller can be designed by including the actuator model (10.152) in the stability analysis and solving for the commanded actuator state that stabilizes the closed loop system. Examples of this approach can be found in [7, 19] for traditional backstepping and [20] for the use of DSC. In the case of traditional backstepping with no exogenous disturbances or model uncertainty, the commanded actuator state will have the form

$$\mathbf{u}_c = \mathbf{u} + \mathbf{T}\frac{\partial \tau}{\partial \mathbf{u}}^{-1} \left[ \dot{\alpha}_2 - \mathbf{J}^{-1}(\eta)\mathbf{s} - \mathbf{K}_z \mathbf{z} \right], \qquad (10.153)$$

where $\alpha_2$ is given by

$$\alpha_2 = \mathbf{M}\dot{\mathbf{v}}_r + \mathbf{C}(\mathbf{v})\mathbf{v}_r + \mathbf{D}(\mathbf{v})\mathbf{v}_r + \mathbf{g}(\eta) - \mathbf{J}^T(\eta)\mathbf{K}_d\mathbf{s} - \mathbf{J}^T(\eta)\mathbf{K}_p\tilde{\eta}, \qquad (10.154)$$

$\mathbf{z}$ is

$$\mathbf{z} = \tau(\mathbf{u}) - \alpha_2, \qquad (10.155)$$

$\mathbf{s}$, $\tilde{\eta}$ and $\mathbf{v}_r$ are defined in (10.92), (10.89) and (10.91), respectively. As above, the terms $\mathbf{K}_d = \mathbf{K}_d^T \in \mathbb{R}^{n \times n}$ and $\mathbf{K}_p = \mathbf{K}_p^T \in \mathbb{R}^{n \times n}$ are control design matrices.

When the actuator states are large ($\mathbf{u} \to \infty$) the slope of $\tau(\mathbf{u})$ becomes small so that $\partial \tau / \partial \mathbf{u}^{-1} \to \infty$. To prevent this term from becoming unbounded in (10.153), the actuator state vector can be thresholded to $u_i = u_{iM}^* \text{sat}\left(u_i/u_{iM}^*\right)$, where $\mathbf{u}_{iM}^* := \mathbf{u}_M/\tanh(1)$, before being used as an input argument of the function $\partial \tau / \partial \mathbf{u}^{-1}$.

When the magnitude of the actuator states is much smaller than that of the actuator saturation limits $|u_i/u_{iM}| \ll 1$, such that $[\partial \tau / \partial \mathbf{u}]^{-1} \approx \mathbf{1}$, the controller behaves like a standard backstepping controller without actuator saturation. However, when the magnitude of the actuator states approaches that of the actuator saturation limits, the product $\mathbf{T}[\partial \tau / \partial \mathbf{u}]^{-1}$ acts like a nonlinear gain function that scales the inputs to the controller. Thus, the control law is *dynamically rescaled* so that it produces a magnitude- and rate-limited bounded signal. When the control input lies within the actuator constraints, the nonlinear scaling performs no action and the nominal control law is used. As the actuator states approach the actuator constraints, the dynamic rescaling modifies the controller inputs and outputs so that the desired control action can be achieved without causing the system to become unstable [20].

Please see [2] for an alternative approach to modeling actuator saturation and rate limits, which can also be used to model systems with asymmetric actuator characteristics, i.e. $\tau(-\mathbf{u}) \neq \tau(+\mathbf{u})$.

## 10.7 Nonlinear Disturbance Observer Based Control

As discussed in Sect. 6.8, unmodeled dynamics (e.g. imprecise models of a plant's inertia tensor, drag, actuator dynamics, etc.) and external disturbances (such as wind, currents or waves) can generally affect the practical implementation of a controller in a real marine system. In Disturbance Observer Based Control (DOBC), the disturbance estimate $\hat{\mathbf{w}}_d$ is typically subtracted from the control input $\tau$ to counteract the actual (unknown) disturbance $\mathbf{w}_d$. The basic structure of a DOBC system is shown in Fig. 10.10.

When the nonlinear dynamics of a marine system are known, or at least partially known, a nonlinear DOBC system can be designed by exploiting the (known)

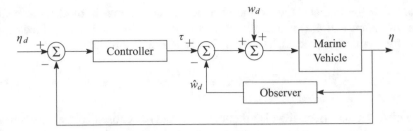

**Fig. 10.10** Structure of a disturbance observer based control system

nonlinear dynamics in the observer design. Here, the nonlinear disturbance observer developed in [3] is first presented, and then customized for use in marine vehicle systems. An important underlying assumption is that the disturbance vector $\mathbf{w}_d$ consists of two types of disturbances $\mathbf{w}_d = \mathbf{w}_{de} + \mathbf{w}_{du}$, where $\mathbf{w}_{de}$ is a vector of unknown external disturbances, such as wind or current, and the term $\mathbf{w}_{du}$ is modeled as a bounded disturbance-like error that arises from model uncertainty. It is also assumed that the components of $\mathbf{w}_d$ and their time derivatives are unknown and time-varying, yet bounded by a positive constant $\rho$, such that

$$\|\dot{\mathbf{w}}_d(t)\| \le \rho, \tag{10.156}$$

where $\| \cdot \|$ represents the 2-norm of a vector or matrix.

Consider the system

$$\begin{aligned} \dot{\mathbf{x}} &= \mathbf{f}(\mathbf{x}) + \mathbf{G}_1(\mathbf{x})\mathbf{u} + \mathbf{G}_2(\mathbf{x})\mathbf{w}_d, \\ \mathbf{y} &= \mathbf{h}(\mathbf{x}), \end{aligned} \tag{10.157}$$

where $\mathbf{x} \in \mathbb{R}^n$, $\mathbf{u} \in \mathbb{R}^m$, $\mathbf{w}_d \in \mathbb{R}^q$ and $\mathbf{y} \in \mathbb{R}^s$ are the state, the control input, the disturbance, and the output vectors, respectively. It is assumed that $\mathbf{f}(\mathbf{x})$, $\mathbf{G}_1(\mathbf{x})$, $\mathbf{G}_2(\mathbf{x})$, and $\mathbf{h}(\mathbf{x})$ are smooth functions of $\mathbf{x}$. As in Sect. 6.8, the disturbances are taken to be generated by an exogenous system of the form

$$\dot{\xi} = \mathbf{A}_\xi \xi, \quad \mathbf{w}_d = \mathbf{C}_\xi \xi, \tag{10.158}$$

where $\xi \in \mathbb{R}^q$ is an internal state variable describing the time evolution of the disturbance, $\mathbf{A}_\xi$ is a dynamics matrix and $\mathbf{C}_\xi$ is an output matrix.

The nonlinear disturbance observer can be found by augmenting the equations of motion (10.157) with the following system

$$\begin{aligned} \dot{\mathbf{z}} &= \left[\mathbf{A}_\xi - \mathbf{L}(\mathbf{x})\mathbf{G}_2(\mathbf{x})\mathbf{C}_\xi\right]\mathbf{z} + \mathbf{A}_\xi \mathbf{p}(\mathbf{x}) \\ &\quad - \mathbf{L}(\mathbf{x})\left[\mathbf{G}_2(\mathbf{x})\mathbf{C}_\xi \mathbf{p}(\mathbf{x}) + \mathbf{f}(x) + \mathbf{G}_1(\mathbf{x})\mathbf{u}\right], \\ \hat{\xi} &= \mathbf{z} + \mathbf{p}(\mathbf{x}), \\ \hat{\mathbf{w}}_d &= \mathbf{C}_\xi \hat{\xi}, \end{aligned} \tag{10.159}$$

where $\mathbf{z}$ is the internal state of the observer, $\mathbf{L}(\mathbf{x})$ is a matrix of state-dependent observer gains and $\mathbf{p}(\mathbf{x})$ is a nonlinear vector function, which is determined during the observer design process. The terms $\mathbf{L}(\mathbf{x})$ and $\mathbf{p}(\mathbf{x})$ are related as

$$\mathbf{L}(\mathbf{x}) = \frac{\partial \mathbf{p}(\mathbf{x})}{\partial \mathbf{x}}. \tag{10.160}$$

The observer will track the disturbance when $\mathbf{L}(\mathbf{x})$ is chosen so that the solution to the equation describing the time evolution of the disturbance estimation error $\mathbf{e}_\xi = \xi - \hat{\xi}$,

$$\dot{\mathbf{e}}_\xi = \left[ \mathbf{A}_\xi - \mathbf{L}(\mathbf{x})\mathbf{G}_2(\mathbf{x})\mathbf{C}_\xi \right] \mathbf{e}_\xi, \tag{10.161}$$

is globally exponentially stable (regardless of the value of the state vector $\mathbf{x}$).

The general solution for $\mathbf{L}(\mathbf{x})$ is beyond the scope of this discussion, but it turns out that for marine vehicles the solution is rather simple. Once again, the equations of motion for a marine vehicle are given by

$$\dot{\eta} = \mathbf{J}(\eta)\mathbf{v},$$

$$\mathbf{M}\dot{\mathbf{v}} + \mathbf{C}(\mathbf{v})\mathbf{v} + \mathbf{D}(\mathbf{v})\mathbf{v} + \mathbf{g}(\eta) = \tau + \mathbf{w}_d.$$

Assuming that the dynamics of the vehicle are much faster than the dynamics over which the disturbance varies, the disturbance will be treated as if though it were constant such that $\mathbf{A}_\xi = \mathbf{0}$. Let $\mathbf{C}_\xi = \mathbf{1}$, where $\mathbf{1}$ is the identity matrix. Rewriting the equation of motion for $\mathbf{v}$ as

$$\dot{\mathbf{v}} = -\mathbf{M}^{-1}\left[ \mathbf{C}(\mathbf{v})\mathbf{v} + \mathbf{D}(\mathbf{v})\mathbf{v} + \mathbf{g}(\eta) \right] + \mathbf{M}^{-1}\tau + \mathbf{M}^{-1}\mathbf{w}_d,$$

and comparing the result with (10.157), it can be seen that here $\mathbf{x} = \mathbf{v}$,

$$\mathbf{f}(\mathbf{v}) = -\mathbf{M}^{-1}\left[ \mathbf{C}(\mathbf{v})\mathbf{v} + \mathbf{D}(\mathbf{v})\mathbf{v} + \mathbf{g}(\eta) \right],$$

$\mathbf{G}_1(\mathbf{v}) = \mathbf{M}^{-1}$ and $\mathbf{G}_2(\mathbf{v}) = \mathbf{M}^{-1}$.

If $\mathbf{L}(\mathbf{v})$ is selected as $\mathbf{L}(\mathbf{v}) = \mathbf{K}_0\mathbf{M}$, where $\mathbf{K}_0 = \mathbf{K}_0^T > 0$ is a diagonal matrix of observer design gains, and $\mathbf{p}(\mathbf{v}) = \mathbf{K}_0\mathbf{M}\mathbf{v}$ the equations for the internal state of the observer (10.159) become

$$\dot{\mathbf{z}}(t) = -\mathbf{K}_0\mathbf{z}(t) - \mathbf{K}_0\left[ -\mathbf{C}(\mathbf{v})\mathbf{v} - \mathbf{D}(\mathbf{v})\mathbf{v} - \mathbf{g}(\eta) + \tau(\mathbf{u}) + \mathbf{K}_0\mathbf{M}\mathbf{v} \right],$$
$$\hat{\mathbf{w}}_d = \mathbf{z}(t) + \mathbf{K}_0\mathbf{M}\mathbf{v}. \tag{10.162}$$

A detailed general survey of DOBC techniques can be found in [4]. The disturbance observer developed in [5] is specifically developed for applications involving the control of surface ships.

## Problems

**10.1** Let us reexamine the use of a tug to guide the path of a cargo ship, as discussed in Sect. 8.1. The equations of motion are given by

$$\dot{x}_1 = x_2,$$
$$\dot{x}_2 = \sin(x_1)\cos(x_1) + u\cos(x_1),$$
$$y = x_1,$$

where $x_1 := \psi$ and $x_2 := r$ are the yaw angle and yaw rate of the cargo ship. The control objective is to stabilize the system at $\mathbf{x} = [\psi \ r]^T = 0$. Use the backstepping approach.

(a) Define suitable error surfaces $z_1$ and $z_2$.
(b) Rewrite the equations of motion in terms of the error surfaces.
(c) Using the candidate Lyapunov function

$$V = \frac{1}{2}z_1^2 + \frac{1}{2}z_2^2,$$

find the stabilizing function and control input that stabilize the system. How does the control input compare with the one found in Example 8.1?

**10.2** Revisit the design of a controller to rapidly position an underwater arm described in Problem 8.7. Here, use the backstepping controller design technique to find a suitable control input.

**10.3** Revisit the design of a controller to regulate the profiling speed of a CTD, as described in Problem 8.8. Here, use the backstepping controller design technique to find a suitable control input.

**10.4** Implement a second order filter in simulation, as discussed in Sect. 10.3.3. Use a Simulink state space block with the matrices $\mathbf{A}$ and $\mathbf{B}$ in (10.110), and obtain the associated $\mathbf{C}$ and $\mathbf{D}$ matrices from the output relation $\mathbf{y}_d = [y_{d1} \ y_{d2}]^T$ when both $y_d(t)$ and $\dot{y}_d(t)$ are output from the state space block for an input signal $r(t)$. Use the variables $\zeta$ and $\omega_n$ in the state space block. Define $\zeta$ and $\omega_n$ at the command line, so that they are available to the Simulink simulation and can be changed to see the resulting effects, without having to reopen the state space block to modify them. Set the initial conditions for the output of the state space block to $\mathbf{y}_d(t = 0) = [0 \ 0]^T$.

Use two sources, a sine wave with an amplitude of 1 and a frequency of 1 rad/sec (i.e. $\sin(t)$) and its time derivative, a cosine wave ($\cos(t)$) with an amplitude of 1 and a frequency of 1 rad/sec. The cosine wave can be created by duplicating the sine wave block and using a phase angle of $\pi/2$. Take the input to the state space matrix to be the sine wave $r(t) = \sin(t)$.

Use a demultiplexing block to separate the state space block output signals $y_d(t)$ and $\dot{y}_d(t)$. Using two scopes, display $y_d(t)$ vs $r(t)$ on one scope and $\dot{y}_d(t)$ vs $\cos(t)$ on the second scope.

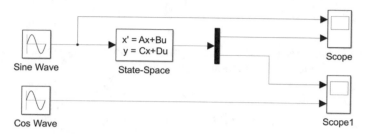

**Fig. 10.11** The Simulink system for Problem 10.4

The Simulink system should look like that shown in Fig. 10.11 when complete.

(a) Let $\zeta = 0.9$ and $\omega_n = 1$ rad/sec. How do the filter outputs $y_d(t)$ and $\dot{y}_d(t)$ compare with the original signals $r(t) = \sin(t)$ and $\dot{r}(t) = \cos(t)$?

(b) Take $\zeta = 0.9$ and $\omega_n = 50$ rad/sec. How do the filter outputs $y_d(t)$ and $\dot{y}_d(t)$ compare with the original signals $r(t) = \sin(t)$ and $\dot{r}(t) = \cos(t)$?

(c) Now, set the initial conditions in the state space block to $\mathbf{y}_d(t = 0) = [0 \quad 1]^T$ and leave all remaining terms the same. Is there a significant improvement? Why?

(d) Remove the cosine source block and replace the sine source block with a step input. Leave the first scope, which displays $y_d(t)$ and $r(t)$ unmodified. Modify the second scope to display $\dot{y}_d(t)$ only. How do the input and output signals compare for this new $r(t)$?

**10.5** Repeat Problem 10.4 using a fourth order filter by implementing (10.108) in Simulink using a state space block. As before, use the variables $\zeta$ and $\omega_n$ in the state space block and define $\zeta$ and $\omega_n$ at the command line, so they are available in the workspace and can also be quickly changed to see the resulting effects. Set the initial conditions for the output of the state space block to $\mathbf{y}_d(t = 0) = [0 \ 0 \ 0 \ 0]^T$.

Let $r(t) = 1(t)$, i.e. a step input with a magnitude of 1.

(a) Let $\zeta = 0.9$ and $\omega_n = 10$ rad/sec. Plot the the filtered output $y_d(t)$ and its first three time derivatives $\dot{y}_d(t)$, $\ddot{y}_d(t)$ and $y_d^{(3)}(t)$.

(b) Take $\zeta = 0.9$ and $\omega_n = 50$ rad/sec. Plot the the filtered output $y_d(t)$ and its first three time derivatives $\dot{y}_d(t)$, $\ddot{y}_d(t)$ and $y_d^{(3)}(t)$. How do the filter outputs compare with your results from part a)?

**10.6** Solve the problem in Example 10.1 using dynamic surface control.

**10.7** Solve the problem in Example 10.2 using dynamic surface control.

**10.8** Revisit the design of a controller to rapidly position an underwater arm described in Problem 8.7. Here, use the dynamic surface control design technique to find a suitable control input.

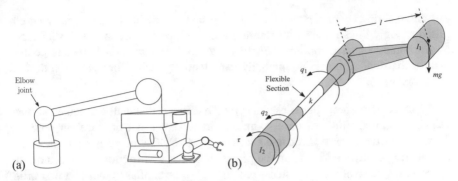

**Fig. 10.12** **a** A davit with flexible elbow joint deploying an ROV, and **b** a simplified model of the joint

**10.9** Revisit the design of a controller to regulate the profiling speed of a CTD, as described in Problem 8.8. Here, use the dynamic surface control design technique to find a suitable control input.

**10.10** Consider the control of a davit for deploying an ROV from the deck of a ship. Model the elbow joint of the davit as a flexible joint (Fig. 10.12). Ignoring damping, the equations of motion are

$$\ddot{q}_1 = -\frac{mgl}{I_1}\sin q_1 - \frac{k}{I_1}(q_1 - q_2),$$
$$\ddot{q}_2 = -\frac{k}{I_2}(q_2 - q_1) + \frac{\tau}{I_2}. \tag{10.163}$$

Note that the effect of gravity makes the system nonlinear.

(a) Define the state variables as

$$x_1 := q_1, \quad x_2 := \dot{q}_1, \quad x_3 := q_2, \quad \text{and} \quad x_4 := \dot{q}_2,$$

and rewrite (10.163) in terms of the state variables.
(b) Let the control input be $u := \tau/I_2$ and the output be $y = x_1$. Without using nonlinear damping (i.e. $n_i(z_i) = 0$) identify the stabilizing functions and control input $u$ that render the system globally exponentially stable for tracking a desired trajectory $y_d(t)$.
(c) Write the closed loop error system corresponding to (10.163).

**10.11** Reconsider the ROV davit studied in Problem 10.10. Suppose that a nonlinear spring force is used to model the joint flexibility, so that the equations of motion are

$$\ddot{q}_1 = -\frac{mgl}{I_1}\sin q_1 - \frac{k}{I_1}\left[(q_1 - q_2) + (q_1 - q_2)^3\right],$$
$$\ddot{q}_2 = -\frac{k}{I_2}\left[(q_2 - q_1) + (q_2 - q_1)^3\right] + \frac{\tau}{I_2}. \tag{10.164}$$

The cubic spring characteristic shown here is characteristic of many systems comprised of rigid links joined with rotating joints, especially when the elasticity can be attributed to gear flexibility [16]. Identify the stabilizing functions and control input $u$ that render the system globally exponentially stable for tracking a desired trajectory $y_d(t)$.

**10.12** An electric propeller is used to provide thrust for a small unmanned surface vehicle. The magnitude of the maximum thrust that it can produce (in both the forward and the reverse directions) is $u_M = 500$ N. Owing to the inertia of the propeller, hydrodynamic drag on the propeller blades and friction in the motor, the dynamic response of the propeller is anticipated to be similar to that of a first order system. From qualitative observations, the time constant is estimated to be about $T = 1$ second.

(a)  Using (10.152) and (10.151), build a Simulink model to simulate the propeller response . Plot the thrust output by the propeller $\tau$ for step input commands of

   (i)  $u_c = 650$ N,
   (ii)  $u_c = 65$ N, and
   (iii)  $u_c = -65$ N.

(b)  Owing to friction in the motor and the limited torque that can be produced with low current, most electric propellers also display a deadband region (sometimes also called a *dead zone*) where no thrust is produced when the commanded thrust is low. Add a deadband element to your simulation so that $\tau = 0$ for $|u| \le 20$ N. Plot the time response of the output thrust $\tau$ and the commanded thrust $u_c$ for $0 \le t \le 20$ secs for the ramp-like input

$$u_c = \begin{cases} 0\,\text{N}, & t < 0 \text{ secs}, \\ 20t\,\text{N}, & 0 \le t < 10 \text{ secs}, \\ 200\,\text{N}, & t \ge 10 \text{ secs}. \end{cases}$$

**10.13** Consider the surge position tracking of a surface vessel in a headwind discussed in Sect. 10.3. Design a DOBC proportional derivative backstepping controller by including the estimated disturbance force $\hat{w}_d$ in the governing equations when deriving the control input $\tau$.

Use the following nonlinear disturbance observer system

$$\dot{z}(t) = -k_0 z(t) - k_0 \left[ -d(u)u + \tau + k_0 x_2 \right],$$
$$\hat{w}_d = z(t) + k_0 x_2,$$

which can be obtained from (10.159), to estimate the wind-induced disturbance force $w_d$.

When the estimated disturbance force $\hat{w}_d$ is included in the governing equations they become

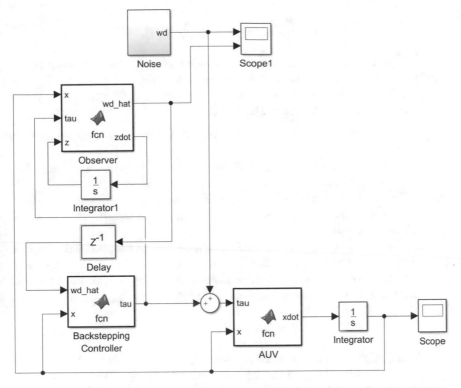

**Fig. 10.13** An example Simulink simulation structure for the backstepping controller and disturbance observer of Problem 10.14

$$\dot{x} = u,$$
$$\dot{u} = -d(u)u + \tau + \hat{w}_d,$$
$$y = x,$$

where $y = x$ is the distance traveled by the vessel along its surge direction, $u$ is its speed, $\tau$ is the normalized thrust generated by the propeller, and $d(u)$ is the normalized nonlinear, speed-dependent drag.

(a) Write out the resulting tracking errors and the system of equations describing the time derivatives of the tracking errors.
(b) Determine the resulting stabilizing function and control input.
(c) Explain how you would implement the nonlinear disturbance observer and controller in simulation and on a real vehicle.

**10.14** Consider the pitch stabilization of an AUV, as discussed in Example 7.4, but in the presence of exogenous disturbances and unmodeled dynamics. The equations of motion can be written as

$$\dot{\theta} = q,$$
$$\dot{q} = c_0 \sin(2\theta) + \tau + w_d(\theta, q, t),$$

where $\tau$ is the control input and $w_d(\theta, q, t)$ represents the effects of unknown exogenous disturbances and unmodeled vehicle dynamics. The magnitude of the time derivative of the disturbance is upper bounded as $\|\dot{w}_d(\theta, q, t)\| \leq \rho$, where $\rho$ is a known constant. Let the state vector be $\mathbf{x} = [\theta \quad q]^T$.

(a) Let $w_d = 0$. Use backstepping to design a control input $\tau$ that stabilizes the state to $\mathbf{x} = 0$.

(b) Write the equations of motion in the form $\dot{\mathbf{x}} = \mathbf{f}(\mathbf{x}) + \mathbf{G}_1(\mathbf{x})\mathbf{u} + \mathbf{G}_2(\mathbf{x})\mathbf{w}_d$ when $\mathbf{u} = \tau$ and $w_d \neq 0$.

(c) Write a Simulink simulation using $c_0 = 9$ and the initial condition $\mathbf{x}(t = 0) = [\pi/2 \quad 0]^T$ to show that for $k_1 = 1$ and $k_2 = 2$ the system stabilizes to near $\mathbf{x} = 0$ in about 6 seconds.

(d) Consider the situation when the AUV is affected by an exogenous disturbance acting about its pitch axis (perhaps the AUV is inspecting a cable in a current and large vortices are being shed ahead of the AUV). Model the disturbance as a first order Markov process (see Sect. 6.8) with the form

$$T_b \dot{w}_d = -w_d + a_n b_n(t),$$

where $w_d$ is the current-generated moment, $T_b = 180$ seconds is a time constant, $b_n(t)$ is zero-mean Gaussian white noise (with $|b_n(t)| \leq 1.0$), and $a_n = 100$ is a constant that scales the amplitude of $b_n$. Build a model for the disturbance in Simulink and plot $w_d(t)$ for $0 \leq t \leq 60$ seconds with an initial condition of $w_d(t = 0) = 0.5$ N-m.

(e) Add the Simulink disturbance model to the Simulink simulation created in part c) above. It should be seen that the backstepping controller is unable to stabilize the system to $\mathbf{x} = 0$ without error.

(f) Assume that the dynamics of the vehicle vary much faster than $w_d(t)$ varies in time, so that the disturbance can be treated as if though it is approximately constant with $\mathbf{A}_\xi = 0$ and $\mathbf{C}_\xi = 1$. Take $p = k_0 x_2$, so that the vector of observer gains is given by $\mathbf{L} = [0 \quad k_0]^T$. Use (10.159) to show that the state equations of the observer can be written as

$$\dot{z}(t) = -k_0 z(t) - k_0 [c_0 \sin(2x_1) + \tau + k_0 x_2],$$
$$\hat{w}_d = z(t) + k_0 x_2.$$

(g) Build an observer block in your Simulink simulation to compute the estimated disturbance $\hat{w}_d(t)$ using your answer to part f) above. Plot $w_d$ and $\hat{w}_d$ together so that you can tune the observer gain $k_0$ to obtain a suitable performance.

(h) Modify your backstepping controller design analysis from part (a) by including the estimated disturbance $\hat{w}_d$ in the equations of motion, as

$$\dot{\theta} = q,$$
$$\dot{q} = c_0 \sin(2\theta) + \tau + \hat{w}_d.$$

Now, when you solve for the backstepping control input, $-\hat{w}_d$ should appear as one of the terms.

(i) Include the modified control input $\tau$ in your Simulink simulation to show that the effects of the disturbance can be significantly mitigated by including the observer. Note that you may need to include a delay block between the observer and the controller in your Simulink model to avoid an algebraic loop. The structure of the controller implemented in Simulink is shown in Fig. 10.13.

# References

1. Bertaska, I., von Ellenrieder, K.: Experimental evaluation of supervisory switching control for unmanned surface vehicles. IEEE J. Oceanic Eng. **44**(1), 7–28 (2019)
2. Chen, M., Ge, S.S., Ren, B.: Adaptive tracking control of uncertain mimo nonlinear systems with input constraints. Automatica **47**(3), 452–465 (2011)
3. Chen, W.-H.: Disturbance observer based control for nonlinear systems. IEEE/ASME Trans. Mechatron. **9**(4), 706–710 (2004)
4. Chen, W.-H., Yang, J., Guo, L., Li, S.: Disturbance-observer-based control and related methods-an overview. IEEE Trans. Industr. Electron. **63**(2), 1083–1095 (2016)
5. Do, K.D.: Practical control of underactuated ships. Ocean Eng. **37**(13), 1111–1119 (2010)
6. Fossen, T.I.: Handbook of Marine Craft Hydrodynamics and Motion Control. Wiley, New York (2011)
7. Fossen, T.I., Berge, S.P.: Nonlinear vectorial backstepping design for global exponential tracking of marine vessels in the presence of actuator dynamics. In: Proceedings of the 36th IEEE Conference on Decision Control, pp. 4237–4242 (1997)
8. Fossen, T.I., Grovlen, A.: Nonlinear output feedback control of dynamically positioned ships using vectorial observer backstepping. IEEE Trans. Control Syst. Technol. **6**(1), 121–128 (1998)
9. Fossen, T.I., Strand, J.P.: Tutorial on nonlinear backstepping: applications to ship control (1999)
10. Johnson, W.R., Van Leer, J.C., Mooers, C.N.K.: A cyclesonde view of coastal upwelling. J. Phys. Oceanogr. **6**(4), 556–574 (1976)
11. Khalil, H.K.: Nonlinear Systems, 3ed edn. Prentice Hall Englewood Cliffs, New Jersey (2002)
12. Pan, Y., Haoyong, Yu.: Dynamic surface control via singular perturbation analysis. Automatica **57**, 29–33 (2015)
13. Sarda, E.I., Qu, H., Bertaska, I.R., von Ellenrieder, K.D.: Station-keeping control of an unmanned surface vehicle exposed to current and wind disturbances. Ocean Eng. **127**, 305–324 (2016)
14. Slotine, J.-J., Li, W.: Adaptive strategies in constrained manipulation. In: Proceedings of the 1987 IEEE International Conference on Robotics and Automation, vol. 4, pp. 595–601. IEEE (1987)
15. Slotine, J.-J.E., Li, W.: Applied Nonlinear Control. Prentice-Hall Englewood Cliffs, New Jersey (1991)
16. Spong, M.W., Hutchinson, S., Vidyasagar, M.: Robot Modeling and Control. Wiley, New York (2006)
17. Swaroop, D., Hedrick, J.K., Yip, P.P., Gerdes, J.C.: Dynamic surface control for a class of nonlinear systems. IEEE Trans. Autom. Control **45**(10), 1893–1899 (2000)

18. Swaroop, D.V.A.H.G., Gerdes, J.C., Yip, P.P., Hedrick, J.K.: Dynamic surface control of non-linear systems. In: American Control Conference, 1997. Proceedings of the 1997, vol. 5, pp. 3028–3034. IEEE (1997)
19. von Ellenrieder, K.D.: Stable backstepping control of marine vehicles with actuator rate limits and saturation. IFAC-PapersOnLine **51**(29), 262–267 (2018)
20. von Ellenrieder, K.D.: Dynamic surface control of trajectory tracking marine vehicles with actuator magnitude and rate limits. Automatica **105**, 433–442 (2019)

# Chapter 11
# Adaptive Control

## 11.1 Introduction

The dynamics of a marine vehicle are generally almost always uncertain. In the first place, it can be very difficult to accurately measure the inertial characteristics (e.g. the exact center of mass, distribution of mass, etc.) and damping (which can depend on temperature and loading) of a marine vehicle (parametric uncertainties). Secondly, even if these characteristics could be measured with high precision, they would likely vary over time as the vehicle is used, or because the system is reconfigured slightly to accommodate specific task requirements.

Marine vehicles are often used to transport loads (e.g. payloads, actuators, manipulators) of various sizes, weights, and inertial distributions. It is very restrictive to assume that the inertial parameters of such loads are well-known and static. If controllers with constant gains are used and the load parameters are not accurately known, the vehicle motion can be inaccurate or unstable.

Marine vehicles are often automatically steered to follow a path or trajectory. However, the dynamic characteristics of the vehicle strongly depend on many uncertain parameters, such as the drag characteristics of its hull, loading, wind conditions and even the operational speed of the vehicle. Adaptive control can be used to achieve good control performance under varying operating conditions, as well as to avoid energy loss due to excessive actuation.

In some cases, the parameter uncertainty must be gradually reduced on-line by an adaptation or estimation mechanism, or it may cause inaccuracy or instability of the controlled marine vehicle. Without the continuous "redesign" of the controller, an initially appropriate controller design may not be able to control the changing plant well. Generally, the basic objective of adaptive control is to maintain consistent performance of a system in the presence of uncertainty or unknown variation in plant parameters.

© Springer Nature Switzerland AG 2021
K. D. von Ellenrieder, *Control of Marine Vehicles*, Springer Series on Naval
Architecture, Marine Engineering, Shipbuilding and Shipping 9,
https://doi.org/10.1007/978-3-030-75021-3_11

Note that ordinary feedback control with a constant gain controller can also reduce the effects of parameter uncertainty. However, a constant gain feedback controller is not an adaptive controller. Here, as in [1], we will take the attitude that an adaptive controller is a controller with adjustable parameters and a mechanism which adjusts those parameters online.

Adaptive control is an approach to the control of systems with constant or slowly-varying uncertain parameters. Adaptive controllers can be thought of as having two loops, an inner feedback loop consisting of the vehicle and controller, and an outer loop which adjusts the control parameters. Here, we will assume that the unknown plant parameters are constant. In practice, adaptive control is often used to handle time-varying unknown parameters. In order for the analysis to be applicable to these practical cases, the time-varying plant parameters must vary considerably slower than the parameter adaptation. Fortunately, this is often satisfied in practice.

## 11.2  Model Reference Adaptive Control

With *model reference adaptive control* (MRAC) techniques, a control system is generally composed of four blocks, as shown in Fig. 11.1: (a) the marine vehicle, (b) a reference model, which specifies the desired output of the system, (c) a feedback control law containing the adjustable parameters, and (d) an adaptation law, which updates the adjustable parameters in the controller.

The structure of the dynamic equations governing the motion of the marine vehicle are assumed to be known, although the exact values of the parameters in the equations may be uncertain (e.g. the inertial and drag terms).

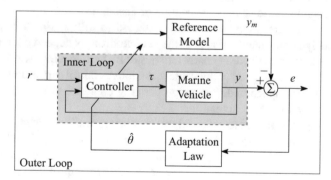

**Fig. 11.1** A model reference adaptive control system. The arrow drawn across the controller indicates that the control parameters are adjusted. The system can be thought of as having an inner loop for control and an outer loop for parameter estimation

A *reference model* is used to specify the ideal response of the adaptive control system to the reference command $r(t)$. It provides the desired ideal plant response that the adaptation mechanism drives the system toward when adjusting the parameters. The selection of an appropriate reference model is part of the control system design and must satisfy two requirements: (a) it should have the desired performance characteristics to perform the given control task, e.g. rise time, settling time, overshoot, etc. (b) the ideal reference behavior should be achievable for the adaptive control system, i.e., it must have the correct order and relative degree for the assumed structure of the plant model.

The controller is parameterized by a number of adjustable terms. When the plant parameters are exactly known, the corresponding controller parameters should make the plant output identical to that of the reference model. When the plant parameters are not known, the adaptation mechanism will adjust the controller parameters so that perfect tracking can be achieved. If the control law is linear in terms of the adjustable parameters, it is said to be a *linearly parameterized control law*. Existing adaptive control designs normally require linear parameterization of the controller in order to obtain adaptation mechanisms with guaranteed stability and tracking convergence.

An *adaptation mechanism* is used to adjust the parameters in the control law until the response of the plant under adaptive control becomes the same as that of the reference model. As the objective of the adaptation mechanism is to make the tracking error converge to zero, the adjustable parameters will not necessary converge to the plant parameters. The basic ideas of MRAC are next illustrated using an example in which a single parameter is adjusted.

**Example 11.1** Consider the use of MRAC for the vertical position tracking of an unmanned aerial vehicle (UAV). Let the mass of the vehicle be $m$, the vertical force input from the propellers be $\tau$ and $g$ be gravity, the dynamics of the UAV in the vertical direction are

$$m\ddot{z} = mg + \tau. \tag{11.1}$$

The UAV is required to carry a small load, such as a camera, with an unknown weight (Fig. 11.2). The control objective is to track a desired time-dependent height $z_d(t)$.

Suppose that we would like the UAV to have the response of an ideal linear second order system with the form

$$\ddot{z}_m + 2\zeta\omega_n\dot{z}_m + \omega_n^2 z_m = \omega_n^2 z_d(t), \tag{11.2}$$

where $z_m$ is the position of the reference model, and the damping ratio $\zeta$ and the natural frequency $\omega_n$ are selected to satisfy a set of desired performance characteristics, such as maximum overshoot

$$M_p = e^{-\pi\zeta/\sqrt{1-\zeta^2}}, \quad 0 < \zeta \leq 1$$

and 2% settling time

**Fig. 11.2** An unmanned
aerial vehicle with
bottom-mounted camera.
Photo provided courtesy of
Prof. Stephen Licht,
University of Rhode Island

$$t_s = \frac{4.0}{\zeta \omega_n}.$$

First, let us assume that the mass of the UAV and its load are precisely known. Take
the control input to be $\tau = m\left(-g + \ddot{z}_m - 2\lambda\dot{\tilde{z}} - \lambda^2\tilde{z}\right)$, where $\tilde{z} := z(t) - z_m(t)$ is
the tracking error and $\lambda > 0$ is a constant control gain, gives the closed loop error
dynamics

$$\begin{aligned}
\dot{\tilde{z}}_1 &= \tilde{z}_2 \\
\dot{\tilde{z}}_2 &= -2\lambda\tilde{z}_2 - \lambda^2\tilde{z}_1,
\end{aligned} \tag{11.3}$$

where $\tilde{z}_1 := \tilde{z}$ and $\tilde{z}_2 := \dot{\tilde{z}}$. This system can be shown to be exponentially stable with
an equilibrium point at $\tilde{z}_2 = \tilde{z}_1 = 0$ (perfect tracking).

Now let us assume that the mass is not precisely known, but that we have an esti-
mate of it $\hat{m}$, which we will use in the control law as $\tau = \hat{m}\left(-g + \ddot{z}_m - 2\lambda\dot{\tilde{z}} - \lambda^2\tilde{z}\right)$.
Then, (11.1) becomes

$$m\ddot{\tilde{z}} = \tilde{m}g - \hat{m}\left(-\ddot{z}_m + 2\lambda\tilde{z}_2 + \lambda^2\tilde{z}_1\right), \tag{11.4}$$

where $\tilde{m} := m - \hat{m}$. Following the approach in [6], we will define a new "combined
error variable" variable

$$s := \tilde{z}_2 + \lambda\tilde{z}_1. \tag{11.5}$$

If we compute $\dot{s} + \lambda s$ and simplify the result using (11.4), we get the following
combined expression for the closed loop tracking errors

$$\begin{aligned}
m\left(\dot{s} + \lambda s\right) &= m\left(\dot{\tilde{z}}_2 + 2\lambda\tilde{z}_2 + \lambda^2\tilde{z}_1\right), \\
&= m\ddot{\tilde{z}} + m\left(-\ddot{z}_m + 2\lambda\tilde{z}_2 + \lambda^2\tilde{z}_1\right), \\
&= \tilde{m}g - \hat{m}\left(-\ddot{z}_m + 2\lambda\tilde{z}_2 + \lambda^2\tilde{z}_1\right) + m\left(-\ddot{z}_m + 2\lambda\tilde{z}_2 + \lambda^2\tilde{z}_1\right), \\
&= \tilde{m}\left(-\ddot{z}_m + 2\lambda\tilde{z}_2 + \lambda^2\tilde{z}_1 + g\right), \\
&= \tilde{m}v.
\end{aligned} \tag{11.6}$$

Consider the following candidate Lyapunov function

$$V = \frac{1}{2}ms^2 + \frac{1}{2\gamma}\tilde{m}^2, \tag{11.7}$$

where $\gamma > 0$ is an adaption gain. Taking the derivative of this along trajectories of the closed loop system gives

$$\dot{V} = sm\dot{s} + \frac{1}{\gamma}\tilde{m}\dot{\tilde{m}} = sm\dot{s} - \frac{1}{\gamma}\tilde{m}\dot{\hat{m}}, \tag{11.8}$$

where it is assumed that $\dot{\tilde{m}} \approx -\dot{\hat{m}}$ because the value of the uncertain mass is either constant or very slowly varying in time (compared to its estimate, which can be updated much more rapidly). Using (11.6), this can be simplified to

$$\dot{V} = s\left[-m\lambda s + \tilde{m}v\right] - \frac{1}{\gamma}\tilde{m}\dot{\hat{m}},$$
$$= -m\lambda s^2 + \tilde{m}\left[sv - \frac{1}{\gamma}\dot{\hat{m}}\right]. \tag{11.9}$$

We can ensure that $\dot{V} < 0, \forall s \neq 0$ by selecting an update law for the mass estimate to be

$$\dot{\hat{m}} = \gamma sv = \gamma s\left(-\ddot{z}_m + 2\lambda\tilde{z}_2 + \lambda^2\tilde{z}_1 + g\right) \tag{11.10}$$

Using Barbalat's Lemma (Sect. 7.8), one can show that $s$ converges to zero. Thus, from (11.5) it can be seen that both the position tracking error $\tilde{z}_1$ and the velocity tracking error $\tilde{z}_2$ converge to zero.

Consider the use of an adaptive controller that permits a UAV to track a desired vertical position containing step-like jumps of $\pm 2.5\,\text{m}$ (Fig. 11.3). Simulation results are shown in Fig. 11.4. For comparison, the same controller with the adaptation mechanism disabled is also shown. The unloaded mass of the UAV is $m = 15\,\text{kg}$. The UAV is carrying a payload (for example a camera) with a mass of $m_p = 1.5\,\text{kg}$, but the control law is implemented without knowledge of the payload mass. The reference model is taken to be an ideal second order system (11.2). The control design specifications are that for each step-like jump the peak overshoot and settling time of the UAV should be $M_p \leq 0.05$ and $t_s \leq 1.8\,\text{s}$, respectively. These design criteria can be satisfied using $\zeta = 1/\sqrt{2}$ and $\omega_n = \pi$. It is also assumed that the UAV can only generate thrust in the downward direction and that the maximum thrust is limited to twice the weight of the UAV. The gain of the parameter update law is $\gamma = 1$ and the control gain $\lambda = 10$. □

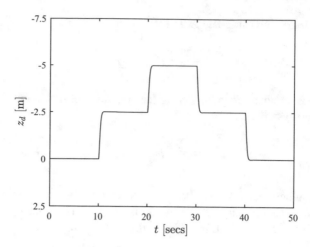

**Fig. 11.3** Desired position $z_d(t)$ of the UAV. The trajectory is plotted in the NED frame, where upwards is in the negative direction

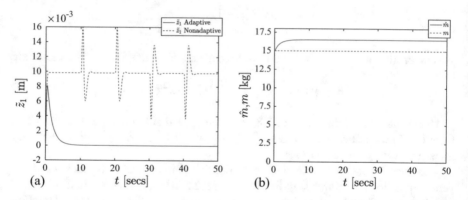

**Fig. 11.4** **a** Vertical position error of the UAV, and **b** mass estimate $\hat{m}$ from the adaptive controller versus actual mass $m$. The initial estimate of the total mass at $t = 0$ is taken to be $\hat{m} = 15\,\mathrm{kg}$. The solid blue lines correspond to the adaptive controller and the dashed red line corresponds to the corresponding non adaptive controller

## 11.3   Adaptive SISO Control via Feedback Linearization

Next we explore the use of adaptive control via feedback linearization for single-input single-output systems, with multiple uncertain parameters. Consider an $n$th order nonlinear system in the form

$$mx^{(n)} + \sum_{i=1}^{n} \theta_i f_i(\mathbf{x}, t) = \tau, \tag{11.11}$$

where $\mathbf{x} = [x \; \dot{x} \; \cdots \; x^{(n-1)}]^T$ is the state vector,

$$x^{(n)} = \frac{d^n x}{dt^n}, \quad \forall n,$$

the $f_i(\mathbf{x}, t)$ are known nonlinear functions, and the parameters $\theta_i$ and $m$ are uncertain constants. Take the output of the system to be $y(t) = x(t)$. Assume that the full state vector $\mathbf{x}$ is measured.

The control objective is to make the output asymptotically track a desired, time-dependent, trajectory $y_d(t)$, despite an uncertain knowledge of the parameters.

As in Example 11.1, a combined error variable $s$ can be defined for the system as

$$s := e^{(n-1)} + \lambda_{n-2} e^{(n-2)} + \cdots + \lambda_0 e, \tag{11.12}$$

where $e := (y - y_d) = (x - x_d)$ is the output tracking error and $\lambda_i > 0$ for all $i \in \{0, \ldots, (n-2)\}$. Let $x_r^{(n-1)} := x_d^{(n-1)} - \lambda_{n-2} e^{(n-1)} - \cdots - \lambda_0 e$ so that $s$ can be rewritten as

$$s := x^{(n-1)} - x_r^{(n-1)}. \tag{11.13}$$

Take the control law to be

$$\tau = m x_r^{(n)} - ks + \sum_{i=1}^{n} \theta_i f_i(\mathbf{x}, t), \tag{11.14}$$

where $k$ is a constant of the same sign as $m$ and $x_r^{(n)}$ is the derivative of $x_r^{(n-1)}$ so that

$$x_r^{(n)} := x_d^{(n)} - \lambda_{n-2} e^{(n-1)} - \cdots - \lambda_0 \dot{e}. \tag{11.15}$$

If the parameters were all well-known, the tracking error dynamics would be

$$m\dot{s} = -ks, \tag{11.16}$$

which would give the exponential convergence of $s$, and by extension the exponential convergence of $e$.

However, since the model parameters are not known, the control law is given by

$$\tau = \hat{m} x_r^{(n)} - ks + \sum_{i=1}^{n} \hat{\theta}_i f_i(\mathbf{x}, t), \tag{11.17}$$

where $\hat{m}$ and $\hat{\theta}_i$ are the estimated values of $m$ and the $\theta_i$, respectively. Then, the closed loop tracking error is given by

$$ m\dot{s} = -ks + \tilde{m}x_r^{(n)} + \sum_{i=1}^{n} \tilde{\theta}_i f_i(\mathbf{x}, t), \tag{11.18} $$

where $\tilde{m} := \hat{m} - m$ and $\tilde{\theta}_i := \hat{\theta}_i - \theta_i$ are the parameter estimation errors. Consider the candidate Lyapunov function

$$ V = |m|s^2 + \gamma^{-1}\left[\tilde{m}^2 + \sum_{i=1}^{n} \tilde{\theta}_i^2\right]. \tag{11.19} $$

Its time derivative is

$$ \dot{V} = 2m|s\dot{s} + 2\gamma^{-1}\left[\tilde{m}\dot{\tilde{m}} + \sum_{i=1}^{n} \tilde{\theta}_i\dot{\tilde{\theta}}_i\right]. \tag{11.20} $$

Using $m\dot{s} = |m|\text{sgn}(m)\dot{s}$ in (11.18), $\dot{V}$ can be rewritten as

$$
\begin{aligned}
\dot{V} &= \left[2\text{sgn}(m)s\right]|m|\text{sgn}(m)\dot{s} + 2\gamma^{-1}\left[\tilde{m}\dot{\tilde{m}} + \sum_{i=1}^{n} \tilde{\theta}_i\dot{\tilde{\theta}}_i\right], \\
&= 2\text{sgn}(m)s\left[-ks + \tilde{m}x_r^{(n)} + \sum_{i=1}^{n} \tilde{\theta}_i f_i(\mathbf{x}, t)\right] + 2\gamma^{-1}\left[\tilde{m}\dot{\tilde{m}} + \sum_{i=1}^{n} \tilde{\theta}_i\dot{\tilde{\theta}}_i\right], \\
&= -2\text{sgn}(m)ks^2 + 2\tilde{m}\left[\text{sgn}(m)sx_r^{(n)} + \gamma^{-1}\dot{\tilde{m}}\right] \\
&\quad + 2\sum_{i=1}^{n} \tilde{\theta}_i\left[\text{sgn}(m)sf_i(\mathbf{x}, t) + \gamma^{-1}\dot{\tilde{\theta}}_i\right].
\end{aligned}
\tag{11.21}
$$

With the assumption above that $m$ and $k$ have the same sign, $\dot{V}$ can be reduced to

$$ \dot{V} = -2|k|s^2, \tag{11.22} $$

which is negative definite for every $s \neq 0$, by selecting the parameter update laws to be

$$
\begin{aligned}
\dot{\hat{m}} &= -\gamma\,\text{sgn}(m)sx_r^{(n)}, \\
\dot{\hat{\theta}}_i &= -\gamma\,\text{sgn}(m)sf_i(\mathbf{x}, t).
\end{aligned}
\tag{11.23}
$$

By using Barbalat's Lemma one can show that the system is globally asymptotically stable, so that the tracking errors converge to zero. It can also be shown that global tracking convergence is preserved if a different adaptation gain $\gamma_i$ is used for each unknown parameter [6], i.e.

$$
\begin{aligned}
\dot{\hat{m}} &= -\gamma_0\,\text{sgn}(m)sx_r^{(n)}, \\
\dot{\hat{\theta}}_i &= -\gamma_i\,\text{sgn}(m)sf_i(\mathbf{x}, t).
\end{aligned}
$$

**Fig. 11.5** Use of a moving
mass system to stabilize the
roll axis of a USV in waves

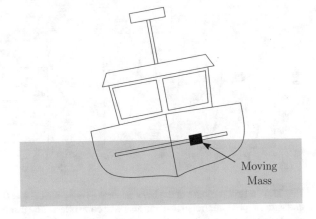

Moving
Mass

**Example 11.2** A classic example of (11.11) is the nonlinear equation for the rolling
motion of an unmanned surface vehicle about its surge axis

$$m\ddot{\phi} + c_d|\dot{\phi}|\dot{\phi} + k_s\phi = \tau, \tag{11.24}$$

where $m$ is the mass moment of inertia (including added mass) about the roll axis, $c$
is a damping coefficient and $k$ is a spring constant related to the buoyancy-induced
righting moment. We will assume that $m$, $c_d$ and $k_s$ are unknown and use an adaptive
controller to stabilize the USV at $\phi = 0$. For example, consider the use of a moving
mass system to stabilize the roll axis of the vehicle in waves (Fig. 11.5).

Comparing (11.11) and (11.24), it can be seen that $\theta_1 = c_d$, $\theta_2 = k_s$, $f_1 = |\dot{\phi}|\dot{\phi}$
and $f_2 = \phi$. The system is second order so that $n = 2$. Taking the state variable to be
$x = \phi$ and $\phi_d = \dot{\phi}_d = \ddot{\phi}_d = 0$, we get $e = \phi$, $\ddot{x}_r = -\lambda_0\dot{\phi}$ and $s = \dot{\phi} + \lambda_0\phi$. Thus,
the control law (11.17) is given by

$$\tau = \hat{m}\ddot{x}_r - ks + \hat{\theta}_1|\dot{\phi}|\dot{\phi} + \hat{\theta}_2\phi.$$

The parameters estimates are found using the update laws

$$\begin{aligned}
\dot{\hat{m}} &= -\gamma_0\, s\ddot{x}_r, \\
\dot{\hat{\theta}}_1 &= -\gamma_1\, s|\dot{\phi}|\dot{\phi}, \\
\dot{\hat{\theta}}_2 &= -\gamma_2\, s\phi.
\end{aligned} \tag{11.25}$$

The roll response of the USV is shown in Fig. 11.6 when the system is uncontrolled,
and controlled with adaptive control and controlled with non adaptive control. The
true values of the system are $m = 100\,\text{kg-m}^2$, $c_d = 8.4\,\text{kg-m}^2$ and $k_s = 17.5\,\text{N-m}$.
The control gains used are $\lambda_0 = 2$, $\gamma_0 = 10$, $\gamma_1 = 10$, $\gamma_2 = 1$, and $k = 5$.     □

As can be seen in Example 11.2, the adaptation mechanism does not necessarily
estimate the unknown parameters exactly, but simply generates values that permit a

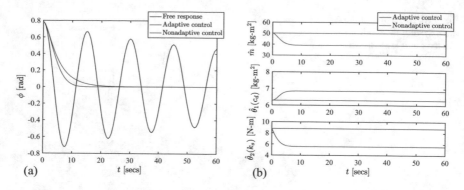

**Fig. 11.6  a** Roll response of the USV when $\phi = \pi/4$ at time $t = 0$ when the USV is uncontrolled, and with adaptive controller and non adaptive controllers. The response with the adaptive controller settles to within $\phi = 0.02$ rad in about 10.6 s, whereas for the non adaptive controller it takes about 17.5 s. **b** Parameter estimates for the adaptive controller and non adaptive controllers. The initial values of the estimates are $\hat{m} = 0.5m$, $\hat{\theta}_1 = 0.75c_d$ and $\hat{\theta}_2 = 0.5k_s$

desired task to be achieved. The practical relevance is that physical effects, such as inertial loading and drag can still be accounted for.

## 11.4   Adaptive MIMO Control via Feedback Linearization

Here, we extend the results of the previous section to multiple input, multiple output systems. Consider the equations of motion (both kinematic and kinetic) of a marine vehicle, (1.58)–(1.59),

$$\dot{\eta} = \mathbf{J}(\eta)\mathbf{v},$$
$$\mathbf{M}\dot{\mathbf{v}} = -\mathbf{C}(\mathbf{v})\mathbf{v} - \mathbf{D}(\mathbf{v})\mathbf{v} - \mathbf{g}(\eta) + \boldsymbol{\tau}.$$

The passivity-based backstepping approach presented in Sect. 10.3 is used in the first part of the controller design. Recall that the body-fixed and Earth-fixed reference trajectories are defined by

$$\dot{\eta}_r := \dot{\eta}_d - \Lambda\tilde{\eta},$$
$$\mathbf{v}_r := \mathbf{J}^{-1}(\eta)\dot{\eta}_r,$$

respectively, where $\tilde{\eta} := \eta - \eta_d$ is the Earth-fixed tracking error and $\Lambda > 0$ is a diagonal design matrix.

The measure of tracking is defined as

$$\mathbf{s} := \dot{\eta} - \dot{\eta}_r = \dot{\tilde{\eta}} + \Lambda\tilde{\eta}.$$

Using (10.93), the equations of motion (both kinematic and kinetic) are

$$\mathbf{M}_\eta(\eta)\ddot{\eta} = -\mathbf{C}_\eta(\mathbf{v}, \eta)\dot{\eta} - \mathbf{D}_\eta(\mathbf{v}, \eta)\dot{\eta} - \mathbf{g}_\eta(\eta) + \mathbf{J}^{-T}(\eta)\tau,$$

where

$$\begin{aligned}
\mathbf{M}_\eta(\eta) &= \mathbf{J}^{-T}(\eta)\mathbf{M}\mathbf{J}^{-1}(\eta),\\
\mathbf{C}_\eta(\mathbf{v}, \eta) &= \mathbf{J}^{-T}(\eta)\left[\mathbf{C}(\mathbf{v}) - \mathbf{M}\mathbf{J}^{-1}(\eta)\dot{\mathbf{J}}(\eta)\right]\mathbf{J}^{-1}(\eta),\\
\mathbf{D}_\eta(\mathbf{v}, \eta) &= \mathbf{J}^{-T}(\eta)\mathbf{D}(\mathbf{v})\mathbf{J}^{-1}(\eta),\\
\mathbf{g}_\eta(\eta) &= \mathbf{J}^{-T}(\eta)\mathbf{g}(\eta).
\end{aligned}$$

The definitions of the virtual reference trajectory and measure of tracking can be used to write the vehicle dynamics as

$$\begin{aligned}
\mathbf{M}_\eta(\eta)\dot{\mathbf{s}} = &-\mathbf{C}_\eta(\mathbf{v}, \eta)\mathbf{s} - \mathbf{D}_\eta(\mathbf{v}, \eta)\mathbf{s}\\
&+\mathbf{J}^{-T}(\eta)\left[\tau - \mathbf{M}\dot{\mathbf{v}}_r - \mathbf{C}(\mathbf{v})\mathbf{v}_r - \mathbf{D}(\mathbf{v})\mathbf{v}_r - \mathbf{g}_\eta(\eta)\right],
\end{aligned} \tag{11.26}$$

where $\dot{\mathbf{v}}_r = \dot{\mathbf{J}}^{-1}(\eta)\dot{\eta}_r + \mathbf{J}^{-1}(\eta)\ddot{\eta}_r = \dot{\mathbf{J}}^{-1}(\eta)\dot{\eta}_r + \mathbf{J}^{-1}(\eta)\left[\ddot{\eta}_d - \Lambda\left(\dot{\eta} - \dot{\eta}_d\right)\right]$.

In order to develop an adaptive controller, it is assumed that the terms $\mathbf{M}$, $\mathbf{C}(\mathbf{v})$, $\mathbf{D}(\mathbf{v})$ and $\mathbf{g}_\eta(\eta)$ have a linear dependence on a constant vector of unknown parameters $\theta$. Thus, we define a known *regressor matrix* $\Phi(\dot{\mathbf{v}}_r, \mathbf{v}_r, \mathbf{v}, \eta)$, such that

$$\mathbf{M}\dot{\mathbf{v}}_r + \mathbf{C}(\mathbf{v})\mathbf{v}_r + \mathbf{D}(\mathbf{v})\mathbf{v}_r + \mathbf{g}_\eta(\eta) = \Phi(\dot{\mathbf{v}}_r, \mathbf{v}_r, \mathbf{v}, \eta)\theta. \tag{11.27}$$

Consider the candidate Lyapunov function

$$V = \frac{1}{2}\tilde{\eta}^T\mathbf{K}_p\tilde{\eta} + \frac{1}{2}\mathbf{s}^T\mathbf{M}_\eta(\eta)\mathbf{s} + \frac{1}{2}\tilde{\theta}^T\Gamma^{-1}\tilde{\theta},$$

where $\tilde{\theta} := \hat{\theta} - \theta$ is the parameter estimation error, $\mathbf{K}_p > 0$ is a diagonal control design matrix and $\Gamma = \Gamma^T > 0$ is a positive definite weighting matrix. Then, the time derivative $\dot{V}$ is

$$\begin{aligned}
\dot{V} &= \tilde{\eta}^T\mathbf{K}_p\dot{\tilde{\eta}} + \mathbf{s}^T\mathbf{M}_\eta(\eta)\dot{\mathbf{s}} + \frac{1}{2}\mathbf{s}^T\dot{\mathbf{M}}_\eta(\eta)\mathbf{s} + \dot{\hat{\theta}}^T\Gamma^{-1}\tilde{\theta},\\
&= -\tilde{\eta}^T\mathbf{K}_p\Lambda\tilde{\eta} + \tilde{\eta}^T\mathbf{K}_p\mathbf{s} + \mathbf{s}^T\mathbf{M}_\eta(\eta)\dot{\mathbf{s}} + \frac{1}{2}\mathbf{s}^T\dot{\mathbf{M}}_\eta(\eta)\mathbf{s} + \dot{\hat{\theta}}^T\Gamma^{-1}\tilde{\theta}, \tag{11.28}\\
&= -\tilde{\eta}^T\mathbf{K}_p\Lambda\tilde{\eta} + \mathbf{s}^T\mathbf{K}_p\tilde{\eta} + \mathbf{s}^T\left[\mathbf{M}_\eta(\eta)\dot{\mathbf{s}} + \frac{1}{2}\dot{\mathbf{M}}_\eta(\eta)\mathbf{s}\right] + \dot{\hat{\theta}}^T\Gamma^{-1}\tilde{\theta}.
\end{aligned}$$

Using (11.26), the term in square brackets becomes

$$\begin{aligned}
\mathbf{s}^T\left[\frac{1}{2}\dot{\mathbf{M}}_\eta(\eta)\mathbf{s} + \mathbf{M}_\eta(\eta)\dot{\mathbf{s}}\right] = &\frac{1}{2}\mathbf{s}^T\left[\dot{\mathbf{M}}_\eta(\eta) - 2\mathbf{C}_\eta(\mathbf{v}, \eta)\right]\mathbf{s} - \mathbf{s}^T\mathbf{D}_\eta(\mathbf{v}, \eta)\mathbf{s}\\
&+\mathbf{s}^T\mathbf{J}^{-T}(\eta)\left[\tau - \mathbf{M}\dot{\mathbf{v}}_r - \mathbf{C}(\mathbf{v})\mathbf{v}_r - \mathbf{D}(\mathbf{v})\mathbf{v}_r - \mathbf{g}_\eta(\eta)\right].
\end{aligned}$$

It can be shown that $\left[\dot{\mathbf{M}}_\eta(\eta) - 2\mathbf{C}_\eta(\mathbf{v}, \eta)\right]$ is antisymmetric, so that

$$s^T \left[ \dot{\mathbf{M}}_\eta(\boldsymbol{\eta}) - 2\mathbf{C}_\eta(\mathbf{v}, \boldsymbol{\eta}) \right] s = 0, \quad \forall s \neq 0.$$

(see [3]). Note that the antisymmetry of $\dot{\mathbf{M}}_\eta(\boldsymbol{\eta}) - 2\mathbf{C}_\eta(\mathbf{v}, \boldsymbol{\eta})$ is analogous to the anti-symmetry of the quantity $\dot{\mathbf{H}} - 2\mathbf{C}$ in the study of multiple link robotic manipulators, where $\mathbf{H}$ is an inertial tensor and $\mathbf{C}$ is a matrix of centripetal and Coriolis terms [6]. With this result, (11.28) becomes

$$
\begin{aligned}
\dot{V} &= -\tilde{\boldsymbol{\eta}}^T \mathbf{K}_p \Lambda \tilde{\boldsymbol{\eta}} - s^T \mathbf{D}_\eta(\mathbf{v}, \boldsymbol{\eta}) s + s^T \mathbf{J}^{-T}(\boldsymbol{\eta}) \mathbf{J}^T(\boldsymbol{\eta}) \mathbf{K}_p \tilde{\boldsymbol{\eta}} \\
&\quad + s^T \mathbf{J}^{-T}(\boldsymbol{\eta}) \left[ \boldsymbol{\tau} - \mathbf{M}\dot{\mathbf{v}}_r - \mathbf{C}(\mathbf{v})\mathbf{v}_r - \mathbf{D}(\mathbf{v})\mathbf{v}_r - \mathbf{g}_\eta(\boldsymbol{\eta}) \right] + \dot{\hat{\theta}}^T \Gamma^{-1} \tilde{\theta}, \\
&= -\tilde{\boldsymbol{\eta}}^T \mathbf{K}_p \Lambda \tilde{\boldsymbol{\eta}} - s^T \mathbf{D}_\eta(\mathbf{v}, \boldsymbol{\eta}) s + \dot{\hat{\theta}}^T \Gamma^{-1} \tilde{\theta} \\
&\quad + s^T \mathbf{J}^{-T}(\boldsymbol{\eta}) \left[ \boldsymbol{\tau} + \mathbf{J}^T(\boldsymbol{\eta}) \mathbf{K}_p \tilde{\boldsymbol{\eta}} - \mathbf{M}\dot{\mathbf{v}}_r - \mathbf{C}(\mathbf{v})\mathbf{v}_r - \mathbf{D}(\mathbf{v})\mathbf{v}_r - \mathbf{g}_\eta(\boldsymbol{\eta}) \right].
\end{aligned}
\tag{11.29}
$$

If the control law is taken as

$$\boldsymbol{\tau} = -\mathbf{J}^T(\boldsymbol{\eta}) \mathbf{K}_p \tilde{\boldsymbol{\eta}} - \mathbf{J}^T(\boldsymbol{\eta}) \mathbf{K}_d s + \boldsymbol{\Phi}\hat{\theta}, \tag{11.30}$$

then (11.29) reduces to

$$
\begin{aligned}
\dot{V} &= -\tilde{\boldsymbol{\eta}}^T \mathbf{K}_p \Lambda \tilde{\boldsymbol{\eta}} - s^T \mathbf{D}_\eta(\mathbf{v}, \boldsymbol{\eta}) s - s^T \mathbf{K}_d s + s^T \mathbf{J}^{-T}(\boldsymbol{\eta}) \boldsymbol{\Phi}\tilde{\theta} + \dot{\hat{\theta}}^T \Gamma^{-1}\tilde{\theta} \\
&= -\tilde{\boldsymbol{\eta}}^T \mathbf{K}_p \Lambda \tilde{\boldsymbol{\eta}} - s^T \left[ \mathbf{D}_\eta(\mathbf{v}, \boldsymbol{\eta}) + \mathbf{K}_d \right] s + \left[ s^T \mathbf{J}^{-T}(\boldsymbol{\eta}) \boldsymbol{\Phi} + \dot{\hat{\theta}}^T \Gamma^{-1} \right] \tilde{\theta}, \\
&= -\tilde{\boldsymbol{\eta}}^T \mathbf{K}_p \Lambda \tilde{\boldsymbol{\eta}} - s^T \left[ \mathbf{D}_\eta(\mathbf{v}, \boldsymbol{\eta}) + \mathbf{K}_d \right] s + \tilde{\theta}^T \Gamma^{-1} \left[ \dot{\hat{\theta}} + \Gamma \boldsymbol{\Phi}^T \mathbf{J}^{-1}(\boldsymbol{\eta}) s \right].
\end{aligned}
\tag{11.31}
$$

The terms $-\tilde{\boldsymbol{\eta}}^T \mathbf{K}_p \Lambda \tilde{\boldsymbol{\eta}}$ and $-s^T \left[ \mathbf{D}_\eta(\mathbf{v}, \boldsymbol{\eta}) + \mathbf{K}_d \right] s$ are negative definite because $\mathbf{K}_p > 0$, $\Lambda > 0$, $\mathbf{K}_d > 0$ and $\mathbf{D}_\eta(\mathbf{v}, \boldsymbol{\eta}) > 0$ (see Remark 10.1). Thus, we can ensure $\dot{V} \leq 0$ for all $s \neq 0$ and for all $\tilde{\boldsymbol{\eta}} \neq 0$ by selecting the update law to be

$$\dot{\hat{\theta}} = -\Gamma \boldsymbol{\Phi}^T \mathbf{J}^{-1}(\boldsymbol{\eta}) s. \tag{11.32}$$

However, note that $\dot{V}$ is still only negative semidefinite, not negative definite, because it does not contain terms that are negative definite in $\tilde{\theta}$. Thus, with $V > 0$ the stability analysis will involve showing that $\lim_{t \to 0} \dot{V} \to 0$, such that $\tilde{\boldsymbol{\eta}} \to 0$ and $s \to 0$ using Barbalat's Lemma (see Example 7.7).

**Example 11.3** Consider the adaptive trajectory tracking control of a fully-actuated unmanned surface vessel called the WAM-V USV16 (see Fig. 11.7 and Table 11.1). It is assumed that the vehicle operates in a three dimensional configuration space (surge–sway–yaw). Detailed information about the vessel can be found in [2, 5]. The hydrodynamic coefficients (added mass and damping) are calculated using the approach in [4].

In the closed loop maneuver simulated here, the position and orientation of the USV are specified as a time-dependent reference trajectory. Before the start of the maneuver, the vehicle is at rest and positioned at the origin of a North-East-Down coordinate system with its bow pointing northward (Fig. 11.8). The USV is commanded to accelerate to a surge speed of 0.5 m/s along a straight line. After 15 s, the

**Fig. 11.7** The WAM-V USV16. Reprinted from [5] ©(2016), with permission from Elsevier

**Fig. 11.8** Desired trajectory $[x_d \ y_d]^T$ (dashed line) and actual trajectory $[x_n \ y_n]^T$ (solid line) of the USV with adaptive MIMO control. At this scale the desired and actual trajectories appear to perfectly overlap

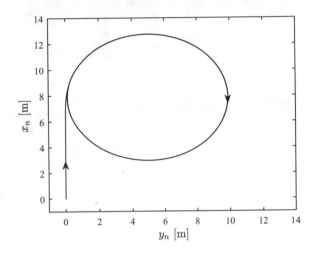

**Table 11.1** Principle characteristics of the WAM-V USV16; w.r.t. is an acronym for the phrase "with respect to"

| Parameter | Value |
| --- | --- |
| Length on the waterline | 3.20 m |
| Draft | 0.15 m |
| Centerline to centerline hull separation | 1.83 m |
| Mass | 180 kg |
| Mass moment of inertia about $z_b$-axis | 250 kg-m² |
| Longitudinal center of gravity w.r.t aft plane | 1.30 m |

USV is commanded to turn in a circle with a radius of 5 m, while maintaining the same surge speed of 0.5 m/s.

The control law in (11.30) is used with

$$\Lambda = \begin{bmatrix} 1 & 0 & 0 \\ 0 & 1 & 0 \\ 0 & 0 & 1 \end{bmatrix}, \quad \mathbf{K}_p = 1000 \begin{bmatrix} 1 & 0 & 0 \\ 0 & 1 & 0 \\ 0 & 0 & 10 \end{bmatrix}, \quad \text{and} \quad \mathbf{K}_d = 500 \begin{bmatrix} 1 & 0 & 0 \\ 0 & 1 & 0 \\ 0 & 0 & 2 \end{bmatrix}.$$

The dynamic model of the USV includes cross-coupled inertial terms, the nonlinear centripetal and Coriolis matrix (including added mass), and both nonlinear and linear drag. The parameter estimates are initialized at zero and no knowledge of the model parameters are used in the controller and adaptation mechanism. Since the configuration (and workspace) of the vehicle is constrained to the horizontal plane, (11.27) becomes

$$\mathbf{M}\dot{\mathbf{v}}_r + \mathbf{C}(\mathbf{v})\mathbf{v}_r + \mathbf{D}(\mathbf{v})\mathbf{v}_r = \Phi(\mathbf{v}, \mathbf{v}_r)\theta.$$

The regressor matrix was selected as

$$\Phi(\mathbf{v}, \mathbf{v}_r) = \begin{bmatrix} \dot{u} & 0 & 0 & uu_r & 0 & rr_r & 0 & 0 & vr_r & u_r & 0 & 0 \\ 0 & \dot{v} & \dot{r} & 0 & 0 & 0 & 0 & ur_r & vr_r & 0 & v_r & r_r \\ 0 & \dot{v} & \dot{r} & 0 & 0 & 0 & uv_r & ur_r & 0 & 0 & v_r & r_r \end{bmatrix},$$

where the subscript $r$ indicates a component of the reference velocity $\mathbf{v}_r :=$ $[u_r \ v_r \ r_r]^T$. The corresponding matrix of adaptation gains is selected to be

$$\Gamma = 500 \begin{bmatrix} 2 & 0 & 0 & 1 & 0 & 1 & 0 & 0 & 1 & 1 & 0 & 0 \\ 0 & 2 & 2 & 0 & 0 & 0 & 0 & 1 & 1 & 0 & 1 & 1 \\ 0 & 2 & 2 & 0 & 0 & 0 & 1 & 1 & 0 & 0 & 1 & 1 \\ 1 & 0 & 0 & 2 & 0 & 0 & 0 & 0 & 0 & 0 & 0 & 0 \\ 0 & 0 & 0 & 0 & 2 & 0 & 0 & 0 & 0 & 0 & 0 & 0 \\ 1 & 0 & 0 & 0 & 0 & 2 & 0 & 0 & 0 & 0 & 0 & 0 \\ 0 & 0 & 1 & 0 & 0 & 0 & 2 & 0 & 0 & 0 & 0 & 0 \\ 0 & 1 & 1 & 0 & 0 & 0 & 0 & 2 & 0 & 0 & 0 & 0 \\ 1 & 1 & 0 & 0 & 0 & 0 & 0 & 0 & 2 & 0 & 0 & 0 \\ 1 & 0 & 0 & 0 & 0 & 0 & 0 & 0 & 0 & 2 & 0 & 0 \\ 0 & 1 & 1 & 0 & 0 & 0 & 0 & 0 & 0 & 0 & 2 & 0 \\ 0 & 1 & 1 & 0 & 0 & 0 & 0 & 0 & 0 & 0 & 0 & 2 \end{bmatrix}.$$

As can be seen in Fig. 11.9, the position errors settle to within less than a centimeter and the parameter updates slowly converge to constant steady state values.  □

The preceding example suggests that it might be possible to construct an effective closed loop adaptive controller without any knowledge of the open loop dynamics of the marine vehicle. Given that the time required to required to formulate and validate, in both simulation and physical experimentation, an accurate open loop dynamic model of a system can be substantial, the potential of trading engineering effort for a more intelligent controller is attractive. However, it should be noted that a good understanding of the dynamics of the vehicle is still essential for selecting the correct control design specification, the structure of the controller (e.g. the regressor matrix and update coefficient matrix), and the design method [1].

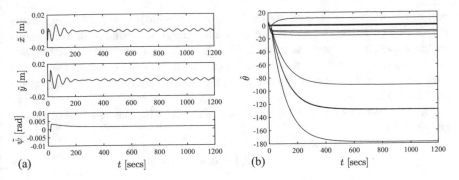

**Fig. 11.9** **a** Pose errors $\tilde{\eta} = [\tilde{x} \ \tilde{y} \ \tilde{\psi}]^T$ of the USV, and **b** convergence of the parameters estimates $\hat{\theta}$. The initial estimate of each parameter at $t = 0$ is zero, i.e. $\hat{\theta}_i = 0, i \in \{1, \ldots, 12\}$

# Problems

**11.1** Most propellers used on marine vehicles are rate-limited (see Sect. 10.6). The time response of the thrust generated can be modeled as a first order system of the form

$$T_d \dot{u} = -u + u_c,$$

where $u_c$ is the commanded thrust, $u$ is the output thrust and $T_d$ is a time constant. In general, $T_d$ is an uncertain parameter, as it can be difficult to measure experimentally, varies with the speed of the vessel and can change over time (e.g. from propeller blade damage caused by unintentional grounding). Use an adaptive control input to provide an estimate of $\hat{T}_d$ via the feedback linearization approach. Let the error variable be $s = u - u_d$, where $u_d(t)$ is the desired thrust, and take the control input to be $u_c = -ks + \hat{T}_d \dot{u}_d + u$.

(a) Take the candidate Lyapunov function to be

$$V = \frac{1}{2}s^2 + \frac{1}{2\gamma}\tilde{T}_d^2,$$

with $\tilde{T}_d = \hat{T}_d - T_d$. Show that the resulting parameter update law is given by

$$\dot{\hat{T}}_d = -\gamma \dot{u}_d s,$$

where $\gamma > 0$ is a constant adaptation gain.
(b) What variables must be measured in order for use of this approach to work? Explain the pros and cons of implementing this approach in practice (i.e. on a real marine vessel).

**11.2** A station-keeping controller for a lightweight unmanned surface vehicle (USV) with a low draft and a large windage area is being developed. The vehicle will be outfitted with an anemometer to measure the apparent wind velocity. The anemometer measurements will be used as an input to a model for estimating the wind forces acting on the above-surface portions of the USV. The estimated wind drag is fedforward as a negative term in the control input to counteract the wind drag. As the USV is frequently reconfigured between missions, the coefficients in the wind drag model are highly uncertain.

Here, we'll focus on only the surge axis of the vehicle. The equation of motion are given by

$$\dot{x}_1 = x_2,$$
$$\dot{x}_2 = -c_0 x_2 |x_2| - c_w u_a |u_a| + \tau,$$

where $x_1$ is the vehicle position along the surge direction, $x_2$ is the speed of the USV along the surge axis, $c_0$ is a drag coefficient (for the underwater portion of the hull), $c_w$ is a drag coefficient for the wind, $u_a$ is the apparent wind velocity along the longitudinal axis of the vehicle ($u_a > 0$ when the USV is in a headwind) and $\tau$ is the control input (the equation for the dynamics has been normalized by the mass of the USV). For station-keeping the control objective is to keep the USV at the origin $(x_1, x_2) = 0$.

Use adaptive feedback linearization to design a station-keeping controller for the USV. Take the control input to be

$$\tau = -ks + \hat{c}_0 x_2 |x_2| + \hat{c}_w u_a |u_a| + \dot{x}_r$$

and define appropriate expressions for the error variable $s$, $x_r$, and the candidate Lyapunov function $V$. What are the resulting update laws for $\hat{c}_0$ and $\hat{c}_w$.

**11.3** Consider the surge speed tracking control of an unmanned surface vehicle. The equation of motion is

$$m\dot{u} = -C_d u |u| + \tau,$$

where $m$ is the vehicle mass, $C_d$ is the drag coefficient and $\tau$ is the thrust from a propeller. The control objective is for the vehicle to track a time-dependent trajectory $u_d(t)$ when $C_d$ and $m$ are uncertain and possibly slowly varying in time. Using model reference adaptive control, take the reference model to be the first order equation

$$\dot{u}_m = -\frac{u_m - u_d}{T_d},$$

where $T_d$ is a time constant. Define the surge-speed velocity tracking error to be $\tilde{u} := u - u_m$. Using partial feedback linearization, the control input is taken to be

$$\tau = \hat{C}_d u |u| + \hat{m} \dot{u}_m - k_p \tilde{u},$$

where $\hat{C}_d$ and $\hat{m}$ are estimates of $C_d$ and $m$, respectively.

(a) Assuming that we have a perfect model of the vehicle, such that $\hat{C}_d = C_d$ and $\hat{m} = m$, substitute the control input $\tau$ into the equation of motion to show that the closed loop system has a stable equilibrium point at $\tilde{u} = 0$.

(b) If we no longer have a perfect model of the vehicle, such that $\hat{C}_d \neq C_d$ and $\hat{m} \neq m$, substitute the control input $\tau$ into the equation of motion to show that the closed loop system is given by

$$m\dot{\tilde{u}} = -\tilde{C}_d u |u| - \tilde{m}\dot{u}_m - k_p \tilde{u},$$

where $\tilde{C}_d := C_d - \hat{C}_d$ and $\tilde{m} := m - \hat{m}$ are the estimation errors.

(c) Take the combined error variable to be $s = \tilde{u}$ and consider the candidate Lyapunov function

$$V = \frac{1}{2}ms^2 + \frac{1}{2\gamma_m}\tilde{m}^2 + \frac{1}{2\gamma_d}\tilde{C}_d^2.$$

(d) Using the result from part (b) above and the definitions of $\tilde{m}$ and $\tilde{C}_d$, show that the time derivative of $\dot{V}$ can be written as

$$\dot{V} = -k_p\tilde{u}^2 - \tilde{m}\left[\frac{1}{\gamma_m}\dot{\hat{m}} + \dot{u}_m\tilde{u}\right] - \tilde{C}_d\left[\frac{1}{\gamma_d}\dot{\hat{C}}_d + u|u|\tilde{u}\right].$$

(e) What are the parameter update laws for $\dot{\hat{m}}$ and $\dot{\hat{C}}_d$ that render $\dot{V} \leq 0$ for all $\tilde{u} \neq 0$.

**11.4** Let us revisit Problem 11.3. Create a Simulink model of the model reference adaptive control system using user defined blocks for the controller, vehicle model (plant) and update mechanism. The reference model can be constructed from either a subsystem composed of simple blocks, or a user defined function. Use the parameter values $m = 35\,\text{kg}$, $C_d = 0.5$, $\gamma_d = 10$, $\gamma_m = 10$, $T_d = 1\,\text{s}$ and $kp = 20$. Let the initial values of the parameter estimates be $\hat{C}_d = 0$ and $\hat{m} = 0$.

(a) Take $u_d = 2\,\text{m/s}$ (constant). Show that the controlled system achieves the desired velocity.

(b) Now, let the desired velocity be a $u_d(t) = 2 + 0.5\cos(2\pi t/20)\,\text{m/s}$. Show that the controlled system is able to track the desired velocity.

(c) Next, let the desired velocity once again be $u_d = 2\,\text{m/s}$ (constant). However, now assume that a constant disturbance of $w_d = 50\,\text{N}$ is acting on the vehicle and change the Simulink model accordingly. Show that the output now has a steady state error. What is its magnitude?

(d) We will add integral control to see if the steady state error can be removed. Define the combined error variable to be

$$s = \tilde{u} + \lambda \int_0^t \tilde{u}\,dt$$

and let the control input be

$$\tau = \hat{C}_d u |u| + \hat{m} \, (\dot{u}_m - \lambda \tilde{u}) - k_p s.$$

As before, let the candidate Lyapunov function be

$$V = \frac{1}{2} m s^2 + \frac{1}{2} \gamma_m \tilde{m}^2 + \frac{1}{2} \gamma_d \tilde{C}_d^2.$$

Show that the time derivative of $\dot{V}$ can now be written as

$$\dot{V} = -k_p s^2 - \tilde{m} \left[ \gamma_m \dot{\hat{m}} + (\dot{u}_m - \lambda \tilde{u}) \, s \right] - \tilde{C}_d \left[ \gamma_d \dot{\hat{C}}_d + u |u| s \right].$$

What are the parameter update laws for $\dot{\hat{m}}$ and $\dot{\hat{C}}_d$ that render $\dot{V} < 0$ for all $\tilde{u} \neq 0$ (negative definite)?

(e) Revise your simulation using the new control law and parameter update rule for $\hat{m}$. Set $\lambda = 2$ and leave the values of all the other coefficients the same. Show that the steady state error is now driven to zero.

**11.5** Consider the automatic launch and recovery of an AUV from a USV. After the AUV is launched, the USV must accurately track a desired trajectory so that it can rendezvous with the AUV at the correct time and place for recovery. When the AUV is launched, the mass of the above surface system changes substantially so that the draft of the USV, and the associated drag and added mass characteristics change. Explore the use of adaptive control for the trajectory tracking of a USV performing an automatic launch. For simplicity, focus on the surge motion of the USV.
  The equations of motion are

$$\dot{x} = u,$$
$$m\dot{u} = -c_0 u |u| + \tau,$$

where $x$ is the surge position, $u$ is the surge velocity, $\tau$ is the control input, $m$ is the uncertain/changing mass, and $c_0$ is the drag coefficient.
  The mass of the AUV is 40 kg, the mass of the USV is 150 kg (without AUV) and the drag coefficient of the USV is $c_0 = 8.5$ kg/m when carrying the AUV, and $c_0 = 6.0$ kg/m after the AUV has been launched.
  A short launch and recovery test sequence is performed in which the system is at zero speed during both the launch and the recovery. The AUV is launched at time $t = 0$ at the position $x(0) = 0$ and recovered at time $t = t_f = 300$ s at the location $x(t_f) = x_f = 600$ m. The desired trajectory is given by

$$x_d(t) = 3x_f \left( \frac{t}{t_f} \right)^2 - 2x_f \left( \frac{t}{t_f} \right)^3.$$

Define the tracking error to be $e = x - x_d$.

(a) Use inverse dynamics feedback linearization to design the control input, by selecting

$$\tau = c_o u|u| + m\ddot{x}_d - k_1 e - k_2 \dot{e}.$$

   (i) Select $k_1$ and $k_2$ to obtain good performance when $m$ corresponds to the mass of the entire system (USV with AUV) and $c_0$ corresponds to the drag coefficient when the USV is carrying the AUV.

  (ii) Create a Simulink simulation and confirm the expected performance of the system when the USV carries the AUV.

 (iii) Modify the simulation so that at $t = t_l = 60\,\text{s}$ the AUV is suddenly dropped. The parameters $c_0$, $m$ and the control gains in the controller $\tau$ should remain fixed, but the mass and drag in the vehicle model should change step-wise at time $t_l$. How well is the controller able to cope with the sudden changes?

(b) Use adaptive feedback linearization to design the control input.

   (i) What are the parameter update laws for $\hat{c}_0$ and $\hat{m}$?

  (ii) Create a new Simulink simulation with the adaptive update law and tune the control parameters to obtain good performance when the AUV is suddenly dropped at $t = t_l = 60\,\text{s}$. How well does the adaptive controller handle the sudden changes?

 (iii) Plot the tracking error and the parameters estimates $\hat{\theta}$ to show their convergence.

**11.6** Revisit Problem 10.10 Consider the control of a davit for deploying an ROV from the deck of a ship. Model the elbow joint of the davit as a flexible joint (Fig. 10.12). Ignoring damping, the equations of motion are

$$I_1 \ddot{q}_1 = -mgl \sin q_1 - k(q_1 - q_2),$$
$$I_2 \ddot{q}_2 = -k(q_2 - q_1) + \tau.$$

Note that the effect of gravity makes the system nonlinear.

(a) Let

$$\eta = \begin{bmatrix} q_1 \\ q_2 \end{bmatrix}, \quad \text{and} \quad v = \begin{bmatrix} \dot{q}_1 \\ \dot{q}_2 \end{bmatrix}.$$

Rewrite the equations of motion in the form of (1.58)–(1.59). Which terms correspond to $J(\eta)$, $M$, $C(v)$, $D(v)$ and $g(\eta)$?

(b) Let $\eta_d(t)$ be the desired reference trajectory. Using your answer from part (a) in (11.27), recommend a suitable regressor matrix $\Phi(\dot{v}_r, v_r, v, \eta)$ and vector of unknown parameters $\theta$?

**11.7** A quadrotor unmanned aerial vehicle is used to drop measurement instrumentation into the sea for water quality testing. Let the mass of the vehicle be $m = 20\,\text{kg}$, the vertical force input from the propellers be $\tau$ and $g$ be gravity, the dynamics of the UAV in the vertical direction are

$$m\ddot{z} = mg + \tau.$$

The mass of the instrumentation is a fifth of the mass of the vehicle $m_i = m/5$. The control objective is for the quadrotor to maintain a fixed altitude after the instrumentation is dropped.

Design a controller using adaptive feedback linearization. Create a simulation of the closed loop system using Matlab/Simulink and tune the controller to achieve a suitable performance.

# References

1. Astrom, K.J., Wittenmark, B.: Adaptive Control, 2nd edn. Dover Publications, Inc., New York (2008)
2. Bertaska, I., von Ellenrieder, K.: Experimental evaluation of supervisory switching control for unmanned surface vehicles. IEEE J. Ocean. Eng. **44**(1), 7–28 (2019)
3. Fossen, T.I.: Handbook of Marine Craft Hydrodynamics and Motion Control. Wiley, Hoboken, 2011
4. Klinger, W.B., Bertaska, I.R., von Ellenrieder, K.D., Dhanak M.R.: Control of an unmanned surface vehicle with uncertain displacement and drag. IEEE J. Ocean. Eng. **42**(2), 458–476 (2017)
5. Sarda, E.I., Qu, H., Bertaska, I.R., von Ellenrieder K.D.: Station-keeping control of an unmanned surface vehicle exposed to current and wind disturbances. Ocean Eng. **127**, 305–324 (2016)
6. Slotine, J.-J.E., Li, W.: Applied Nonlinear Control. Prentice-Hall, Englewood Cliffs (1991)

# Chapter 12
# Sliding Mode Control

## 12.1 Introduction

Consider the dynamic equations of a marine vehicle (Fig. 12.1) in body-fixed coordinates (1.58)–(1.59), which are reproduced here for convenience

$$\dot{\eta} = \mathbf{J}(\eta)\mathbf{v},$$
$$\mathbf{M}\dot{\mathbf{v}} = -\mathbf{C}(\mathbf{v})\mathbf{v} - \mathbf{D}(\mathbf{v})\mathbf{v} - \mathbf{g}(\eta) + \boldsymbol{\tau} + \mathbf{w}_d.$$

It should be noted that dynamic models of marine vehicles generally tend to be fairly uncertain. The drag properties $\mathbf{D}(\mathbf{v})$ and actuator capabilities (in generating $\boldsymbol{\tau}$) may change if the vehicle is operated at off-design speeds, or from the wear and tear that occurs with normal usage. Further, the vehicle may have been designed to operate in one configuration, but a user may later add or remove components of the system, significantly changing the inertial properties represented by $\mathbf{M}$ and $\mathbf{C}(\mathbf{v})$, as well as the drag term $\mathbf{D}(\mathbf{v})$. Lastly, there are often unpredictable exogenous (externally-generated) disturbances that affect the vehicle's performance, which are unknown, but can be bounded. Here, the disturbances and model uncertainty will be combined and represented in the equations of motion (1.59) as the vector $\mathbf{w}_d$.

From (1.59) it can be seen that the dynamics of a marine vehicle tend to be nonlinear. Model uncertainties and exogenous disturbances, tend to strongly affect nonlinear systems, thus it is important to find control approaches that are *robust* to these uncertainties. Sliding mode control is an important robust control technique. The main advantages of sliding mode control include:

(1) robustness to uncertain plant model parameters, unmodeled dynamics and exogenous disturbances,
(2) finite-time convergence, and
(3) reduced order dynamics.

Several marine vehicle trajectory tracking control designs are based on the use of sliding mode control (SMC) to provide a fast response and to guarantee that tracking errors are bounded and robust to unmodeled dynamics and exogenous disturbances. SMC is generally based upon maintaining a sliding surface constraint through the use

© Springer Nature Switzerland AG 2021
K. D. von Ellenrieder, *Control of Marine Vehicles*, Springer Series on Naval Architecture, Marine Engineering, Shipbuilding and Shipping 9,
https://doi.org/10.1007/978-3-030-75021-3_12

**Fig. 12.1** 3 DOF
maneuvering coordinate
system definitions

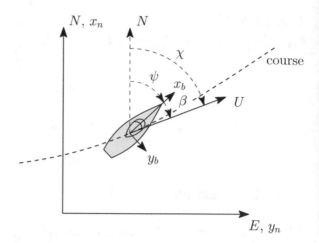

of high frequency switching, which is usually accomplished by injecting a non-linear discontinuous term into the control input.

The order of a sliding mode controller is categorized according to the relative degree $r_s$ between the sliding surface $s = 0$ and the injected discontinuous term. For example, a first order sliding mode controller (standard SMC) has $r_s = 1$, such that the discontinuous control input explicitly appears in the first total time derivative $\dot{s}$.

A potential problem with first order SMC designs is that, while the discontinuous control input enables the system to reject disturbances and improves robustness to model uncertainty, it can lead to an undesirable chattering effect, even with the use of some of the standard chattering mitigation techniques that are presented below.

The use of higher order sliding mode (HOSM) control design techniques, where $(r_s \geq 2)$, can be used to alleviate the chattering problem. In addition, HOSM techniques can also be used for the design of real-time differentiators (for example, to take the time derivatives of a measured signal or of a backstepping virtual control input) and observers (for estimating system states or disturbances) [15, 16, 23]. Higher order sliding mode controllers have convergence and robustness properties similar to those of first order sliding mode controllers, but they produce a continuous control signal (without chatter). Effectively, the approach still uses a non-linear discontinuous term, as is done in first order sliding mode techniques, but the relative degree $r_s$ between $s$ and the discontinuous term is increased, so that the control signal is continuous. HOSM controllers are often used in combination with HOSM differentiators/observers, as computation of the control input can require knowledge of state derivatives and disturbances.

Note that since the highest order terms in a sliding mode system are discontinuous, modified forms of the nonlinear Lyapunov stability analysis methods presented in Chap. 7 can be applied in some limited cases, i.e. when $r_s = 2$ [17, 18], but more generally the stability of such systems must be understood in the sense of Filippov [9]. Such analyses are beyond the scope of this text. Here the focus is on implementation and so the main conditions required for the stability of the given sliding mode systems will only be stated. Interested readers can refer to the citations provided to more fully study the stability of sliding mode systems.

We will explore the application of sliding mode techniques, as well as HOSM, to marine vehicles. However, before diving into the more complex multivariable systems needed to represent marine vehicles, we will focus on a simple, linear system that illustrates the basic concepts.

## 12.2 Linear Feedback Control Under the Influence of Disturbances

Let us start by exploring the effects of uncertainty and external disturbance on a linear controller.

**Example 12.1** Consider depth control of a Lagrangian float, as discussed in Example 4.8. Let the depth be represented by $x_1 = z$ and the vertical descent/ascent rate be $x_2 = \dot{z}$. Then the equations of motion can be written as

$$\begin{aligned} \dot{x}_1 &= x_2, \\ \dot{x}_2 &= \tau + w_d(x_1, x_2, t), \end{aligned} \tag{12.1}$$

where $\tau$ is the control input and $w_d$ is a bounded disturbance (e.g. from a current), with bounds $L \geq |w_d|$, where $L > 0$ is a positive constant.

The control objective is to design a feedback control law that drives $x_1 \to 0$ and $x_2 \to 0$ asymptotically. Achieving asymptotic convergence in the presence of the unknown disturbance is challenging. For example, consider a linear control law of the form $\tau = -k_1 x_1 - k_2 x_2$. When $w_d = 0$, the closed loop system can be written in the state space form

$$\dot{\mathbf{x}} = \frac{d}{dt} \begin{bmatrix} x_1 \\ x_2 \end{bmatrix} = \begin{bmatrix} 0 & 1 \\ -k_1 & -k_2 \end{bmatrix} \begin{bmatrix} x_1 \\ x_2 \end{bmatrix} = \mathbf{A}\mathbf{x}. \tag{12.2}$$

The closed loop system is stable when the matrix $\mathbf{A}$ is Hurwitz (when the real part of all eigenvalues of $\mathbf{A} < 0$). The eigenvalues of $\mathbf{A}$ can be computed by solving the equation

$$\det(\lambda \mathbf{1} - \mathbf{A}) = \begin{vmatrix} \lambda & -1 \\ k_1 & \lambda + k_2 \end{vmatrix} = \lambda(\lambda + k_2) + k_1 = 0, \tag{12.3}$$

for $\lambda$, where $\mathbf{1}$ is the identity matrix and $\det(\cdot)$ is the determinant. Solving (12.3) gives the two eigenvalues

$$\lambda_{1,2} = -\frac{k_2}{2} \pm \frac{1}{2}\sqrt{k_2^2 - 4k_1}. \tag{12.4}$$

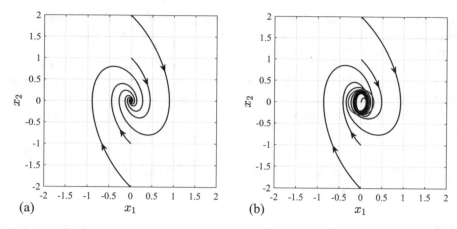

**Fig. 12.2** Phase portrait of multiple solution trajectories of (12.1) for $k_1 = 2$ and $k_2 = 1$ when **a** $w_d = 0$ and **b** $w_d = 0.25 \sin(1.5t)$. When $w_d = 0$, the controller asymptotically drives the closed loop system to the origin of its state space. Instead, when $w_d \neq 0$ the controller drives the states to a bounded region of convergence (represented approximately by the red circle)

In order for both eigenvalues to have negative real parts, we require $k_2 > 0$ and $k_1 > k_2^2/4$. A plot of the solution trajectories of (12.1) with $w_d = 0$ in the state space of the system is shown in Fig. 12.2a. When $w_d \neq 0$, the solution trajectories are ultimately bounded (see Sect. 7.6), that is they do not necessary approach the origin, but instead converge to a region, or "ball", of radius $\delta$ near the origin (Fig. 12.2b).                                                                                      □

Here, we will use the methods presented in Sects. 7.4–7.7 to examine the stability of the closed loop systems. As discussed in Sect. 7.2.2, Lyapunov functions are continuously differentiable, non-negative energy-like functions that always decrease along the solution trajectories of the system. We start by identifying a candidate Lyapunov function (CLF) $V(x_1, x_2)$, where

(a)  $V$ is a radially unbounded function of the state variables, i.e.

$$\lim_{|x_1|, |x_2| \to \infty} V(x_1, x_2) \to \infty, \tag{12.5}$$

and

(b)  the time derivative of $V$ is negative definite,

$$\dot{V} < 0, \forall x_1, x_2 \neq 0. \tag{12.6}$$

**Example 12.2** Let explore the closed loop stability of (12.1) with linear control input $\tau = -k_1 x_1 - k_2 x_2$ using the CLF

$$V = \frac{1}{2}x_1^2 + \frac{1}{2}x_2^2. \tag{12.7}$$

From (12.7), it can be seen that this function is radially unbounded, i.e. that

$$\lim_{|x_1|,|x_2| \to \infty} V \to \infty.$$

Taking the time derivative of (12.7), it will be shown that control gains can be selected so that $\dot{V} < 0, \forall x_1, x_2 \neq 0$. First, we see that

$$
\begin{aligned}
\dot{V} &= x_1 \dot{x}_1 + x_2 \dot{x}_2, \\
&= x_1 x_2 - x_2(-k_1 x_1 - k_2 x_2 + w_d), \\
&= (1 - k_1) x_1 x_2 - k_2 x_2^2 + x_2 w_d.
\end{aligned}
\tag{12.8}
$$

This can be further simplified as

$$
\begin{aligned}
\dot{V} &\leq (1 - k_1) \left[ \frac{1}{2} x_1^2 + \frac{1}{2} x_2^2 \right] - k_2 x_2^2 + \left[ \frac{1}{2} x_2^2 + \frac{1}{2} w_d^2 \right], \\
&\leq -\frac{(k_1 - 1)}{2} x_1^2 - \left[ \frac{(k_1 - 1) + 2k_2 - 1}{2} \right] x_2^2 + \frac{1}{2} w_d^2,
\end{aligned}
\tag{12.9}
$$

by using Young's Inequality.

Let

$$\mu_1 := \frac{(k_1 - 1)}{2} \quad \text{and} \quad \mu_2 := \left[ \frac{(k_1 - 1) + 2k_2 - 1}{2} \right] \tag{12.10}$$

where we take $\mu_1 > 0$ and $\mu_2 > 0$ to ensure stability. Then define

$$\mu := \min \{\mu_1, \mu_2\} \tag{12.11}$$

and

$$C_d := \sup_{\{x_1, x_2, t\}} \left| \frac{1}{2} w_d^2 \right|. \tag{12.12}$$

With these definitions, (12.9) becomes

$$\dot{V} \leq -\mu_1 x_1^2 - \mu_2 x_2^2 + C_d. \tag{12.13}$$

It can be shown that $\dot{V}$ is upper-bounded by $-2\mu V + C_d$ when $\mu$ is either $\mu_1$ or $\mu_2$. For example, let $\mu = \mu_1$. Then (12.13) becomes

$$\dot{V} \leq -\mu_1 \left( x_1^2 + \frac{\mu_2}{\mu_1} x_2^2 \right) + C_d = -\mu \left( x_1^2 + \frac{\mu_2}{\mu} x_2^2 \right) + C_d. \tag{12.14}$$

Using the definition of $V$ in (12.7), and noting that $\mu_2/\mu > 1$, gives

$$\mu \left( x_1^2 + \frac{\mu_2}{\mu} x_2^2 \right) > \mu \left( x_1^2 + x_2^2 \right) = 2\mu V. \tag{12.15}$$

Therefore,

$$- \mu \left( x_1^2 + \frac{\mu_2}{\mu} x_2^2 \right) + C_d < -2\mu V + C_d, \tag{12.16}$$

so that

$$\dot{V} \leq -2\mu V + C_d. \tag{12.17}$$

The same can be shown to be true when $\mu = \mu_2 = \min\{\mu_1, \mu_2\}$.

Equation (12.17) can be integrated to show that

$$V \leq \left( V_0 - \frac{C_d}{2\mu} \right) e^{-2\mu t} + \frac{C_d}{2\mu}, \tag{12.18}$$

where $V_0$ is the value of $V$ at time $t = 0$. Note that, as discussed in Sect. 7.7, this system is uniformly globally practically exponentially stable (USPES), see Theorem 7.13.                                                                                     □

## 12.3  First Order Sliding Mode Control

We have seen that when exogenous disturbances and uncertainties are present, the closed loop trajectories of the system do not reach a specific equilibrium point, but instead asymptotically approach a finite region (ball) in the state space. Let's see if we can *reverse engineer* the desired closed loop dynamics of our system (12.1) in order to drive it to a desired equilibrium point when $w_d \neq 0$, instead.

When no disturbances are present, a stable linear controller for the state $x_1$ would have the form

$$\dot{x}_1 = -\lambda x_1, \quad \lambda > 0. \tag{12.19}$$

Since, $x_2 := \dot{x}_1$, the general solution of (12.19) is

$$\begin{aligned} x_1(t) &= x_1(0)e^{-\lambda t}, \\ x_2(t) &= \dot{x}_1(t) = -\lambda x_1(0)e^{-\lambda t}, \end{aligned} \tag{12.20}$$

where both $x_1(t)$ and $x_2(t)$ exponentially converge to zero. Ideally, nonzero disturbances $w_d(x_1, x_2, t) \neq 0$ will not affect our desired solution.

To design a controller that can achieve this desired solution we can introduce a new variable $s$ in the state space of the system as

$$s = x_2 + \lambda x_1, \quad \lambda > 0. \tag{12.21}$$

We seek a control input $\tau$ that can drive $s$ to zero in finite time. To find this control, we take the time derivative of (12.21), and use (12.1) to introduce the control input from the dynamics of the closed loop system into the resulting equation

$$\dot{s} = \dot{x}_2 + \lambda \dot{x}_1 = \tau + w_d(x_1, x_2, t) + \lambda x_2. \tag{12.22}$$

This expression can then be used to identify a suitable (CLF) to stabilize the closed loop system. Let

$$V = \frac{1}{2}s^2. \tag{12.23}$$

In order to ensure that the closed loop system converges to its equilibrium point in finite time, we also require the CLF to also (simultaneously) satisfy the inequality

$$\dot{V} \leq -\alpha V^{1/2}, \quad \alpha > 0. \tag{12.24}$$

The inequality can be solved using separation of variables, so that

$$\int_{V_0}^{V} \frac{dV}{V^{1/2}} \leq -\int_0^t \alpha dt,$$
$$2\left[V^{1/2} - V_0^{1/2}\right] \leq -\alpha t, \tag{12.25}$$
$$V^{1/2} \leq -\frac{\alpha t}{2} + V_0^{1/2},$$

where, as above, $V_0$ is the value of $V$ at time $t = 0$. The Lyapunov function $V = V(x_1(t), x_2(t))$ becomes zero in the finite time $t_r$, which is bounded by

$$t_r \leq \frac{2V_0^{1/2}}{\alpha}. \tag{12.26}$$

Differentiating (12.23) we can find a control input $\tau$ so that $\dot{V} \leq 0$,

$$\dot{V} = s\dot{s} = s\left[\lambda x_2 + w_d(x_1, x_2, t) + \tau\right] \tag{12.27}$$

by taking $\tau = -\lambda x_2 + v$ to get

$$\dot{V} = s\left[v + w_d(x_1, x_2, t)\right] = sv + sw_d(x_1, x_2, t) \leq |s|L + sv. \tag{12.28}$$

Let $v = -\rho \operatorname{sgn}(s)$, $\rho > 0$, where, as shown in Fig. 12.3,

$$\operatorname{sgn}(s) := \begin{cases} +1, & s > 0, \\ -1, & s < 0. \end{cases} \tag{12.29}$$

Note that strictly speaking one can take

$$\operatorname{sgn}(0) \in [-1, 1], \tag{12.30}$$

(see [23]). However, it is common in engineering practice to use $\operatorname{sgn}(0) = 0$, as in [10].

**Fig. 12.3** A plot of the
function $-\mathrm{sgn}(s)$

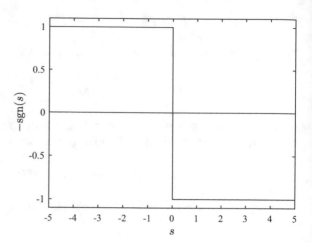

Then, noting that $\mathrm{sgn}(s)s = |s|$, (12.28) can be rewritten as

$$\dot{V} \le L|s| - \rho\,\mathrm{sgn}(s)s = (L - \rho)|s|, \tag{12.31}$$

where $L$ is an upper bound for the magnitude of the disturbance. Using (12.23)
and (12.24), $\dot{V}$ can also be written as

$$\dot{V} \le -\frac{\alpha}{\sqrt{2}}|s|. \tag{12.32}$$

Matching the right hand sides of inequalities (12.31) and (12.32) gives

$$(L - \rho) = -\frac{\alpha}{\sqrt{2}}, \tag{12.33}$$

so that the control gain can be selected as

$$\rho = L + \frac{\alpha}{\sqrt{2}}, \tag{12.34}$$

where the first component $L$ compensates for the bounded disturbance and the second
component determines the sliding surface reaching time $t_r$, which increases as $\alpha$
increases. The resulting control law is

$$\tau = -\lambda x_2 - \rho\,\mathrm{sgn}(s). \tag{12.35}$$

**Remark 12.1** The surface $s = 0$ (which is the straight line $x_2 + \lambda x_1 = 0$ in two
dimensions) is called a *sliding surface*. Equation (12.24), which is equivalent to

$$s\dot{s} \leq -\frac{\alpha}{\sqrt{2}}|s|, \tag{12.36}$$

is called a *reaching condition*. When the reaching condition is met, the solution trajectory of the closed loop system is driven towards the sliding surface and remains on it for all subsequent times.

**Definition 12.1** (*sliding mode controller*) A control law that drives the state variables to the sliding surface in finite time $t_r$ and keeps them on the surface thereafter in the presence of a bounded disturbance is called a sliding mode controller. An ideal sliding mode is taking place when $t > t_r$.

**Example 12.3** Let us revisit control of the Lagrangian float discussed in Example 12.1. Consider the response of system (12.1) when the control input is instead taken to be $\tau = -\lambda x_2 - \rho \text{sgn}(s)$ with $\lambda = 1.5$, $\rho = 2$ and $w_d(x_1, x_2, t) = 0.25 \sin(1.5t)$ and initial conditions $x_1(0) = 0$, $x_2(0) = 2$. From Eq. (12.23) it can be seen that

$$V_0 = \frac{1}{2}[x_2(0) + \lambda x_1(0)]^2 = 2. \tag{12.37}$$

If we take $L = 0.25$, then from (12.33) we have

$$\alpha = \sqrt{2}(\rho - L) = 2.475. \tag{12.38}$$

From (12.26), we see that the estimated *reaching time* is

$$t_r \leq \frac{2V_0^{1/2}}{\alpha} = \frac{2\sqrt{2}}{2.475} = 1.143 \text{ s.} \tag{12.39}$$

The time responses of $x_1$, $x_2$ and $s$ are plotted in Fig. 12.4a. It can be seen that for times between $t = 0$ and slightly earlier than $t = t_r$, the sliding surface $s \neq 0$, so that the closed loop system is in its reaching phase. At just before $t = t_r$ the value of $x_2$ changes discontinuously as $s \to 0$ and the system enters the sliding phase. It can be seen that both $x_1$ and $x_2$ are driven to zero. The corresponding phase portrait of the system is shown in Fig. 12.4b. By comparing Fig. 12.4a, b it can be seen that the curved branch of the trajectory in the phase plane corresponds to the reaching phase and the straight line portion (which moves along the line $x_2 = -1.5x_1$) corresponds to the sliding phase. The solution trajectory of the closed loop system reaches the desired equilibrium point $(x_1, x_2) = (0, 0)$, despite the nonzero disturbance acting upon it.

For comparison purposes, the phase portrait is shown in Fig. 12.5 for the same initial conditions used in Fig. 12.2a, b.

Thus far, the sliding mode control approach seems fantastic. However, there is a catch. In numerical and practical implementations of the control law (12.35) the $\text{sgn}(s)$ term rapidly switches between $\text{sgn}(s) = +1$ and $\text{sgn}(s) = -1$ as the controller approaches the sliding mode phase, i.e. as $s \to 0$. Owing to the finite size of the time step in numerical simulations and to time delays or other imperfections in switching

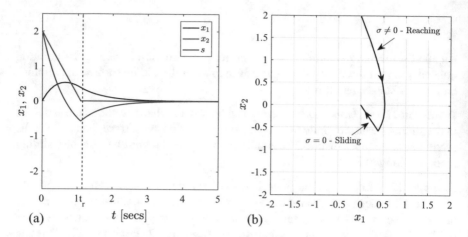

**Fig. 12.4** **a** Time response and **b** phase portrait of the system in (12.1) when the control input is taken to be $\tau = -\lambda x_2 - \rho\,\mathrm{sgn}(s)$ with $\lambda = 1.5$, $\rho = 2$ and $w_d(x_1, x_2, t) = 0.25\sin(1.5t)$ and initial conditions $x_1(0) = 0$, $x_2(0) = 2$

**Fig. 12.5** Phase portrait of the sliding mode controller with $\lambda = 1.5$ and $\rho = 2$ and disturbance $w_d = 0.25\sin(1.5t)$

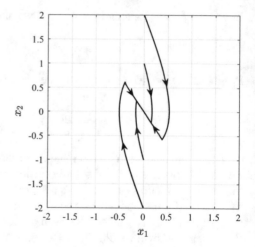

devices found in real implementations, the sign of $s$ will continue to switch between positive and negative values after the reaching phase, so that the solution trajectory of the system zig-zags across the sliding surface, rather than truly sliding along it, in a condition known as *chattering* (Fig. 12.6a). This is can also be seen in the control input signal, which will also tend to rapidly oscillate in amplitude (Fig. 12.6b). In an ideal sliding mode, the frequency of the zig-zag motion would approach infinity and its amplitude would approach zero. However, in practical implementations this high frequency switching is undesirable, as it can result in low control accuracy, high heat losses in electrical circuits and the rapid wear of moving mechanical parts.   $\square$

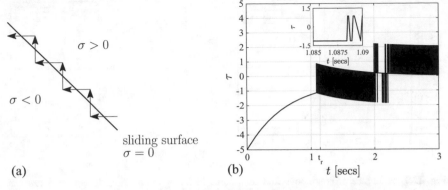

**Fig. 12.6 a** Zig-zag motion of the solution trajectory across the sliding surface $s = 0$ exhibited during chattering. **b** Chattering of the control input when using the sliding mode controller with $\lambda = 1.5$ and $\rho = 2$ and disturbance $w_d = 0.25 \sin(1.5t)$. The inset for times $1.085 \leq t \leq 1.090$ illustrates how the control signal rapidly switches in amplitude during chattering

## 12.4   Chattering Mitigation

The chattering problem discussed in the previous section can be overcome by replacing the sgn($s$) function in (12.35) with a continuous approximation $f(s)$. In principle, the approximate function could be any function that is continuous, sigmoidal $\lim_{s \to \pm\infty} f(s) \to \pm 1$, odd $f(-s) = -f(s)$, zero at zero $f(0) = 0$ and strictly increasing. In practice, it is common to use a saturation function [24]

$$f(s) = \text{sat}\left(\frac{s}{\Phi}\right) := \begin{cases} \dfrac{s}{\Phi}, & \left|\dfrac{s}{\Phi}\right| < 1, \\ \text{sgn}\left(\dfrac{s}{\Phi}\right), & \left|\dfrac{s}{\Phi}\right| \geq 1, \end{cases} \tag{12.40}$$

a hyperbolic tangent function [10]

$$f(s) = \tanh\left(\frac{s}{\Phi}\right), \tag{12.41}$$

or a simple sigmoidal function [23], such as

$$f(s) = \frac{\dfrac{s}{\Phi}}{\dfrac{|s|}{\Phi} + 1}, \tag{12.42}$$

where $\Phi > 0$ is a parameter that stretches the region along the $s$ direction around the origin $s = 0$ (Fig. 12.7).

In the case of the saturation function (Fig. 12.8), it can be shown that the stretched region $|s| < \Phi$ functions like a boundary layer [24]. From (12.32) it can be seen that

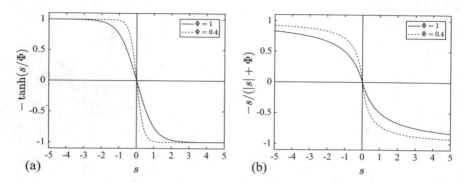

**Fig. 12.7** Examples of continuous functions suitable for approximating the $-\text{sgn}(s)$ function in (12.35) with $\Phi = 1$ and $\Phi = 0.4$: **a** the hyperbolic tangent function (12.41), and **b** the sigmoidal function (12.42). It can be seen that as $\Phi$ decreases the output of each function is "squeezed" about $s = 0$

**Fig. 12.8** Use of the saturation function (12.40) for approximating the $-\text{sgn}(s)$ function in (12.35). The region from $-\Phi \le 0 \le \Phi$ is the boundary layer

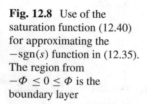 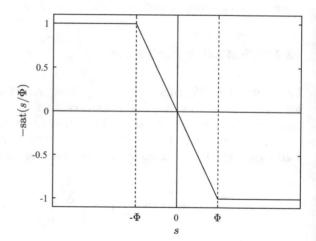

$\dot{V} < 0$ for $s \ne 0$ (negative definite) so that the boundary layer region is attractive. At the edges of the boundary layer, $|s| = \Phi$ so that $V_{BL} = \Phi^2/2$. Thus, modifying (12.25) as

$$\int_{V_0}^{V_{BL}} \frac{dV}{V^{1/2}} \le -\int_0^{t_{BL}} \alpha \, dt \quad \Rightarrow \quad 2\left[V_{BL}^{1/2} - V_0^{1/2}\right] \le -\alpha t_{BL},$$

it can be shown that the boundary layer is reached in the finite time $t_{BL}$,

$$t_{BL} \le \frac{\sqrt{2}(s_0 - \Phi)}{\alpha},$$

where, from (12.23), $s_0 = \sqrt{2V_0}$.

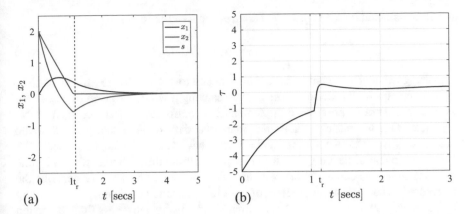

**Fig. 12.9** The **a** time response and **b** control input of system (12.1) when $\tau = -\lambda x_2 - \rho\mathrm{sat}(s/\Phi)$, where $\Phi = 0.05$, $\lambda = 1.5$, $\rho = 2$ and $w_d(x_1, x_2, t) = 0.25\sin(1.5t)$ and initial conditions $x_1(0) = 0$, $x_2(0) = 2$

Once inside the boundary layer, the solution trajectories will remain within the region $B = \{\mathbf{x} : |s(\mathbf{x}, t)| \leq \Phi\}$ for all future times. However, within the boundary layer it is no longer guaranteed that $s \to 0$ for nonzero disturbances. Instead, the system will reach some bounded region around $s = 0$, the size of which depends upon $\Phi$ and $\lambda$ (see [24]). Although $s$ and the state variables no longer converge to the origin, the performance of the system can be very close to that of a true sliding mode controller.

**Example 12.4** Returning to the control of the Lagrangian float discussed in Examples 12.1–12.3 above, we now implement the continuous saturation function (12.40) in the control law as $\tau = -\lambda x_2 - \rho\mathrm{sat}(s/\Phi)$, where $\Phi = 0.05$, $\lambda = 1.5$, $\rho = 2$ and $w_d(x_1, x_2, t) = 0.25\sin(1.5t)$ and initial conditions $x_1(0) = 0$, $x_2(0) = 2$, as before. Comparing Figs. 12.9a, and 12.4a, it can be seen that the time response of the system is almost the same. However, while the control input resulting from use of the discontinuous function exhibits substantial chattering (Fig. 12.6b), the control input obtained with the saturation function is smooth (Fig. 12.9b). □

## 12.5 Equivalent Control

It is shown above that for times $t \geq t_r$ the trajectory of the closed loop system is on the sliding surface where $s = x_2 + \lambda x_1 = 0$. Once on the sliding surface, the trajectory remains there because of the sliding mode controller, so that $s = \dot{s} = 0$. Let us examine the condition $\dot{s} = 0$ more closely. From (12.27), we see that

$$\dot{s} = \lambda x_2 + w_d(x_1, x_2, t) + \tau = 0, \quad s(t \geq t_r) = 0 \tag{12.43}$$

would be satisfied by a an *equivalent control* input of the form

$$\tau_{eq} = -\lambda x_2 - w_d(x_1, x_2, t). \tag{12.44}$$

This suggests that the control law, which needs to be applied to ensure that the system remains on the sliding surface after reaching it would be given by $\tau_{eq}$. However, it would not be possible to implement this control law in practice, as the term $w_d(x_1, x_2, t)$ represents exogenous disturbances and model uncertainty, which we do not know. Given that the sliding mode control law (12.35) developed in Sect. 12.3 above is capable of keeping the closed loop system on the sliding surface after the reaching time, one can argue that we have identified a control law that acts like the desired $\tau_{eq}$. Thus, the average effect of the high frequency switching in our sliding mode control law is $\tau_{eq}$. We can approximate $\tau_{eq}$ by using a low pass filtered version of our sliding mode control law, which averages the effects of the high frequency switching caused by the sgn($s$) term as

$$\hat{\tau}_{eq} = -\lambda x_2 - \rho \text{LPF}\left[\text{sgn}(s)\right], \quad t \geq t_r. \tag{12.45}$$

The low pass filter can be implemented as a first order differential equation of the form

$$\begin{aligned} T\dot{z} &= -z + \text{sgn}(s) \\ \hat{\tau}_{eq} &= -\lambda x_2 - \rho z, \end{aligned} \tag{12.46}$$

where $T \in [0, 1)$ is a filter time constant (which should be as small as possible, but larger than the sampling time of the computer). Comparing (12.44) and (12.45), it can be seen that the term $\text{LPF}\left[\text{sgn}(s)\right]$ in (12.45) acts like a disturbance observer, which can be used to provide an estimate of the combined effects of exogenous disturbances and model uncertainty.

## 12.6  Summary of First Order Sliding Mode Control

Based on our results thus far, we can state the following observations:

(1) First Order sliding mode control design consists of two main tasks:

    (a) Design of a first order sliding surface $s = 0$.
    (b) Design of the control $\tau$, which drives the sliding variable $s$ in (12.21) to zero use the sliding variable dynamics in (12.22).

(2) Application of the sliding mode control law (12.35) in (12.1) yields the closed loop system dynamics

$$\begin{aligned} \dot{x}_1 &= x_2, \\ \dot{x}_2 &= -\lambda x_2 - \rho \text{sgn}(s) + w_d(x_1, x_2, t). \end{aligned} \tag{12.47}$$

When the system is in sliding mode ($t \geq t_r$), the closed loop system is driven by its equivalent dynamics (12.44) so that (12.47) becomes

$$\dot{x}_1 = x_2,$$
$$\dot{x}_2 = \underbrace{-\lambda x_2 - w_d(x_1, x_2, t)}_{\tau_{eq}} + w_d(x_1, x_2, t) = -\lambda x_2 = -\lambda \dot{x}_1, \qquad (12.48)$$

which implies that

$$\dot{x}_1 = x_2,$$
$$x_2 = -\lambda x_1.$$

Thus, once the sliding mode condition is reached, the dynamics of the system are first order (12.48), whereas the original system (12.1) is second order. This reduction of order is known as a *partial dynamical collapse*.

(3) In the sliding mode the closed loop system dynamics do not depend on the disturbance term $w_d(x_1, x_2, t)$. However, the upper limit of $w_d$ must be taken into account when selecting the controller gain.

(4) In practical implementations, it may be necessary to use a continuous approximation of the discontinuous function $\text{sgn}(s)$ in the control law to avoid chattering.

## 12.7 Stabilization Versus Tracking

Recall the definitions of stabilization and tracking, as given in Example 1.2. The problem of designing a control input that drives the closed loop system to an equilibrium point is known as a *stabilization problem*. However, many important marine vehicle control applications require one to design a control law that tracks a desired trajectory $y_d$, i.e. to design a control input which solves a *tracking problem*.

Let the output of the system be $y$. To ensure the system is stable, it is important to identify its relative degree (see Sect. 8.6.1). This can be thought of as the number of times $r$ the output must be differentiated before the control input explicitly appears in the input–output relation. For example, if we take $y = x_1$ for system (12.1),

$$\dot{x}_1 = x_2,$$
$$\dot{x}_2 = \tau + w_d(x_1, x_2, t), \qquad (12.49)$$

it can be seen that $\ddot{y} = \ddot{x}_1 = \dot{x}_2 = \tau + w_d(x_1, x_2, t)$. Thus, the relative degree of system (12.1) is $r = 2$. When the order of the system is the same as its relative degree, it does not have any (hidden) internal dynamics (e.g. dynamics which cannot be directly controlled using the control input $\tau$). When the relative degree of the system is less than its order, additional analysis is necessary to ensure that the internal dynamics of the system are stable.

Recent work that utilizes one or more first order sliding mode control terms for the trajectory tracking control of marine systems includes [1, 8, 13, 21, 27, 29]. An

example of the use of first order sliding mode control for station keeping is provided
in the following example.

**Example 12.5** Consider the first order sliding mode station keeping control of a
marine vehicle. Here, following [22], we will use passivity based backstepping
(Sect. 10.3.2) to derive the control law, but with modified forms of the reference
velocity and measure of tracking.

As before, let the Earth-fixed tracking error be

$$\tilde{\eta} := \eta - \eta_d. \tag{12.50}$$

Now, the modified Earth-fixed reference velocity, will be defined as

$$\dot{\eta}_r := \dot{\eta}_d - 2\Lambda\tilde{\eta} - \Lambda^2 \int_0^t \tilde{\eta} dt, \tag{12.51}$$

where $\Lambda > 0$ is a diagonal control design matrix. The corresponding body-fixed
reference velocity is

$$\mathbf{v}_r := \mathbf{J}^{-1}(\eta)\dot{\eta}_r. \tag{12.52}$$

Let the measure of tracking be

$$\mathbf{s} := \dot{\eta} - \dot{\eta}_r = \dot{\tilde{\eta}} + 2\Lambda\tilde{\eta} + \Lambda^2 \int_0^t \tilde{\eta} dt. \tag{12.53}$$

In order to implement a first order sliding mode controller, the passivity-based control
law (10.104) is modified to get

$$\tau = \mathbf{M}\dot{\mathbf{v}}_r + \mathbf{C}(\mathbf{v})\mathbf{v}_r + \mathbf{D}(\mathbf{v})\mathbf{v}_r + \mathbf{g}(\eta) - \mathbf{J}^T(\eta)\mathbf{K}_d\Phi^{-1}\mathbf{s}, \tag{12.54}$$

where $\mathbf{K} = \mathbf{K}^T > 0$ is a positive definite diagonal gain matrix and $\Phi = \Phi^T > 0$
defines the thickness of the boundary layer.                                          □

## 12.8   SISO Super–Twisting Sliding Mode Control

Once again, consider the system (12.49), where the output is taken to be $y = x_1$.
Here, we will modify our definition of the sliding variable in (12.21) to solve an
output tracking problem. Let $\tilde{x}_1 := (y - y_d) = (x_1 - y_d)$ and

$$s := \dot{\tilde{x}}_1 + \lambda\tilde{x}_1 = \tilde{x}_2 + \lambda\tilde{x}_1, \quad \lambda > 0. \tag{12.55}$$

We seek a control input that drives $s$ to $s = 0$ in finite time and keeps it there, such
that $\dot{s} = 0$. Taking the time derivative of (12.55) gives

$$\begin{aligned}
\dot{s} &= \ddot{\tilde{x}}_1 + \lambda\dot{\tilde{x}}_1, \\
&= \ddot{y} - \ddot{y}_d + \lambda\dot{y} - \lambda\dot{y}_d, \\
&= \dot{x}_2 - \ddot{y}_d + \lambda x_2 - \lambda\dot{y}_d,
\end{aligned} \tag{12.56}$$

Using (12.49), this can be reduced to

$$\dot{s} = \underbrace{-\ddot{y}_d - \lambda\dot{y}_d + \lambda x_2 + w_d(x_1, x_2, t)}_{\gamma(x_1, x_2, t)} + \tau = 0, \tag{12.57}$$

where $\gamma(x_1, x_2, t)$ is a *cumulative* disturbance term, which is upper-bounded by a positive constant $M$, e.g. $|\gamma(x_1, x_2, t)| \leq M$.

When $\gamma(x_1, x_2, t) = 0$ using the control input

$$\tau = -\lambda|s|^{1/2}\text{sgn}(s), \quad \lambda > 0 \tag{12.58}$$

in (12.56) makes compensated dynamics in the sliding mode become

$$\dot{s} = -\lambda|s|^{1/2}\text{sgn}(s), \quad s(0) = s_0. \tag{12.59}$$

This can be integrated to get

$$|s|^{1/2} - |s_0|^{1/2} = -\frac{\lambda}{2}t. \tag{12.60}$$

Taking $s = 0$, we find that the corresponding reaching time would be given by

$$t_r \geq \frac{2}{\lambda}|s_0|^{1/2}. \tag{12.61}$$

Unfortunately, when $\gamma(x_1, x_2, t) \neq 0$, these relations no longer hold. Therefore, we ask if it might be possible to add another term to the controller to cancel the effects of cumulative disturbance.

Assuming that the time derivative of the cumulative error can be upper bounded, i.e. $\dot{\gamma}(x_1, x_2, t) \leq C$, a modified form of the controller above,

$$\begin{aligned}
\tau &= -\lambda|s|^{1/2}\text{sgn}(s) + z, \quad \lambda = 1.5\sqrt{C} \\
\dot{z} &= -b\text{sgn}(s), \quad b = 1.1\,C
\end{aligned} \tag{12.62}$$

makes the compensated $s$-dynamics become

$$\begin{aligned}
\dot{s} &= \gamma(x_1, x_2, t) - \lambda|s|^{1/2}\text{sgn}(s) + z, \\
\dot{z} &= -b\text{sgn}(s).
\end{aligned} \tag{12.63}$$

The term $z$ acts like an equivalent control term, which produces a low pass filtered estimate of the cumulative disturbance. When it starts following the disturbance term, the sliding variable dynamics correspond to (12.58) so that the remaining term $\lambda|s|^{1/2}\text{sgn}(s)$ in the control law can drive $s$ to zero in finite time.

Based on these results, we make the following observations about super-twisting sliding mode control

(1) From (12.57), it can be seen that, using the definition of $s$ in (12.55), the relative degree between $s$ and $\tau$ is 1, but the relative degree between $s$ and the discontinuous injected term $b\text{sgn}(s)$ is $r_s = 2$. Thus, the super twisting control input $\tau$ is continuous because both $\lambda|s|^{1/2}\text{sgn}(s)$ and $z = -\int b\text{sgn}(s)dt$ are continuous. The high frequency switching in the $\text{sgn}(s)$ term is smoothed by the integral.

(2) Use of the term $z$ and the augmented system $\dot{z} = -b\text{sgn}(s)$ are mandatory (very important) for ensuring continuity of the control function while simultaneously canceling the effects of the (unknown) cumulative disturbance.

(3) Once the cumulative disturbance is canceled, the super-twisting controller uses a nonlinear sliding manifold given by $\dot{s} = -\lambda|s|^{1/2}\text{sgn}(s)$ to drive both drive both $\dot{s}$ and $s$ to zero in finite time (Fig. 12.10). This is different from the conventional sliding mode controller presented in Sect. 12.3, which uses a linear sliding surface that can only drive $s$ to zero in finite time.

(4) If we define our state variables to be the output tracking errors $\tilde{x}_1$ and $\tilde{x}_2 := \dot{\tilde{x}}_1$, we can rewrite the sliding variable (12.55) in the form $s = \tilde{x}_2 + \lambda\tilde{x}_1$. The fact that $\dot{s}$ is driven to zero in finite time, implies that $s \to 0$ in finite time, so that when the system is in sliding mode, its compensated dynamics are

$$\tilde{x}_2 = -\lambda\tilde{x}_1$$
$$\dot{s} = 0. \tag{12.64}$$

Therefore, we have complete dynamical collapse in the sliding mode—the second order uncompensated dynamics of the original system in (12.1) are reduced

**Fig. 12.10** The super-twisting sliding surface given by $\dot{s} = -1.5|s|^{1/2}\text{sgn}(s)$

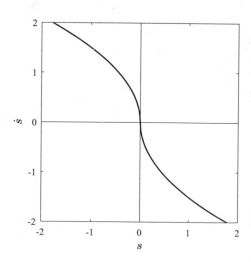

to the algebraic equations $\tilde{x}_2 = \tilde{x}_1 = 0$ in finite time. From a control engineering standpoint, this is an important result, as it guarantees that any inner loop dynamics or parasitic dynamics can be eliminated by the control input.

## 12.9  MIMO Super-Twisting Sliding Modes

Here, a multivariable generalization of super-twisting sliding mode techniques, which is developed in [20], is presented. Consider a multi-input, multi-output dynamic system of the form

$$\dot{\mathbf{x}} = \mathbf{f}(\mathbf{x}, t) + \mathbf{G}(\mathbf{x}, t)\boldsymbol{\tau} + \mathbf{w}_d, \tag{12.65}$$

where $\mathbf{x} \in \mathbb{R}^n$ is the state, $\boldsymbol{\tau} \in \mathbb{R}^m$ is the control input, $\mathbf{w}_d \in \mathbb{R}^n$ represents bounded, but unknown disturbances and model uncertainty, $\mathbf{f}(\mathbf{x}, t) \in \mathbb{R}^n$ is a known vector function and $\mathbf{G}(\mathbf{x}, t) \in \mathbb{R}^n \times \mathbb{R}^m$ is a known nonsingular matrix function (which could be a matrix of actuator terms), $\det(\mathbf{G}(\mathbf{x}, t)) \neq 0$. Here, we consider the case when the system is fully-actuated so that $m = n$.

Let $\mathbf{s}(\mathbf{x}, t) \in \mathbb{R}^n$ be a vector of sliding surfaces, which can be taken to be a tracking error (for design of a control input) or a the error of a state estimate, in the case of an observer or differentiator. The objective is to drive $\mathbf{s}(\mathbf{x}, t) \to 0$ in finite time.

For example, if we take the output to be $\mathbf{y} = \mathbf{x}$, then $\mathbf{s} = \mathbf{x} - \mathbf{y}_d(t)$, where $\mathbf{y}_d(t) \in \mathbb{R}^n$ is the desired trajectory of the sytem. The time derivative of $\mathbf{s}$ is

$$\begin{aligned}
\dot{\mathbf{s}} &= \dot{\mathbf{x}} - \dot{\mathbf{y}}_d(t) \\
&= \mathbf{f}(\mathbf{x}, t) + \mathbf{G}(\mathbf{x}, t)\boldsymbol{\tau} + \mathbf{w}_d - \dot{\mathbf{y}}_d(t), \\
&= \mathbf{f}(\mathbf{x}, t) + \mathbf{G}(\mathbf{x}, t)\boldsymbol{\tau} + \boldsymbol{\gamma}(\mathbf{s}, t),
\end{aligned} \tag{12.66}$$

where $\boldsymbol{\gamma}(\mathbf{s}, t)$ is a cumulative disturbance term, which includes the combined, unknown, but bounded effects of uncertainty and exogenous disturbance.

Consider the feedback linearizing control law $\boldsymbol{\tau} = \mathbf{G}(\mathbf{x}, t)^{-1}[\bar{\boldsymbol{\tau}} - \mathbf{f}(\mathbf{x}, t)]$ and take

$$\begin{aligned}
\bar{\boldsymbol{\tau}} &= -k_1 \frac{\mathbf{s}}{\|\mathbf{s}\|^{1/2}} - k_2\mathbf{s} + \mathbf{z}, \\
\dot{\mathbf{z}} &= -k_3 \frac{\mathbf{s}}{\|\mathbf{s}\|} - k_4\mathbf{s} + \mathbf{z},
\end{aligned} \tag{12.67}$$

where $k_1, \ldots, k_4$ are scalar constants. Using this control law in (12.66) yields the closed loop system

$$\begin{aligned}
\dot{\mathbf{s}} &- -k_1 \frac{\mathbf{s}}{\|\mathbf{s}\|^{1/2}} - k_2\mathbf{s} + \mathbf{z} + \boldsymbol{\gamma}(\mathbf{s}, t) \\
\dot{\mathbf{z}} &= -k_3 \frac{\mathbf{s}}{\|\mathbf{s}\|} - k_4\mathbf{s} + \mathbf{z} + \boldsymbol{\phi}(t),
\end{aligned} \tag{12.68}$$

when $\boldsymbol{\phi}(t) = 0$. Here, following [20], the term $\boldsymbol{\phi}(t) = 0$ is added to the equation for $\dot{\mathbf{z}}$ to maintain compatibility of the approach with the more general form of the super-twisting algorithm presented in [17].

The terms $\boldsymbol{\gamma}(\mathbf{s}, t)$ and $\boldsymbol{\phi}(t)$ are assumed to satisfy

$$\|\boldsymbol{\gamma}(\mathbf{s}, t)\| \le \delta_1 \tag{12.69}$$
$$\|\boldsymbol{\phi}(t)\| \le \delta_2 \tag{12.70}$$

for known scalar bounds $\delta_1, \delta_2 > 0$.

Define

$$
\begin{aligned}
k_3^{\Omega} &:= 3\delta_2 + \frac{2\delta_2^2}{k_1^2}, \\
k_4^{\Omega} &:= \frac{\left(\frac{3}{2}k_1^2 k_2 + 3\delta_2 k_2\right)^2}{k_3 k_1^2 - 2\delta_2^2 - 3\delta_2 k_1^2} + 2k_2^2 + \frac{3}{2}k_2\delta_1,
\end{aligned}
\tag{12.71}
$$

and

$$
\begin{aligned}
k_3^{\Psi} &:= \frac{\frac{9}{16}(k_1\delta_1)^2}{k_2(k_2 - 2\delta_1)} + \frac{\frac{1}{2}k_1^2\delta_1 - 2k_1^2 k_2 + k_2\delta_2}{(k_2 - 2\delta_1)}, \\
k_4^{\Psi} &:= \frac{\alpha_1}{\alpha_2(k_2 - 2\delta_1)} + \frac{2k_1^2\delta_1 + \frac{1}{4}k_2\delta_1^2}{(k_2 - 2\delta_1)},
\end{aligned}
\tag{12.72}
$$

where

$$
\begin{aligned}
\alpha_1 &:= \frac{\frac{9}{16}(k_1\delta_1)^2(k_2 + \frac{1}{2}\delta_1)^2}{k_2^2}, \\
\alpha_2 &:= k_2(k_3 + 2k_1^2 - \delta_2) - \left(2k_3 + \frac{1}{2}k_1^2\right)\delta_1 + \frac{\frac{9}{16}(k_1\delta_1)^2}{k_2}.
\end{aligned}
\tag{12.73}
$$

It has been shown that when the scalar constants $k_1, \ldots, k_4$ are selected to satisfy the inequalities

$$
\begin{aligned}
k_1 &> \sqrt{2\delta_2}, \\
k_2 &> 2\delta_1, \\
k_3 &> \max(k_3^{\Omega}, k_3^{\Psi}), \\
k_4 &> \max(k_4^{\Omega}, k_4^{\Psi}),
\end{aligned}
\tag{12.74}
$$

the variables $s$ and $\dot{s}$ are forced to zero in finite time and remain zero for all subsequent time.

## 12.10  Higher Order Sliding Mode Differentiation

In Sect. 12.8 it is shown that when the relative degree of a system is $r = 2$, second order sliding (super-twisting) sliding mode control can be used to drive the sliding surface and its first total time derivative to zero $s = \dot{s} = 0$ and also provide a continuous control signal. Similarly, higher order sliding mode (HOSM) techniques for $(r \geq 2)$ can be used to design of controllers, as well as signal differentiators and observers [15, 16, 23].

Provided the relative degree $r$ of a system is known, HOSM can be used to drive the sliding surface and its higher total time derivatives up to the $(r - 1)$th derivative to zero in finite time $s = \dot{s} = \cdots = s^{(r-1)} = 0$ with a smooth control signal. At the same time, the derivatives required for implementation of the controller can be simultaneously computed in real time using a HOSM differentiator. We first consider differentiation, then tracking.

To construct a HOSM differentiator, let the input signal $f(t)$ consist of a base signal $f_0(t)$, whose $k$th time derivative has a known Lipschitz constant $L > 0$ (see Definition 7.1), and bounded noise. A HOSM differentiator can be used to obtain real-time estimates of the base signal's derivatives $\dot{f}_0(t), \ddot{f}_0(t), \ldots, f_0^{(k)}(t)$

$$
\begin{aligned}
\dot{z}_0 &= -\lambda_k L^{1/(k+1)} |z_0 - f(t)|^{k/(k+1)} \operatorname{sgn}(z_0 - f(t)) + z_1, \\
\dot{z}_1 &= -\lambda_{k-1} L^{1/k} |z_1 - \dot{z}_0|^{(k-1)/k} \operatorname{sgn}(z_1 - \dot{z}_0) + z_2, \\
&\vdots \\
\dot{z}_{k-1} &= -\lambda_1 L^{1/2} |z_{k-1} - \dot{z}_{k-2}|^{1/2} \operatorname{sgn}(z_{k-1} - \dot{z}_{k-2}) + z_k, \\
\dot{z}_k &= -\lambda_0 L \operatorname{sgn}(z_k - \dot{z}_{k-1}).
\end{aligned}
\tag{12.75}
$$

When the gains $\lambda_0, \lambda_1, \ldots, \lambda_k > 0$ are properly chosen, the derivatives of the base signal are obtained in finite time as

$$
\begin{aligned}
f_0(t) &= z_0, \\
\dot{f}_0(t) &= z_1, \\
&\vdots \\
f_0^{(k-2)}(t) &= z_{k-1}, \\
f_0^{(k-1)}(t) &= z_k.
\end{aligned}
\tag{12.76}
$$

If $\varepsilon$ is an upper bound for the magnitude of the noise in the signal $f(t)$, the accuracy of the $i$th derivative is given by

$$
|z_i(t) - f_0^{(i)}(t)| = O(\varepsilon^{(k+1-i)/(k+1)}).
\tag{12.77}
$$

As discussed in [23], the differentiator can be tuned by first selecting a $\lambda_0 > 1$ and then tuning the lower order derivative terms; possible choices are given there for $k \leq 5$ as

$$
\lambda_0 = 1.1, \lambda_1 = 1.5, \lambda_2 = 3, \lambda_3 = 5, \lambda_4 = 8, \lambda_5 = 12,
$$

or

$$
\lambda_0 = 1.1, \lambda_1 = 1.5, \lambda_2 = 2, \lambda_3 = 5, \lambda_4 = 5, \lambda_5 = 8.
$$

As shown in [15, 23], the differentiator can also be written in the following non-recursive form

$$
\begin{aligned}
\dot{z}_0 &= -\tilde{\lambda}_k L^{1/(k+1)} |z_0 - f(t)|^{k/(k+1)} \mathrm{sgn}(z_0 - f(t)) + z_1, \\
\dot{z}_1 &= -\tilde{\lambda}_{k-1} L^{2/(k+1)} |z_0 - f(t)|^{(k-1)/(k+1)} \mathrm{sgn}(z_0 - f(t)) + z_2, \\
&\;\;\vdots \\
\dot{z}_{k-1} &= -\tilde{\lambda}_1 L^{k/(k+1)} |z_0 - f(t)|^{1/(k+1)} \mathrm{sgn}(z_0 - f(t)) + z_k, \\
\dot{z}_k &= -\tilde{\lambda}_0 L \mathrm{sgn}(z_0 - f(t)),
\end{aligned}
\tag{12.78}
$$

where, as shown in [23], $\tilde{\lambda}_k = \lambda_k$ and $\tilde{\lambda}_i = \lambda_i \tilde{\lambda}_{i+1}^{i/(i+1)}$ for $i = k - 1, k - 2, \ldots, 0$.

The use of general $n$th order HOSM techniques for feedback control, as well as detailed stability analyses, is presented in [15, 23].

## 12.11   An HOSM Controller–Observer

The design of an HOSM controller is often coupled with the design of an HOSM observer, as computation of the control input often requires knowledge of state derivatives and disturbances. It has been shown that when an HOSM observer is coupled with an HOSM controller, the order of the observer must be one higher than the order of the controller so that the control input is continuous [6, 14]. Here, we discuss the design of the coupled modified super-twisting controller and third order sliding mode observer proposed by [19].

Consider a second order system of the form

$$
\begin{aligned}
\dot{x}_1 &= x_2, \\
\dot{x}_2 &= f(x_1, x_2, t) + g(x_1, x_2, t)\tau + w_d(x_1, x_2, t), \\
y &= x_1(t),
\end{aligned}
\tag{12.79}
$$

where $y(t) \in \mathbb{R}$ is the output, $\tau$ is the control input, $f(x_1, x_2, t) \in \mathbb{R}$ and $g(x_1, x_2, t) \in \mathbb{R}$ are known functions, and $w_d(x_1, x_2, t) \in \mathbb{R}$ contains both unmodeled dynamics and external disturbances. With the assumptions that $w_d(x_1, x_2, t)$ is Lipschitz (Definition 7.1), that the magnitude of its time derivative is upper bounded by a positive constant $|\dot{w}_d| < \rho$, and that $g(x_1, x_2, t) \neq 0$ for all $x_1, x_2, t$, a third order sliding mode observer, which is based on the non-recursive form of the HOSM differentiator in (12.78), can be used to estimate $x_2$ and $w_d$.

The control problem is to track a desired trajectory $x_{1d}$ in the presence of uncertainty and exogenous disturbances. Further, it is assumed that state measurements of $x_2$ are unavailable.

Define the estimation errors to be $\tilde{x}_1 := x_1 - \hat{x}_1, \tilde{x}_2 := x_2 - \hat{x}_2$ and $\tilde{x}_3 = w_d - \hat{x}_3$. An observer is constructed using feedback linearization with $k = 2$, by taking the time derivatives of the estimation errors to be

$$
\begin{aligned}
\dot{\hat{x}}_1 &= \hat{x}_2 + \tilde{\lambda}_2|\tilde{x}_1|^{2/3}\mathrm{sgn}(\tilde{x}_1),\\
\dot{\hat{x}}_2 &= \hat{x}_3 + \tilde{\lambda}_1|\tilde{x}_1|^{1/3}\mathrm{sgn}(\tilde{x}_1) + f(x_1, x_2, t) + g(x_1, x_2, t)\tau,\\
\dot{\hat{x}}_3 &= +\tilde{\lambda}_0\mathrm{sgn}(\tilde{x}_1),
\end{aligned}
\tag{12.80}
$$

such that the closed loop estimation error system is

$$
\begin{aligned}
\dot{\tilde{x}}_1 &= \tilde{x}_2 - \tilde{\lambda}_2|\tilde{x}_1|^{2/3}\mathrm{sgn}(\tilde{x}_1),\\
\dot{\tilde{x}}_2 &= \tilde{x}_3 - \tilde{\lambda}_1|\tilde{x}_1|^{1/3}\mathrm{sgn}(\tilde{x}_1),\\
\dot{\tilde{x}}_3 &= \dot{w}_d - \tilde{\lambda}_0\mathrm{sgn}(\tilde{x}_1).
\end{aligned}
\tag{12.81}
$$

The closed loop system of estimation errors has the form (12.78), such that the estimation errors converge to zero in a finite time $T_0$ for a suitable selection of the observer gains $\tilde{\lambda}_1, \tilde{\lambda}_2, \tilde{\lambda}_3$ (see Sect. 12.10 above).

The observer, is paired with a second order, modified super twisting controller [18]. In the implementation proposed by [19], the sliding surface is defined as

$$
s := \hat{e}_2 + c_1 e_1
\tag{12.82}
$$

where the constant $c_1 > 0$ and the tracking errors are defined as $e_1 := \hat{x}_1 - x_{1d}$ and $\hat{e}_2 := \hat{x}_2 - x_{2d}$. The controller is designed to drive $s$ and $\dot{s}$ to zero. Note that with this choice of $s$, the controller will drive the error between the state estimate $\hat{x}_2$ and the time derivative of the desired trajectory $x_{2d} = \dot{x}_{1d}$ to zero. This can be advantageous for marine vehicle control systems in which the position measurement is reliable, but estimates of the velocity are unavailable or noisy/intermittent.

Taking the time derivative of (12.82) and using (12.79)–(12.80) gives

$$
\begin{aligned}
\dot{s} = c_1\left(\tilde{x}_2 + \hat{x}_2 - \dot{x}_{1d}\right) &+ \int_0^t \tilde{\lambda}_0\mathrm{sgn}(\tilde{x}_1)\mathrm{d}t\\
&+ \tilde{\lambda}_1|\tilde{x}_1|^{1/3}\mathrm{sgn}(\tilde{x}_1) + f(x_1, x_2, t) + g(x_1, x_2, t)\tau - \dot{x}_{2d}.
\end{aligned}
\tag{12.83}
$$

Let the control input be

$$
\begin{aligned}
\tau = \frac{1}{g(x_1, x_2, t)}\Bigg[&-c_1\left(\hat{x}_2 - \dot{x}_{1d}\right) - \int_0^t \tilde{\lambda}_0\mathrm{sgn}(\tilde{x}_1)\mathrm{d}t - \tilde{\lambda}_1|\tilde{x}_1|^{1/3}\mathrm{sgn}(\tilde{x}_1)\\
&- f(x_1, x_2, t) + \dot{x}_{2d} - k_1|s|^{1/2}\mathrm{sgn}(s) - k_2 s\\
&- \int_0^t k_3\mathrm{sgn}(s)\mathrm{d}t - \int_0^t k_4 s\,\mathrm{d}t\Bigg],
\end{aligned}
\tag{12.84}
$$

where $k_i > 0$ for $i = 1, \ldots, 4$, so that

$$
\dot{s} = c_1\tilde{x}_2 - k_1|s|^{1/2}\mathrm{sgn}(s) - k_2 s - \int_0^t k_3\mathrm{sgn}(s)\mathrm{d}t - \int_0^t k_4 s\,\mathrm{d}t.
\tag{12.85}
$$

Using the definition of $\tilde{x}_2$, together with (12.82) and (12.85), the closed loop error system can be written as

$$
\begin{aligned}
\dot{x}_1 &= \tilde{x}_2 + s - c_1 e_1 + x_{2d}, \\
\dot{s} &= c_1 \tilde{x}_2 - k_1 |s|^{1/2} \mathrm{sgn}(s) - k_2 s + v, \\
\dot{v} &= -k_3 \mathrm{sgn}(s) - k_4 s.
\end{aligned}
\tag{12.86}
$$

The observer gains are generally selected, such that the dynamics of the observer are much faster than those of the controlled system. For times greater than the finite convergence time of the observer, $t \geq T_0$, the estimation error $\tilde{x}_2 = 0$ so that the closed loop error system is

$$
\begin{aligned}
\dot{x}_1 &= s - c_1 e_1 + x_{2d}, \\
\dot{s} &= -k_1 |s|^{1/2} \mathrm{sgn}(s) - k_2 s + v, \\
\dot{v} &= -k_3 \mathrm{sgn}(s) - k_4 s,
\end{aligned}
\tag{12.87}
$$

where the last two equations correspond to the closed loop error system of a modified super twisting controller [18]. Then, $s$ and $\dot{s}$ are driven to zero in finite time (as are the associated tracking errors) provided that the controller gains $k_1, \ldots, k_4$ are suitably selected to satisfy the inequality

$$
4k_3 k_4 > (8k_3 + 9k_1^2)k_2^2.
\tag{12.88}
$$

## 12.12   An HOSM Controller-Observer for Marine Vehicles

Recent studies that use higher order sliding modes for the trajectory tracking control of marine vehicles include [4, 5, 11, 28]. In [28] a third order sliding mode observer is coupled with a first order sliding mode controller for the trajectory tracking control of underwater vehicles—online estimation of the equivalent control [25] is used to reduce chattering. Ianagui and Tannuri [11] couple a partially HOSM observer with a super twisting controller (second order sliding mode) for the dynamic positioning control of ships (including low speed trajectory tracking). In [4, 5] an adaptive super twisting controller is coupled with the third order sliding mode observer proposed by [14] for the control of a snake-like underwater vehicle.

In this section, a trajectory tracking HOSM controller-observer for marine vehicles is described. The development is based on [26].

Consider a second order dynamical system of the form

$$
\begin{aligned}
\dot{\boldsymbol{\eta}} &= \mathbf{J}(\boldsymbol{\eta})\mathbf{v}, \\
\mathbf{M}\dot{\mathbf{v}} &= \mathbf{N}(\mathbf{v}, \boldsymbol{\eta}) + \boldsymbol{\tau} + \mathbf{w}_{dE}, \\
\mathbf{y} &= \boldsymbol{\eta}.
\end{aligned}
\tag{12.89}
$$

**Table 12.1**  Variables used in (12.89)

| Term | Dimension | Description |
|------|-----------|-------------|
| $\mathbf{M}$ | $\mathbb{R}^n \times \mathbb{R}^n$ | Inertia tensor |
| $\mathbf{v}$ | $\mathbb{R}^n$ | Velocity/angular rate vector |
| $\boldsymbol{\eta}$ | $\mathbb{R}^n$ | Position/attitude vector |
| $\mathbf{N}(\mathbf{v}, \boldsymbol{\eta})$ | $\mathbb{R}^n$ | Uncertain nonlinear function |
| $\boldsymbol{\tau}$ | $\mathbb{R}^n$ | Control input |
| $\mathbf{w}_{dE}$ | $\mathbb{R}^n$ | Unknown exogenous disturbance |
| $\mathbf{y}$ | $\mathbb{R}^n$ | System output |

The terms appearing in these equations are defined in Table 12.1, where $n$ is the dimension of the vehicle configuration space, i.e. the number of vehicle degrees of freedom (DOF). The objective is to design a control law so that a fully-actuated system can track a desired trajectory $\boldsymbol{\eta}_d$ with exogenous disturbances and system uncertainties.

**Remark 12.2** In general, a marine craft with actuation in all DOFs, such as an underwater vehicle, requires a $n = 6$ DOF model for model-based controller and observer design, while ship and semi-submersible control systems can be designed using an $n = 3$, or 4 DOF model. In 6 DOF $\boldsymbol{\eta}$ is a composite vector of translations in 3 DOF and Euler angle rotations in 3 DOF. Thus, in 6 DOF, its dimensions are often denoted as $\mathbb{R}^3 \times \mathcal{S}^3$. In 3 DOF, $\boldsymbol{\eta}$ often represents $x$ (North) and $y$ (East) translations in a North-East-Down (NED) reference frame with an angular rotation of $\psi$ about the downward axis and would have dimensions $\mathbb{R}^2 \times \mathcal{S}$.

**Remark 12.3** Generally, for marine vessels, the nonlinear term $\mathbf{N}(\mathbf{v}, \boldsymbol{\eta})$ in (12.89) is

$$\mathbf{N}(\mathbf{v}, \boldsymbol{\eta}) = -\mathbf{C}(\mathbf{v})\mathbf{v} - \mathbf{D}(\mathbf{v})\mathbf{v} - \mathbf{g}(\boldsymbol{\eta}), \tag{12.90}$$

where $\mathbf{C}(\mathbf{v}) \in \mathbb{R}^{n \times n}$ is a tensor of centripetal and Coriolis acceleration terms, $\mathbf{D}(\mathbf{v}) \in \mathbb{R}^{n \times n}$ is a tensor of hydrodynamic damping terms and $\mathbf{g}(\boldsymbol{\eta}) \in \mathbb{R}^n$ is a vector of hydrostatic forces/moments.

**Assumption 12.1** In many applications a coordinate transformation matrix $\mathbf{J}(\boldsymbol{\eta})$ is used to convert the representation of vectors between a body-fixed coordinate system and an inertial coordinate system. This transformation matrix is assumed to be non-singular, e.g. it is assumed that the pitch angle $\theta$ of the vehicle is bounded such that $|\theta| < \pi/2$.

**Assumption 12.2** The trajectory $\boldsymbol{\eta}_d$ and its derivatives $\dot{\boldsymbol{\eta}}_d$ and $\ddot{\boldsymbol{\eta}}_d$ are smooth and bounded.

**Assumption 12.3** The uncertain nonlinear function $\mathbf{N}(\mathbf{v}, \boldsymbol{\eta})$ is related to the system model $\hat{\mathbf{N}}(\mathbf{v}, \boldsymbol{\eta})$ as

$$\mathbf{N}(\mathbf{v}, \boldsymbol{\eta}) = \hat{\mathbf{N}}(\mathbf{v}, \boldsymbol{\eta}) + \mathbf{w}_{dN}, \tag{12.91}$$

where $\mathbf{w}_{dN}$ is the model uncertainty.

**Assumption 12.4** The total disturbance vector $\mathbf{w}_d$ includes the effects of both exogenous disturbances $\mathbf{w}_{dE}$ and model uncertainty $\mathbf{w}_{dN}$,

$$\mathbf{w}_d := \mathbf{w}_{dE} + \mathbf{w}_{dN}. \tag{12.92}$$

The components of $\mathbf{w}_d$ and their time derivatives are unknown and time-varying, yet bounded. There exists a positive constant $\rho$, such that

$$\|\dot{\mathbf{w}}_d(t)\| \leq \rho, \tag{12.93}$$

where $\| \bullet \|$ represents the 2-norm of a vector or matrix.

**Assumption 12.5** Full state feedback of the pose and velocity/angular rates in the NED reference frame is available.

Let

$$\dot{\boldsymbol{\xi}}_\eta := \mathbf{M}\mathbf{J}^{-1}(\boldsymbol{\eta})\dot{\boldsymbol{\eta}} \quad \text{and} \quad \boldsymbol{\xi}_v := \mathbf{M}\mathbf{v},$$

where Assumption 12.1 is used to ensure that $\mathbf{J}$ can be inverted. Then, (12.89) can be rewritten as

$$\dot{\boldsymbol{\xi}}_\eta = \boldsymbol{\xi}_v, \quad \text{and} \quad \dot{\boldsymbol{\xi}}_v = \hat{\mathbf{N}}(\mathbf{v}, \boldsymbol{\eta}) + \boldsymbol{\tau} + \mathbf{w}_d.$$

The HOSM observer developed in [19] is used to estimate the disturbance $\mathbf{w}_d$. Let the tracking surface estimation error, the velocity surface estimation error and the disturbance estimation error be defined as

$$\tilde{\boldsymbol{\xi}}_\eta := \boldsymbol{\xi}_\eta - \hat{\boldsymbol{\xi}}_\eta, \quad \tilde{\boldsymbol{\xi}}_v := \boldsymbol{\xi}_v - \hat{\boldsymbol{\xi}}_v, \quad \tilde{\boldsymbol{\xi}}_d := \mathbf{w}_d - \hat{\mathbf{w}}_d, \tag{12.94}$$

respectively. We use the third order sliding mode observer given by

$$\dot{\hat{\boldsymbol{\xi}}}_\eta = \hat{\boldsymbol{\xi}}_v + \lambda_1 |\tilde{\boldsymbol{\xi}}_\eta|^{2/3}\mathrm{sgn}(\tilde{\boldsymbol{\xi}}_\eta),$$
$$\dot{\hat{\boldsymbol{\xi}}}_v = \hat{\mathbf{w}}_d + \lambda_2 |\tilde{\boldsymbol{\xi}}_\eta|^{1/3}\mathrm{sgn}(\tilde{\boldsymbol{\xi}}_\eta) + \hat{\mathbf{N}}(\mathbf{v}, \boldsymbol{\eta}) + \boldsymbol{\tau}, \tag{12.95}$$
$$\dot{\hat{w}}_d = \lambda_3 \mathrm{sgn}(\tilde{\boldsymbol{\xi}}_\eta),$$

where $|\tilde{\boldsymbol{\xi}}_\eta|$ is the length of $\tilde{\boldsymbol{\xi}}_\eta$, $\mathrm{sgn}(\cdot)$ is extended element-wise to vector arguments, and $\lambda_i > 0, i \in \{1, 2, 3\}$ are observer design constants.

Then, the estimation error dynamics are

$$\dot{\tilde{\xi}}_\eta = -\lambda_1 |\tilde{\xi}_\eta|^{2/3} \text{sgn}(\tilde{\xi}_\eta) + \tilde{\xi}_v,$$
$$\dot{\tilde{\xi}}_v = -\lambda_2 |\tilde{\xi}_\eta|^{1/3} \text{sgn}(\tilde{\xi}_\eta) + \tilde{\xi}_d, \qquad (12.96)$$
$$\dot{\tilde{\xi}}_d = -\lambda_3 \text{sgn}(\tilde{\xi}_\eta) + \dot{\mathbf{w}}_d.$$

The dynamics of the estimation error (12.96) have the form of the non-recursive exact robust differentiator in [15], where it is proved that the estimation errors of this system will converge in finite time $t \geq T_0$ when the gains are suitably chosen [15, 17].

The sliding surface for a second order sliding mode controller is given by

$$\mathbf{s}(\mathbf{v}, \boldsymbol{\eta}) = \dot{\mathbf{s}}(\mathbf{v}, \boldsymbol{\eta}) = 0, \qquad (12.97)$$

where $\mathbf{s}(\mathbf{v}, \boldsymbol{\eta})$ is the sliding variable. We select the sliding manifold defined by

$$\mathbf{s} := c_1 \mathbf{e}_1 + \mathbf{e}_2 = 0, \qquad (12.98)$$

where $c_1 > 0$ is a constant,

$$\mathbf{e}_1 := \boldsymbol{\eta} - \boldsymbol{\eta}_d, \qquad \mathbf{e}_2 := \mathbf{M}(\mathbf{v} - \mathbf{v}_d) = \boldsymbol{\xi}_v - \mathbf{M}\mathbf{v}_d,$$

and

$$\mathbf{v}_d := \mathbf{J}^{-1}(\boldsymbol{\eta}) \dot{\boldsymbol{\eta}}_d.$$

Taking the time derivative of $\mathbf{s}$ and using (12.89) and (12.95) gives

$$\begin{aligned}
\dot{\mathbf{s}} &= c_1(\dot{\boldsymbol{\eta}} - \dot{\boldsymbol{\eta}}_d) + \mathbf{M}(\dot{\mathbf{v}} - \dot{\mathbf{v}}_d), \\
&= c_1(\mathbf{J}\mathbf{v} - \dot{\boldsymbol{\eta}}_d) + \dot{\tilde{\xi}}_v + \mathbf{M}\dot{\hat{v}} - \mathbf{M}\dot{\mathbf{v}}_d, \qquad (12.99) \\
&= c_1 \mathbf{J}\mathbf{M}^{-1}\tilde{\xi}_v + c_1(\mathbf{J}\hat{\mathbf{v}} - \dot{\boldsymbol{\eta}}_d) + \tilde{\xi}_d + \hat{\mathbf{w}}_d + \hat{\mathbf{N}}(\mathbf{v}, \boldsymbol{\eta}) + \boldsymbol{\tau} - \mathbf{M}\dot{\mathbf{v}}_d,
\end{aligned}$$

where $\hat{\mathbf{v}}$ is computed as $\hat{\mathbf{v}} = \mathbf{M}^{-1}\hat{\xi}_v$. In order to ensure that the time derivative of the sliding function is driven to zero, we select $\boldsymbol{\tau}$ such that

$$\dot{\mathbf{s}} = c_1 \mathbf{J}\mathbf{M}^{-1}\tilde{\xi}_v + \tilde{\xi}_d - k_1|\mathbf{s}|^{1/2}\text{sgn}(\mathbf{s}) - k_2\mathbf{s} - \int_0^t k_3\text{sgn}(\mathbf{s})\mathrm{d}t - \int_0^t k_4\mathbf{s}\mathrm{d}t, \qquad (12.100)$$

where the constants $k_i > 0$, $i \in \{1, \ldots, 4\}$ are controller design gains. Thus, we take the control input to be

$$\tau = -c_1(\mathbf{J}\hat{\mathbf{v}} - \dot{\boldsymbol{\eta}}_d) - \hat{\mathbf{w}}_d - \hat{\mathbf{N}}(\mathbf{v}, \boldsymbol{\eta}) + \mathbf{M}\dot{\mathbf{v}}_d - k_1|\mathbf{s}|^{1/2}\text{sgn}(\mathbf{s}) - k_2\mathbf{s}$$
$$- \int_0^t k_3\text{sgn}(\mathbf{s})dt - \int_0^t k_4\mathbf{s}dt. \tag{12.101}$$

**Remark 12.4** The term

$$\dot{\mathbf{v}}_d = \frac{d\mathbf{J}^{-1}}{dt}\dot{\boldsymbol{\eta}}_d + \mathbf{J}^{-1}\ddot{\boldsymbol{\eta}}_d$$

is found in the control input $\tau$. Assumption 12.2 is sufficient to ensure $\tau$ is continuous.

Then, using the definition of **s** in (12.98), the closed loop control system can be written as

$$\dot{\boldsymbol{\eta}} = \mathbf{J}\mathbf{M}^{-1}[\mathbf{s} - c_1\mathbf{e}_1 + \mathbf{M}\mathbf{v}_d]$$
$$\dot{\mathbf{s}} = c_1\mathbf{J}\mathbf{M}^{-1}\tilde{\boldsymbol{\xi}}_v + \tilde{\boldsymbol{\xi}}_d - k_1|\mathbf{s}|^{1/2}\text{sgn}(\mathbf{s}) - k_2\mathbf{s} + \boldsymbol{v} \tag{12.102}$$
$$\dot{\boldsymbol{v}} = -k_3\text{sgn}(\mathbf{s}) - k_4\mathbf{s}.$$

The estimation error of the observer converges to zero in finite time, so that the closed loop system is given by the dynamics

$$\dot{\boldsymbol{\eta}} = \mathbf{J}(\boldsymbol{\eta})\mathbf{M}^{-1}[\mathbf{s} - c_1\mathbf{e}_1 + \mathbf{M}\mathbf{v}_d]$$
$$\dot{\mathbf{s}} = -k_1|\mathbf{s}|^{1/2}\text{sgn}(\mathbf{s}) - k_2\mathbf{s} + \boldsymbol{v} \tag{12.103}$$
$$\dot{\boldsymbol{v}} = -k_3\text{sgn}(\mathbf{s}) - k_4\mathbf{s}. \tag{12.104}$$

Equations (12.103) and (12.104) are the modified super twisting SMC developed in [18], where it is shown that the tracking errors are asymptotically stable when the controller gains are properly selected.

**Remark 12.5** The observer coefficients $\lambda_i$ in (12.95), the coefficient $c_1$ in (12.98) and the control gains $k_i$ in (12.101) could instead be formulated as diagonal design matrices with dimensions $\mathbb{R}^n \times \mathbb{R}^n$, e.g. $\Lambda_i = \Lambda_i^T > 0, i \in \{1, \ldots, 3\}$, $\mathbf{C}_1 = \mathbf{C}_1^T > 0$ and $\mathbf{K}_i = \mathbf{K}_i^T > 0, i \in \{1, \ldots, 4\}$, which could allow more precise tuning through the independent selection of the corresponding coefficient/gain for each component.

**Example 12.6** As originally presented in [26], here the performance of the observer-controller given in (12.95) and (12.101) is investigated using a 3 DOF simulation of a fully-actuated unmanned surface vehicle (USV) following a time-dependent, nominally 40 m × 40 m square-shaped trajectory in the NED reference frame. This particular class of surface vessel has a very low draft, which permits it to operate as a fully-actuated vessel, without the need for extraordinarily large actuators. On-water physical experiments using this vessel in a fully-actuated configuration have been presented in [2, 3, 22]. The trajectory is constructed so that the vehicle maintains a constant surge (forward) speed of 1 m/s. The vehicle starts from rest at the origin of the NED frame. At each corner the commanded heading angle increases step-wise by

**Fig. 12.11** The smooth
reference trajectory $\eta_d(t)$

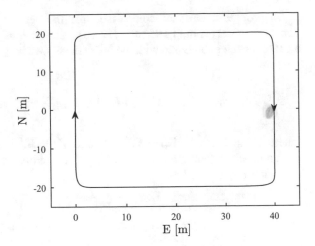

**Fig. 12.12** Top view of the
simulated USV. Reprinted
from [22] ©(2016), with
permission from Elsevier

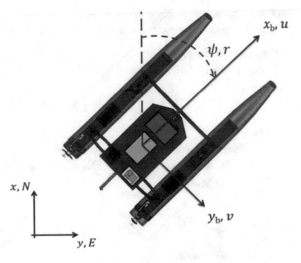

$+\pi/2$. To ensure its smoothness, the reference trajectory $\eta_d(t)$ tracked by the USV is obtained by passing the nominal trajectory through a fourth order linear filter, which is comprised of a cascaded pair of second order filters (Fig. 12.11). Each second order filter has a damping ratio of $\xi = 1.0$ and a natural frequency of $\omega_n = 1.0$ rad/s.

The USV simulated is a catamaran with an overall length of 4 m, an overall beam of 2.44 m and a mass of 180 kg (Fig. 12.12). The maneuvering coefficients and physical characteristics of the USV are obtained from [22].

Exogenous disturbances are modeled as first order Markov processes [12] of the form $T_b \dot{b} = -b + a_n w_n$, where $b \in \mathbb{R}^3$ is a vector of bias forces and moments, $T_b \in \mathbb{R}^{3 \times 3}$ is a diagonal matrix of time constants, $w_n \in \mathbb{R}^3$ is a vector of zero-mean Gaussian white noise, and $a_n \in \mathbb{R}^{3 \times 3}$ is a diagonal matrix that scales the amplitude of $w_n$.

To provide a basis of comparison for the performance of the higher order controller-observer in (12.95) and (12.101), simulations were also conducted using a PID controller combined with the NDO developed by [7]. The PID control law has the form

$$\tau_{\text{PID}} = -\hat{\mathbf{d}} - \mathbf{J}^T(\eta)\left(\mathbf{K}_p\tilde{\eta} + \mathbf{K}_d\dot{\tilde{\eta}} + \mathbf{K}_i\int_0^t \tilde{\eta}\mathrm{d}t\right), \qquad (12.105)$$

where $\mathbf{K}_p = \mathbf{K}_p^T > 0$, $\mathbf{K}_d = \mathbf{K}_d^T > 0$ and $\mathbf{K}_i = \mathbf{K}_i^T > 0$ are diagonal controller design matrices (and elements of $\mathbb{R}^n \times \mathbb{R}^n$). The disturbance estimate for the PID controller is computed as

$$\hat{\mathbf{d}} = \mathbf{q}(t) + \mathbf{K}_0\mathbf{M}\mathbf{v} \qquad (12.106)$$
$$\dot{\mathbf{q}}(t) = -\mathbf{K}_0\mathbf{q}(t) - \mathbf{K}_0\left[\mathbf{N}(\mathbf{v})\mathbf{v} + \tau + \mathbf{K}_0\mathbf{M}\mathbf{v}\right],$$

where $\mathbf{K}_0 = \mathbf{K}_0^T > 0$, $\mathbf{K}_0 \in \mathbb{R}^n \times \mathbb{R}^n$.

Three simulation cases were studied using each controller:

(1)  no disturbances;
(2)  continuous disturbances with $T_b = 10^3 \cdot \mathbf{1}_{6\times6}$, $b_0 = -a_n$, where

$$a_n = [1.75\ 2.35\ 2.35\ 0.25\ 1.20\ 1.20]^T;$$

and
(3)  the continuous disturbance simulated in Case (2) combined with a step disturbance having a magnitude of 450 N along the direction $\psi = +\pi/4$, starting at $t = 40$ s (Fig. 12.13).

**Fig. 12.13**  Case (3) disturbances in body-fixed frame. Jumps in disturbance magnitudes reflect rapid turning at corners of square trajectory

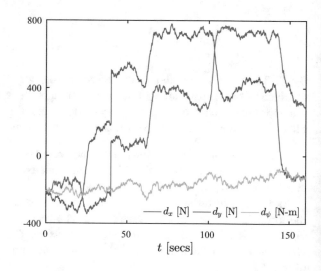

The controller-observers were manually tuned using simulations of Case (3). The manually-tuned parameters are:

(a) $\lambda_1 = 10, \lambda_2 = 250, \lambda_3 = 750, c_1 = 1, k_1 = 100, k_2 = 75, k_3 = 50$ and $k_4 = 35$ for the HOSM controller-observer; and

(b) $\mathbf{K}_p = 4000 * \text{diag}([1\ 1\ 2]), \mathbf{K}_d = 2000 * \text{diag}([1\ 1\ 2])$ and $\mathbf{K}_i = 250 * \mathbf{1}$ and $\mathbf{K}_0 = 100 * \mathbf{1}$ for the PID controller and NDO in (12.106).

The integral of time multiplied by absolute error (ITAE) of the USV's trajectory tracking error is provided in Table 12.2 for simulation Cases (1)–(3). Based on the ITAE, the HOSM controller-observer performed better than the PID controller. Of special note is the fact that the ITAE of the HOSM controller-observer appears to decrease in the presence of the disturbances simulated.

The tracking errors of the PID controller and the HOSM controller for simulation Case (3) are shown in Figs. 12.14 and 12.15, respectively. The tracking error of the PID controlled system exhibits large peaks as the USV passes each corner of the square-shaped trajectory. In comparison, the HOSM controlled system appears to be

**Table 12.2** ITAE of pose

| Case | Control | N (m-s) | E (m-s) | $\psi$ (rad-s) |
|------|---------|---------|---------|----------------|
| 1 | **PID** | 86.6 | 81.0 | 95.6 |
|   | **HOSM** | 51.5 | 76.2 | 19.7 |
| 2 | **PID** | 86.3 | 80.9 | 95.6 |
|   | **HOSM** | 13.0 | 12.4 | 2.21 |
| 3 | **PID** | 86.3 | 80.9 | 95.6 |
|   | **HOSM** | 7.32 | 10.3 | 2.51 |

**Fig. 12.14** PID controller Case (3) trajectory tracking error

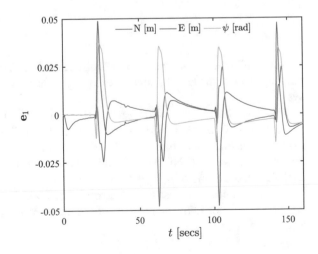

**Fig. 12.15** HOSM
controller Case (3) trajectory
tracking error

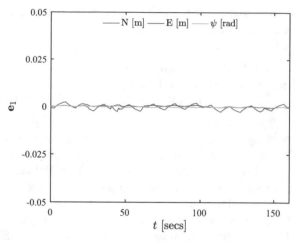

**Fig. 12.16** Case (3) NDO
disturbance estimation error.
Upper inset: magnified view
of $\tilde{\xi}_d$ for $100 \leq t \leq 120$ s.
Lower inset: corresponding
disturbance **d**

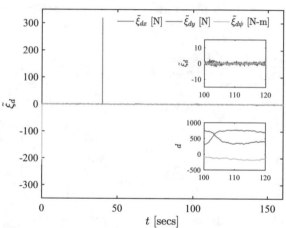

largely unaffected by the sharp turns required at each corner. Generally, the trajectory
tracking error of the HOSM controlled system is less than that of the PID controlled
system.

The disturbance estimation errors from simulation Case (3) for the NDO
in (12.106) and for the HOSM observer are shown in Figs. 12.16 and 12.17, respectively. The estimation errors of the HOSM observer are roughly three to five times
larger than those of the NDO in (12.106). However, as can be seen from the insets in
each figure, the errors of both estimators are small in comparison to the magnitudes
of the disturbances. The estimation errors of both observers exhibit a spike at 40 s,
which corresponds to the activation time of the step disturbance.                          □

**Fig. 12.17** Case (3) HOSM
disturbance estimation error.
Upper inset: magnified view
of $\tilde{\xi}_d$ for $100 \leq t \leq 120$ s.
Lower inset: corresponding
disturbance **d**

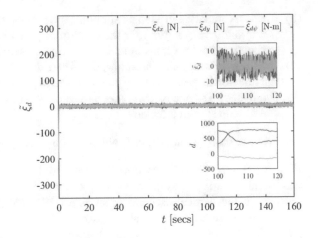

## Problems

**12.1** Sliding mode techniques make extensive use of the sgn function. To see the
interesting relationship between sgn$(x)$ and $|x|$, do the following.

(a)  Sketch the functions sgn$(x)$ versus $x$ and $|x|$ versus $x$ next to one another.
(b)  Using the sketches show that

$$\frac{d|x|}{dx} = \text{sgn}(x).$$

(c)  Using the sketch argue that $x/|x| = \text{sgn}(x)$.

**12.2** Consider the pitch stabilization of an AUV, as discussed in Example 7.4, but in
the presence of exogenous disturbances and unmodeled dynamics. The equations of
motion can be written as

$$\dot{\theta} = q,$$
$$\dot{q} = c_0 \sin(2\theta) + \tau + w_d(\theta, q, t),$$

where $\tau$ is the control input and $w_d(\theta, q, t)$ represents the effects of unknown exogenous disturbances and unmodeled vehicle dynamics. The magnitude of $w_d(\theta, q, t)$
is upper bounded by a known constant $L$.

(a)  First using either inverse dynamics (Sect. 8.2.1) or backstepping to obtain a
linearized system, design a first order sliding mode controller to stabilize the
pitch angle to $\theta = 0$ in finite time when $w_d = 0.25 \sin(2\pi t/10) + 2(\theta/\pi)^2$.
Provide an estimate for $L$. Take the initial condition to be $[\theta_0 \ q_0]^T = [\pi/4 \ 0]^T$,
give an estimate for the reaching time $t_r$.

(b) Create a simulation in Matlab/Simulink and plot the time response of the system (i.e. $\theta$ and $q$ versus time) for the conditions given in part (a) above. Also plot the control input $\tau$ versus time. Is chattering present?

(c) Use the nonlinear function $\tanh(s/\Phi)$, where $\Phi > 1$, as an approximation for the sgn function in the control law you designed in part (a). Let $\Phi = 10$. Is chattering eliminated? Try one or two other values of $\Phi$—how is the performance of the closed loop system affected?

**12.3** Repeat parts (a) and (b) of Problem 1 above using a super-twisting controller.

**12.4** A pressure sensor is used to measure the depth of an automatic underwater water column profiling system. As the measurements of other instrumentation are sensitive to the vertical ascent/descent rate of the profiler (e.g. conductivity and temperature sensors), it is desirable to use the depth measurements to also estimate the vertical velocity. In addition, the measurement includes some unknown noise and an unknown weak dynamic effect associated with the depth of the sensor. Using Matlab/Simulink, create a signal of the form

$$d(t) = d_0(t) + w_d(d, t),$$

where $d_0(t) = 100 + 50\sin(2\pi t/100)$ and $w_d(d, t) = d/200 + 0.2\sin(2\pi t/5)$. Take the initial conditions of the depth profiler to be $[d_0(0)\ \dot{d}_0(0)]^T = [50\ 0]^T$.

(a) In Matlab/Simulink implement a second order recursive HOSM differentiator to estimate $d_0(t)$ and $\dot{d}_0(t)$. What is the order of the approximate measurement error for this differentiator?

(b) In Matlab/Simulink implement a third order recursive HOSM differentiator to estimate $d_0(t)$ and $\dot{d}_0(t)$. What is the order of the approximate measurement error for this differentiator? How does the performance of this differentiator compare with the one implemented in part (a) above?

**12.5** Consider the backstepping PID trajectory tracking control of a surface vessel discussed in Example 10.3. The equations of motion of the system are

$$\begin{aligned}
\dot{x} &= u, \\
\dot{u} &= -d(u)u + \tau + w_d, \\
y &= x,
\end{aligned}$$

where $y = x$ is the distance traveled by the vessel along its surge direction, $u$ is its speed, $\tau$ is the thrust generated by the propeller, $d(u)$ is a nonlinear, speed-dependent drag and $w_d$ represents the effects of exogenous disturbances (e.g. wind/waves) and unmodeled dynamics.

Let the tracking error be defined as $e := y - y_d$, where $y_d(t)$ is the desired time-dependent trajectory to be tracked. In Example 10.3, the backstepping control design process yielded the backstepping PID control law and stabilizing functions

$$\alpha_1 = -k_1 z_1,$$
$$\alpha_2 = -k_2 z_2 - z_1 + \dot{y}_d + \dot{\alpha}_1,$$
$$\tau = -k_3 z_3 - z_2 + d(u)u + \dot{\alpha}_2,$$

where $k_1 > 0, k_2 > 0$ and $k_3 > 0$ are the integral, proportional, derivative and control gains, respectively. The associated closed loop tracking errors are defined as

$$z_1 := \int_0^t e \, dt,$$
$$z_2 := e - \alpha_1,$$
$$z_3 := u - \alpha_2.$$

Rather than compute the time derivatives of the stabilizing functions directly (Sect. 10.3.3), or use Dynamic Surface Control to estimate the time derivatives of the stabilizing functions (Sect. 10.5), let us explore the use of a second order HOSM differentiator to estimate them.

(a) Create a Matlab/Simulink simulation containing a second order non-recursive HOSM differentiator using the coefficients recommended in Sect. 12.10. Explore the use of the differentiator to compute the derivative of a signal $f_0(t) = \sin(2\pi t/10)$. Add a little noise to $f_0(t)$ and verify that it still performs well.

(b) Using (10.110) with $\zeta = 0.9$ and $\omega_n = \pi$, create a state space block in Matlab/Simulink block to generate a smooth desired trajectory $y_d(t)$ with a smooth first time derivative $\dot{y}_d(t)$ when the input reference trajectory is given by

$$r(t) = \begin{cases} 0 \text{ m}, & 0 \le t < 10 \text{ s} \\ 5 \text{ m}, & 10 \le t < 30 \text{ s} \\ 15 \text{ m}, & 30 \le t < 60 \text{ s} \\ 20 \text{ m}, & 60 \le t < 90 \text{ s} \\ 30 \text{ m}, & 90 \le t \text{ s}. \end{cases}$$

Plot $r(t)$ and $y_d(t)$ on the same graph. How do they compare?

(c) Expand your simulation by duplicating (copying) your differentiator from part (a), so that one differentiator can be used to compute $\dot{\alpha}_1$ from the signal $\alpha_1$ and the second can be used to compute $\dot{\alpha}_2$ from the signal $\alpha_2$. Add blocks to your Simulink model for the controller, boat (plant), and disturbance terms. Take $d(u) = |u|$ and $w_d = 0.1 \sin(2\pi t/10) + 0.05u$. Configure the system to track the $y_d(t)$ generated in part (b) above. Tune the controller gains to get suitable performance.

**12.6** Consider the situation in which we would like to simultaneously estimate and differentiate the values of two signals from possibly noisy measurements. Let the nominal values of the two signals be

$$f_1(t) = 2 \sin(2\pi t/2),$$
$$f_2(t) = 1.5 \sin(2\pi t).$$

Rather than create two separate differentiators, use the super-twisting MIMO HOSM techniques presented in Sect. 12.9 to estimate $f_1(t)$, $\dot{f}_1(t)$, $f_2(t)$ and $\dot{f}_2(t)$ from measurements.

Create a Matlab/Simulink simulation with a user defined block in which the error system (12.68) is implemented as a user-defined block with inputs $\mathbf{s}$ and $\mathbf{z}$, and outputs $\dot{\mathbf{s}}$ and $\dot{\mathbf{z}}$. Take the disturbance terms $\gamma(\mathbf{s}, t)$ and $\phi(t)$ to be zero. Define the vector version of the sliding surface to be $\mathbf{s} := \hat{\mathbf{f}}(t) - \mathbf{f}_m(t)$, where $\hat{\mathbf{f}}(t)$ is the estimate of the measured signal $\mathbf{f}_m(t) = [f_1(t) \quad f_2(t)]^T$ and create it in your Simulink model. The initial conditions can be taken as $\mathbf{s} = 0$ and $\mathbf{z} = 0$. Compare the estimated values of $\hat{\mathbf{f}}(t)$ and $\dot{\hat{\mathbf{f}}}(t)$ with the original signals.

**Note:** The observer will drive $\mathbf{s}$ to zero. In order to avoid singularities in the numerical implementation of (12.68) one must add a small number (e.g. $1 \times 10^{-4}$ or so) to the values of $\mathbf{s}$ in the denominator terms in (12.68).

**12.7** In Sect. 9.6 surge speed and yaw rate PID controllers are designed to enable a surface vessel to perform path following control. Modify the design of the surge speed and yaw rate controllers given to use super-twisting control, instead of PID control. Using the same terms already defined in Sect. 9.6, select the sliding surface for the surge speed controller to be

$$\sigma_u := (m - X_{\dot{u}})\dot{\tilde{u}} + k_x|\tilde{u}|^{1/2}\mathrm{sgn}(\tilde{u}), \tag{12.107}$$

where $k_x > 0$ is a control gain, and let the sliding surface for the yaw rate controller be

$$\sigma_r := \dot{\tilde{\psi}} + k_r\tilde{r}$$

where $k_r > 0$ is a control design parameter.

(a) When using a super twisting yaw rate controller, should the term $-k_\psi\tilde{\psi}$ still be included in the virtual control input

$$r_d = -\frac{\Delta}{(e^2 + \Delta^2)}\dot{e} - \dot{\beta} - k_\psi\tilde{\psi}?$$

   Why or why not?
(b) Give the control inputs $\tau_x$ and $\tau_\psi$ for super twisting control.
(c) Give the cumulative disturbance term for each controller.
(d) Show how to estimate the finite reaching times when $\sigma_u = 0$ and when $\sigma_r = 0$.

**12.8** Let us revisit the trajectory tracking control of a surface vessel discussed in Problem 12.5. The equations of motion of the system can be expressed as

$$\ddot{x} + \dot{x}|\dot{x}| = \tau + w_d,$$

where $y = x$ is the system output (position), $\tau$ is the thrust generated by the propeller and $w_d$ represents the effects of exogenous disturbances (e.g. wind/waves) and unmodeled dynamics.

Generate a smooth desired trajectory $y_d(t)$ with a smooth first time derivative $\dot{y}_d(t)$ using (10.110) with $\zeta = 0.9$ and $\omega_n = \pi$, when the input reference trajectory is given by

$$r(t) = \begin{cases} 0 \text{ m}, & 0 \le t < 10 \text{ s} \\ 5 \text{ m}, & 10 \le t < 30 \text{ s} \\ 15 \text{ m}, & 30 \le t < 60 \text{ s} \\ 20 \text{ m}, & 60 \le t < 90 \text{ s} \\ 30 \text{ m}, & 90 \le t \text{ s}. \end{cases}$$

(a) Create a Matlab/Simulink simulation of the vessel using the HOSM Controller–Observer presented in Sect. 12.11 to track the desired position of the vessel when $w_d = 0$. Tune the controller gains appropriately.

(b) Now add the disturbance signal $w_d = 0.1 \sin(2\pi t/10) + 0.05u$ to your simulation. Retune the controller gains slightly, if necessary. Using (12.80), integrate the $\dot{\hat{x}}_3$ to obtain an estimate of $w_d$. How does it compare with the disturbance signal you generated? Would this signal be suitable for use in disturbance observer based control?

# References

1. Ashrafiuon, H., Nersesov, S., Clayton, G.: Trajectory tracking control of planar underactuated vehicles. IEEE Trans. Autom. Control **62**(4), 1959–1965 (2017)
2. Bell, Z.I., Nezvadovitz, J., Parikh, A., Schwartz, E.M., Dixon, W.E.: Global exponential tracking control for an autonomous surface vessel: an integral concurrent learning approach. IEEE J. Ocean. Eng. (2018)
3. Bertaska, I., von Ellenrieder, K.: Experimental evaluation of supervisory switching control for unmanned surface vehicles. IEEE J. Ocean. Eng. **44**(1), 7–28 (2019)
4. Borlaug, I.-L.G., Pettersen, K.Y., Gravdahl, J.T.: Trajectory tracking for an articulated intervention AUV using a super-twisting algorithm in 6 DOF. IFAC-PapersOnLine **51**(29), 311–316 (2018)
5. Borlaug, I.L.G., Gravdahl, J.T., Sverdrup-Thygeson, J., Pettersen, K.Y., Loria, A.: Trajectory tracking for underwater swimming manipulators using a super twisting algorithm. Asian J. Control (2018)
6. Chalanga, A., Kamal, S., Fridman, L.M., Bandyopadhyay, B., Moreno, J.A.: Implementation of super-twisting control: super-twisting and higher order sliding-mode observer-based approaches. IEEE Trans. Ind. Electron. **63**(6), 3677–3685 (2016)
7. Chen, W.-H., Ballance, D.J., Gawthrop, P.J., O'Reilly, J.: A nonlinear disturbance observer for robotic manipulators. IEEE Trans. Ind. Electron. **47**(4), 932–938 (2000)
8. Elmokadem, T., Zribi, M., Youcef-Toumi, K.: Terminal sliding mode control for the trajectory tracking of underactuated autonomous underwater vehicles. Ocean Eng. **129**, 613–625 (2017). Jan
9. Filippov, A.F.: Differential Equations with Discontinuous Righthand Sides: Control Systems, vol. 18. Springer Science & Business Media, Berlin (2013)

10. Fossen, T.I.: Handbook of Marine Craft Hydrodynamics and Motion Control. Wiley, Hoboken (2011)
11. Ianagui, A.S.S., Tannuri, E.A.: High order sliding mode control and observation for DP systems. IFAC-PapersOnLine **51**(29), 110–115 (2018)
12. Jialu, D., Xin, H., Krstić, M., Sun, Y.: Robust dynamic positioning of ships with disturbances under input saturation. Automatica **73**, 207–214 (2016)
13. Karkoub, M., Hsiu-Ming, W., Hwang, C.-L.: Nonlinear trajectory-tracking control of an autonomous underwater vehicle. Ocean Eng. **145**, 188–198 (2017)
14. Kumari, K., Chalanga, A., Bandyopadhyay, B.: Implementation of super-twisting control on higher order perturbed integrator system using higher order sliding mode observer. IFAC-PapersOnLine **49**(18), 873–878 (2016)
15. Levant, A.: Higher-order sliding modes, differentiation and output-feedback control. Int. J. Control **76**(9–10), 924–941 (2003)
16. Levant, A.: Homogeneity approach to high-order sliding mode design. Automatica **41**(5), 823–830 (2005)
17. Moreno, J.A.: Lyapunov function for Levant's second order differentiator. In: Proceedings of 51st IEEE Conference on Decision and Control, pp. 6448–6453. IEEE (2012)
18. Moreno, J.A., Osorio, M.: A Lyapunov approach to second-order sliding mode controllers and observers. In: Proceedings of 47th IEEE Conference on Decision and Control, pp. 2856–2861 (2008)
19. Muñoz, F., Bonilla, M., Espinoza, E.S., González, I., Salazar, S., Lozano, R.: Robust trajectory tracking for unmanned aircraft systems using high order sliding mode controllers-observers. In: IEEE International Conference on Unmanned Aircraft Systems (ICUAS), pp. 346–352 (2017)
20. Nagesh, I., Edwards, C.: A multivariable super-twisting sliding mode approach. Automatica **50**(3), 984–988 (2014)
21. Qiao, L., Yi, B., Defeng, W., Zhang, W.: Design of three exponentially convergent robust controllers for the trajectory tracking of autonomous underwater vehicles. Ocean Eng. **134**, 157–172 (2017)
22. Sarda, E.I., Qu, H., Bertaska, I.R., von Ellenrieder, K.D.: Station-keeping control of an unmanned surface vehicle exposed to current and wind disturbances. Ocean Eng. **127**, 305–324 (2016)
23. Shtessel, Y., Edwards, C., Fridman, L., Levant, A.: Sliding Mode Control and Observation. Springer, Berlin (2014)
24. Slotine, J.-J.E., Li, W.: Applied Nonlinear Control. Prentice-Hall, Englewood Cliffs (1991)
25. Utkin, V.I.: Sliding Modes in Optimization and Control Problems. Springer, New York (1992)
26. von Ellenrieder, K.D., Henninger, H.C.: A higher order sliding mode controller-observer for marine vehicles. IFAC-PapersOnLine **52**(21), 341–346 (2019)
27. Wang, N., Lv, S., Zhang, W., Liu, Z., Er, M.J.: Finite-time observer based accurate tracking control of a marine vehicle with complex unknowns. Ocean Eng. **145**, 406–415 (2017)
28. Yan, Y., Shuanghe, Y.: Sliding mode tracking control of autonomous underwater vehicles with the effect of quantization. Ocean Eng. **151**, 322–328 (2018)
29. Yan, Z., Haomiao, Y., Zhang, W., Li, B., Zhou, J.: Globally finite-time stable tracking control of underactuated UUVs. Ocean Eng. **107**, 132–146 (2015)

# Index

© Springer Nature Switzerland AG 2021
K. D. von Ellenrieder, *Control of Marine Vehicles*, Springer Series on Naval
Architecture, Marine Engineering, Shipbuilding and Shipping 9,
https://doi.org/10.1007/978-3-030-75021-3

Printed in the United States
by Baker & Taylor Publisher Services